I0064761

# Biofuels and Renewable Energy

# Biofuels and Renewable Energy

Editor: Damian Price

www.callistoreference.com

**Callisto Reference,**
118-35 Queens Blvd., Suite 400,
Forest Hills, NY 11375, USA

Visit us on the World Wide Web at:
www.callistoreference.com



ISBN: 978-1-63239-991-5 (Hardback)



**Cataloging-in-Publication Data**

Biofuels and renewable energy / edited by Damian Price.
p. cm.
Includes bibliographical references and index.
ISBN 978-1-63239-991-5
1. Biomass energy. 2. Renewable energy sources. I. Price, Damian.
TP339 .B56 2018
333.953 9--dc23

# Table of Contents

# Preface

This book is a vital tool for all researching or studying the field of biofuel and renewable energy as it gives incredible insights into emerging trends and concepts related to it. Biofuel is the new form of green energy, which is generated from plants, and plant related activities, like agriculture, cellulosic biomass, bioethanol fermentation, etc. It can be derived from domestic waste, commercial and even industrial waste. From theories to research to practical applications, case studies related to all contemporary topics of relevance to this field have been included in this book. It is a vital tool for all researching and studying this field.

The information shared in this book is based on empirical researches made by veterans in this field of study. The elaborative information provided in this book will help the readers further their scope of knowledge leading to advancements in this field.

Finally, I would like to thank my fellow researchers who gave constructive feedback and my family members who supported me at every step of my research.

**Editor**

1

# Enzymatic saccharification of Tapioca processing wastes into biosugars through immobilization technology

Nurul Aini Edama, Alawi Sulaiman*, Siti Noraida Abd. Rahim

*Tropical Agro-Biomass Research Group, Faculty of Plantation and Agrotechnology, Universiti Teknologi MARA, 40450 Shah Alam, Selangor, Malaysia*

**HIGHLIGHTS**

➢Significance of cassava solid wastes and wastewater for production of biosugars.
➢Cassava-oriented biosugars as an economically-feasible source for second-generation bioethanol production.
➢Unique advantages of enzyme immobilization technologies in increasing the overall viability of cassava waste treatment industry.

**ABSTRACT**

Cassava is very popular in Nigeria, Brazil, Thailand and Indonesia. The global cassava production is currently estimated at more than 200 million tons and the trend is increasing due to higher demand for food products. Together with food products, huge amounts of cassava wastes are also produced including cassava pulp, peel and starchy wastewater. To ensure the sustainability of this industry, these wastes must be properly managed to reduce serious threat to the environment and among the strategies to achieve that is to convert them into biosugars. Later on, biosugars could be converted into other end products such as bioethanol. The objective of this paper is to highlight the technical feasibility and potentials of converting cassava processing wastes into biosugars by understanding their generation and mass balance at the processing stage. Moreover, enzyme immobilization technology for better biosugar conversion and future trends are also discussed.

**Keywords:**
Enzymatic Saccharification
Tapioca Starch Processing
Tapioca Agrowastes
Biosugars
Immobilization Technology

## 1. Introduction

Cassava is one of the major crops in the world and is ranked sixth as the most important food crop (Rattanachomsri et. al., 2009). The major world producers of cassava products are Nigeria, Brazil, Thailand and Indonesia (Hermiati et. al., 2012). Wuttiwai (2009) reported that the global production of cassava in 2007 was 228.14 million tons. Currently, Thailand is the largest exporter of cassava products in the world which represents more than 80% of its global trade (Hermiati et. al., 2012; Wuttiwai, 2009). Among the important cassava root products are tapioca starch and flour. Tapioca starch (95% starch) could be used in many industries such as food, textile, chemical and pharmaceutical (Hermiati et. al., 2011). There are seven major stages involved in the production of starch from cassava root including washing, chopping and grinding, fibrous residue separation, dewatering and protein separation, dehydration, drying and packaging (Chavalparit and Ongwandee, 2009). During these processing stages, large amounts of solid as well as liquid wastes (starchy wastewater) are produced. The solid wastes include cassava pulp (also known as cassava bagasse) and cassava peel (Hermiati et al., 2011).

Pandey et. al. (2000) reported that processing of 250 to 300 tons of cassava root results in approximately 1.16 tons of cassava peels, 280 tons of cassava pulp and 2655 $m^3$ of starchy wastewater. The wastewater contains high amount of starch and fibers. If not properly handled, these wastes and wastewater could pose serious threats to the environment (Djuma'ali et al., 2011) such as strong unpleasant smell as a result of microbial decomposition of organic matters to volatile matters and contamination of rivers and underground water resources located in the vicinity of the industry (Hermiati et. al., 2012). Looking from another perspective, starch and cellulose in tapioca waste materials such as peel and fiber are potential sources of

* Corresponding author
E-mail address: dr_alawi@salam.uitm.edu.my (A. Sulaiman).

carbohydrates which could be converted into different kinds of chemicals or bioproducts such as biofuels, biochemical and biomaterials (Hermiati et. al., 2012; Olaniwoninu and Odunfa, 2012).

Utilization of organic waste materials for renewable energy is currently one of the hottest topics in the market. One of the demanded intermediate products for renewable energy (e.g. bioethanol and biobuthanol) is fermentable sugar. Therefore, the need for cheap biosugars supply such as glucose has recently increased remarkably and one of the potential sources to generate a considerable amount of biosugars would be the cassava processing wastes. In better words, cassava wastes such as peel, bagasse and starchy wastewater can be utilized to produce biosugars since they contain starch, cellulose and hemi-cellulose (Srinorakutara et al., 2006). As discussed earlier the source of cassava wastes is abundant and therefore their utilization is economically feasible and promising because of their availability and continuous supply by tapioca industries as well as consistent quality and characteristics. At the same time, this strategy could also help to address the problems related to wastes treatment and disposal.

Hence, this mini review paper will highlight the cassava industry in terms of wastes generation and mass balance at the processing stage, potential of biosugars production from cassava wastes. Moreover, the potential application of enzyme immobilization technology for better biosugar conversion is briefly discussed.

## 2. Industrial processing of tapioca starch and its agrowastes

Figure 1 shows the stages involved in the processing of cassava roots into tapioca starch with its mass balance. As presented, it is clear that large amounts of solid and liquid wastes are produced along with tapioca starch. The liquid wastewater, however is produced in much greater amount i.e more than 3 times of cassava roots mass (Virunanon et. al., 2012). The liquid wastewater usually contains about 1% solids and is generally treated using open lagoon systems. On the other hand, the solid wastes are generally discarded in the environment as landfill without any treatment and therefore can cause serious problems to the environment (Pandey et. al., 2000). The basic chemical compositions of tapioca processing wastes are shown in Table 1.

**Table 1**
Chemical composition of tapioca agrowastes.

| Parameters | Unit | Solid Agrowastes content | | Cassava Wastewater[c] |
|---|---|---|---|---|
| | | Cassava Peel[a] | Cassava Pulp[b] | |
| Starch | g/100g | 42.6-64.6 | 66-68.89 | 14.98-25.5 |
| Fibers | g/100g | 11.7-12.5 | 21.10-27.75 | - |
| Ash | g/100g | 5.0-6.4 | 1.50-1.70 | - |
| Protein | g/100g | 1.6-8.2 | 1.52-1.55 | - |
| pH | - | - | - | 6.21 |
| TDS[d] | g/L | - | - | 13.52 |
| COD | mg/L | - | - | 20433 |
| BOD | mg/L | - | - | 9750 |
| Cyanide | mg/L | - | - | 64.0 |
| Glucose | g/100g | - | - | 5.62-6.09 |

[a] Oboh (2006) & Obadina et. al. (2006).
[b] Woiciechowski et. al. (2002).
[c] Virunanon et. al., 2012, Nasr et. al., 2013 & Akponah and Akpomie, 2012.
[d] Total Disolved Solid.

### *2.1 Cassava Peel*

In the processing of cassava roots, the roots are normally peeled to remove two outer covering layers i.e the thin brown outer layer and slightly thicker but softer and leathery parenchymatous inner layer (Kongkiattikajorn and Sornvoraweat, 2011). This process is usually being done manually and the peels constitute 20-35% of the total weight of the roots (Olanbiwoninu and Odunfa, 2012). Ubalua (2007) reported that the peels are usually discarded as waste and are let to decompose naturally leading to major environmental concerns such as bad smell and river and underground water contamination.

### *2.2 Cassava Pulp (Bagasse)*

Together with cassava peel, huge amounts of cassava pulp are also produced (Woiciechowski et. al., 2002). Cassava pulp is a fibrous byproduct of the cassava processing industry and is generally used as low-value animal feed. According to Rattanachomsri et. al. (2009), cassava pulp contains about 50-70% starch on dry weight basis and 20-30% fibers which are mainly composed of cellulose and other non-starch polysaccharides. Therefore, cassava pulp is an ideal substrate for the bioconversion into value-added product due to its high organic and low ash contents. Cassava pulp also offers other advantages such as easier hydrolysis process, low collection cost and lack of competition with other industrial uses (Rattanachomsri et. al., 2009). In a study, Hermiati et. al., (2012) reported that, besides starch, some other carbohydrates such as cellulose, galactan, xylan, ramnan, arabinan and mannan are also present in cassava pulp.

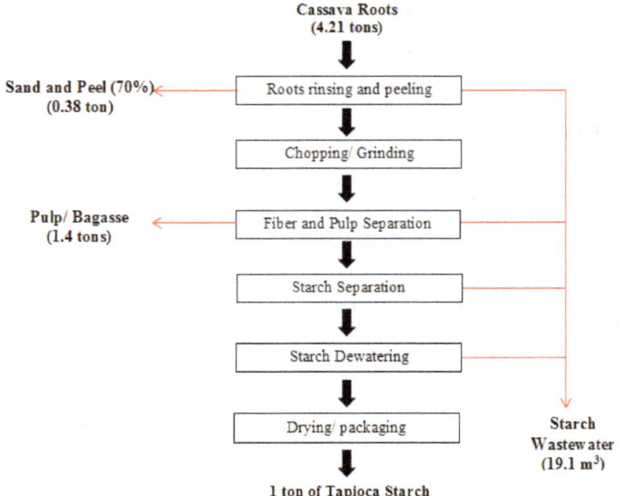


**Fig.1.** Processes involved in tapioca starch production and mass balance (Source: Chavalparit and Ongwandee, 2009 with modifications).

### *2.3 Tapioca Starch Wastewater*

As clearly illustrated in Figure 1, in the tapioca starch manufacturing process, starchy wastewater is produced at the root rinsing and peeling, fiber and pulp separation, starch separation and starch dewatering processes. In the conventional tapioca starch wastewater treatment process, the wastewater is treated using a series of open lagoon treatment system before being discharged into the river (Aisien et. al., 2010). In an investigation, Sangyoka et. al. (2007) elaborated that, 1 kg of fresh cassava roots leads to the production of about 0.2 kg of starch, 0.4-0.9 kg of cake and about 5-7 liters of wastewater. The generated wastewater has a very high chemical oxygen demand (COD), biochemical oxygen demand (BOD) and total solid (TS). A study by Kamaraj (2006) reported that the total BOD, COD and TS in the cassava starch wastewater were at 6820, 10363 and 5290 mg/L, respectively. Therefore, the treatment and disposal of this wastewater require high cost and impose additional burden on the tapioca starch manufacturers.

## 3. Potential utilization of tapioca agrowates for biosugars production

Biosugars are one of the important intermediate products that can be produced from tapioca or cassava processing wastes (Ayoola et. al., 2012). As mentioned earlier, one of the main waste feedstock generated by the tapioca processing industry is cassava peels. Due to the presence of high amounts of starch, this feedstock could be well used to produce biosugars (Yoonan and Kongkiattikajorn, 2004; Ubalua, 2007; Olanbiwoninu and Odunfa, 2012). Yoonan and Kongkiattikajorn (2004) compared the production of glucose from cassava peels using both diluted acid and enzymatic hydrolysis processes. They argued that enzymatic saccharification using amylase followed by amyglucosidase led to better results. They also found sequential treatment by alpha amylase and amyglucosidase more promising than the other hydrolytic enzymes i.e. cellulose, xylanase and pecticnase (Yoonan and Kongkiattikajorn, 2004).

Another waste feedstock that could be used in biosugars production is cassava bagasse. According to Woiciechowski et. al. (2002), cassava bagasse could be ideally converted into reducing sugars (mainly glucose) due to its richness in organic materials mainly starch. To obtain reducing sugars from cassava bagasse, this waste must also undergo acid treatment or enzymatic hydrolysis. Woiciechowski et. al. (2002) also reported that reducing sugar recovery from cassava bagasse by using enzymatic hydrolysis stood at about 97%. This rate was only slightly higher than that of acid hydrolysis (95%). But the major weakness of using acid hydrolysis is that the equipment used must be designed to withstand corrosive conditions as well as high temperature and pressure. These lead to higher production cost which could jeopardize the economic viability of the whole process (Yoonan and Kongkiattikajorn, 2004). Beside economic issues, enzymatic hydrolysis is also preferred over acid treatment due to the prevention of unwanted browning compounds (Srinorakutara et. al., 2006).

According to Chotineeranat et. al. (2004), sachharification of cassava pulp using mixed enzymes (alpha-amylase and glucoamylase) compared to using a single enzyme resulted in higher yields of reducing sugar at 9.88 mg/mL. More specifically, combined application of alpha-amylase and glucoamylase led to about 4 and 1.3 times higher biosugars yield than by alpha-amylase and glucoamylase alone. This is ascribed to the fact that alpha-amylase supplies the non-reducing end groups to glucoamylase (Chotineeranat et. al., 2004).

Akaracharanya et. al. (2011) also reported that glucose levels climbed up by increasing glucoamylase levels from 0.4 U/g (11.8 g glucose/L) to 2.0 U/g and that the maximum glucose yield of 22.6 g/L was achieved when cassava pulp underwent enzymatic saccharification process using both alpha amylase and glucoamylase enzymes. In an earlier study, Ubalua (2007) reported the multi-enzyme hydrolysis of cassava residues by cellulase and pectinase at 28 °C for 1 hour, followed by alpha-amylase treatment at 100 °C for 2 hours and finally glucoamylase treatment at 60 °C for 4 hours. In that investigation, it was found out that cassava residues with an initial non-water-soluble carbohydrate concentration of 11% (w/v), could yield 122.4 g/L of reducing sugars. The significantly increased amount of produced reducing sugars was achieved owing to the synergistic action of cellulase, alpha-amylase and glucoamylase (Ubalua, 2007).

## 4. Enzyme Immobilization

Immobilization can be defined as the limitation of movement of biocatalyst, such as enzymes, through the application of chemical and or physical treatment (Abdelmajeed et. al., 2012). Some other researchers have defined immobilization as a biocatalyst which has been confined or localized so that it could be reused continuously (Murty et. al., 2002). Abdelmajeed et. al. (2012) and Sheldon (2007) studied and compared free and immobilized enzyme systems and concluded that the latter offers several advantages as it allows enzymes to be easily reused multiple times for the same reaction. Immobilization also results in longer enzyme half-life and less degradation and moreover, it provides a straightforward method for controlling reaction rate as well as reaction start and termination times. Immobilization also prevents the contamination of the substrate with enzyme/protein or other compounds, which consequently results in decreased purification costs. These benefits attributed to enzyme immobilization have highlighted the applicability of this technology in a wide range of biotechnologies including biosugars and biofuel industry (Abdelmajeed et. al., 2012).

Currently there are different methods commonly used to immobilize enzymes which involve different degree of complexity and efficiency. The common methods to immobilize enzymes are listed in Table 2 which include adsorption, covalent binding, cross-linking, entrapment, and encapsulation (Murty et al., 2002; Abdelmajeed et. al., 2012). In recent years, using immobilized enzymes in starch processing has been extensively studied to obtain high purity sugars from starch. Currently in the cassava industry, biosugars such as glucose is commonly produced by processing the cassava starch. However, the tendency of using cassava for producing biosugars would affect supply of cassava as food. Therefore, the utilization of cassava wastes from the tapioca industry combined with the application of modern technologies such as enzyme immobilization could offer promising outcomes in this vast industry. (Hermiati et. al., 2012).

Despite the fact that to date there has been an increasing number of studies on the utilization of the cassava wastes for biosugar production, however, most of them have applied free enzyme systems. To the best of our knowledge, so far just a couple of studies have reported the utilization of cassava wastes for biosugars production by using enzyme immobilization technology. Recently, Abdul Rahim et. al. (2013) investigated enzyme encapsulation for glucose production by using cassava starch as substrate. In their study, multi-enzymes (α-amylase, cellulase and glucoamylase) were successfully encapsulated within the alginate-clay beads and the beads were reused 5 times with remaining relative activity of 12%. A few years before, Baskar et. al. (2008) investigated enzyme immobilization through gel entrapment. They immobilized α-amylase enzyme through this technique in order to produce glucose from cassava starch. These few examples highlight the necessity of using enzyme immobilization technology as a replacement for the free enzyme system for glucose production from cassava waste to increase efficiency and lower production cost.

## 5. Future trends

In dealing with global warming and other environmental crises caused by the increasing utilization of fossil fuels and consequent emission of greenhouse gases, renewable energy resources has been increasingly highlighted (Ayoola et. al., 2012). One of such renewable alternatives is bioethanol obtained from waste-oriented biosugars. As mentioned earlier, cassava roots processing wastes could be used for producing biosugars which could be then fermented into bioethanol (Ubalua, 2007, Hermiati et. al., 2012). In a study conducted by Oyeleke et. al. (2012) revealed that the yield obtained through enzymatic production of bioethanol from cassava peel was higher than that of sweet potato peels. This could be due to the presence of higher amount of carbohydrate (starch) in cassava peel which could be fermented into bioethanol.

Other ways to utilize cassava wastes such as pulp would be for the production of various fermentation products, including citric acid by *Candida lipolytica*, Lactic acid by *Lactobacilli*, and glutamic acid or xanthan gum (Rattanachomsri et. al., 2009; Hermiati et. al., 2012). There were also some reports indicating that the cassava pulp could be used as a medium for the production of amylase enzyme as well as for preparing composite materials (Hermiati et. al., 2012). In an investigation, Pandey et. al., (2000) managed to generate fumaric acid; an important organic acid, by using cassava pulp.

Fumaric acid has a wide range of applications as an intermediate in chemical synthesis involving esterification reactions. It is non-toxic and non-hygroscopic in nature and due to these properties is also used as an acidulant in food and pharmaceutical industries (Pandet et. al., 2000).

As mentioned earlier, starch recovery from cassava roots is commonly achieved by wet processing which generates a large volume of high strength liquid stream containing volatile fatty acid (VFA). These VFAs i.e. acetic, propionic and butyric acids are the intermediates for the production of biohydrogen (Reunsang et. al., 2006; Sangyoka et. al., 2007). Biohydrogen is expected to receive more attention in near future as a promising alternative energy due to the fact that its combustion is clean and non-polluting.

Cassava wastewater also can be used as fertilizer, herbicide, insecticide, biosurfactant and *etc.* (Aisien et. al., 2010) and last but not least, Virunanon et. al. (2012) reported this wastewater can be utilized to replace the fresh water used in fermentation practices to produce bioethanol form other resources. This not only results in saving water and preventing the discharge of polluting wastewater, but also in an increased ethanol yield, due to the starch content of the cassava wastewater.

## 6. Conclusions

The applications of tapioca or cassava wastes such as peel, pulp as well as the processing wastewater containing high amounts of carbohydrates in generating valuable products such as biosugars and bioethanol were discussed. Moreover, since the conversion process of tapioca wastes mostly involves enzymatic processes, therefore, the application of immobilized enzyme technology in order to enhance sugar (glucose) yield was recommended. Finally, the utilization of tapioca wastes to produce biosugars and the other valuable product such as bioethanol and biohydrogen seems like a critical step in order to achieve a fully-economized process while preserving the environment in the developing countries.

**Table 2.**
Methods of enzyme immobilization.

| Methods | Definition | Advantages | Disadvantages | Ref. |
|---|---|---|---|---|
| **Physical Adsorption** | Nonspecific physical binding between the enzyme protein and the surface of the matrix brought about by mixing a concentrated solution of enzyme with the solid | Cheap, fast, simple processes, no chemical changes to support or enzymes are necessary, reversible immobilization | Leakage of the enzyme from the support, the possible steric hindrance by the support and the nonspecific binding. | Abdelmajeed *et. al.*, 2012 |
| **Entrapment** | Enzyme is free in solution, but restricted in movement by the lattice structure of a gel | Allow free diffusion of low molecular weight substrates and the reaction products | The support acts as a barrier to mass transfer. Only low molecular weight substrates can diffuse rapidly in the enzyme | Abdelmajeed *et. al.*, 2012 |
| **Cross-linking** | Intermolecular cross linking of protein, either to other protein molecules/ polymerized gel or to functional groups on an insoluble support intra matrix | The enzyme easily separated from the reaction mixture and reused | Very low immobilization yields, the absence of mechanical properties and the poor stability | Abdelmajeed *et. al.*, 2012 |
| **Covalent binding** | The formation of covalent bonds between a support material and some functional groups of the amino acid residues on the surface of the enzyme | The strength of the bonds and the consequent stability of immobilization | High costs and low yields, as the enzyme conformation and of course activity is be strongly influenced by the covalent binding | Abdelmajeed *et. al.*, 2012 |
| **Encapsulation** | Enveloping the biological components within various forms of semi permeable membranes, usually microcapsules varying from 10-100 μm in diameter. | Large proteins cannot pass out of or into the capsule, but small substrates and products can pass freely across the semi permeable membrane | Acute diffusion problem | Park et. al., 2010 |

## References


Abdul Rahim, S. N., Sulaiman A., Ku Hamid, K. H., 2013. Performance encapsulated enzymes within calcium alginate-clay beads in a stirred bioreactor for biosugar production. Adv. Environ. Biol. 7, 3783-3788.

Abdulmajeed, N. A., Khelil, O. A., Danial, E. N., 2013. Immobilization technology for enhancing bio-products industry. Afr. J. Biotechnol. 11, 13528-13539.

Aisien, F. A., Aguye, M. D., Aisien, E. T., 2010. Blending of ethanol produced from cassava waste water with gasoline as source of automobile fuel. Electron. J. Environ. Agric. Food. Chem. 9, 946-950.

Akponah, E., Akpomie, O. O., 2012. Optimization of bio-ethanol production from cassava effluent using Saccharomyces cerevisiae. Afr. J. Biotechnol. 11, 8110-8116.

Ayoola, A. A., Adeeyo, O. A., Efeovbokhan, V. C., Ajileye, O., 2012. A comparative study on glucose production from sorghum bicolor and manihot esculenta species in Nigeria. Intl. J. Sci. Technol. 2, 353-357.

Baskar, G., Muthukumaran, C., Renganathan, S., 2008. Optimization of enzymatic hydrolysis of manihot esculenta root starch by immobilized α-amylase using reponse surface methodology. Intl. J. Chem. Biomol. Eng. 13, 364-368.

Chavalparit, O., Ongwandee M., 2009. Clean technology for the tapioca starch industry in Thailand. J. Clean. Prod. 17, 105-110.

Chotineeranat, S., Pradistsuwana, C., Siritheerasas, P., Tantratian, S., 2004. Reducing sugar production from cassava pulp using enzymes and ultra-filtration 1: enzymatic hydrolyzation. J. Sci. Res. Chulalongkorn Univ. 29, 119-128.

Djuma'ali, Soewarno, N., Sumarno, Primarini, D., Sumaryono, W., 2011. Cassava pulp as a biofuel feedstock of an enzymatic hydrolysis process. J. Makara Teknol. 15, 183-192.

Gaewchingduang, S., Pengthemkeerati, P., 2010. Enhancing efficiency for reducing sugar from cassava bagasse by pretreatment. J. Word Acad. Sci. Eng. Technol. 46, 727-730.

Hermiati, E., Azuma, J., Mangunwidjaja, D., Sunarti, T., Suparno, O., Prasetya, B., 2012. Hydrolysis of carbohydrates in cassava pulp and tapoca flour under microwave irradiation. Indones. J. Chem. 11, 238-245.

Hermiati, E., Mangunwidjaja, D., Sunarti, T., Suparno, O., Prasetya, B., 2012. Potential utilization of cassava pulp for ethanol production in Indonesia. J. Sci. Res. Essays, 7, 100-106.

Kamaraj, A., 2006. Biofuel production from tapioca starch industry wastewater using a hybrid anaerobic reactor. Energy Sustain. Dev. 10, 73-77.

Kongkiattikajorn, J., Sornvoraweat, B., 2011. Comparative study of bioethanol production from cassava peels by monoculture and co-culture of yeast. Kasetsart J.: Nat. Sci. 45, 268-274.

Mahmoud, D. A. R., Helmy, W. A., 2009. Potential application of immobilization technology in enzyme and biomass production. J. Appl. Sci. Res. 5, 2466-2476.

Murty, V., Bhat, J., Muniswaran, P., 2002. Hydrolysis of oils by using immobilized lipase enzyme: A review. Biotechnol. Bioprocess Eng. 7, 57-66.

Nasr, M., Tawfik, A., Ookawara, S., Suzuki, M., 2013. Biological hydrogen production from starch wastewater using a novel up-flow anaerobic staged reactor. BioResour. 8, 4951-4968.

Obadina, A. O., Oyewole, O. B, Sanni, L. O., Abiola, S. S., 2006. Fungal enrichment of cassava peels protein. Electron. Afr. J. Biotechnol. 5, 302-304.

Oboh, G., 2006. Nutrient enrichment of cassava peels using mixed culture of Saccharomyces carevisae and Lactobacillus spp solid media fermentation techniques. Electron. J. Biotechnol. 9, 46-49.

Olanbiwoninu, A. A., Odunfa, S. A., 2012. Enhancing the production of reducing sugars from cassava peels by pretreatment methods. Intl. J. Sci. Technol. 2, 650-657.

Pandey, A., Sccol, C. R., Nigam, P., Soccol, V. T., Vandenberghe, L. P. S., Mohan, R., 2000. Biotechnological potential of agro-industrial residues. II: cassava bagasse. Bioresour. Technol. 74, 81-87.

Park, B. W., Yoon, D. Y., Kim, D. S., 2010. Recent progress in bio-sensing techniques with encapsulated enzyme. Biosens. Biotechnol. 26, 1-10.

Rattanachomsri, U., Tanapongpipat, S., Eurwilaichitr, L., Champreda, V., 2009. Simultaneous non-thermal saccharification of cassava pulp by multi-enzyme activity and ethanol fermentation by Candida tropicalis. J. Biosci. Bioeng. 107, 488-493.

Reunsang, A., Sangyoka, S., Imai, T., Chaiprasert, P., 2006. Biohydrogen production from cassava starch manufacturing wastewater. Asian J. Energy Env. 7, 367-377.

Sangyoka, S., Reungsang, A., Samart, M., 2007. Repeated batch fermentative for biohydrogen production from cassava starch manufacturing wastewater. Pak. J. Biol. Sci. 10, 1782-1789.

Sheldon, R. A., 2007. Enzyme Immobilization: The Quest for Optimum Performance. Adv. Synth. Catal. 349, 1289-1307.

Silva, R. D. N., Quintino, F. B., Monteiro, V. N., Asquieri, E. R., 2010. Production of glucose and fructose syrups from cassava (Manihot esculenta Crantz) starch using enzymes produced by microorganisms isolated from Brazilian Cerrado soil. Ciênc. Tecnol. Aliment. 30, 2466-2476.

Srinorakutara, T., Kaewvimol, L., Saengow, L., 2006. Approach of cassava waste pretreatments for fuel ethanol production in Thailand. J. Sci. Res. Chulalongkorn Univ. 31, 77-84.

Virunanon, C., Ouephanit, C., Burapatana, V., Chulalaksananukul, W., 2012. Cassava pulp enzymatic hydrolysis process as a preliminary step in bio-alcohols production from waste starchy resources. J. Clean. Prod. 1-7.

Woiciechowski, A. L., Nitsche, S., Pandey, A., Ricardo, C., 2002. Acid and enzymatic hydrolysis to recover reducing sugars from cassava bagasse: an economic study. J. Braz. Arch. Biol. Technol. 45, 393-400.

Wongskeo, P., Rangsunvigit, P., Chavadej, S., 2012. Production of glucose from the hydrolysis of cassava residue using bacteria isolates from Thai higher termies. J. Word Acad. Sci. Eng. Technol. 64, 353-35.

Wuttiwai, P., 2009. Cassava solves world's crisis. The 9th International Students Summit on Food, Agriculture and Environment in the New Century. Tokyo, Japan.

Yoonan, K., Kongkiattikajorn, J., 2004. A study of optimal conditions for reducing sugars production from cassava peels by diluted acid and enzymes. Kasetsart J.: Nat. Sci. 38, 29-35.

# 2

# Microwave-assisted methyl esters synthesis of Kapok (*Ceiba pentandra*) seed oil: parametric and optimization study

Awais Bokhari, Lai Fatt Chuah, Suzana Yusup*, Junaid Ahmad, Muhammad Rashid Shamsuddin, Meng Kiat Teng

*Biomass Processing Laboratory, Centre of Biofuel and Biochemical Research (CBBR), Chemical Engineering Department, Universiti Teknologi PETRONAS, Bandar Seri Iskandar, 32610 Seri Iskandar, Perak, Malaysia.*

## HIGHLIGHTS

- Kapok oil methyl ester was produced by microwave-assisted technique.
- About 14 to 37 fold lesser reaction time using microwave than mechanical stirring.
- Optimization and parametric study by using response surface methodology.
- The properties of the produced fuels met EN 14214 and ASTM D 6751.

## GRAPHICAL ABSTRACT

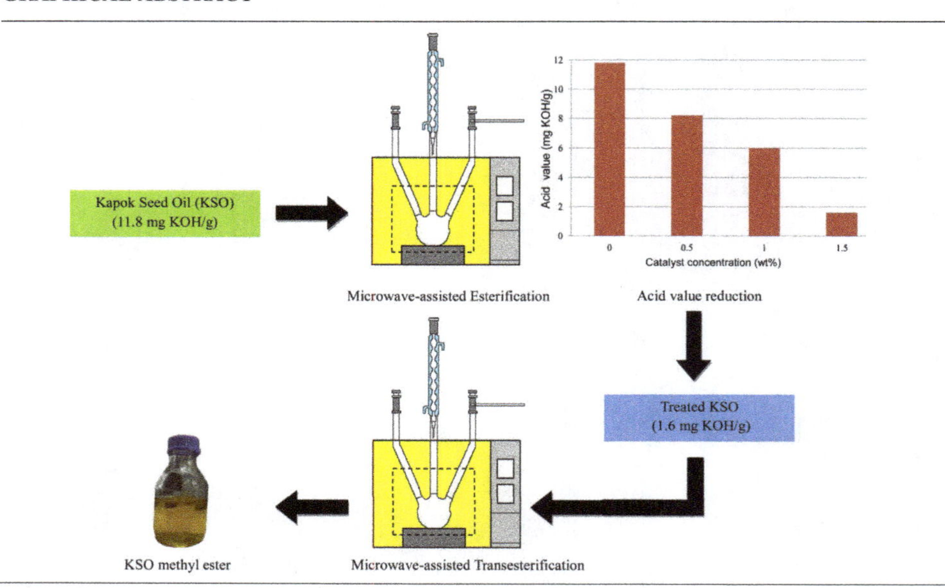


## ABSTRACT

The depleting fossil fuel reserves and increasing environmental concerns have continued to stimulate research into biodiesel as a green fuel alternative produced from renewable resources. In this study, Kapok (*Ceiba pentandra*) oil methyl ester was produced by using microwave-assisted technique. The optimum operating conditions for the microwave-assisted transesterification of Kapok seed oil including temperature, catalyst loading, methanol to oil molar ratio, and irradiation time were investigated by using Response Surface Methodology (RSM) based on Central Composite Design (CCD). A maximum conversion of 98.9 % was obtained under optimum conditions of 57.09 °C reaction temperature, 2.15 wt% catalyst (KOH) loading, oil to methanol molar ratio of 1:9.85, and reaction time of 3.29 min. Fourier Transform Infra-Red (FT-IR) spectroscopy was performed to verify the conversion of the fatty acid into methyl esters. The properties of Kapok oil methyl ester produced under the optimum conditions were characterized and found in agreement with the international ASTM D 6751 and EN 14214 standards.

**Keywords:**
Microwave
Kapok (*Ceiba pentandra*) seed oil
Biodiesel
Response surface methodology

* Corresponding author
E-mail address: drsuzana_yusuf@petronas.com.my

## 1. Introduction

The fast growing population and industrialization have relentlessly increased the demand for energy (Moradi et al., 2015). Generally, energy demands are mostly met by non-renewable resources, such as petroleum, natural gas, and coal. It is well documented that these resources have serious negative environmental impacts particularly due to the emissions of nitrogen oxides, sulphur oxides, unburned hydrocarbons, and particulate matters (Chuah et al., 2015a; Jaber et al., 2015). Renewable fuels such as bioethanol and biodiesel are amongst the best alternatives to fossil fuels (Atabani et al., 2015).

Biodiesel is advantageous as a replacement for conventional diesel because up to a certain inclusion rate, it does not require any modifications in diesel engines. This green fuel is defined as the mono-alkyl esters of long chain fatty acids derived from triglycerides of vegetable oils and animal fat (Fadhil, 2013). These triglycerides cannot be used directly in diesel engines due to the high viscosity of the oils and their low volatility resulting in incomplete combustion and carbon depositions. Thus, they need to be converted into biodiesel *via* a number of processes such as transesterification (Sharma et al., 2008). In the transesterification reaction, triglycerides react with an alcohol (methanol or ethanol) in the presence of a catalyst (Chuah et al., 2015b). The catalyst could be either acidic or basic depending on the free fatty acid content of the oil feedstock (Georgogianni et al., 2009). Stoichiometrically, one mole of triglycerides reacts with three moles of methanol to produce three moles of methyl ester and one mole of glycerin (Chuah et al., 2015c).

Malaysia is diversifying its biodiesel feedstock towards non-edible plant oils, such as *Ceiba pentandra* (Kapok), *Calophyllum inophyllum* (bintangor laut / nyamplung / penaga laut), *Jatropha curcas* (Jatropha), *Ricinus communis* (Castor), *Hevea brasiliensis* (Rubber), and waste from palm oil processing. These local non-edible plant oils have drawn attention as biodiesel feedstock due to their potential and abundant supply. *Ceiba pentandra* locally known as Kapok or Kekabu is grown in Malaysia, India and other parts of Asia. It grows naturally in humid and sub-humid tropical regions. Kapok pod contains 17% fiber that is mainly utilized in making pillows and mattresses, whereas the seeds are traditionally considered as waste (Ong et al., 2013). Kapok seeds make up about 25 - 28 wt% of each pod with an average potential oil yield of 1280 kg/ha annually (Yunus Khan et al., 2015). The most common method for extracting oil from the Kapok seeds is mechanical expeller (Vedharaj et al., 2013). The use of Kapok seeds is well in line with the purpose of the second generation biodiesel production, i.e., utilization of non-edible feedstock to avoid direct conflict with human food (Lee et al., 2011).

Various non-edible oils have been used for biodiesel production through transesterification of triglycerides by using different methods, such as mechanical stirring, supercritical procedure (Ong et al., 2013), ultrasonic techniques (Ji et al., 2006), hydrodynamic cavitation (Chuah et al., 2015b), and microwave (Lee et al., 2010). However, only a few studies have reported on biodiesel production from Kapok seed oil (KSO). Among them was the study recently conducted by Yunus Khan et al. (2015) who investigated the fuel properties of a biodiesel obtained from the blends of *Ceiba pentandra* and *Nigella sativa* by mechanical stirring. Sivakumar et al. (2013) also studied the effect of molar ratio of methanol to Kapok oil, temperature, time, and catalyst concentration on biodiesel production process by using mechanical stirring method. In a different study, Vedharaj et al. (2013) reported that biodiesel derived from Kapok oil emitted higher nitrogen oxides compared to diesel fuel. The authors further strived to reduce the nitrogen oxides by using urea based selective non-catalytic reduction system, which was retrofitted in the exhaust pipe (Vedharaj et al., 2014).

Transesterification reaction assisted by mechanical stirring has been widely used for biodiesel production. However, there are a number of problems associated with this technique, i.e., long reaction time, non-uniform heat distribution, and large energy requirements. These drawbacks have rendered researchers to find alternative methods. Microwave-assisted biodiesel production has been investigated using various oil feedstock, e.g., *Camelina sativa* oil, rice bran oil, *pongamia pinnata*, tallow, yellow horn oil, castor oil, used cooking oil, palm oil, coconut oil, soybean oil, and *Jatropha curcas* (Motasemi and Ani, 2012; Yunus khan et al., 2014). This technique has advantages over the mechanical stirring method including shorter reaction time, efficient heating, and facilitated separation of glycerol from biodiesel (Motasemi and Ani, 2012). Despite its advantages, microwave-assisted transesterification has never been attempted on raw KSO.

Therefore in the present study, the design of experiment and optimization of microwave-assisted transesterification of Kkapok oil into biodiesel was conducted by incorporating four reaction parameters, namely methanol to oil molar ratio, catalyst concentration, temperature, and reaction time using response surface methodology (RSM) and four-way analysis of variance (ANOVA). The response (i.e., methyl ester conversion) was fitted by a quadratic polynomial regression model using least square analysis in a five-level-four-factor central composite design (CCD). The quality of Kapok oil methyl ester produced was investigated according to the ASTM D 6751 and EN 14214 standards.

## 2. Materials and methods

KSO was purchased from the East Jawa Province, Indonesia. Solvent and chemical used in the experiments, i.e., anhydrous methanol, 95 % sulphuric acid ($H_2SO_4$), and potassium hydroxide (KOH) were analytical grade. All chemicals were purchased from Merck (Malaysia) except for $H_2SO_4$ which was purchased from Sigma Aldrich (Malaysia).

### *2.1. Kapok seed oil characterization*

The properties of KSO including acid value, saponification value, iodine value, density, kinematic viscosity, and flash point were analyzed. All analysis were performed by following the AOCS, DIN and ASTM methods (Chuah et al., 2015a).

### *2.2. Pretreatment of kapok seed oil*

Due to high acid value, the KSO was pretreated to ensure a high conversion rate into methyl esters. To reduce the acid value of the KSO, acid esterification reaction was performed under microwave irradiations by reacting KSO with methanol (1:6) in the presence of sulphuric acid (1.5 wt.%) as catalyst at 60 °C for 5 min. The acid value of the KSO was measured after the pretreatment to ensure that it was lower than 2 % before proceeding with the alkali-catalyzed transesterification reaction (Ramadhas et al., 2005).

### *2.3. Experimental design*

The experimental arrays were designed by CCD The reaction variables and their respective ranges are shown in **Table 1**. The independent input process variables were primarily classified in terms of low and high levels. The factors were further distributed into versatile points called axial, center and factorial points. The axial points were coded by the CCD as -2 and +2. The low and high lever factor points were designated as -1 and +1. Whereas the centre points were coded as 0 and the repeated experimental arrays were designed on the centre points.

**Table 1.**
Design Parameters for transesterification process.

| Process parameters | -2 | -1 | 0 | +1 | +2 |
|---|---|---|---|---|---|
| Methanol to oil (molar ratio) | 2 | 6 | 10 | 14 | 18 |
| Catalyst loading (wt%) | 0 | 1 | 2 | 3 | 4 |
| Temperature (°C) | 25 | 40 | 55 | 70 | 85 |
| Time (min) | 0.5 | 2 | 3.50 | 5 | 6.5 |

### *2.4. Transesterification and microwave configuration*

Transesterification was performed according to the experimental design set by RSM and following the procedure described by Yusup and Khan (2010). A total of 30 runs were performed under different combinations of process parameters. In each run, 50 g of KSO was mixed with a specified amount of methanol-KOH solution in a 500 mL three-neck round-bottom reactor, and was heated and stirred in a 3000 W microwave (**Fig. 1**) for a specified time period set by the design of expert. Upon completion of the reaction, the mixture was transferred into a separating funnel and was left for 6 h for complete separation. Two layers of immiscible phases were obtained. The upper layer consisted of Kapok oil methyl ester (KOME), whereas the lower layer included the by-product and residues consisting of glycerol, excess methanol and un-reacted catalyst. KOME was separated and washed with warm deionized water to remove residual catalyst. Rotary evaporator was then employed to remove the residual water in KOME.

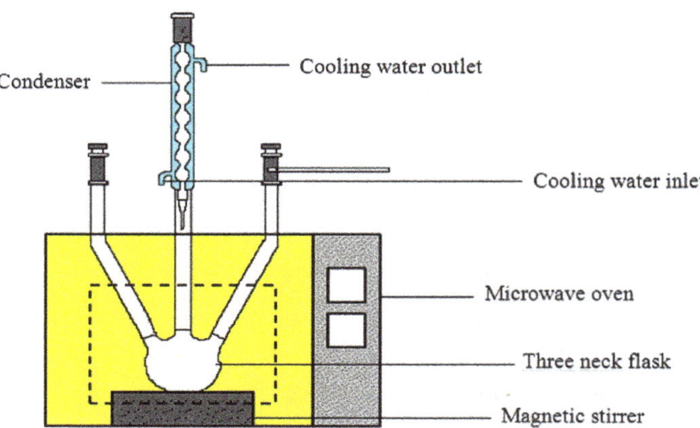


**Fig.1.** Experimental setup

### 2.5. *Fatty acid methyl ester analysis*

The produced KOME was analyzed by using Gas Chromatography (GC) to determine the fatty acid methyl ester (FAME) conversion achieved in each experimental run based on the EN 14013 standard method (Knothe, 2006). The Agilent-Technologies 7890A model GC was used for the FAME determination. The GC system was equipped with a variable split flow injector, a temperature programmable oven, a flame ionization detector, and a capillary column coated with methylpolysiloxane (DB-23) (60 x 0.25 mm; film thickness 0.25μm).

Temperature program included holding for 2 min at 100 °C, heating at 10 °C/min until 200 °C, heating at 5 °C/min until 240 °C and holding for 7 min. Helium was used as the carrier gas at a flow rate of 4 mL/min. Hydrogen and air were used at flow rates of 50 and 400 mL/min, respectively, for flame. The injector temperature and detector temperature were set at 250 °C. The volume of the sample injected was 1 μL. All the experiments were conducted in three replicates and the reported values are averages of the individual runs and the inaccuracy percentage was less than 2% of the average value. The properties of the purified biodiesel were analysed according to both ASTM and EN standards.

### 2.6. *Fourier Transform Infra-Red Spectroscopy*

Fourier transform infra-red (FTIR) spectroscopy was used to analyze the conversion of the KSO into KOME.

## 3. Results and discussion

### 3.1. *Kapok seed oil characterization*

Physiochemical analysis was performed on KSO to analyze the quality of the feedstock. The results were given in **Table 2**. The acid value is an important parameter to indicate the quality, age and purity degree of an oil during processing and storage. Oil samples possessing acid values > 4 mg KOH/g require a two-step processing, e.g. acid esterification followed by alkaline transesterification. The results of the present study revealed that the acid value of the KSO was 11.8 mg KOH/g, which was higher than the set point of 4 mg KOH/g. Saponification is a process by which the fatty acids in the glycerides of oil are hydrolysed by an alkali. The results obtained revealed that the saponification value of the KSO was 194 mg KOH/g. The density (at 20 °C), kinematic viscosity (at 40 °C) and flash point of the KSO were 0.91 g/cm$^3$, 36.21 mm$^2$/s and 210 °C, respectively.

**Table 2.**
Physiochemical properties of the kapok seed oil.

| Analysis | Crude kapok seed oil |
|---|---|
| Acid value (mg KOH/g) | 11.80 |
| Saponification value (mg KOH/g) | 194.00 |
| Iodine value (g $I_2$/100 g) | 102.40 |
| Density at 20 °C (g/cm$^3$) | 0.91 |
| Kinematic viscosity at 40 °C (mm$^2$/s) | 36.21 |
| Flash point (°C) | 210 |

### 3.2. *Pretreatment of kapok seed oil*

It can be observed from **Figure 2** that a significant reduction in acid value from 11.8 to 1.6 mg KOH/g was achieved by using 1.5 wt% of $H_2SO_4$ as a catalyst, methanol to oil ratio molar ratio of 6:1, and reaction time of 5 min at 60 °C.

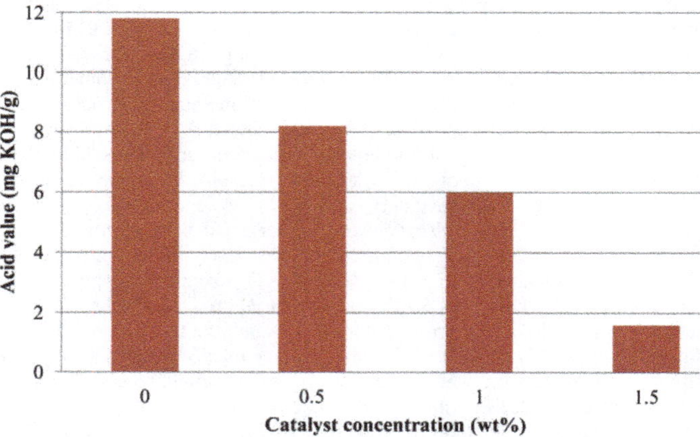


**Fig.2.** Acid value reduction after pretreatment.

### 3.3. *Optimization study of transesterification process variables*

RSM was used to optimize the four process variables, i.e., methanol to oil molar ratio, catalyst concentration, reaction temperature, and reaction time. Thirty runs at points were set based on the Design Expert. The response FAME conversion was calculated by the GC-FID and the results were compared with the predicted response (**Table 3**). The FAME conversion was observed in the range of 37.40 to 98.98 %. More specifically, the lowest conversion of 37.40 % was associated with the 10:1 methanol to oil molar ratio in the absence of catalyst at 55 °C for 3.5 min of reaction time, while the maximum FAME conversion of 99.98 % was obtained at 10:1 methanol to oil molar ration in the presence of 2 wt% catalyst at 55 °C for 3.50 min of reaction time.

### 3.4. *ANOVA analysis of base transesterification*

**Table 4** shows the ANOVA results of the base transesterification. P-value of this model was <0.0001 showing that the model was significant. P-value represents the significance of the model and F-value represents the most influencing factor in a study (Lee et al., 2005). The significance of the reaction parameters with regard to FAME conversion was in the order of catalyst concentration > temperature > reaction time > methanol to oil molar ratio.

The FAME was the response of the process variables in this study and the factor methanol to oil molar ratio (A), catalyst concentration (B), reaction temperature (C), and reaction time (D) were the process variables. The $R^2$ value was measured at 0.9085 (**Table 4**), revealing that the experimental data validated 90.85 % of the model. The regression analysis resulted in a response surface equation for the output response, i.e., FAME conversion.

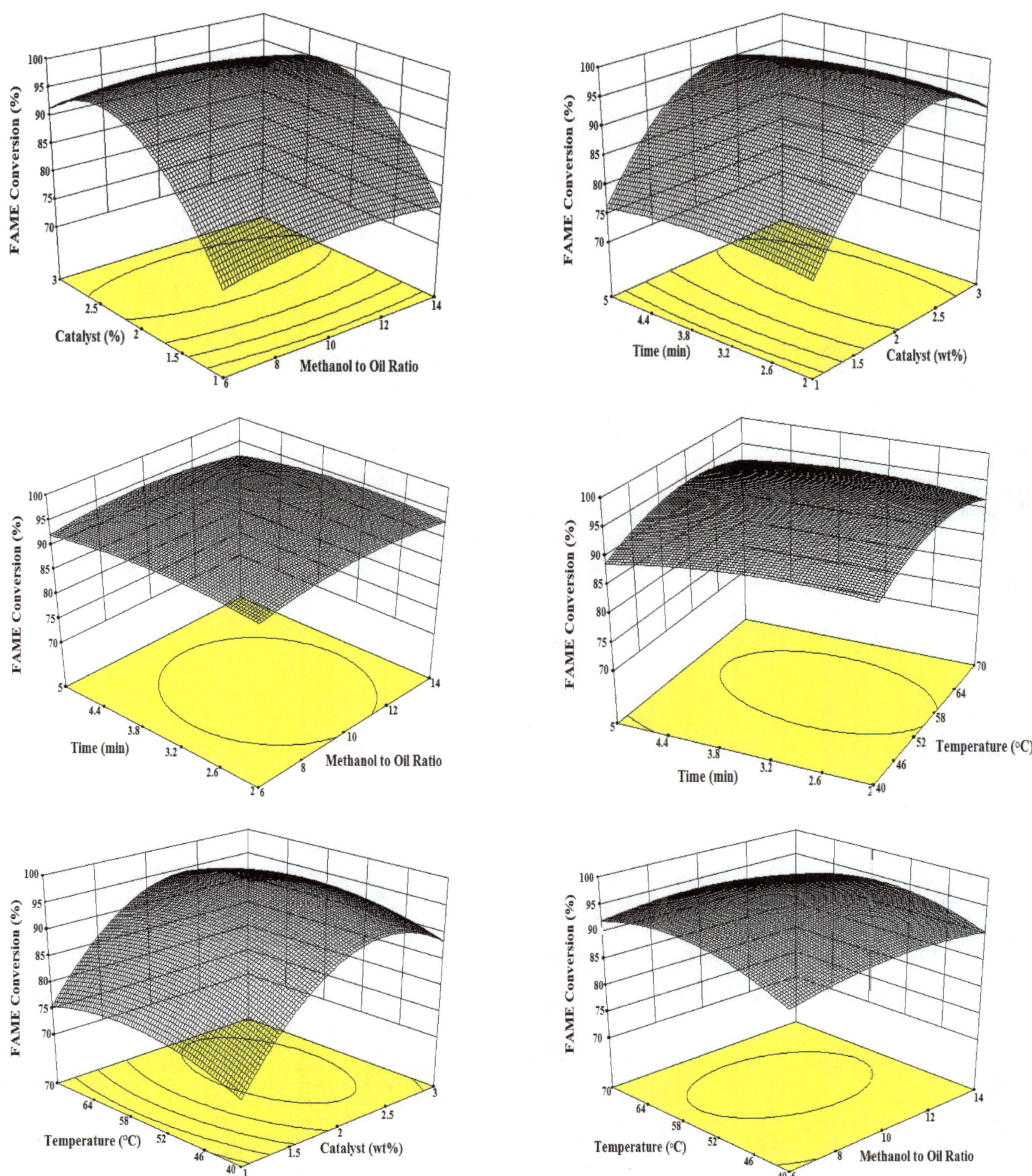


**Fig.3.** 3-D plots of process variables with respect to fatty acid methyl ester conversion.

**Table 3.**
Transesterification experimental designed and response of fatty acid methyl ester conversion.

| Run | Methanol to oil molar ratio | Catalyst concentration (wt%) | Temperature (°C) | Time (min) | Experimental FAME conversion (%) |
|---|---|---|---|---|---|
| 1 | 6.00 | 1.00 | 40.00 | 2.00 | 74.14 |
| 2 | 6.00 | 1.00 | 70.00 | 2.00 | 77.57 |
| 3 | 6.00 | 1.00 | 40.00 | 5.00 | 72.52 |
| 4 | 6.00 | 1.00 | 70.00 | 5.00 | 76.34 |
| 5 | 6.00 | 3.00 | 40.00 | 2.00 | 79.75 |
| 6 | 6.00 | 3.00 | 70.00 | 2.00 | 80.25 |
| 7 | 6.00 | 3.00 | 40.00 | 5.00 | 75.00 |
| 8 | 6.00 | 3.00 | 70.00 | 5.00 | 80.96 |
| 9 | 14.00 | 1.00 | 40.00 | 2.00 | 76.03 |
| 10 | 14.00 | 1.00 | 70.00 | 2.00 | 77.56 |
| 11 | 14.00 | 1.00 | 40.00 | 5.00 | 73.87 |
| 12 | 14.00 | 1.00 | 70.00 | 5.00 | 77.27 |
| 13 | 14.00 | 3.00 | 40.00 | 2.00 | 79.70 |
| 14 | 14.00 | 3.00 | 70.00 | 2.00 | 77.52 |
| 15 | 14.00 | 3.00 | 40.00 | 5.00 | 75.10 |
| 16 | 14.00 | 3.00 | 70.00 | 5.00 | 75.32 |
| 17 | 10.00 | 2.00 | 25.00 | 3.50 | 82.69 |
| 18 | 10.00 | 2.00 | 85.00 | 3.50 | 85.94 |
| 19 | 10.00 | 2.00 | 55.00 | 0.50 | 91.65 |
| 20 | 10.00 | 2.00 | 55.00 | 6.50 | 93.89 |
| 21 | 10.00 | 0.00 | 55.00 | 3.50 | 37.40 |
| 22 | 10.00 | 4.00 | 55.00 | 3.50 | 72.77 |
| 23 | 2.00 | 2.00 | 55.00 | 3.50 | 89.65 |
| 24 | 18.00 | 2.00 | 55.00 | 3.50 | 92.87 |
| 25 | 10.00 | 2.00 | 55.00 | 3.50 | 98.80 |
| 26 | 10.00 | 2.00 | 55.00 | 3.50 | 98.90 |
| 27 | 10.00 | 2.00 | 55.00 | 3.50 | 98.96 |
| 28 | 10.00 | 2.00 | 55.00 | 3.50 | 98.98 |
| 29 | 10.00 | 2.00 | 55.00 | 3.50 | 98.02 |
| 30 | 10.00 | 2.00 | 55.00 | 3.50 | 98.64 |

**Table 4.**
ANOVA analysis of the Transestrification experiment.

| Source | Sum of Squares | DF | Mean Square | F value | P value |
|---|---|---|---|---|---|
| Model | 4269.52 | 14 | 04.97 | 10.64 | <0.0001 significant |
| A (Methanol) | 0.22 | 1 | 0.22 | 7.554E-003 | 0.9319 |
| B (Catalyst) | 330.34 | 1 | 330.34 | 11.52 | 0.0040 |
| C (Temperature) | 22.39 | 1 | 22.39 | 0.78 | 0.3908 |
| D(Time) | 5.66 | 1 | 5.66 | 0.20 | 0.6630 |
| AB | 9.73 | 1 | 9.73 | 0.34 | 0.5688 |
| AC | 7.21 | 1 | 7.21 | 0.25 | 0.6234 |
| AD | 0.35 | 1 | 0.35 | 0.012 | 0.9137 |
| BC | 3.69 | 1 | 3.69 | 0.13 | 0.7249 |
| BD | 1.92 | 1 | 1.92 | 0.067 | 0.7994 |
| CD | 6.40 | 1 | 6.40 | 0.22 | 0.6434 |
| $A^2$ | 176.87 | 1 | 176.87 | 6.17 | 0.0253 |
| $B^2$ | 3680.06 | 1 | 3680.06 | 128.34 | <0.0001 |
| $C^2$ | 501.42 | 1 | 501.42 | 17.49 | 0.0008 |
| $D^2$ | 128.19 | 1 | 128.19 | 4.47 | 0.0516 |
| Residual | 430.12 | 15 | 28.67 | | |
| Lack of Fit | 429.46 | 10 | 42.95 | 325.18 | < 0.0001 significant |
| Pure Error | 0.66 | 5 | 0.13 | | |
| $R^2$ = 0.9085 | $R^2_{adj}$ = 0.8231 | | Adeq. precision= 14.19 | | |

This equation represented a second order polynomial regression model as shown in (**Eq. 1**).

$$\text{FAME Conversion (\%)} = +98.72 + (0.095\times A) + (3.71\times B) + (0.97\times C) - (0.49\times D) - (0.78\times A\times B) - (0.67\times A\times C) - (0.15\times A\times D) - (0.48\times B\times C) - (0.35\times B\times D) + (0.63\times C\times D) - (2.54\times A^2) - (11.58\times B^2) - (4.28\times C^2) - (2.16\times D^2) \quad \text{(Eq. 1)}$$

*3.5. Parametric analysis*

**Figure 3** depicts the 3-D plots of the transesterification process parameters with respect to the response (i.e., FAME conversion). Excess methanol was used to shift the reaction towards equilibrium. Methyl esters conversion was increased by increasing methanol ratio up to 10, but further increases in this parameter led to no remarkable effect on the response and hindered the glycerol separation. Catalyst loading was observed as the most significant variable effecting the FAME conversion. Maximum conversion of 98.98 % was achieved using 2.0 wt% of the catalyst and further increases in catalyst loading decreased the methyl esters conversion. Reaction rate was increased by increasing the reaction temperature up to 55 °C. The maximum FAME conversion was achieved within minimum reaction time of 3.5 min under microwave irradiations. Hence, the microwave-assisted transesterification method investigated herein proved to be effective in terms of enhancing methyl esters conversion and decreasing reaction time.

*3.6. Fatty acid methyl ester profile of Kapok oil methyl ester*

The fatty acid profile of KOME is presented in **Table 5** and compared with those of other studies conducted on KSO and some non-edible oil feedstock. As shown, the KSO contained higher amounts of unsaturated fatty acids than saturated ones. This attribute of non-edible oil feedstock would result in favourable cold flow properties, but on the other hand, lead to poor oxidation stability. On the contrary, palm oil methyl ester or soybean methyl ester do not possess suitable cold flow properties, but have a good oxidation stability (Ma and Hanna, 1999).

*3.7. Fourier Transform Infra-Red analysis*

The FT-IR analysis was also performed on the KOME. **Figure 4** shows the FT-IR spectrum of KOME confirming the successful ransesterification of the KSO. The band range from 2854 – 3008 $cm^{-1}$ represents the asymmetric stretching of the methyl group. The bands appearing in the range of the 1436 - 1741 $cm^{-1}$ are associated with aldehyde, ketone, and fatty acids. Symmetric stretching and vibration of the hydroxyl group could be observed in the range of the 1169 - 1246 $cm^{-1}$ (Zhang et al., 2012).

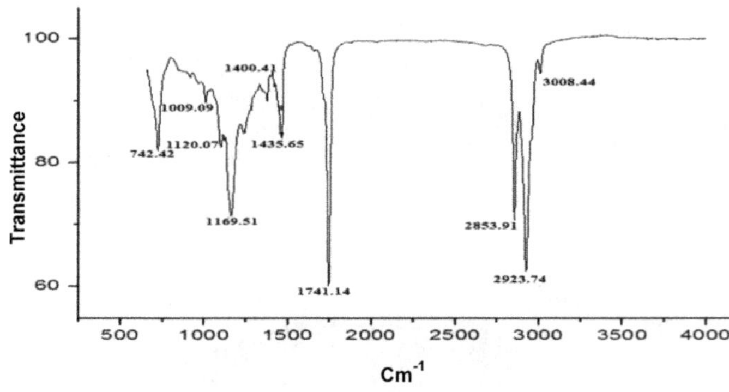


**Fig.4.** FT-IR spectrum of the Kapok oil methyl ester.

*3.8. Optimized conditions and fuel properties*

Optimum conditions for KOME production using microwave-assisted method conducted in the present study and mechanical stirring procedure (Sivakumar et al., 2013; Yunus Khan et al., 2015) are presented in **Table 6**. The optimum conditions were determined by numerical optimization tool in the Design Expert 8.0 software. Basically, numerical optimization selects the most appropriate and optimized values between the ranges of input variables and its associated output response designated arrays which were designed by CCD. Comparison of the optimized values of the current work, i.e., microwave assisted technology and those of Sivakumar et al. (2013) and Yunus Khan et al. (2015) who investigated the mechanical stirring method reveals that biodiesel production using the microwave assisted technology required less reaction time and temperature.

**Table 5.**
Kapok seed oil fatty acid profile

| Fatty acid | Kapok oil methyl ester (This study) | Kapok oil methyl ester Sivakumar et al. (2013) | Kapok oil methyl ester Yunus Khan et al. (2015) | Jatropha oil methyl ester Lee et al. (2011) | Rubber seed oil methyl ester Ahmad et al. (2014) | Rapeseed methyl ester Kusdiana and Saka (2001) |
|---|---|---|---|---|---|---|
| Palmitic (C16:0) | 25.75 | 23.20 | 20.80 | 11.00 | 9.89 | 3.49 |
| Stearic (C18:0) | 5.43 | 5.68 | 2.70 | 4.00 | 9.90 | 0.85 |
| Oleic (C18:1) | 24.32 | 29.69 | 20.10 | 22.00 | 24.89 | 64.40 |
| Linoleic (C18:2) | 42.46 | 35.11 | 38.10 | 53.00 | 35.00 | 22.30 |
| Linolenic (C18:3) | - | - | 1.70 | 8.00 | 16.78 | 8.23 |
| Arachidic (C20:0) | 2.04 | 1.89 | 0.50 | - | - | - |

**Table 7** depicts the fuel properties of the produced KOME with microwave-assisted and mechanical stirring methods. All the fuel properties (except oxidation stability) were in agreement with the international standards, i.e., ASTM D 6751 and EN 14214. Compared to the studies in which mechanical stirring method was used, the present study led to improved fuel properties. For instance, cetane number, which plays a significant role in fuel ignition, was significantly higher in the present study that those of the previous investigations on KOME production using mechanical stirring (Sivakumar et al., 2013). Oxidation stability and cold flow properties of KOME showed significant improvements as compared to the previous studies as well (Yunus Khan et al., 2015). More specifically, the oxidation stability of the KOME herein study was measured at 3.69 h, whereas Yunus Khan et al. (2015) reported a much lower value of 1.14 h. The cloud, pour, and cold filter plugging points of the KOME produced in the present study were also improved compared to those of the mechanical-stirring based study performed by Yunus Khan et al. (2015); 2, 0, and 3 °C, compared to 3, 5, and 4 °C, respectively.

**Table 6.**
Numerical optimization for biodiesel production from Kapok seec oil.

| Optimum conditions | **This study** | Sivakumar et al. (2013) | Yunus Khan et al. (2015) |
|---|---|---|---|
| | **Microwave** | Mechanical stirring | Mechanical stirring |
| Oil to methanol ratio | 1:9.85 | 1:6 | 1:4 |
| Catalyst loading (wt%) | 2.15 | 1 | 1 |
| Temperature (°C) | 57.09 | 65 | 60 |
| RPM | - | 600 | 700 |
| Time (min) | 3.29 | 45 | 120 |
| Predicted conversion (%) | 99.07 | - | - |
| Actual conversion (%) | 98.90 | 99.5 | - |

**Table 7.**
Fuel properties of Kapok oil methyl ester.

| Properties | **This study** | Sivakumar et al. (2013) | Yunus Khan et al. (2015) | Methods | ASTM D 6751 | EN 14214 |
|---|---|---|---|---|---|---|
| | **Microwave** | Mechanical stirring | Mechanical stirring | | | |
| Density 25 °C (kg $m^{-3}$) | 874 | - | 885 (at 15 °C) | ASTM D 5002 | - | 0.86-0.90 |
| Cloud point (°C) | 2 | 1 | 3 | ASTM D 97 | - | - |
| Pour point (°C) | 0 | - | 5 | ASTM D 2500 | - | - |
| Cold filter plugging point (°C) | 3 | - | 4 | ASTM D 6371 | - | - |
| Flash point (°C) | 149 | 169 | 202.5 | ASTM D 93 | ≥ 93 | ≥ 120 |
| Kinematic viscosity 40 °C ($mm^2s^{-1}$) | 1.90 | 4.17 | 4.42 | ASTM D 445 | 1.9 – 6.0 | 3.5 – 5.0 |
| Oxidative stability (h) | 3.69 | - | 1.14 | EN 14112 | - | ≥ 6 |
| Moisture content (wt%) | 0.03 | 0.03 | - | ASTM D 2709 | < 0.05 | < 0.03 |
| Acid value (mg KOH $g^{-1}$) | 0.3 | 0.04 | 0.16 | Cd 3d-63 | < 0.8 | < 0.5 |
| Cetane number | 57.08 | 47 | - | ASTM D 613 | ≥ 47 | ≥ 51 |
| Higher heating value (MJ $kg^{-1}$) | 39.7 | - | 39.4 | ASTM D 4868 | - | - |
| Free glycerin (wt%) | 0.016 | - | - | ASTM D 6584 | ≤ 0.020 | ≤ 0.020 |
| Total glycerin (wt%) | 0.24 | - | - | ASTM D 6584 | ≤ 0.240 | ≤ 0.240 |
| Ester Content (wt%) | 98.9 | 99.5 | - | EN 14103 | - | ≥ 96.5 |

## 4. Conclusions

High free fatty acid content of the KSO was significantly reduced by acid esterification. RSM was used to optimize the process variables for base transesterification. The optimum operating conditions corresponding to 98.90 % KOME conversion were 57.09 °C, 2.15 wt% KOH catalyst loading, 1:9.85 molar ratio of oil to methanol, and 3.29 min of reaction time. A significant reduction (~ 14 - 37 folds) in the optimum reaction time for transesterification was achieved; i.e., from 45 to 120 min for the mechanical stirring approach to 3.29 min for the microwave-assisted approach. The conversion of the fatty acid into methyl ester was verified by The FT-IR analysis and the fuel properties of KOME met the ASTM D 6751 and EN14214 standards.

## Acknowledgments

This research was conducted under ERGS Grant (No. 0153AB-169), MyRA Grant (No. 0153AB-J19), and PRGS Grant (No. 0153AB-K19). The authors would like to thank Universiti Teknologi PETRONAS, Public Service Department of Malaysia, and Marine Department Malaysia for their supports.

## References

Atabani, A.E., Badruddin, I., Masjuki, H.H., Chong, W.T., Lee, K., 2015. Pangium edule Reinw: A Promising Non-edible Oil Feedstock for Biodiesel Production. Arab. J. Sci. Eng. 40, 583-594.

Chuah, L.F., Abd Aziz, A.R., Yusup, S., Bokhari, A., Klemeš, J.J., Abdullah, M.Z., 2015a. Performance and emission of diesel engine fuelled by waste cooking oil methyl ester derived from palm olein using hydrodynamic cavitation. Clean Techn. Environ. Policy. 1-13. DOI: 10.1007/s10098-015-0957-2.

Chuah, L.F., Yusup, S., Abd Aziz, A.R., Bokhari, A., Abdullah, M.Z., 2015b. Cleaner production of methyl ester using waste cooking oil derived from palm olein using a hydrodynamic cavitation reactor. J. Cleaner Prod. DOI:10.1016/j.jclepro.2015.06.112.

Chuah, L.F., Yusup, S., Abd Aziz, A.R., Bokhari, A., Klemeš, J.J., Abdullah, M.Z., 2015c. Intensification of biodiesel synthesis from waste cooking oil (Palm Olein) in a Hydrodynamic Cavitation Reactor: Effect of operating parameters on methyl ester conversion. Chem. Eng. Process. Process Intensif. 95, 235-240.

Fadhil, A., 2013. Biodiesel Production from Beef Tallow Using Alkali-Catalyzed Transesterification. Arab. J. Sci. Eng. 38, 41-47.

Georgogianni, K.G., Katsoulidis, A.K., Pomonis, P.J., Manos, G., Kontominas, M.G., 2009. Transesterification of rapeseed oil for the production of biodiesel using homogeneous and heterogeneous catalysis. Fuel Process. Technol. 90, 1016-1022.

Jaber, R., Shirazi, M.M.A., Toufaily, J., Hamieh, A.T., Noureddin, A., Ghanavati, H., Ghaffari, A., Zenouzi, A., Karout, A., Ismail, A.F., Tabatabaei, M., 2015. Biodiesel wash-water reuse using microfiltration: toward zero-discharge strategy for cleaner and economized biodiesel production. Biofuel Res. J. 2, 148-151.

Ji, J., Wang, J., Li, Y., Yu, Y., Xu, Z., 2006. Preparation of biodiesel with the help of ultrasonic and hydrodynamic cavitation. Ultrasonics. 44, e411-e414.

Knothe, G., 2006. Analyzing biodiesel: standards and other methods. J. Amer. Oil Chem. Soc. 83, 823-833.

Lee, H.V., Yunus, R., Juan, J.C., Taufiq-Yap, Y.H., 2011. Process optimization design for jatropha-based biodiesel production using response surface methodology. Fuel Process. Technol. 92, 2420-2428.

Lee, J.Y., Yoo, C., Jun, S.-Y., Ahn, C.Y., Oh, H.M., 2010. Comparison of several methods for effective lipid extraction from microalgae. Bioresour. Technol. 101, S75-S77.

Lee, K.T., Matlina Mohtar, A., Zainudin, N.F., Bhatia, S., Mohamed, A.R., 2005. Optimum conditions for preparation of flue gas desulfurization absorbent from rice husk ash. Fuel. 84, 143-151.

Ma, F., Hanna, M.A., 1999. Biodiesel production: a review. Bioresour. Technol. 70, 1-15.

Moradi, G.R., Mohadesi, M., Ghanbari, M., Moradi, M.J., Hosseini, S., Davoodbeygi, Y., 2015. Kinetic comparison of two basic heterogenous catalysts obtained from sustainable resources for transesterification of waste cooking oil. Biofuel Res. J. 2, 236-241.

Motasemi, F., Ani, F.N., 2012. A review on microwave-assisted production of biodiesel. Renew. Sustain. Energy Rev. 16, 4719-4733.

Ong, L.K., Effendi, C., Kurniawan, A., Lin, C.X., Zhao, X.S., Ismadji, S., 2013. Optimization of catalyst-free production of biodiesel from Ceiba pentandra (kapok) oil with high free fatty acid contents. Energy. 57, 615-623.

Ramadhas, A.S., Jayaraj, S., Muraleedharan, C., 2005. Biodiesel production from high FFA rubber seed oil. Fuel. 84, 335-340.

Sharma, Y.C., Singh, B., Upadhyay, S.N., 2008. Advancements in development and characterization of biodiesel: A review. Fuel. 87, 2355-2373.

Sivakumar, P., Sindhanaiselvan, S., Gandhi, N.N., Devi, S.S., Renganathan, S., 2013. Optimization and kinetic studies on biodiesel production from underutilized *Ceiba Pentandra* oil. Fuel. 103, 693-698.

Vedharaj, S., Vallinayagam, R., Yang, W.M., Chou, S.K., Chua, K.J.E., Lee, P.S., 2013. Experimental investigation of kapok (Ceiba pentandra) oil biodiesel as an alternate fuel for diesel engine. Energy Convers. Manage.75, 773-779.

Vedharaj, S., Vallinayagam, R., Yang, W.M., Saravanan, C.G., Chou, S.K., Chua, K.J.E., Lee, P.S., 2014. Reduction of harmful emissions from a diesel engine fueled by kapok methyl ester using combined coating and SNCR technology. Energy Convers. Manage. 79, 581-589.

Yunus Khan, T.M., Atabani, A.E., Badruddin, I.A., Ankalgi, R.F., Mainuddin Khan, T.K., Badarudin, A., 2015. Ceiba pentandra, Nigella sativa and their blend as prospective feedstocks for biodiesel. Ind. Crops Prod. 65, 367-373.

Yunus khan, T.M., Atabani, A.E., Badruddin, I.A., Badarudin, A., Khayoon, M.S., Triwahyono, S., 2014. Recent scenario and technologies to utilize non-edible oils for biodiesel production. Renew. Sustain. Energy Rev. 37, 840-851.

Yusup, S., Khan, M.A., 2010. Base catalyzed transesterification of acid treated vegetable oil blend for biodiesel production. Biomass Bioenergy. 34, 1500-1504.

Zhang, Q., Liu, C., Sun, Z., Hu, X., Shen, Q., Wu, J., 2012. Authentication of edible vegetable oils adulterated with used frying oil by Fourier Transform Infrared Spectroscopy. Food Chem. 132, 1607-1613.

# 3

# Process analysis of superheated steam pre-treatment of wheat straw and its relative effect on ethanol selling price

Dave Barchyn, Stefan Cenkowski*

*Department of Biosystems Engineering, University of Manitoba, Winnipeg, Manitoba, Canada.*

## HIGHLIGHTS

- ➢ *Examined superheated steam as a method of pre-treatment for lignocellulosic wheat straw.*
- ➢ *Glucose yield from material pre-treated with superheated steam was measured at 47%.*
- ➢ *Preliminary economic evaluation of the pre-treatment method showed that using superheated steam would produce ethanol as economically as if steam explosion was used.*

## GRAPHICAL ABSTRACT

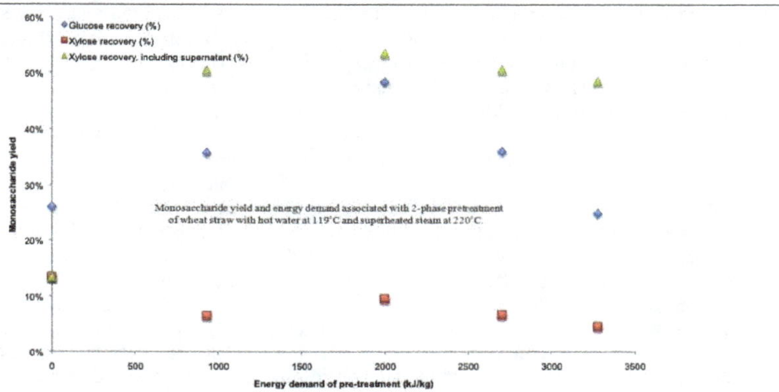


## ABSTRACT

Existing bioethanol operations rely on starch-based substrates, which have been criticized for their need to displace food crops in order to be produced. As an alternative to these first generation biofuels, the use of agricultural residues is being considered to create more environmentally-benign second generation, or cellulosic biofuels. Recalcitrance of these substrates to fermentation requires extensive pre-treatment processes, which often consume more energy than can be extracted from the ethanol that they produce, so one of the priorities in developing cellulosic ethanol is an effective and efficient pre-treatment method. This study examines the use of superheated steam (SS) as a process medium by which wheat straw lignocellulosic material is pre-treated. Following enzymatic hydrolysis, it was found that 47% of the total glucose could be liberated from the substrate, and the optimal conditions for pre-treatment were 15 min in hot water (193 kPa, 119°C) followed by 2 min in SS. Furthermore, a preliminary relative economic analysis showed that the minimum ethanol selling price (MESP) was comparable to that obtained from steam explosion, a similar process, while energy consumption was 22% less. The conclusion of the study is that SS treatment stands to be a competitive pre-treatment technology to steam explosion.

**Keywords:**
Superheated steam
Bioethanol
Lignocellulose
Wheat straw
Pre-treatment

## 1. Introduction

Growing concern surrounding the scarcity of fossil fuels has spurred research into alternative renewable sources for high energy density liquid fuels, such as biologically-derived ethanol. Existing operations rely heavily on sugar- and starch-based ethanol production from dedicated crops such as corn in the United States and sugarcane in Brazil, though the former has been criticized in terms of the net energy balance achieved by the conversion

* Corresponding author
E-mail address: stefan.cenkowski@umanitoba.ca

| Abbreviation | Definition |
|---|---|
| AOAC | Association of Official Agricultural Chemists |
| AFFCO | Association of American Feed Control Officials |
| HPLC | High performance liquid chromatography |
| MESP | Minimum ethanol selling price |
| NREL | National Renewable Energy Laboratory |
| SS | Superheated steam |

process (Shapouri et al., 2002). Increasingly, development of biologically-derived fuels has encountered resistance due to the 'food versus fuel' debate, proponents of which state that displacing food crops for the purpose of cultivating fuel crops is ethically questionable (Pimental, 2003). Among the manifestations of these concerns is the shift from sugar and starch substrates (referred to as 1st generation biofuels) to cellulosic substrates (2nd generation biofuels), obtained from agricultural residues and byproducts, forgoing the need to displace existing food crops for the purpose of fuel production.

Lignocellulosic agricultural residues such as wheat straw represent a promising resource in terms of abundance, feedstock cost, and environmentally benign production (Saha et al., 2005; Brodeur et al., 2011), and can be successfully hydrolyzed and subsequently fermented into ethanol by a variety of organisms (Lynd et al., 2005). While initially attractive, one drawback inherent in the utilization of wheat straw is its relatively high lignin content when compared to corn or sugarcane (Kaparaju and Felby, 2010). Lignin is problematic in that it renders the substrate recalcitrant to hydrolysis and fermentation, resulting in low yields from raw material. Consequently, intensive pre-treatment prior to hydrolysis is required to achieve yields suitable for economical large-scale fuel production (Mosier, et al., 2005). As the pre-treatment can represent a large proportion of the overall conversion cost, it is currently a barrier to widespread adoption in industry (Wyman, 2007).

Several effective pre-treatment technologies have been developed, with varying effectiveness across a variety of substrates. These include, among others, physical, liquid hot water, steam explosion, acid hydrolysis, lime, and wet oxidation pretreatment. A summary of the effects and conditions of various pre-treatment methods are shown in Table 1 (Talebnia et al., 2010). The energy-intensive nature of these pre-treatment methods contributes substantially to their prohibitive costs (Eggeman and Elander, 2005). In addition to the process economics, the large energy expenditures involved in the pre-treatment shift the energy balance more unfavorably for lignocellulosic feedstocks (von Blottnitz and Curran, 2007). The overall conversion process stands to benefit greatly from improvements in the pre-treatment component.

**Table 1.**
Effects of pre-treatment methods on sugar yields from wheat straw.

| Pre-treatment | | Condition | Effect |
|---|---|---|---|
| Physical | | | |
| | Ball milling | Hammer milling at 11.4 kWh/t, 2 hours | Degree of saccharification increased from 17.7% to 61.1% [a] |
| Thermal | | | |
| | Liquid hot water | 80°C, 5-10 mins followed by 195°C, 6-12 min | Increased recovery of hemicellulose and cellulose to 70% and 93%, respectively [b] |
| | Steam explosion | 230°C, 1 min | Increased recovery of hemicellulose and cellulose to 83.7% and 93.5%, respectively [c] |
| Chemical | | | |
| | Acid hydrolysis | 170°C, sulfuric acid | 98% of theoretical glucose yield following enzymatic hydrolysis [d] |
| | Lime | 50°C -135°C, 1 - 24h | Reducing sugar yield increased by factor of 10 [e] |
| Oxidative | | | |
| | Wet oxidation | 10 bar pressure, 170°C | Enzymatic convertibility of cellulose increased to 85% [f] |

[a] Pedersen and Meyer, 2009 [b] Perez et al., 2008 [c] Beltrame et al., 1992
[d] Kootstra et al., 2009 [e] Chang et al., 1998 [f] Bjerre et al., 1996

This study examines the use of superheated steam (SS) as a means of pre-treating wheat straw prior to enzymatic hydrolysis without the use of chemical catalysts and at a potentially reduced energy expenditure. Superheated steam has been successfully implemented into industrial processes such as food processing and drying and biomass decontamination and has led to substantial increases in energy efficiency due to high penetration and energy delivery (Cenkowski et al., 2007; Pronyk et al., 2004). Thermal processes such as steam explosion have been used successfully for pre-treatment of lignocellulosic biomass, though in addition to generation of toxic by-products, it requires the addition of acid catalyst for best results (Alviraet al., 2010). Pretreatment with SS could provide a more energy efficient process, without the need for acid catalysis or the generation of inhibitory compounds.

Economic feasibility of producing ethanol from lignocellulosic substrates has been hampered by the intensive nature of the pre-treatment processes, and the pursuit of more efficient pre-treatment technologies is an important factor (Wyman, 2007).

The objectives addressed in this paper are the effectiveness of SS pre-treatment of lignocellulosic wheat straw in terms of glucose yield from enzymatic hydrolysis, and a relative analysis of the effect that using this pre-treatment method would have on the selling price of ethanol produced.

## 2. Thermal pre-treatment

Typical thermal pre-treatments targeting delignification of lignocellulosic substrates such as steam explosion operate at temperatures of 180 to 230°C (Talebnia et al., 2010). Additionally, depolymerization of lignin occurs at approximately 180°C, and thermal pre-treatments resulting in increased digestibility have been shown to be effective at temperatures ranging from 170 to 270°C (Agbor et al., 2011). The primary goal of an ideal pre-treatment process is to render 100% of the fermentable material accessible to the microorganisms for conversion. The 6-carbon sugars are left intact in the biomass, while the more thermosensitive 5-carbon sugars are recovered from the supernatant (Alvira et al., 2010).

Increased yield has been shown to correlate with the severity of the pre-treatment process (Li et al., 2007), often represented in terms of severity factor, $S_o$, a function of time and temperature (Overend and Chornet, 1987). However, increasing the pre-treatment temperature can also lead to deleterious effects, such as sugar degradation (decreasing the total theoretical yield), and generation of inhibitory compounds (impeding microbial conversion) (Jönsson et al., 2013). For this reason, the temperature selected for pre-treatment with SS in this study was 220°C.

### *2.1. Pre-treatment methodology*

The raw biomass, wheat straw, was obtained from Biovalco Inc. The theoretical yields of cellulose and hemicellulose from the raw biomass were 0.437 g/g and 0.240 g/g, respectively. A compositional analysis of the raw biomass prior to pre-treatment is shown in Table 2. The analysis was performed by the University of Saskatchewan College of Agriculture and Bioresources in accordance with Association of Official Agricultural Chemists (AOAC, 2011) and Association of American Feed Control Officials (AFFCO, 2009) standards.

**Table 2.**
Compositional analysis of raw biomass.

| Parameter | Composition (%) | Method |
|---|---|---|
| Moisture | 4.82 | AOAC 930.15 |
| Dry matter | 95.18 | AOAC 930.15 |
| Lignin | 10.63 | AOAC 973.18 |
| ADF* | 54.37 | AOAC 973.18 |
| NDF** | 78.4 | AOAC 2002.04 |
| Cellulose | 43.74 | cellulose = ADF - lignin |
| Hemicellulose | 24.03 | hemicellulose = NDF - ADF |

* ADF = acid detergent fibre
** NDF = neutral detergent fibre

Initial tests indicated that the effect of SS pre-treatment was enhanced by the use of hot water prior to exposure to SS. Biomass treated using only SS did not show significant improvement in glucose or xylose yields. For this analysis, the samples were subjected to 15 min of hot water treatment followed by 2, 5, or 10 min of SS treatment, using the following steps:

1) Prior to pre-treatment, the raw biomass samples were ground and sieved to <355 μm.
2) Samples were then boiled in pressurized hot water (193 kPa, 119°C) for 15 min using a Lagostina 3L pressure cooker.
3) Following hot water treatment, the samples were placed into the steam chamber where SS at atmospheric pressure was passed through them for the prescribed duration of time in a batch process.

Temperature stability occurred at 217°C, and this was the temperature used for the calculation of the severity factor of each treatment. Following the pre-treatment, the samples were subjected to enzymatic hydrolysis to verify the degree to which the celluloses and hemicelluloses had been rendered accessible for saccharification. The degree to which the sugars were converted is indicative of the potential for fermentation. Hydrolysis was carried out in accordance with the laboratory analytical procedure released by the National Renewable Energy Laboratory (NREL) for the enzymatic saccharification of lignocellulosic biomass (Selig et al., 2008). The enzymes used were lyophilized powder cellulase (from *Aspergillus niger*, 1.3 FPU/mg) and β-glucosidase (from *Prunus amygdalus*, 7.4 pNPGU/mg) supplied by Sigma-Aldrich. Incubation was carried out at 50°C +/- 1°C for a duration of 96 h followed by high performance liquid chromatography (HPLC) analysis for the presence of glucose and xylose. In addition to the hydrolysate, the supernatant from the hot water phase of the treatment was similarly analyzed. The model of HPLC used was the Breeze 2 system from Waters (Mississauga, ON). The system consisted of an HPX-87H column from Biorad (Hercules, CA) with Micro-guard Cation H+ guard column, and a 2414 refractive index detector. The HPLC system used 5 mM sulfuric acid as the mobile phase at a flow rate of 0.6 mL/min and a column temperature of 45°C.

## 3. Effect on economy of processing

Proper assessment of a pre-treatment technology must include an economic analysis in order to evaluate its feasibility. Existing literature generally evaluates economic viability on the basis of a minimum ethanol selling price (MESP) resulting from production costs to compare different technologies (Eggeman and Elander, 2005). While this is a useful tool for evaluation, it is susceptible to uncertainties arising from the proprietary nature of ethanol corporate cost data, as well as the economies of scale (Gallagher et al., 2005). Being a thermal process, SS pre-treatment of wheat straw will be compared to steam explosion, a well-developed technology for pre-treatment of lignocellulose. The main differences between the two processes are a) an acid catalyst is used in the first phase of steam explosion processing, b) SS pre-treatment occurs at atmospheric pressure, and c) steam is passed through the biomass during SS pre-treatment as opposed to the explosive decompression of steam and biomass used in steam explosion.

### *3.1. Methodology*

An equation specific to the economics of cellulosic ethanol was proposed to calculate the cost of ethyl alcohol production (Soloman et al., 2007) and is expressed as follows (Eq. 1):

$$C_A = C_B/95 + C_K + C_L + C_E + C_M + C_O - P_P \qquad \text{Eq. 1}$$

Where
$C_A$ = cost of ethanol production ($/gal)
$C_B$ = cost of biomass feedstock ($/dry short ton)
$C_K$ = cost of capital investment
$C_L$ = cost of labor
$C_E$ = cost of energy
$C_M$ = cost of raw materials
$C_O$ = other costs
$P_P$ = price of excess electric power byproduct to be sold

The following assumptions were made in order to employ this equation: Estimates pertaining to the cost of raw material are generally inconsistent, and are in the range of $10/Mg to $25/Mg (Kaylen et al., 2000; Walsh, 1997). An extensive study of lignocellulosic biomass harvest systems concluded that, depending on crop yield, the cost of raw wheat straw can be estimated as being between $11.26 and $14.01 per Mg, which is the assumption made for this study (Thorsell et al., 2004). An average sized (50 Mgal/year) ethanol plant typically requires approximately $65-$100 million in capital costs and $45-$60 million in annual operating costs (Urbanchuk, 2006). These costs are assumed to encompass the cost of labor and raw materials for this evaluation, and be similar between the two processes. The effect on MESP discussed is in the context of modifying the pre-treatment procedure while keeping all other parameters constant.

Energy costs associated with SS processing are generally evaluated based on the moisture removed from the biomass during the treatment (Berghel and Renstrom, 2002). Processing of wet, pressed agricultural pulps requires approximately 2900 kJ/kg of water removed (Mujumdar, 2006), and the assumption made for this study was that this is an appropriate number to use for the pre-treatment of wheat straw, for the moisture content prior to SS processing is 70-80%. Table 3 shows the energy consumed during the SS processing for each of the treatments. Steam explosion, by contrast, consumes approximately 1800 kJ/kg of biomass (Zhu et al., 2010; Zhu and Pan, 2010).

**Table 3.**
Energy demand associated with SS treatment.

| Treatment | Change in moisture content (kg/kg) | Corresponding energy demand* (kJ/kg) |
|---|---|---|
| 15 min. HW | 0 | 0 |
| 15 min. HW + 2 min. SS | 0.368 | 1068 |
| 15 min. HW + 5 min. SS | 0.611 | 1772 |
| 15 min. HW + 10 min. SS | 0.809 | 2348 |

HW = hot water treatment
*due to SS processing

Similarly to the methodology used in SS processing, steam explosion occurs in 2 steps, the first being an acid impregnation to catalyze delignification. In order to achieve 80% of theoretical ethanol conversion from wheat straw, the raw material was soaked in 0.9% w/w $H_2SO_4$ at 45°C for 18 h prior to steam explosion (Ballesteros et al., 2006). The corresponding step in SS pre-treatment is soaking in hot water at 193 kPa (119°C) for 15 min.

In order to develop an estimate for comparison of the magnitude of energy consumption associated with these thermal processes, the severity factor for each shall be considered. The severity factor is expressed as follows (Eq. 2) (Overend and Chornet, 1987):

$$logR_O = log[t \times exp((T - 100)/14.75)] \qquad \text{Eq. 2}$$

Where
$logR_o$ = severity factor
t = time (min)
T = temperature (°C)

The ratio of severity factors between acid impregnation and hot water prior to SS to steam explosion are 0.42 and 0.52, respectively, so for the purposes of evaluation, it is assumed that the energy expenditure for the treatments are 760 kJ/kg and 930 kJ/kg, respectively.

Power generation from the combustion of excess biomass is assumed to generate energy based on a lower heating value of wheat straw of 16 MJ/kg (Jenkins et al., 1998) and an electrical conversion efficiency of 25% (Energy and environmental analysis, 2008) and sold back to the grid at 0.04$/kWh (Eggeman and Elander, 2005).

## 4. Results and discussion

HPLC analysis of the hydrolyzed samples showed that the glucose yields increased significantly as a result of SS pre-treatment. Xylose however, was shown to decrease following pre-treatment, likely due to being solubilized during the hot water phase of the treatment (Brodeur et al., 2011). Furthermore, glucose yield decreased with increasing time subjected to SS treatment, indicating the possibility of glucose degradation, or the formation of inhibitory compounds (Talebniaet al., 2010). The greatest degree to which the celluloses were hydrolyzed occurred with a treatment of 15 min hot water and 2 min of SS. The glucose and xylose conversion as a result of treatment with SS are shown in Figure 1. The conversion was calculated based on the observed glucose and xylose contents in the hydrolysate compared to the theoretical yield of gluose and xylose from the cellulose and hemicellulose, respectively.

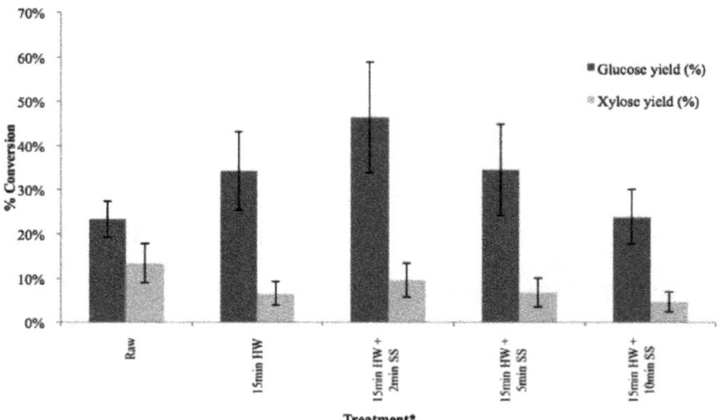


**Fig.1.** Glucose and xylose conversion as a result of enzymatic hydrolysis.
*HW indicates treatment in hot water at 193 kPa, 119°C, SS indicates treatment in superheated steam at 220°C.

While the treatment was successful in increasing the glucose yield from the substrate, there was a marked decrease in the xylose production. In order to ascertain the total xylose that could potentially be recovered from the substrate, the water used in the hot water phase of the pre-treatment was analyzed with the HPLC for the presence of glucose, xylose, and cellobiose that could potentially have been solubilized during that phase of the process.

Results of the analysis showed that in the treatment water, there was an average concentration of 16.6 mM of xylose present, representing a yield of 43.9%. No glucose was detected, and only trace amounts of cellobiose. Taking this into account, the yield of xylose as a result of the combined hot water and SS treatment is shown in Figure 2.

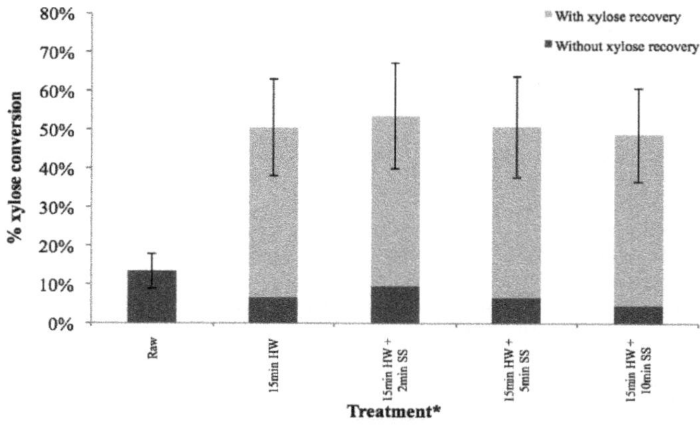


**Fig.2.** Xylose conversion with and without recovery from hot water.
*HW indicates treatment in hot water at 193 kPa, 119°C, SS indicates treatment in superheated steam at 220°C.

There was a substantial increase in xylose recovery if the fraction solubilized in the hot water phase of the treatment was accounted for. This would significantly improve the total yield of sugars of the overall treatment.

The trend observed was that glucose yield tended to increase up to 2 min of SS pre-treatment, after which it declined. Xylose yield came primarily from the hot water phase of the treatment. The purpose of the pre-treatment process was to maximize the total sugar yield, so based on the results obtained, the optimal configuration of SS pre-treatment was 15 min in hot water followed by 2 min in SS. This corresponds to a theoretical ethanol production of approximately 0.103g ethanol/g biomass, or 0.136g ethanol/g biomass with xylose recovery.

Referring back to Eq. 1, and operating under the assumptions listed above, an estimate can be put forth for the MESP of ethanol derived from wheat straw using SS and steam explosion for pre-treatment. The assumed price of $H_2SO_4$ is $100/ton ($0.11/kg) (Bout and Shewchuk, 2013). Table 4 shows the cost breakdown of the conversion of wheat straw to ethanol using steam explosion and SS pre-treatment based on the assumptions listed above. All costs are adjusted per unit volume of ethanol produced. Introduction of xylose recovery is assumed to increase the capital and operating costs by 3%.

**Table 4.**
Cost breakdown of ethanol production.

| Pre-treatment method | Steam explosion | Superheated steam | Superheated steam with xylose recovery |
|---|---|---|---|
| Raw materials | $0.06 | $0.10 | $0.07 |
| Capital cost | $0.44 | $0.44 | $0.45 |
| Operating cost | $0.28 | $0.28 | $0.29 |
| Energy cost* | $0.22 | $0.30 | $0.22 |
| Other cost** | $0.01 | $0.00 | $0.00 |
| Energy revenue | $0.20 | $0.34 | $0.26 |
| **MESP** | **$0.80** | **$0.77** | **$0.78** |

* energy cost is assumed to be $0.07 / kWh
** cost of $H_2SO_4$

While the MESP of ethanol produced with SS pre-treatment is lower than that using steam explosion, a significant portion is due to the increased revenue associated with the energy generated from excess biomass combustion, a result of the lower glucose conversion efficiency, and consequently, more excess biomass. However, if the efficiency of the pre-treatment increases, under the same assumptions, the MESP would decrease, shown in Figure 3.

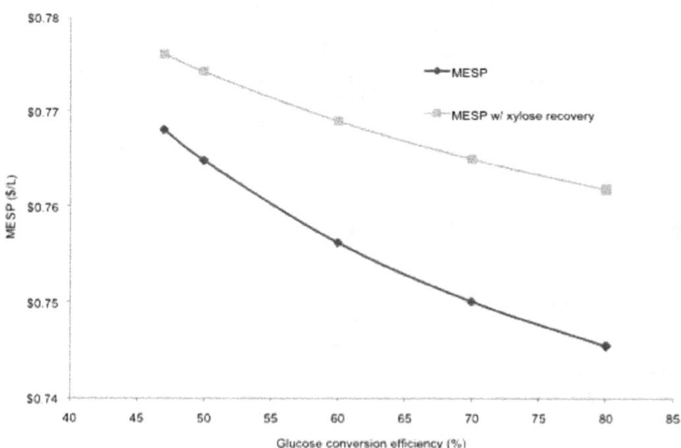


**Fig.3.** Effect of glucose conversion efficiency on MESP of wheat straw pre-treated with SS.

As the goal of this study was not to conduct a comprehensive economic evaluation of the production of cellulosic ethanol using SS pre-treated wheat straw, the preceding analysis yields a MESP not representative of values found in existing literature (Eggeman and Elander, 2005; Tao, et al., 2011). The MESP developed were meant to be used in relative terms, and not absolute. The results are, however, conclusive in that they indicate that pre-treatment with SS is not only technically feasible, but produces a similar result in terms of economic viability to existing pre-treatment technologies. More importantly, SS pre-treatment stands to improve the energy efficiency of the pre-treatment process, as it was shown to consume only 78% of the energy of steam explosion pre-treatment per kg of biomass.

## 4. Conclusion

Pre-treatment of lignocellulosic wheat straw using SS has similar performance characteristics to developed technologies such as steam explosion in terms of economic viability. Furthermore, it is a promising technology in that it consumes less energy than steam explosion, reducing the energy requirements of the pre-treatment step, which is critical to developing viable cellulosic ethanol biorefineries at an industrial scale. Improvements can be made to the SS pre-treatment in order to improve the overall yield from the biomass, which is still comparatively low with regards to other pre-treatment technologies (Conde-Mejia et al., 2012). Hypothetically, increasing the process efficiency to 80% through optimization and addition of catalysts could further decrease the energy consumption by approximately 38%, with only a modest increase in process cost associated with catalyst (0.01$/L ethanol) if the chemical demand is similar to steam explosion. The conclusion drawn from this study is that pre-treatment of wheat straw using a SS process stands to be competitive with existing technology from a technical and economic standpoint, and can be improved upon through optimization.

## 5. Acknowledgments

The authors would like to acknowledge the support of BioFuelNet Canada in the preparation of this work.

## References


AAFCO, 2009. Procedures manual. Association of American Feed Control Officials, Inc. AAFCO.

Agbor, V., Cicek, N., Sparling, R., Berlin, A., Levin, D., 2011. Biomass pretreatment: Fundamentals toward application . Biotechnol. Adv. 29(6), 675-685.

Alvira, P., Tomas-Pejo, E., Ballesteros, M., & Negro, M., 2010. Pretreatment technologies for an efficient bioethanol production process based on enzymatic hydrolysis: a review. Bioresour. Technol. 101(13), 4851-4861.

AOAC, 2011. Official Methods. Journal of AOAC International .

Ballesteros, I., Negro, M. J., Olivia, J., Cabanas, A., Manzanares, P., Ballesteros, M., 2006. Ethanol production from steam-explosion pretreated wheat straw. Appl. Biochem. Biotechnol. 130, 496-508.

Beltrame, P., Carniti, P., Visciglio, A., Focher, B., Marzetti, A., 1992. Fractionation and bioconversion of steam-exploded wheat straw. Bioresour. Technol. 39(2), 165-171.

Berghel, J., & Renstrom, R., 2002. Basic design criteria nad corresponding results performance of a pilot-scale fluidized superheated atmospheric condition steam dryer. Biomass Bioenergy. 23(2), 103-112.

Bjerre, A., Olesen, A., Fernqvist, T., Ploger, A., Schmidt, A., 1996. Pretreatment of wheat straw using combined wet oxidation and alkaline hydrolysis resulting in convertible cellulose and hemicellulose. Biotechnol. Bioeng. 49(5), 568-577.

Bout, J., Shewchuk, S., 2013. 2013 Agriculture, chemical and fertilizer outlook. Toronto, Canada.

Brodeur, G., Yau, E., Badal, K., Collier, J., Ramachandran, K., Ramakrishnan, S., 2011. Chemical and physiochemical pretreatment of lignocellulosic biomass: a review. Enzyme Res. 1-18.

Cenkowski, S., Pronyk, C., Zmidzinska, D., Muir, W., 2007. Decontamination of food products with superheated steam. J. Food eng. 83(1), 68-75.

Chang, V., Nagwani, M., Holtzapple, M., 1998. Lime pretreatment of crop residues bagasse and wheat straw. Appl. Biochem. Biotechnol. 74(3), 135-159.

Conde-Mejia, C., Jimenez-Gutierrez, A., El-Halwagi, M., 2012. A comparison of pretreatment methods for bioethanol production from lignocellulosic materials. Process Saf. Environ. Prot. 90(3), 189-202.

Eggeman, T., Elander, R., 2005. Process and economic analysis of pretreatment technologies. Bioresour. technol. 96(18), 2019-2025.

Energy and environmental analysis, 2008. Technology characterization: steam turbines. Environmental protection agency, Combined heat and power partnership. Washington, DC: USEPA.

Gallagher, P., Brubaker, H., Shapouri, H., 2005. Plant size: capital cost relationships in the dry mill ethanol industry. Biomass Bioenergy. 28(6), 565-571.

Jönsson, L., Alriksson, B., Nilvebrant, N.O., 2013. Bioconversion of lignocellulose: inhibitors and detoxification . Biotechnol. Biofuels. 6(1), 16.

Jenkins, B., Baxter, L., Miles, T. J., Miles, T., 1998. Combustion properties of biomass. Fuel Process. Technol. 54(1), 17-46.

Kaparaju, P., & Felby, C., 2010. Characterization of lignin during oxidative and hydrothermal pre-treatment of wheat straw and corn stover. Bioresour. Technol. 101(9), 3175-3181.

Kaylen, M., VanDyne, D., Choi, Y., Blase, M., 2000. Economic feasibility of producing ethanol from lignocellulosic feedstocks. Bioresour. Technol. 72 (1), 19-32.

Kootstra, A., Beeftink, H., Scott, E., Sanders, J., 2009. Comparison of dilute mineral and organic acid pretreatment for enzymatic hydrolysis of wheat straw. Biochem. Eng. J. 46(2), 126-131.

Li, J., Henriksson, G., Gellerstedt, G., 2007. Lignin depolymerization/repolymerization and its critical role for deligniWcation of aspen wood by steam explosion . Bioresour. Technol.98(16), 3061-3068.

Lynd, L., VanZyl, W., McBride, J., Laser, M., 2005. Consolidated bioprocessing of cellulosic biomass: an update. Curr. Opin. Biotechnol. 16 (5), 577-583.

Mosier, N., Wyman, C., Dale, B., Elander, R., Lee, Y., Holtzapple, M., et al., 2005. Features of promising technologies for pretreatment of lignocellulosic biomass. Bioresour. Technol. 96(6), 673-686.

Mujumdar, A., 2006. Superheated steam drying. Handbook of industrial drying. 2, 1071-1086.

Overend, R., Chornet, E., 1987. Fractionation of lignocellulosics by steam-aqueous pretreatments. Philos. Trans. R. Soc. Lond. 321, 523–536.

Pedersen, M., Meyer, A., 2009. Influence of substrate particle size and wet oxidation on physical surface structures and enzymatic hydrolysis of wheat straw. Biotechnol. prog. 25(2), 399-408.

Perez, J., Ballesteros, I., M., B., Suez, F., Negro, M., Manzanares, P., 2008. Optimizing liquid hot water pretreatment conditions to enhance sugar recovery from wheat straw for fuel ethanol production. Fuel. 87(17), 3640-3647.

Pimental, D.,2003. Ethanol fuels: energy balance, economics, and environmental impacts are negative. Nat. Resour. Res. 12(2), 127-134.

Pronyk, C., Cenkowski, S., Muir, W., 2004. Drying Foodstuffs with Superheated Steam. Drying Technol. 22 (5), 899-916.

Saha, B., Iten, L., Cotta, M., Wu, Y., 2005. Dilute acid pretreatment, enzymatic saccharification and fermentation of wheat straw to ethanol . Process Biochem. 40(12), 3693-3700.

Schell, D., Torget, R., Power, A., Walter, P., Grohmann, K., Hinman, N., 1991. A technical and economic analysis of acid-catalyzed steam explosion and dilute acid pretreatments using wheat straw and aspen wood chips. Appl. Biochem. Biotechnol. 28(1), 87-97.

Selig, M., Weiss, N., Ji, Y., 2008. Enzymatic saccharification of lignocellulosic biomass: Laboratory Analytical Procedure (LAP): Issue Date, 3/21/2008. National Renewable Energy Laboratory.

Shapouri, H., Duffield, J., Wang, M., 2002. The energy balance of corn ethanol: an update (No. 34075). United States Department of Agriculture, Economic Research Service.

Soloman, B., Barnes, J., Halvorsen, K., 2007. Grain and cellulosic ethanol: history, economics, and energy policy. Biomass Bioenergy. 31(6), 416-425.

Talebnia, F., Karakashev, D., Angelidaki, I., 2010. Production of bioethanol from wheat straw: an overview on pretreatment, hydrolysis and fermentation. Bioresour. Technol. 101(13), 4744-4753.

Tao, L., Aden, A., Elander, R., Pallapolu, V., Lee, Y., Garlock, R., et al., 2011. Process and technoeconomic analysis of leading pretreatment technologies for lignocellulosic ethanol production using switchgrass. Bioresour. Technol. 102(24), 11105-11114.

Thorsell, S., Epplin, F., Huhnke, R., & Taliaferro, C., 2004. Economics of a coordinated biorefinery feedstock harvest system: lignocellulosic biomass harvest cost. Biomass Bioenerg. 27(4), 327-337.

Urbanchuk, J., 2007. Contribution of the ethanol industry to the economy of the United States. Prepared for the Renewable Fuels Association. LECG, LLC.

von Blottnitz, H., Curran, M., 2007. A review of assessments conducted on bio-ethanol as a transportation fuel from a net energy, greenhouse gas, and environmental life cycle perspective. J. cleaner prod.15(7), 607-619.
Walsh, M., 1998. US Bioenergy crop economic analyses: status and needs. Biomass Bioenergy. 14(4), 341-350.
Wyman, C., 2007. What is (and is not) vital to advancing cellulosic ethanol. Trends Biotechnol. 25(4), 153-157.
Zhu, J., Pan, X., 2010. Woody biomass pretreatment for cellulosic ethanol production: Technology and energy consumption evaluation. Bioresour. Technol. 101(13), 4992-5002.
Zhu, W., Zhu, J., Gleisner, R., Pan, X., 2010. On energy consumption for size-reduction and yields from subsequent enzymatic saccharification of pretreated lodgepole pine. Bioresour. Technol. 101(8), 2782-2792.

# 4

# Improved microbial conversion of de-oiled Jatropha waste into biohydrogen via inoculum pretreatment: process optimization by experimental design approach

Gopalakrishnan Kumar[1,*], Péter Bakonyi[2], Periyasamy Sivagurunathan[3], Nándor Nemestóthy[2], Katalin Bélafi-Bakó[2], Chiu-Yue Lin[4]

[1] *Center for Materials Cycles and Waste Management Research, National Institute for Environmental Studies, Tsukuba, Japan.*

[2] *Research Institute on Bioengineering, Membrane Technology and Energetics, University of Pannonia, Egyetem u. 10, 8200 Veszprém, Hungary.*

[3] *Department of Environmental Engineering, Daegu University, Republic of Korea.*

[4] *Department of Environmental Engineering and Science, Feng Chia University, 40724 Taichung, Taiwan.*

## HIGHLIGHTS

- *Enhanced hydrogen fermentation via heat-treated inoculum and statistical optimization.*
- *Production was increased by nearly 4 folds through using pretreated inoculum.*
- *Peak hydrogen production rate of 1.42 ± 0.03 L H2/L-d was achieved.*
- *An insight into the microbial aspects of the process was achieved bt PCR-DGGE.*

## GRAPHICAL ABSTRACT

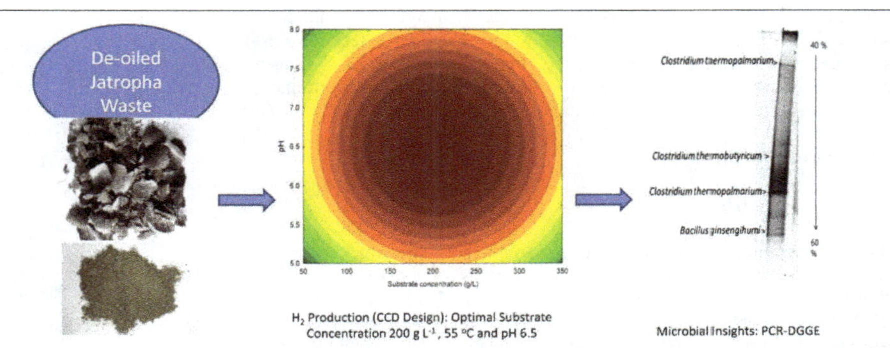


## ABSTRACT

In this study various pretreatment methods of sewage sludge inoculum and the statistical process optimization of de-oiled jatropha waste have been reported. Peak hydrogen production rate (HPR) and hydrogen yield (HY) of 0.36 L $H_2$/L-d and 20 mL $H_2$/g Volatile Solid (VS) were obtained when heat shock pretreatment (95 °C, 30 min) was employed. Afterwards, an experimental design was applied to find the optimal conditions for $H_2$ production using heat-pretreated seed culture. The optimal substrate concentration, pH and temperature were determined by using response surface methodology as 205 g/L, 6.53 and 55.1 °C, respectively. Under these circumstances, the highest HPR of 1.36 L $H_2$/L-d was predicted. Verification tests proved the reliability of the statistical approach. As a result of the heat pretreatment and fermentation optimization, a significant (~ 4 folds) increase in HPR was achieved. PCR-DGGE results revealed that *Clostridium* sp. were majorly present under the optimal conditions.

**Keywords:**
Biohydrogen
De-oiled Jatropha waste
Experimental design
Inoculum pretreatment
Optimization
Pretreatment

* Corresponding author
E-mail address: kumar.gopal@nies.go.jp

## 1. Introduction

The facts of ever increasing energy consumption, diminishing reservoirs of fossil fuels and threatening environmental problems awaked the scientists to find acceptable fuel alternatives both from the ecological and energetic points of views. As a result, various green energy carriers have been proposed, among which hydrogen is a promising candidate due to its unique characteristics (Akil and Jayanthi, 2014). However, the majority of hydrogen is currently derived from non-renewable sources such as methane conversion and oil/naphtha (Ewan and Allen, 2005). Consequently, clean technologies should be developed to make hydrogen a more attractive energy carrier. For this purpose, biological approaches are among the emerging opportunities (Hallenbeck, 2009). In fact, the dark fermentative way to produce hydrogen is one of the most extensively studied fields and is currently the most promising one when practicality is considered (Han et al., 2012; Wang et al., 2012; Diamantis et al., 2013; Sarma et al., 2013). However, utilization of abundant and inexpensive lignocellulosic wastes, such as de-oiled Jatropha waste (DJW) needs adequate microbial consortia (Fan et al., 2006). To obtain such bacterial populations, the seed pretreatment is often a key step. Nevertheless, the most appropriate technique to enrich the hydrogen-producing strains has to be found specifically for each case due to the diversities in the microbial population structure of seed sources of different origins (Mohammadi et al., 2011; Wang et al., 2011). In addition to the importance of seed pretreatment, it is also known that biotechnological hydrogen formation is influenced by the environmental conditions applied. Among them, temperature, pH and substrate concentration are most crucial (Hawkes et al., 2007; Wang and Wan, 2009a). Therefore, these factors must carefully be optimized to improve hydrogen generation e.g. by the experimental design approach (Wang and Wan, 2009b).

In our previous study, it was shown that the DJW could be successfully converted into hydrogen and methane by untreated, mixed microbial flora (Kumar and Lin, 2013). In this work, firstly, the selection of feasible seed pretreatment was attempted. Subsequently, the enhancement of hydrogen production was aimed by optimizing the substrate (i.e. DJW) concentration, pH and temperature using the experimental design approach.

## 2. Materials and methods

### 2.1. Inoculum and DJW

The seed inoculum was collected from a municipal wastewater treatment plant located in central Taiwan. The main features of the inoculum were previously described elsewhere (Kumar and Lin, 2013). The cellulose (14.1%), hemicellulose (24.2%) and lignin (30.4) contents of the DJW were measured by the FIBERTEC$^{TM}$ analyzer as indicated in our previous study (Kumar et al., 2012).

### 2.2. Inoculum pretreatment methods

Several methods were tested for inoculum pretreatment. The heat shock treatment was conducted by heating the sludge in a water bath at 95 °C for 30 min. The acid pretreatment was performed by adjusting the pH of the sludge to 3.0 for 24 h. During base pretreatment, the pH of the sludge was increased to 11.0 and maintained for 1 d. The chemical treatment was carried out by adding $KNO_3$ (1000 mg/L) to the sludge. The combination of acid and heat pretreatment involved acidification and subsequent heat shock, meanwhile the base and heat strategy was also established by exposing the sludge firstly to base and then to heat, according to the procedure mentioned above.

### 2.3. Experiment I: selection of efficient inoculum pretreatment method

Batch vials (holding capacity of 225 mL) with a working volume of 150 mL were used for the fermentation. The bottles contained 30 mL of seed sludge (pretreated by various ways), dried DJW substrate to a final concentration of 50 g/L, 5 mL of the nutrient solution and drops of either 1N HCl or NaOH to adjust the pH of the solution to neutral. The final working volume was made up with tap water. The composition of the nutrient solution used can be found elsewhere (Kumar and Lin, 2013). The vials were purged with argon for 5 min to provide a fully anaerobic environment. Afterwards, the batch bottles were placed in a reciprocal air-bath shaker at 150 rpm agitation rate and temperature of 55 °C. The volume and composition of the biogas formed were measured periodically.

### 2.4. Experiment II: statistical parameter optimization

The batch tests were performed by following the procedure given in Section 2.3, except that the batch vials used had a total and working volume of 125 mL and 60 mL, respectively. This change was made in order to reduce the amount of chemicals and substrate needed. The initial measurement conditions – substrate concentration, temperature and pH – were set according to the experimental design matrix (**Table 1**). A five-level central composite design (CCD) and response surface methodology (RSM) were used to optimize substrate concentration, pH and temperature. After optimization, verification experiments were conducted in parallel.

**Table 1.**
Central composite design for optimizing the hydrogen production rate.

| Substrate concentration | | Temperature | | Initial pH | | HPR* |
|---|---|---|---|---|---|---|
| $X_1$ code | $X_1$ (g/L) | $X_2$ code | $X_2$ (°C) | $X_3$ code | $X_3$ (1) | (L $H_2$/L-d) |
| -1 | 100 | -1 | 45 | -1 | 5.5 | 0.33 |
| -1 | 100 | -1 | 45 | 1 | 7.5 | 0.35 |
| -1 | 100 | 1 | 65 | -1 | 5.5 | 0.36 |
| -1 | 100 | 1 | 65 | 1 | 7.5 | 0.35 |
| 1 | 300 | -1 | 45 | -1 | 5.5 | 0.47 |
| 1 | 300 | -1 | 45 | 1 | 7.5 | 0.48 |
| 1 | 300 | 1 | 65 | -1 | 5.5 | 0.46 |
| 1 | 300 | 1 | 65 | 1 | 7.5 | 0.45 |
| -1.682 | 32 | 0 | 55 | 0 | 6.5 | 0.32 |
| 1.682 | 368 | 0 | 55 | 0 | 6.5 | 0.34 |
| 0 | 200 | -1.682 | 38 | 0 | 6.5 | 0.34 |
| 0 | 200 | 1.682 | 72 | 0 | 6.5 | 0.38 |
| 0 | 200 | 0 | 55 | -1.682 | 4.8 | 0.37 |
| 0 | 200 | 0 | 55 | 1.682 | 8.2 | 0.5 |
| 0 | 200 | 0 | 55 | 0 | 6.5 | 1.3 |
| 0 | 200 | 0 | 55 | 0 | 6.5 | 1.34 |
| 0 | 200 | 0 | 55 | 0 | 6.5 | 1.38 |
| 0 | 200 | 0 | 55 | 0 | 6.5 | 1.4 |
| 0 | 200 | 0 | 55 | 0 | 6.5 | 1.38 |

* HPR: Hydrogen production rate

### 2.5. Analytical procedures

The soluble metabolic products (SMPs) including volatile fatty acids (VFAs) and alcohols (e.g. ethanol an butanol) were determined by GC-FID (Shimadzu GC-14, Japan). The volume of biogas was measured by a glass syringe at room temperature and 1 atm. The gas composition was analyzed by GC-TCD (China Chromatograph 8700T). All the measurement conditions are available in our previous work (Kumar and Lin, 2013). Kinetic analysis using the modified Gompertz equation was performed as previously reported (Kumar and Lin, 2013). Microbial community analysis (PCR-DGGE) was conducted by following the protocol described by Sivagurunathan et al., (2014).

## 3. Results and discussion

### 3.1. Influence of various pretreatment methods on hydrogen yield and formation rate

In **Figure 1A**, it can be seen that both biogas and hydrogen generation were highly dependent on the applied seed sludge pretreatment, demonstrating that selection of appropriate pretreatment is a key step for hydrogen generation. The obtained production performances are listed in **Table 2**, where it is shown that heat treatment was the most efficient method in comparison with the other methods employed. On the other hand, chemical, base and acid pretreatments as well as the control produced not only hydrogen but also methane. Thus, these methods were not suitable for efficient hydrogen production due to the co-generation of methane (Wang et al., 2012).

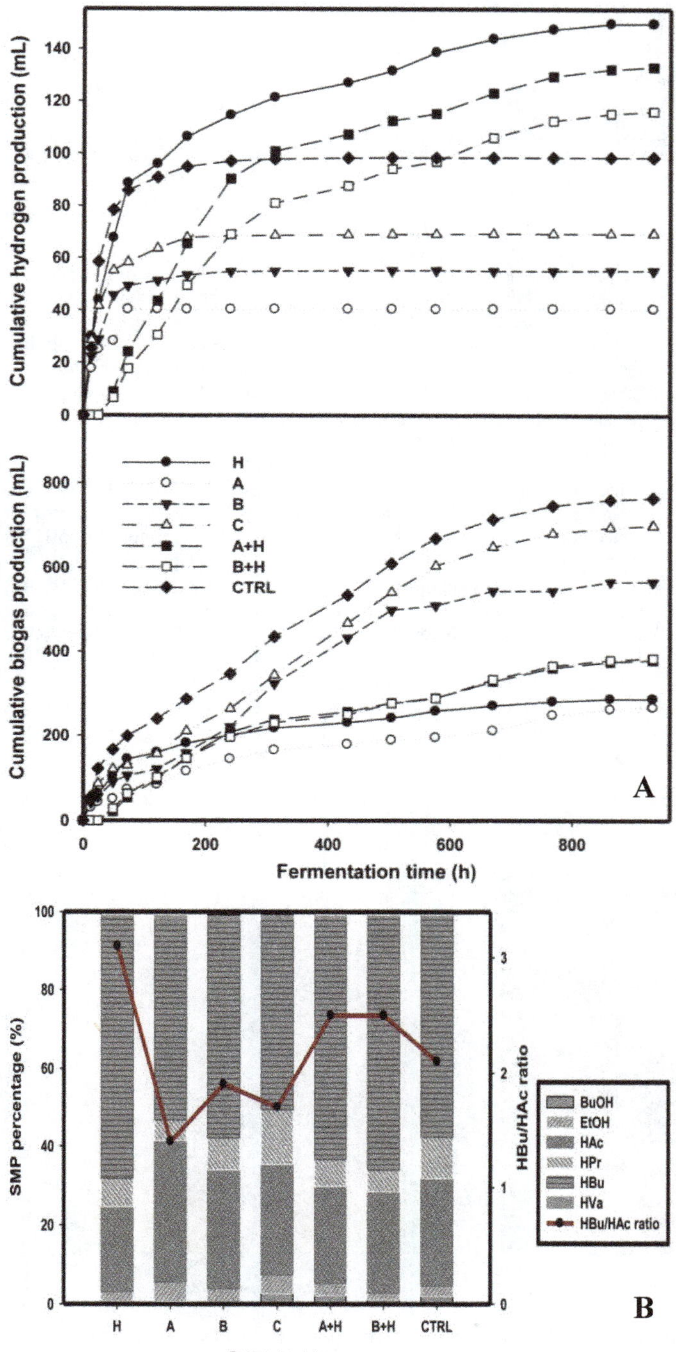


**Fig.1.** A: Progress curves of biogas and hydrogen fermentation for the various pretreatment methods, and B: SMP distribution of hydrogen fermentation for the various pretreatment methods (BuOH: butanol; EtOH: ethanol: HAc: acetate; HPr: propionate; HBu: butyrate; and HVa: valerate).

The reason assumed beyond the superiority of heat-associated treatment is that it preferentially enhances the growth of the endospore-forming *Clostridium* species, which are considered as good hydrogen-producers (Lin et al., 2006; Li and Fang, 2007). The kinetic analyses of the various pretreatment methods are tabulated in **Table 2**, where it could be observed that treatment conditions significantly influenced the bacterial lag phase time. This can be explained by the distinct adaptation capabilities of the microbes to the changes in the environmental conditions caused by the pretreatments applied.

### *3.2. SMP distribution*

Monitoring the SMPs during anaerobic fermentation is suggested to be used to judge the bioreactor's performance and consequently the appropriateness of the available hydrogen producing cultures (Dabrock et al., 1992; Khanal et al., 2004).

The concentrations of VFAs and solvents formed in the hydrogen evolution step for different sludge pretreatment methods are summarized in **Figure 1B**, where it is indicated that the SMPs were dominated by butyric- and acetic acids accounting for 76.6-90.7% of the total SMPs.

The other side products such as propionate, valerate, and alcohols showed less significant contributions (9.3-23.4%) to the total SMPs. The high HBu/HAc ratio indicated an efficient biohydrogen generation system, since efficient hydrogen production is usually associated with high HBu production. Similar findings were reported in other studies as well (Lin and Chang, 1999; Chen et al., 2002; Lin et al., 2006; Demirbas, 2007; Hawkes et al., 2007).

According to **Figure 1B**, it can be pointed out that the highest HBu/HAc ratio (3.1) could be attained via applying heat pretreatment. These research findings demonstrate that metabolic activity (e.g. SMPs release) of the microbes is related to the pretreatment method used to promote the growth of the reliable hydrogen-evolving organisms.

### *3.3. Optimization of process parameters affecting biohydrogen production*

The main operational parameters in biohydrogen production are substrate concentration, temperature and pH. In order to evaluate their effects on hydrogen fermentation of DJW, central composite design and response surface methodology were employed.

During these experimental runs, the heat-pretreated sludge was used. The experimental design matrix along with the corresponding response values of the dependent variable HPR are shown in **Table 1**.

The levels of the independent variables were chosen based on the preliminary experimental results. Analysis of Variance (ANOVA) was carried out by using Statistica 8 software to get the significance of each process variable. The results of ANOVA are listed in **Table 3**. The impacts of the parameters scoped were ranked based on the obtained *P-values*.

Basically, a smaller *P-value* stands for higher influence and only factors having *P-values* <0.05 can be considered as statistically important ones (Guo et al., 2009).

Accordingly, as it can be seen in **Table 3**, all the input variables studied in this work could affect biohydrogen formation in a statistically significant manner. As a result of the statistical evaluation, a mathematical model describing the hydrogen production, more specifically the HPR in connection with substrate concentration, pH and temperature could be established (**Eq. 1**):

$$HPR = 1.356 + 0.037X_1 - 0.345X_1^2 + 0.004X_2 - 0.334X_2^2 + 0.017X_3 - 0.308X_3^2 - 0.009X_1X_2 - 0.001X_1X_3 - 0.006X_2X_3. \quad \text{(Eq. 1)}$$

In the present study, according to the ANOVA results, the $R^2$ value of 0.986 suggested that there was a reliable agreement between the experimental data and the values predicted by the model (Chong et al., 2009). Using the results obtained the contour plots could be constructed with the two-dimensional projections of the fitted three-dimensional surfaces (**Fig. 2**).

**Table 2.**
Biogas production performance under various pretreatment conditions.

| Pretreatments | Final pH | Total biogas (mL) | Cumulative $H_2$ (mL) | Cumulative $CH_4$ (mL) | $HPR_{max}$ (L $H_2$/L-d) | $HY_{max}$ (mL $H_2$/g VS) | $MPR_{max}$ ( L $CH_4$/L-d) | $MY_{max}$ (mL $CH_4$/g VS) | λ (h) | |
|---|---|---|---|---|---|---|---|---|---|---|
| | | | | | | | | | $H_2$ | $CH_4$ |
| CTRL | 6.8 | 764.5 | 98.5 | 258.5 | 0.35 | 13.1 | 0.09 | 34.5 | 0.4 | 114 |
| A | 6.7 | 267.0 | 40.5 | 86.1 | 0.12 | 5.4 | 0.02 | 11.5 | 6 | 66 |
| B | 6.5 | 566.0 | 55.3 | 242.0 | 0.18 | 7.4 | 0.12 | 32.3 | 5 | 159 |
| C | 6.4 | 699.0 | 69.3 | 239.9 | 0.23 | 9.2 | 0.09 | 33.0 | 3 | 120 |
| A+H | 5.4 | 379.5 | 133.2 | ND | 0.25 | 17.8 | ND | ND | 29 | ND |
| B+H | 5.8 | 384.0 | 116.0 | ND | 0.29 | 15.5 | ND | ND | 27 | ND |
| H | 5.4 | 287.5 | 149.5 | ND | 0.36 | 20.0 | ND | ND | 37 | ND |

A : Acid B : Base C : Chemical H : Heat treatment HY : expressed in terms of $VS_{added}$ ND : not detected

**Table 3.**
ANOVA table for hydrogen production rate (HPR).

| Response variable: HPR (L $H_2$/L-d) | | | | | |
|---|---|---|---|---|---|
| *Factor* | *SS* | *df* | *MS* | *F-value* | *P-value* |
| $X_1$ | 0.019 | 1 | 0.019 | 11.608 | 0.027 |
| $X_1^2$ | 1.622 | 1 | 1.622 | 1014.016 | <0.001 |
| $X_2$ | <0.001 | 1 | <0.001 | 0.150 | 0.718 |
| $X_2^2$ | 1.524 | 1 | 1.524 | 952.582 | <0.001 |
| $X_3$ | 0.004 | 1 | 0.004 | 2.392 | 0.197 |
| $X_3^2$ | 1.292 | 1 | 1.292 | 807.396 | <0.001 |
| $X_1X_2$ | 0.001 | 1 | 0.001 | 0.383 | 0.570 |
| $X_1X_3$ | <0.001 | 1 | <0.001 | 0.008 | 0.934 |
| $X_2X_3$ | <0.001 | 1 | <0.001 | 0.195 | 0.681 |
| Pure Error | 0.006 | 4 | 0.002 | - | - |
| Total SS | 3.506 | 18 | - | - | - |

SS : sum of squares df : degree of freedom MS : mean square

These graphs were intended to illustrate the effects of two independent variables on HPR while keeping the level of the third factor at its center value (**Fig. 2A-C**). As it appears in **Figure 2A-C**, all the contour plots demonstrated a round ridge running around the center point implying that the interactions between the factors had only low importance (Kim et al., 2004). This is confirmed by the data presented in **Table 3** as well, where it can be observed that the interactive effects ($X_1X_2$, $X_1X_3$, $X_2X_3$) are insignificant (*P-value*>0.05).

Moreover, it is to be concluded that the maximum HPR value could be achieved within the design boundaries. Hence, the optimal conditions yielding the highest HPR could be derived by RSM as follows: substrate concentration of 205 g/L, initial pH of 6.53 and temperature of 55.1 °C, where peak HPR value of 1.36 L $H_2$/L-d was estimated.

Although experimental design methods have definite benefits, the statistically estimated optimum conditions need to be validated (Mu et al., 2006). Under the predicted optimal conditions, the HPR value was found as 1.42±0.03 L$H_2$/L-d. This proved that the actual optimal HPR satisfactorily matched its statistically-forecasted value (1.36 L $H_2$/L-d) with only 4% difference between them. Therefore, the results obtained confirmed the feasibility of using the experimental design and RSM for biohydrogen process development.

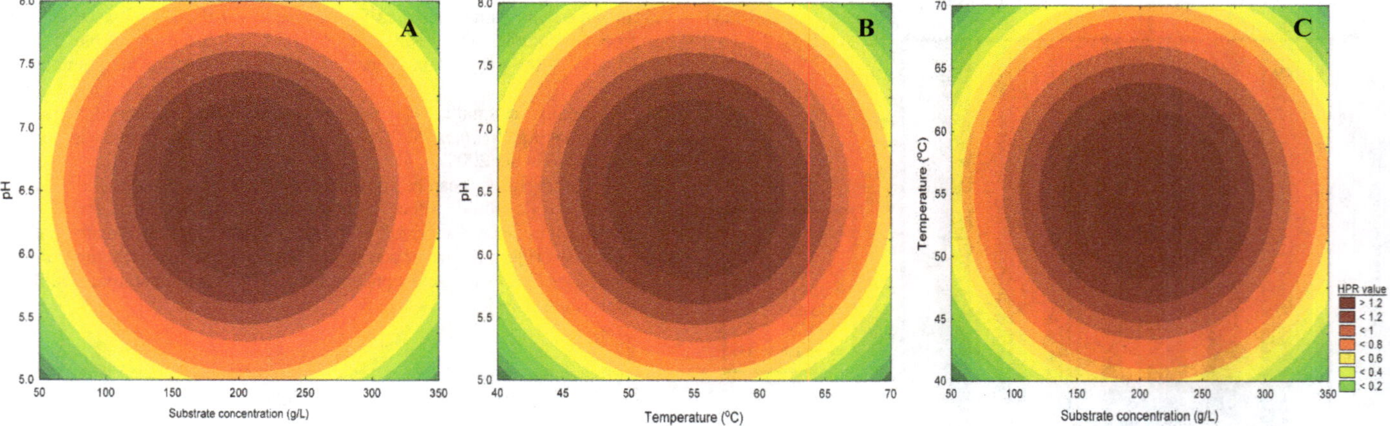


**Fig.2.** (A) Contour plots showing the effects of initial pH and substrate concentration on hydrogen production rate (HPR), (temperature: 55 °C), (B) Contour plots showing the effects of initial pH and temperature on hydrogen production rate (HPR), (substrate concentration: 200 g/L), (C) Contour plots showing the effects of temperature and substrate concentration on hydrogen production rate (HPR), (pH: 6.5).

**Table 4.**
Comparison of the results with relevant literature data.

| Substrates | Inoculum | Range studied | | | Optimal conditions | Hydrogen production index | References |
|---|---|---|---|---|---|---|---|
| | | Substrate Concentration index | Temperature (°C) | pH | | | |
| Glucose | Potato & soybean oil soil, HT-Compost | 1.5-44.8 g COD/L | NA | 4.5-7.5 | SC:7.5 g COD/L, T:NS, pH:5.5 | HPR:74.7 mL/L-h | Van Ginkel et al., 2001 |
| POME | *C. butyricum* EB6 | 60-100 g COD/L | 32-42 | 5.3-6.7 | SC: 94 g COD/L, T: 36 °C, pH:6.05 | HPR: 849.5 mL/h | Chong et al., 2009 |
| Sucrose | HT-sludge | 10-30 g COD/L | NA | 5.5-8.5 | SC:20g COD/L, T:NS, pH:7.5 | HPR :745 mL/L-h | Wang et al., 2006 |
| Mushroom waste | HT-cow dung | NA | NA | NA | SC: 20 g COD/L, T: 55 °C, pH:8.0[a] | HY: 0.68 mol/g COD | Lay et al., 2012 |
| DJW | HT-Sludge | 40-240 g/L | 45-65 | 5.5-7.5 | SC: 205 g/L, T:55 °C, pH:6.5 | HPR: 59.2 mL/L-h | This study |

DJW : De-oiled Jatropha Waste; T : temperature (in °C); POME : palm oil mill effluent; NA : not available; HT : heat treated; a : given values; SC : substrate concentration (specified)

A comparison with other works focusing on substrate concentration, temperature and pH for enhanced biohydrogen production is given in **Table 4.** As a conclusion, it can be pointed out that most of these investigations were performed using the classic, so-called "one-factor-at-a-time" method. On the other hand, only a few studies were dedicated to the experimental design. The optimum value obtained in the current research is quite comparable with the results of other reports employing complex (lignocellulosic) substrate materials.

### *3.4. Microbial community pattern*

The DGGE profile obtained using the bacterial primer set EUB968gc–UNIV1392r revealed the structural composition of the microbial communities based on the V6 region of the 16s rRNA. The DGGE profile of the hydrogen producing microbial community using DJW under optimal condition are depicted in **Figure 3A.** As shown, four individual distinct bands patterns representing four different strains were observed. The phylogenetic tree distribution was established using the bootstrap neighbor joining method (Saitou and Nei, 1987) as shown in **Figure 3B**.

The microbial load in the seed inocula enriched with de-oiled Jatropha waste was made up by *Clostridium sp.*, and *Lactobacillus sp.* According to the major bands shown in **Figure 3A**, *Clostridium sp.* was the most abundant followed by *Bacillus sp.* The bacterial species *C. thermopalmarium, C. thermobutyricum,* and *B. ginsengihumi* were previously reported as potential hydrogen producing bacteria ( Wiegel et al., 1989; Geng et al., 2010; Walton et al., 2010). *Clostridium Sp.*, a low G+ C content bacterium, is known to generate hydrogen along with butyrate and acetate as major SMPs (Chen et al., 2002; Levin et al., 2004).

In addition, *C. thermopalamarium* was reported to efficiently utilize cellulose (a major component of the DJW) to produce hydrogen. Moreover, *Bacillus sp.* was also reported to produce lactic acid from hemicellulose extracts (Walton et al., 2010). The inoculum source i.e. the sewage sludge selected in the present study was a rich source of hemicelluloytic and cellulolytic bacteria (Sleat et al., 1984).

Under the optimum conditions employed in the present study, *Clostridium sp.* and *Bacillus sp.* were the predominant microorganisms. while the former was more abundant. *Clostridium* species are known hydrogen producers with butyrate type fermentation in which acetic acid and alcohols are the minor metabolites (Levin et al., 2004). This was in agreement with the results of the SMPs analysis, suggesting that butyrate was the major metabolite observed during the fermentation, followed by acetic acid and ethanol. The PCR-DGGE based sequence analysis also revealed the presence of dominant butyrate-mediated hydrogen-producing bacteria present in the mixed cultures.

## 4. Conclusion

Various sludge pretreatment methods were tested for biohydrogen production using de-oiled Jatropha waste as feedstock. It turned out that among the methods studied, heat pretreatment was most effective in order to enrich efficient hydrogen-producing microorganisms. Moreover, central composite design was successfully used for the optimization of the most crucial operational parameters. The distribution pattern of the SMPs showed that the fermentation mainly followed the acetate-butyrate pathway. Moreover, PCR-DGGE results revealed that *Clostridium* sp. were majorly present under the optimal conditions.

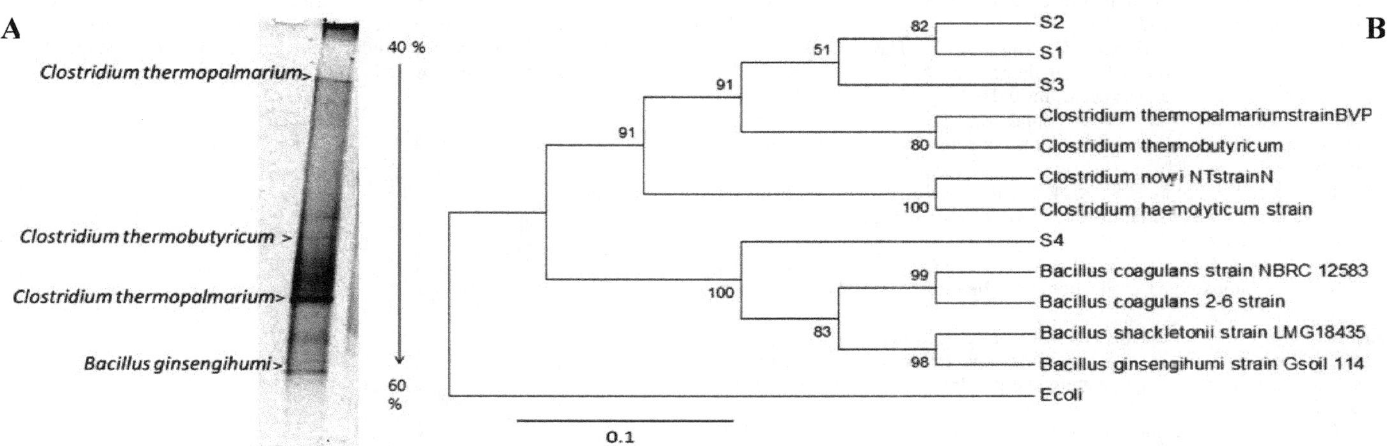


**Fig.3.** A: The DGGE profile of the hydrogen-producing microbial community under optimum conditions using DJW, and B: Phylogenetic tree showing the relatedness of the sequences identified in the mixed cultures. The tree was constructed based on maximum composite likelihood method using the neighbor-joining algorithm with 1,000 bootstrapping. *E.coli* was selected as the outgroup species. The scale bar represents 0.1 substitutions per nucleotide position. Numbers at the nodes are the bootstrap values.

**Acknowledgements**

The authors gratefully acknowledge the financial support by Taiwan's Bureau of Energy (grant no. 102-D0616), Taiwan's National Science Council (NSC-102-2221-E-035 -002 -MY3 and NSC-102-2622E-035-016-CC1). We also thank Hua Neng Environmental Protection and Energy Technology Ltd., Taiwan, for providing us the DJW. This work was also supported by the European Union and financed by the European Social Fund in the framework of the TAMOP-4.2.2/A-11/1/KONV-2012-0071 project and the János Bolyai Research Scholarship of the Hungarian Academy of Sciences.

**References**

Akil, K., Jayanthi, S., 2014. The biohydrogen potential of distillery wastewater by dark fermentation in an anaerobic sequencing batch reactor. Int. J. Green Energy. 11, 28-39.

Chen, C.C., Lin, C.Y., Lin, M.C., 2002. Acid-base enrichment enhances anaerobic hydrogen production process. Appl. Microbiol. Biotechnol. 58, 224-228.

Chong, M.L., Abdul Rahman, N.A., Rahim, R.A., Aziz, S.A., Shirai, Y., Hassan, M.A., 2009. Optimization of biohydrogen production by *Clostridium butyricum* EB6 from palm oil mill effluent using response surface methodology. Int. J. Hydrogen Energy. 34, 7475-7482.

Dabrock, B., Bahl, H., Gottschalk, G., 1992. Parameters affecting solvent production by Clostridium pasteurianum. Appl. Env. Microbiol. 58, 1233-1239.

Demirbas, A., 2007. Progress and recent trends in biofuels. Prog. Energy. Combust. Sci. 33, 1-18.

Diamantis, V., Khan, A., Ntougias, S., Stamatelatou, K., Kapagiannidis, A.G., Aivasidis, A., 2013. Continuous biohydrogen production from fruit wastewater at low pH conditions. Bioproc. Biosyst. Eng. 36, 965-974.

Ewan, B.C.R., Allen, R.W.K., 2005. A figure of merit assessment of the routes to hydrogen. Int. J. Hydrogen Energy. 30, 809-819.

Fan, Y.T., Zhang, G.S., Guo, X.Y., Xing, Y., Fan, M.H., 2006. Biohydrogen-production from beer lees biomass by cow dung compost. Biomass Bioenergy. 30, 493-496.

Geng, A., He, Y., Qian, C., Yan, X., Zhou, Z., 2010. Effect of key factors on hydrogen production from cellulose in a co-culture of Clostridium thermocellum and Clostridium thermopalmarium. Bioresour. Technol. 101, 4029-4033.

Guo, W.Q., Ren, N.Q., Wang, X.J., Xiang, W.S., Ding, J., You, Y., Liu, B.F., 2009. Optimization of culture conditions for hydrogen production by Ethanoligenens harbinense B49 using response surface methodology. Bioresour. Technol. 100, 1192-1196.

Hallenbeck, P.C., 2009. Fermentative hydrogen production: Principles, progress, and prognosis. Int. J. Hydrogen Energy. 34, 7379-7389.

Han, J., Lee, D., Cho, J., Lee, J., Kim, S., 2012. Hydrogen production from biodiesel byproduct by immobilized Enterobacter aerogenes. Bioproc. Biosyst. Eng. 35, 151-157.

Hawkes, F.R., Hussy, I., Kyazze, G., Dinsdale, R., Hawkes, D.L., 2007. Continuous dark fermentative hydrogen production by mesophilic microflora: Principles and progress. Int. J. Hydrogen Energy. 32, 172-184.

Khanal, S.K., Chen, W.H., Li, L., Sung, S., 2004. Biological hydrogen production: Effects of pH and intermediate products. Int. J. Hydrogen Energy. 29, 1123-1131.

Kim, S.H., Han, S.K., Shin, H.S., 2004. Feasibility of biohydrogen production by anaerobic co-digestion of food waste and sewage sludge. Int. J. Hydrogen Energy. 29, 1607-1616.

Kumar, G., Lay, C.H., Chu, C.Y., Wu, J.H., Lee, S.C., Lin, C.Y., 2012. Seed inocula for biohydrogen production from biodiesel solid residues. Int. J. Hydrogen Energy. 37, 15489-15495.

Kumar, G., Lin, C.Y., 2013. Bioconversion of de-oiled Jatropha Waste (DJW) to hydrogen and methane gas by anaerobic fermentation: Influence of substrate concentration, temperature and pH. Int. J. Hydrogen Energy. 38, 63-72.

Lay, C.H., Sung, I.Y., Kumar, G., Chu, C.Y., Chen, C.C., Lin, C.Y., 2012. Optimizing biohydrogen production from mushroom cultivation waste using anaerobic mixed cultures. Int. J. Hydrogen Energy. 37, 16473-16478.

Levin, D.B., Pitt, L., Love, M., 2004. Biohydrogen production: prospects and limitations to practical application. Int. J. Hydrogen Energy, 29, 173-185.

Li, C., Fang, H.H.P., 2007. Fermentative hydrogen production from wastewater and solid wastes by mixed cultures. Crit. Rev. Env. Sci. Technol. 37, 1-39.

Lin, C.Y., Chang, R.C., 1999. Hydrogen production during the anaerobic acidogenic conversion of glucose. J. Chem. Technol. Biotechnol. 74, 498-500.

Lin, C.Y., Hung, C.H., Chen, C.H., Chung, W.T., Cheng, L.H., 2006. Effects of initial cultivation pH on fermentative hydrogen production from xylose using natural mixed cultures. Proc. Biochem. 41, 1383-1390.

Mohammadi, P., Ibrahim, S., Mohamad Annuar, M.S., Law, S., 2011. Effects of different pretreatment methods on anaerobic mixed microflora for hydrogen production and COD reduction from palm oil mill effluent. J. Clean. Prod. 19, 1654-1658.

Mu, Y., Wang, G., Yu, H.Q., 2006. Response surface methodological analysis on biohydrogen production by enriched anaerobic cultures. Enzy. Microb. Technol. 38, 905-913.

Sarma, S.J., Brar, S.K., Le Bihan, Y., Buelna, G., 2013. Bio-hydrogen production by biodiesel-derived crude glycerol bioconversion: A techno-economic evaluation. Bioproc. Biosyst. Eng. 36, 1-10.

Sivagurunathan, P., Sen, B., Lin, C.Y., 2014. Batch fermentative hydrogen production by enriched mixed culture: Combination strategy and their microbial composition. J. Biosci. Bioeng. 117, 222-228.

Sleat, R., Mah, R.A., Robinson, R., 1984. Isolation and characterization of an anaerobic, cellulolytic bacterium, Clostridium cellulovorans sp. nov. Appl. Env. Microbiol. 48, 88-93.

Van Ginkel, S., Sung, S., Lay, J.J., 2001. Biohydrogen production as a function of pH and substrate concentration. Env. Sci. Technol. 35, 4726-4730.

Walton, S.L., Bischoff, K.M., Van Heiningen, A.R.P., Van Walsum, G.P., 2010. Production of lactic acid from hemicellulose extracts by Bacillus coagulans MXL-9. J. Indust. Microbiol. Biotechnol. 37, 823-830.

Wang, C.H., Lin, P.J., Chang, J.S., 2006. Fermentative conversion of sucrose and pineapple waste into hydrogen gas in phosphate-buffered culture seeded with municipal sewage sludge. Proc. Biochem. 41, 1353-1358.

Wang, J., Wan, W., 2009a. Factors influencing fermentative hydrogen production: A review. Int. J. Hydrogen Energy. 34, 799-811.

Wang, J., Wan, W., 2009b. Kinetic models for fermentative hydrogen production: A review. Int. J. Hydrogen Energy. 34, 3313-3323.

Wang, Y.Y., Ai, P., Hu, C.X., Zhang, Y.L., 2011. Effects of various pretreatment methods of anaerobic mixed microflora on biohydrogen production and the fermentation pathway of glucose. Int. J. Hydrogen Energy. 36, 390-396.

Wang, H., Zhi, Z., Wang, J., Ma, S., 2012. Comparison of various pretreatment methods for biohydrogen production from cornstalk. Bioproc. Biosyst.Eng. 35, 1239-1245.

Wiegel, J., Kuk, S.U., Kohring, G.W., 1989. Clostridium thermobutyricum sp. nov., a moderate thermophile isolated from a cellulolytic culture, that produces butyrate as the major product. Int. J. Syst. Bacteriol. 39, 199-204.

# 5

# Kinetic comparison of two basic heterogenous catalysts obtained from sustainable resources for transesterification of waste cooking oil

G.R. Moradi[1,*], M. Mohadesi[2], M. Ghanbari[1], M.J. Moradi[1], Sh. Hosseini[1], Y. Davoodbeygi[1]

[1]*Catalyst Research Center, Chemical Engineering Department, Faculty of Engineering, Razi University, Kermanshah, Iran.*

[2]*Chemical Engineering Department, Kermanshah University of Technology, Kermanshah, Iran.*

## HIGHLIGHTS

➢DM water treatment precipitates as a novel catalyst for biodiesel production.

➢Reusability/stability of the new catalyst was higher than the calcinated waste mussel shells.

➢After five times of reusing the new catalyst, only 6.15% of CaO was extracted by methanol.

➢Reaction in presence of the new catalyst was faster than the calcinated mussel shells.

## GRAPHICAL ABSTRACT

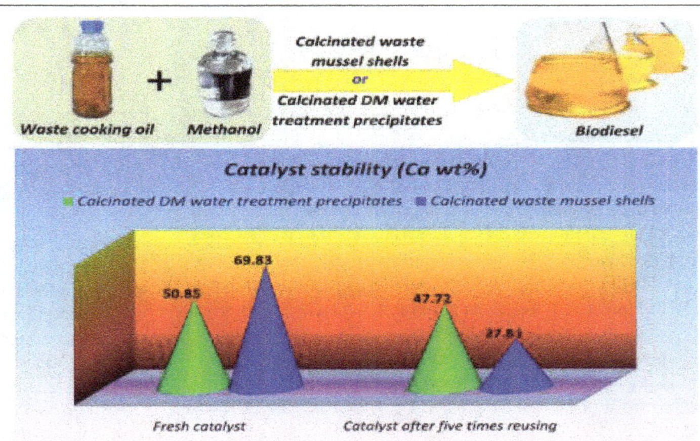


## ABSTRACT

Alkaline earth metal oxides are appropriate catalysts for biodiesel production and among them, CaO and MgO are known for possessing the best efficiency. In this study, catalysts synthesized from economical and sustainable resources were used for biodiesel production. More specifically, waste mussel shells and demineralized (DM) water treatment precipitates as calcium and magnesium carbonate sources, were converted into calcium and magnesium oxides at temperatures above 900 °C. Methanol and waste cooking oil were reacted in a 250 mL two-necked flask at 24:1 and 22.5:1 ratios in presence of 12 and 9.08 wt% of mussel shell-based and DM water treatment precipitates-based catalysts, respectively. The effects of temperature (328, 333, 338, 343 and 348 K) and time (1, 3, 5, 7 and 8 h) at a stirrer speed of 350 rpm on the conversion of the oil into biodiesel were investigated. The results obtained indicated a pseudo-first order kinetics for the transesterification reaction using both catalysts. The activation energies in the presence of the DM water treatment precipitates and mussel shell catalysts were measured at 77.09 and 79.83 kJ.mol$^{-1}$, respectively. Accordingly, the DM water treatment precipitates catalyst resulted in a faster reaction due to its lower activation energy value. Moreover, the catalysts were reused five times and the results obtained showed that the methanol-driven extraction of CaO contained in the DM water treatment precipitates catalyst was lower than the waste mussel shell catalyst proving the higher stability of the new heterogeneous catalyst i.e. the calcinated DM water treatment precipitates.

**Keywords:**
Waste Cooking Oil biodiesel
Heterogeneous Catalyst
DM water treatment precipitates
Kinetics
Catalyst reusability
Catalyst stability

* Corresponding author
E-mail address: gmoradi@razi.ac.ir

## 1. Introduction

Due to the increasing demand for fuel and energy and limited fossil fuel resources, a great deal of attention has been paid to alternative renewable fuels (Liu et al., 2011; Rezaei et al. 2013). Biodiesel, also known as fatty acid methyl ester (FAME), is obtained through the transesterification reaction between oil and alcohol in the presence of a suitable catalyst (Kouzu et al., 2008; Yin et al., 2012). Biodiesel in comparison with the fossil-based diesel fuel, not only has higher lubricity, flash point, oxygen content and cetane number but also results in lower hazardous emissions when combusted because of its low sulfur content (Kouzu et al., 2008; Omar and Amin, 2011; Lin and Cheng, 2012; Yin et al., 2012). Biodiesel can be derived from vegetable oils, animal fats, and waste cooking oil (Endalew et al., 2011; Taufiq-Yap et al., 2011).

Transesterification reaction occurs between triglycerides and methanol leading to the production of FAME and glycerol as byproduct (Di Serio et al., 2008; Veljkovic et al., 2009; Endalew et al., 2011; Sharma et al., 2011; Taufiq-Yap et al., 2011). The typical reaction scheme for the transesterification is presented in the following equation (**Eq. 1**).

$$\begin{array}{l} R_1COOCH_2 \\ \quad | \\ R_2COOCH \\ \quad | \\ R_3COOCH_2 \end{array} + 3CH_3OH \xleftrightarrow{\text{Catalyst}} \begin{array}{l} HOCH_2 \\ \quad | \\ HOCH \\ \quad | \\ HOCH_2 \end{array} + \begin{array}{l} R_1COOCH_3 \\ \\ R_2COOCH_3 \\ \\ R_3COOCH_3 \end{array}$$

$$\text{Triglyceride} + 3\text{MeOH} \xleftrightarrow{\text{Catalyst}} \text{Glycerol} + \text{Methyl Esters(Biodiesel)} \qquad \text{(Eq. 1)}$$

Compared to homogenous catalysts, basic and acidic heterogeneous catalysts have the advantages of easy and cheap separation and regeneration (Birla et al., 2012; Gaikwad and Gogate, 2105). Overall, basic heterogeneous catalysts are preferred because of their higher activity compared to acidic ones (Endalew et al., 2011). Among the basic heterogeneous catalysts, calcium oxide (CaO) has attracted the most attention because of its high catalytic activity, regenerability/reusability, and that there are plenty of relatively inexpensive resources for its production (waste shells, egg shells, etc.). Moreover, CaO is not sensitive to small amounts of FFA and moisture, and is therefore, suitable for waste cooking oils (Veljkovic et al., 2009; Boey et al., 2011).

Regardless of the type of the catalyst used, establishing the reaction kinetics is necessary in order to obtain more in-depth information such as reactor configuration, reaction time and optimum process temperature. Veljkovic et al. (2009) studied the kinetics of sunflower oil transesterification with methanol over CaO as catalyst. They found out that during the first stage of the process, mass transfer of triglyceride controlled the reaction and in the latter stage, the chemical reaction became the rate determining factor (Veljkovic et al., 2009). In a different study, Dossin et al. (2006) investigated the kinetics of ethyl acetate methanolysis catalyzed by magnesium oxide (MgO) as heterogeneous catalyst. They used a three step 'Eley-Ridel' mechanism in liquid phase and reported that methanol adsorption was the rate determining step in that reaction (Dossin et al., 2006). Birla et al. (2012) reported a first order kinetics for the transesterification of waste cooking oil in the presence of snail shell as a heterogeneous base catalyst. They determined the activation energy of 79 $kJ.mol^{-1}$ and the frequency factor of $2.98\times10^{10}$ $min^{-1}$ (Birla et al., 2012). Recently, Pukale et al. (2015) investigated the effect of ultrasound on kinetics of transesterification of waste cooking oil using heterogeneous solid catalyst. They reported an activation energy of 64.241 $kJ.mol^{-1}$ in presence of $K_3PO_4$ catalyst. Moreover, they obtained a higher yield in the presence of ultrasound as compared to the conventional approach under similar conditions (Pukale et al., 2015).

The cost of raw material typically accounts for about 70–80 % of the total cost of biodiesel production. Therefore, there is a need to develop technologies for producing biodiesel from non-edible and waste oil resources using highly efficient catalysts (Gole and Gogate, 2012a; Maddikeri et al., 2012; Gaikwad and Gogate, 2105). On such basis, the present study was set to reduce the cost of biodiesel production from waste cooking oil using CaO and MgO catalysts generated from inexpensive resources. More specifically, waste mussel shells containing lots of calcium carbonate was converted into CaO by calcination at 1050 °C as elaborated in our previous study (Rezaei et al., 2013). Moreover, a novel catalyst was also prepared herein by conversion of DM water treatment precipitates into CaO/MgO by calcination at different temperature values. Both catalysts were then compared by taking into account the effects of temperature (328, 333, 338, 343 and 348 K) and time (1, 3, 5, 7 and 8 h) at a stirrer speed of 350 rpm on the conversion of the oil into biodiesel. Finally, the reusability/stability of the new catalyst was compared with that of the calcinated waste mussel shells by re-using for five times.

## 2. Materials and methods

### *2.1. Materials*

Waste cooking oil obtained from a local restaurant and methanol 99.5 % (Merck, Germany) were used in this study for esterification and transesterification reactions. Sulfuric acid 97 % (Merck, Germany) was used as catalyst in the esterification reaction. Potassium hydroxide 85 % (Merck, Germany) was used for determination of acidic number. Waste mussel shell and DM water treatment precipitates were used as the source for generating the catalysts used in the transesterification reaction. *n*-hexane 95 % (Merck, Germany) and methyl laurate (methyl dodecanoate) >99.7 % supplied by Sigma Aldrich were used in the gas chromatography (GC) analyses for determining the produced biodiesel purity.

### *2.2. Catalysts preparation and characterization*

Mussel shell and DM water treatment precipitates were ground by a mortar. Obtained powders were sieved to separate micro-particles (125-250 μm), and were then dried at 110 °C for 18 h. Finally waste mussel shell powder was calcinated at 1050 °C accordingly to our previous findings (Rezaei et al., 2013) and DM water treatment precipitates powder was calcinated at different temperatures of 800, 900 and 1000 °C for 2 h to determine the optimal temperature value leading to the highest conversion of calcium and magnesium carbonates into CaO/MgO catalyst.

The crystalline phases of the DM water treatment precipitates calcinated at different temperature values were characterized by X-ray diffraction analysis (XRD). XRD analysis was performed through Cu Ka radiation. The data showing the intensity was plotted in a chart based on 2-Theta in a range of 10–80° with a step of 0.06°. X-ray florescence (XRF) (Spectro Xepor 03 plus) analysis was applied to determine the chemical elements composition of the catalyst after the 2 h calcination.

### *2.3. Esterification reaction*

Since the acidic number of the waste cooking oil was measured at 2 mg KOH/g oil, an esterification stage was carried out to achieve an acidic number below 1 mg KOH/g (Gole and Gogate, 2012b). Esterification reaction was performed with 5 wt% $H_2SO_4$ as catalyst, with a methanol to oil molar ratio of 18:1 at 65 °C for 5 h, as previously described by Encinar et al. (2011). After the esterification, the acidic number of the waste cooking oil decreased to 0.49 mg KOH/g oil.

### *2.4. Transesterification reaction procedure*

Transesterification reaction was conducted in a 250 mL two necked flask equipped with a thermometer and a condenser. The mixture of waste cooking oil, methanol and catalyst was mixed using a stirrer at 350 rpm. Optimal values of methanol to oil molar ratio and catalyst concentration were obtained from our previous studies (Rezaei et al., 2013; Davoodbeygi, 2013). To determine the kinetics, the mentioned reaction was carried out at different reaction temperatures of 328, 333, 338, 343, 348 K and different reaction times of 1, 3, 5, 7 and 8 h. After the reaction, the catalyst was first separated by centrifugation and then, glycerol and the produced biodiesel were separated using a separation funnel. For increasing the purity of the produced biodiesel, the product was washed several times with water (90 °C). Finally, the produced biodiesel was dried in the oven at 110 °C for 2 h.

*2.5. FAME analysis*

FAME characterization was carried out by a HP 6890 gas chromatograph with a flame ionization detector (FID). The capillary column was a BPX-70 with a length of 120 m, a film thickness of 0.25 μm and an internal diameter of 0.25 mm. Nitrogen was used as the carrier gas and also as an auxiliary gas for FID. 1 μL of the sample was injected by a 6890 Agilent Series Injector. The inlet temperature was at 50 °C, which was heated up to 230 °C. Methyl laurate (C12:0) was added as an internal reference into each biodiesel sample prior to GC analysis to determine biodiesel purity using the following equations (**Eqs. 2 and 3**) as described by Rezaei et al. (2013).

$$\text{Purity}(\%) = \frac{\text{area of all FAME}}{\text{area of reference}} \times \frac{\text{weight of reference}}{\text{weight of biodiesel sample}} \times 100 \quad \text{(Eq. 2)}$$

$$\text{Conversion}(\%) = \frac{\text{area of all FAME}}{\text{area of reference}} \times \frac{\text{weight of reference}}{\text{weight of biodiesel sample}} \times \frac{\text{weight of biodiesel producted}}{\text{weight of oil used}} \times 100 \quad \text{(Eq. 3)}$$

*2.6. Kinetics of process*

The effects of temperature and time were investigated to determine the reaction kinetics. Since the amounts of catalysts used were enough to convert the oil into FAME, thus, the reverse reactions could be ignored. Moreover, the changes in catalysts concentration changes during the reaction could also be overlooked (Zhang et al., 2010). Assuming that the transesterification reaction is carried out in one step, transesterification reaction rate can be calculated through the following equation (**Eq. 4**) (Vujicic et al., 2010):

$$-r_{TG} = -\frac{d[TG]}{dt} = \frac{d[ME]}{dt} = k'.[TG].[ROH]^3 \quad \text{(Eq. 4)}$$

Where $r_{TG}$ represents the rate at which triglycerides are used, [TG], [ME] and [ROH] are triglyceride, methyl ester and methanol concentrations, respectively, and $k'$ is the reaction constant. Because of the high methanol to oil molar ratio, methanol concentration changes during the reaction can be neglected and the reaction can be considered as a pseudo-first order reaction (Freedman et al., 1984; Zhang et al., 2010; Singh and Fernando, 2007). Therefore, the reaction rate could be expressed as follows (**Eq. 5**):

$$-r_{TG} = -\frac{d[TG]}{dt} = k.[TG] \quad \text{(Eq. 5)}$$

In which $k$ is the reaction constant which is equal to $k'.[ROH]^3$ (**Eq. 6**).

$$-\ln(1 - X_{ME}) = k.t \quad \text{(Eq. 6)}$$

Where $X_{ME}$ is the methyl ester conversion.

*2.7. Activation energy determination*

Arrhenius equation establishes a relation between reaction rate constant ($K$), temperature ($T$) and activation energy ($E_a$) as follows (**Eq. 7**):

$$k = k_0 \exp\left(\frac{-E_a}{RT}\right) \quad \text{(Eq. 7)}$$

Where $k_0$ and $R$ are the frequency factor and universal gas constant, respectively. This equation could be rewritten as follows (**Eq. 8**):

$$\ln(k) = -\frac{E_a}{RT} + \ln(k_0) \quad \text{(Eq. 8)}$$

By plotting the diagram of ln ($k$) vs. $1/T$, the slope is equal to $-E_a/R$ and the intercept will be ln ($k_0$).

*2.8. Reusability of catalyst*

DM water treatment precipitates catalyst obtained under optimal conditions was re-used for five times. Before each reusing, catalyst was washed by methanol and the residual methanol was vaporized, by heating on a stirrer at 90 °C (100 rpm). XRF analysis was used to compare the major available elements in the catalyst re-used for five times and the fresh catalyst.

## 3. Results and discussion

*3.1. Catalyst preparation and characterization*

As illustrated in **Figure 1**, the DM water treatment precipitates catalyst at calcination temperature of 800 °C showed the peaks related to $CaCO_3$, CaO, and MgO. Apparently, at this temperature $MgCO_3$ must have been converted into MgO completely because no peak associated with $MgCO_3$ was observed. At calcination temperatures above 900 °C, the major peaks were related to CaO and MgO only, which indicated full conversion of $CaCO_3$ and $MgCO_3$ to CaO and MgO, respectively. Since the change of calcination temperature from 900 to 1000 °C did not lead to any further favorable changes, therefore, it could be concluded that the optimum calcination temperature for the production of the DM water treatment precipitates catalyst was 900 °C.

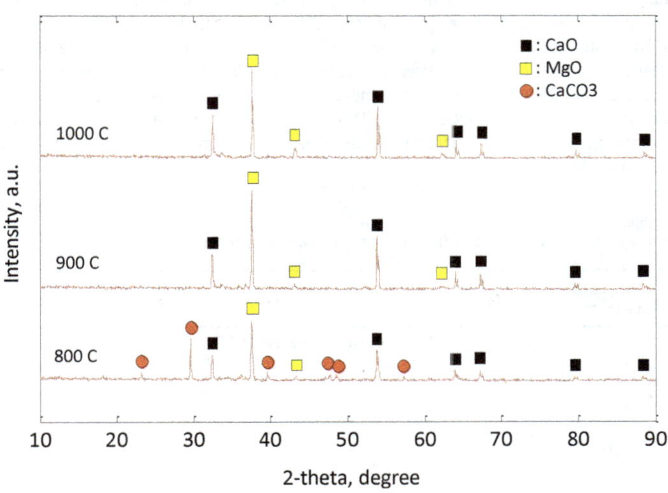


**Fig.1.** The XRD patterns of the DM water treatment precipitates catalyst calcinated at 800, 900 and 1000 °C.

The results of the XRF analysis on the fresh catalyst and the catalyst after five times of reusing in the transesterification reaction are presented in **Table 1**. As seen, about 60 wt% of the fresh calcinated DM water treatment precipitates catalyst composed of two elements of Ca (50.85 wt%) and Mg

(9.34 wt%), which are in existence with CaO and MgO molecules, respectively. The results of the XRF analysis of the calcinated waste mussel shell are also presented in the same table (Rezaei et al., 2013).

**Table 1.**
The weight percentages of the major elements in the fresh and five-time used DM water treatment precipitates catalyst based on the XRF analysis.

| Catalyst type | Element | Fresh catalyst | Catalyst after 5 times of reusing | Reference |
|---|---|---|---|---|
| Calcinated DM water treatment precipitates | Ca, wt% | 50.85 | 47.72 | Present study |
| | Mg, wt% | 9.34 | 9.52 | |
| Calcinated waste mussel shells | Ca, wt% | 69.83 | 27.81 | Rezaei et al. (2013) |

The major concern pertaining to heterogeneous catalysts used in the Transestrification reaction is that the active part of the catalyst e.g. CaO is extracted by methanol. This in turn reduces the stability/reusability of the catalyst in the subsequent reactions. According to the XRF analysis results, 60.17% of the CaO contained in the calcinated waste mussel shells catalyst was extracted by methanol after five times of reusing in transesterification reaction (Rezaei et al., 2013), while only 6.15% of the CaO contained in the calcinated DM water treatment precipitates catalyst was extracted after five times of reusing. Therefore, the latter resulted in significantly higher biodiesel conversion (data not shown).

### *3.2. Temperature and time effect*

According to our previous study (Davoodbeygi, 2013) at stirring speeds lower than 350 rpm, diffusion problem was the rate-limiting step; while high stirring speeds led to saponification. So, the stirrer speed of 350 rpm was considered for all experiments in the present study. **Table 2** shows conversion (%) of oil into FAME and different times internals and temperatures (i.e. 328, 333, 338, 343 and 348 K) using waste mussel shell and DM water treatment precipitates catalysts. As presented in the table, at all the investigated temperatures, the trends observed for FAME production over time using both catalysts were similar. With increasing the temperature from 328 to 343 K, the conversion rate was improved from 28 and 30% to 78 and 82% using the DM water treatment precipitates and the waste mussel shell catalysts, respectively. Similar trend was also reported by Pukale et al. (2015). Such an increase in biodiesel yield by increasing reaction temperature could be attributed to the enhanced solubility of methanol in oil phase as a result of temperature increase. Further increasing of the temperature had no significant effect on biodiesel yield.

**Table 2.**
Conversion (%) of oil into FAME at different times and temperatures.

| Time (h) \ t (K) | 328 | 333 | 338 | 343 | 348 |
|---|---|---|---|---|---|
| **Waste mussel shell catalyst** | | | | | |
| 1 | 8 | 14 | 21 | 25 | 27 |
| 3 | 18 | 26 | 48 | 56 | 58 |
| 5 | 23 | 34 | 60 | 66 | 71 |
| 7 | 27 | 44 | 67 | 78 | 81 |
| 8 | 30 | 46 | 73 | 82 | 83 |
| **DM water treatment precipitates catalyst** | | | | | |
| 1 | 8 | 16 | 26 | 26 | 27 |
| 3 | 15 | 27 | 45 | 57 | 59 |
| 5 | 20 | 38 | 52 | 67 | 69 |
| 7 | 28 | 48 | 65 | 78 | 79 |
| 8 | 32 | 51 | 72 | 83 | 84 |

### *3.3. Reaction kinetics determination*

The kinetics of transesterification reaction of waste cooking oil with methanol in the presence of the waste mussel shell catalyst (12 wt%) and DM water treatment precipitates catalyst (9.08 wt%) at the temperatures of 328, 333, 338, 343 and 348 K under the stirring speed of 350 rpm (methanol to oil molar ratio of 24:1 and 22.5:1, respectively) were investigated in the present study. The exponential trend of methyl ester conversion changes *vs.* time at different temperatures indicated a pseudo-first order kinetics for the transesterification reaction. By fitting the experimental data *via* temperatures in the Equation 6, a good relationship between $-\ln(1-X_{ME})$ and $T$ was obtained. These results for the mussel shell and DM water treatment precipitates catalysts at the temperatures of 328, 333, 338, 343 and 348 K are presented in **Figure 2a** and **b**, respectively. Also the $K$ and $R^2$ values for each temperature are presented in **Table 3**. As seen in **Figure 2** and **Table 3**, the kinetic rate constant increased with increasing the temperature. Also at high temperature values, the difference between the rate constants was low.

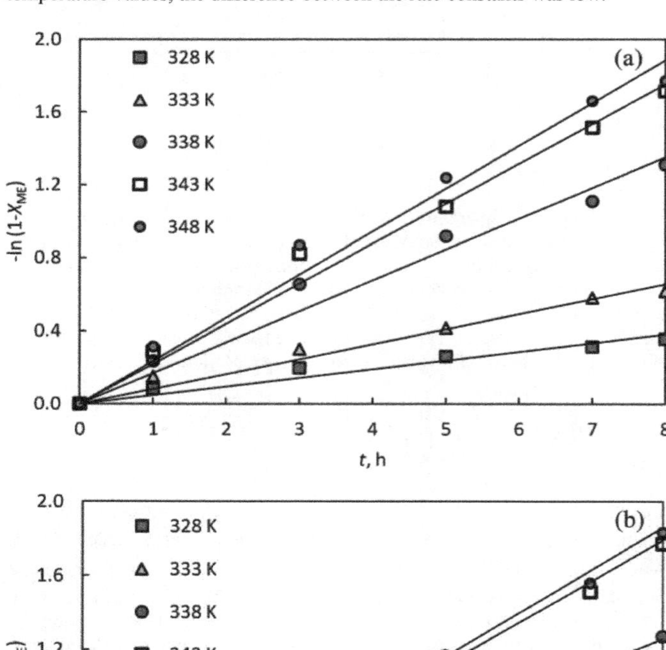


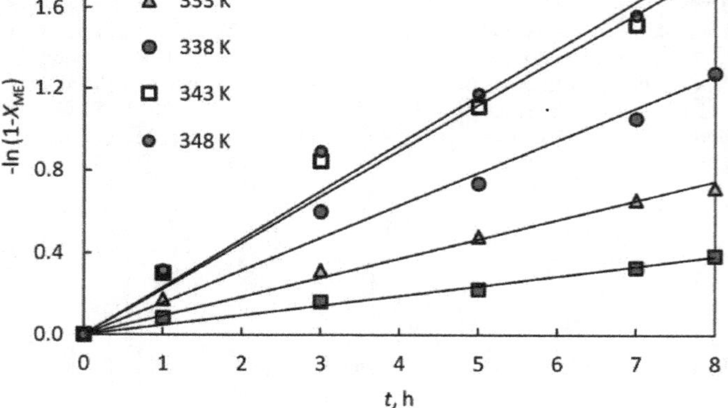


**Fig.2.** Plot of $-\ln(1-X_{ME})$ via $t$ at different temperatures, a) mussel shell, b) DM water treatment precipitates

Considering the reaction constant variations *via* temperature, and as shown in **Figure 3**, the transesterification reaction activation energy was obtained through fitting ln ($k$) data *vs.* $1/T$ with a high accuracy (using **Eq. 8**). For the transesterification reaction in the presence of the waste mussel shell and DM water treatment precipitates catalysts, the activation energy ($E_a$) were equal to 79.83 and 77.09 kJ.mol$^{-1}$, respectively.

**Table 3.**
Transesterification reaction constant rate at different temperatures.

| T (K) | K, $h^{-1}$ | $R^2$ |
|---|---|---|
| **Mussel shell catalyst** | | |
| 328 | 0.0476 | 0.94 |
| 333 | 0.0819 | 0.97 |
| 338 | 0.1690 | 0.97 |
| 343 | 0.2193 | 0.98 |
| 348 | 0.2359 | 0.98 |
| **DM water treatment precipitates catalyst** | | |
| 328 | 0.0478 | 0.98 |
| 333 | 0.0932 | 0.98 |
| 338 | 0.1574 | 0.96 |
| 343 | 0.2240 | 0.98 |
| 348 | 0.2326 | 0.98 |

Usually, the activation energy for the transesterification of oil using base catalysts is in the range of 33.6–84 kJ.mol$^{-1}$ (Freedman et al., 1986). The lower activation energy of the transesterification in the presence DM water treatment precipitates catalyst could be ascribed to the presence of the MgO molecules. This resulted in a faster reaction and therefore, it could be concluded that the DM water treatment precipitates catalyst would be more efficient for transesterification reaction. Moreover, the frequency factor ($k_0$) for the reactions catalyzed by waste mussel shell and DM water treatment precipitates catalysts were found to be at $2.81\times10^{11}$ and $1.08\times10^{11}$ h$^{-1}$, respectively.

*3.4. Kinetics model accuracy*

However The rate constants in the presence of the waste mussel shell and DM water treatment precipitates catalysts were obtained as $k = 2.81\times10^{+11}\exp\left(-\frac{96018}{T}\right)$ and $k = 1.08\times10^{+11}\exp\left(-\frac{92728}{T}\right)$, respectively. By placing these into the Equation 6 and after simplification, the methyl ester conversion in the presence of the catalysts as a function of temperature (K) and time (h), could be achieved as follows (**Eqs. 9** and **10**):

$$X_{\text{ME, musselshell}} = 1-\exp\left(-2.81\times10^{+11}\exp\left(-\frac{96018}{T}\right).t\right) \quad \text{(Eq. 9)}$$

$$X_{\text{ME, DM water unitprecipitation}} = 1-\exp\left(-1.08\times10^{+11}\exp\left(-\frac{92728}{T}\right).t\right) \quad \text{(Eq. 10)}$$

The Equations 9 and 10 were in a good agreement with the experimental data and their mean relative errors (MRE) with the experimental data were measured at 12.33 and 11.68 %, respectively.

## 4. Conclusions

Using economical and environment-friendly waste-oriented materials would play an important role in economizing and consequently expanding biodiesel production and use all around the world. In this study, DM water treatment precipitates were used as basic heterogeneous catalysts for producing biodiesel from waste cooking oil and methanol and was compared to the catalyst obtained through calcination of waste mussel shells (Rezaei et al., 2013). Since the variations in the methyl ester conversion were exponential at all the studied temperature values, therefore, a pseudo-first order kinetics for the transesterification reaction using both catalysts was considered. The activation energies in the presence of the calcinated DM water treatment precipitates and waste mussel shell catalysts were measured at 77.09 and 79.83 kJ.mol$^{-1}$, respectively. Hence, the DM water treatment precipitates catalyst resulted in a faster reaction and could be a better option for industrial biodiesel production. Moreover, the DM water treatment precipitates catalyst had a higher efficiency through five times of reusing and just a small portion (6.15%) of the CaO was extracted by methanol while the loss reported for the calcinated waste mussel shell catalyst stood at 60.17% (Rezaei et al., 2013). Thus, the significantly less methanol-driven extraction rate of CaO from the DM water treatment precipitates marks it as an economical and stable heterogeneous catalyst for biodiesel production from waste cooking oil.

## 5. Acknowledgments

The authors are grateful for the analytical support provided by the Mahidasht Agro-Industry Co. (Nazgol).

## References

Birla, A., Singh, B., Upadhyay, S.N., Sharma, Y.C., 2012. Kinetics studies of synthesis of biodiesel from waste frying oil using a heterogeneous catalyst derived from snail shell. Bioresour. Technol. 106, 95-100.

Boey, P.L., Maniam, G.P., Hamid, S.A., 2011. Performance of calcium oxide as a heterogeneous catalyst in biodiesel production: A review. Chem. Eng. J. 168, 15-22.

Davoodbeygi, Y., 2013. Kinetic of transesterification reaction for biodiesel production using natural catalysts. M.Sc. thesis, Razi University.

Di Serio, M., Tesser, R., Pengmei, L., Santacesaria, E., 2008. Heterogeneous catalysts for biodiesel production. Energ. Fuels. 22, 207-217.

Dossin, T.F., Reyniers, M.F., Marin, G.B., 2006. Kinetics of heterogeneously MgO-catalyzed transesterification. Appl. Catal., B. 61, 35-45.

Encinar, J.M., Sanchez, N., Martinez, G., Garcia, L., 2011. Study of biodiesel production from animal fats with high free fatty acid content. Bioresour. Technol. 102, 10907-10914.

Endalew, A.K., Kiros, Y., Zanzi, R., 2011. Inorganic heterogeneous catalysts for biodiesel production from vegetable oils. Biomass Bioenerg. 35, 3787-3809.

Freedman, B., Butterfield, R.O., Pryde, E.H., 1986. Transesterification kinetics of soybean oil. J. Am. Oil Chem. Soc. 63, 1375-1380.

Freedman, B., Pryde, E.H., Mounts, T.L., 1984. Variables affecting the yields of fatty esters from transesterified vegetable oils. J. Am. Oil Chem. Soc. 61, 1638-1643.

Gaikwad, N.D., Gogate, P.R., 2015. Synthesis and application of carbon based heterogeneous catalysts for ultrasound assisted biodiesel production. Green Process. Synth. 4, 17-30.

Gole, V.L., Gogate, P.R., 2012a. A review on intensification of synthesis of biodiesel from sustainable feed stock using sonochemical reactors. Chem. Eng. Process. 53, 1-9.

Gole, V.L., Gogate, P.R., 2012b. Intensification of synthesis of biodiesel from nonedible oils using sonochemical reactors. Ind. Eng. Chem. Res. 51, 11866-11874.

Kouzu, M., Kasuno, T., Tajika, M., Yamanaka, S., Hidaka, J., 2008. Active phase of calcium oxide used as solid base catalyst for transesterification of soybean oil with refluxing methanol. Appl. Catal., A. 334, 357-365.

Lin, C.Y., Cheng, H.H., 2012. Application of mesoporous catalysts over palm-oil biodiesel for adjusting fuel properties. Energ. Convers. Manage. 53, 128-134.

Liu, H., Su, L., Liu, F., Li, C., Solomon, U.U., 2011 Cinder supported $K_2CO_3$ as catalyst for biodiesel production. Appl. Catal., B. 106, 550-558.

Maddikeri, G.L., Pandit, A.B., Gogate, P.R., 2012. Intensification approaches for biodiesel synthesis from waste cooking oil: a review. Ind. Eng. Chem. Res. 51, 14610-14628.

Omar, W.N.N.W., Amin, N.A.S., 2011. Optimization of heterogeneous biodiesel production from waste cooking palm oil via response surface methodology. Biomass Bioenerg. 35, 1329-1338.

Pukale, D.D., Maddikeri, G.L., Gogate, P.R., Pandit, A.B., Pratap, A.P., 2015. Ultrasound assisted transesterification of waste cooking oil using heterogeneous solid catalyst. Ultra. Sonochem. 22, 278-286.

Rezaei, R., Mohadesi, M., Moradi, G.R., 2013. Optimization of biodiesel production using waste mussel shell catalyst. Fuel. 109, 534-541.

Sharma, Y.C., Singh, B., Korstad, J., 2011. Latest developments on application of heterogenous basic catalysts for an efficient and eco friendly synthesis of biodiesel: A review. Fuel. 90, 1309-1324.

Singh, A.K., Fernando, S.D., 2007. Reaction kinetics of soybean oil transesterification using heterogeneous metal oxide catalysts. Chem. Eng. Technol. 30, 1716-1720.

Taufiq-Yap, Y.H., Lee, H.V., Hussein, M.Z., Yunus, R., 2011. Calcium-based mixed oxide catalysts for methanolysis of Jatropha curcas oil to biodiesel. Biomass Bioenerg. 35, 827-834.

Veljkovic, V.B., Stamenkovic, O.S., Todorovic, Z.B., Lazic, M.L., Skala, D.U., 2009. Kinetics of sunflower oil methanolysis catalyzed by calcium oxide. Fuel. 88, 1554-1562.

Vujicic, D.J., Comic, D., Zarubica, A., Micic, R., Boskovi, G., 2010. Kinetics of biodiesel synthesis from sunflower oil over CaO heterogeneous catalyst. Fuel. 89, 2054-2061.

Yin, X., Ma, H., You, Q., Wang, Z., Chang, J., 2012. Comparison of four different enhancing methods for preparing biodiesel through transesterification of sunflower oil. Appl. Energ. 91, 320-325.

Zhang, L., Sheng, B., Xin, Z., Liu, Q., Sun, S., 2010. Kinetics of transesterification of palm oil and dimethyl carbonate for biodiesel production at the catalysis of heterogeneous base catalyst. Bioresour. Technol. 101, 8144-8150.

# 6

# Integrated volarization of spent coffee grounds to biofuels

Mebrahtu Haile*

*Land resource management and environmental protection department, college of dry land agriculture and natural resource, Mekelle University, Ethiopia*

## HIGHLIGHTS

- Oil extraction was studied for different Solvents
- Two-step biodiesel production method due to high FFA
- Pelletized fuel of the over left solid and glycerin
- Biofuel without growing plants and/or converting food to fuel.

## GRAPHICAL ABSTRACT

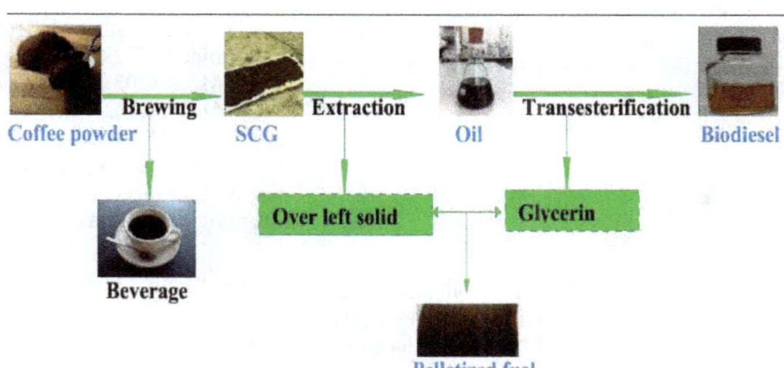


## ABSTRACT

Biodiesel is a renewable energy source produced from natural oils and fats, and is being used as a substitute for petroleum diesel. The aim of this study was to investigate the potential of using spent coffee grounds for biodiesel production and its by-products to produce pelletized fuel, which is expected to help the biodiesel production process achieve zero waste. For this experiment, spent coffee grounds sample was collected from Kaldis coffee, Addis Ababa, Ethiopia. Extraction of the spent coffee grounds oil was then conducted using n-hexane, ether and mixture of isopropanol to hexane ratio (50:50 %vol), and resulted in oil yield of 15.6, 17.5 and 21.5 %w/w respectively. A two-step process was used in biodiesel production with conversion of about 82 %w/w. The biodiesel quality parameters were evaluated using the American Standard for Testing Material (ASTM D 6751). The major fatty acid compositions found by Gas chromatography were linoleic acid (37.6%), palmitic acid (39.8%), oleic (11.7%), and stearic acid (8.6%). In addition, solid waste remaining after oil extraction and glycerin ratio (glycerin content from 20-40%) was evaluated for fuel pellet (19.3-21.6 MJ/Kg) applications. Therefore, the results of this work could offer a new perspective to the production of biofuel from waste materials without growing plants and/or converting food to fuel.

**Keywords:**
Spent coffee ground
Spent coffee ground oil
Transesterification
Biodiesel
Pelletized fuel

## 1. Introduction

Biodiesel, a mixture of long-chain fatty acid alkyl esters obtained from renewable feedstock, such as vegetable oil or animal fats, for use in compression ignition engines (Murugesa et al., 2009). Recently, biodiesel has become more attractive because it is biodegradable and non-toxic in nature, environmentally friendly and derived from renewable resources. It can also be produced from any material that contains fatty acids, either linked to other molecules or present as free fatty acids (Balat and Balat, 2009).

However, a global debate has now emerged because this fuel is derived primarily from soybean oil or other cereals and using food to produce fuel is not reasonable considering the increase in world population. In order to overcome this problem, industries use waste vegetable oil and grease and animal fats from poultry to produce biodiesel Nebel and Mittelbach, 2006).

* Corresponding author
E-mail address: mebrahtu.haile@yahoo.com

Therefore, it would be very useful to look for new raw sustainable materials for biodiesel production that do not involve the use of cereals and land to grow. In this work, production of biodiesel from spent coffee grounds as well as fuel pellets from its by-products has been carried out.

Coffee is one of the world's most widely consumed beverages. It is the most important cash or export crop in Ethiopia, providing approximately 30.6% of Ethiopia's foreign exchange earnings in 2010-2011 (Bureau of African Affairs, 2012). According to the report in (Mebrahtu et al., 2013) Ethiopia is producing an estimated 9.804 million 60-kg bags of coffee and half of it is consumed domestically.

Spent coffee grounds are the main coffee industry residues with a worldwide annual generation of 6 million tons and 235,296 tons in Ethiopia, obtained from the treatment of coffee powder with hot water to prepare instant coffee (Mebrahtu et al., 2013; Tokimoto et al., 2005). Considering this huge amount of coffee residue produced all over the world, the reutilization of this material is a relevant subject. Some attempts for reutilization of SCG have been made. However, none of these attempts have yet been routinely implemented, and most of these residues remain unutilized, being discharged to the environment where they cause severe contamination and environmental pollution problems due to the toxic nature (presence of caffeine, tannins, and polyphenols) (Leifa et al., 2000).

The biodiesel from spent coffee grounds possesses better stability than biodiesel from other sources, due to its high antioxidant content (Campo et al., 2007; Yanagimoto et al., 2004). SCG is also considered as easily available and an inexpensive adsorbent for the removal of cationic dyes in wastewater treatments (Franca et al., 2009).

## 2. Materials and methods

### *2.1. Materials and chemicals*

This work used spent coffee grounds (coffee Arabica) collected from Kaldis coffee shop (Addis Ababa, Ethiopia). All analytical grade chemicals and solvents were obtained from the department of chemical engineering (Addis Ababa University, Ethiopia). Experimental and laboratory work was undertaken in the department of chemical engineering (Addis Ababa University), Ethiopian Petroleum Supply Enterprise (EPSE), Ethiopian Health and Nutrition Research Institute and Geological Survey of Ethiopia Central Laboratory.

### *2.2. Experimental*

#### *2.2.1. Moisture content Determination*

The spent coffee grounds (SCGs) were allowed to air dry for several days and then characterized to evaluate its moisture content by repeated cycles of oven drying at 105 °C followed by cooling in a desiccator over silica gel (0% relative humidity) and weighing until a constant weight. The moisture content was determined as in Equation 1;

$$M = \left(\frac{W1 - W2}{W1}\right) x\ 100\ \% \qquad (1)$$

Where M, $W_1$, $W_2$ are moisture content, initial mass and final mass

#### *2.2.2. Spent coffee grounds (SCGs) oil extraction*

Different solvents were tested to evaluate their suitability for oil extraction from oven dried SCG (at 105 °C). 300 g of the dried SCG sample was placed in a 2 L round bottom flask of the soxhlet apparatus and extracted for 4-8 h using n-hexane, ether and mixture of isopropanol to hexane ratio (50:50) as a solvent. The extraction procedure was stopped when three consecutive measurements of the solvent refraction index were constant and close to the pure solvents' value. The oil was separated from the solvents using a rotary evaporator (Buchi RE111 Rotavapor, BUCHI, Flawil, Switzerland). The solvents were reused in the next batch of extraction. The oil yield was determined as in equation 2;

$$\text{SCG oil content} = \frac{W_o}{Ws}\ x\ 100\ \% \qquad (2)$$

Where: $W_o$ = weight (g) of oil extracted and
$W_s$ = weight (g) of sample (dry base)

#### *2.2.3. Physicochemical characterization of SCGs oil*

Prior to transesterification, oil quality properties of the SCG oil were determined. These properties included saponification value, peroxide value (ASTM D3703), water and sediment (ASTM D2709), iodine value (EN14111), acid value (ASTM D974) density (ASTM D1298), kinematic viscosity (ASTM D445), flash point (ASTM D93), cloud point (ASTM D97) and higher heating value (HHV) (ASTM D240). The oil properties were analyzed in accordance with ASTM D 6751 and EN 14214. All experiments were run in triplicate and mean values were reported.

#### *2.2.4. Two-step Biodiesel Production Process*

##### *2.2.4.1. Acid-catalyzed esterification*

The SCG oil was heated to 54 °C to homogenize the oil. The reaction was conducted in a 1 L round-bottomed flask attached with a reflux condenser and a thermometer placed in an oil bath. The oil is mixed with methanol (a molar ratio of alcohol to free fatty acids of 20:1) and significant quantities of HCl (10 wt% of total fatty acid content). The reactor was stirred at about 600 RPM, at a temperature of 54 °C for 90 minutes (Santori et al., 2012). Then the reaction product mixture was poured into a separating funnel and allowed to settle for 24 h. The top layer, which is comprised of unreacted methanol and water, was removed.

##### *2.2.4.2. Base-catalyzed transesterification*

The transesterification of the oil was carried out under ambient pressure in a 1 L two-necked round bottom reactor equipped with a thermometer, a hot plate with magnetic stirrer, and a reflux condenser. The reaction was carried out at 54 °C for 90 minutes with 1 wt% of KOH and methanol-to-oil molar ratio of 9:1 (Kondamudi et al., 2008). The resulting product was cooled to room temperature without any agitation and transferred to a separatory funnel for glycerol and methyl ester separation. It was left overnight to allow separation by gravity. After the two phases have separated, the upper phase was collected and the excess alcohol in it removed using a vacuum evaporator operated at 80 °C.

#### *2.2.5. Purification*

The resulting methyl ester obtained was purified by successive washing with warm (55 °C) deionized water to remove residual catalyst, glycerol, methanol and soap. A small quantity of sulfuric acid was used in the second washing to neutralize the remaining soap and catalyst. Finally, the SCG oil methyl ester was dried over anhydrous sodium sulfate ($Na_2SO_4$) to remove residual water. A filtration process followed to remove solid traces. The dried methyl ester was then bottled and kept for characterization studies.

*2.2.6. Characterization of SCG Biodiesel*

Physical and chemical properties of SCG biodiesel esters were characterized using internationally accepted standards. Density (15 °C) (ASTM D1298), Kinematic viscosity (40 °C) (ASTM D445), Gross calorific value (ASTM D240), Cloud point (ASTM D97), Iodine number (EN14111), water and sediment (ASTM D2709), Ash content (ASTM D874), acid value (ASTM D974), Carbon residue (ASTM D189), flash point (ASTM D93), Distillation, 90% recovery (ASTM D86) and Copper corrosion (ASTM D130) were used.

*2.2.7.Fatty acid composition of SCG methyl ester*

Fatty acid composition of the synthesized alkyl ester was determined by gas chromatography. Gas chromatography (DANI GC 1000) equipped with flame ionization detector (FID) was employed during fatty acid determination. The GC was calibrated by injecting standards at varying concentrations. The samples were injected (1μL) one by one in a DANI GC 1000, equipped with a capillary column of EC TM-5 (25 m x 0.53 mm x 1 μm). The GC oven was primarily kept at 50 °C for 2 min, and then heated at 4 °C /min up to 250 °C, where it was kept for 15 min, and a pressure of 1.25 Bar was applied. The carrier gas used was nitrogen at (1 ml/min).

*2.2.8. Glycerin and solid waste remaining after biodiesel production for Pellets*

The by-products of spent coffee ground biodiesel were analyzed as pellets. The solid waste remaining after oil extraction were mixed with glycerin content (20-40%) by weight ratio and blended by hand in a large mixing bowl. The glycerin and solid waste ratio were then mixed to produce a crude unfinished material. The pellet mixture (approximately 12g) is placed inside a rolled piece of newspaper wrapping, and the ends are folded down so thatboth ends of the cylinder are covered. No adhesive is used, and until the pellet is compressed this will remain prone to unwrap. The major quality properties of the pellet like heating value (Bomb calorimeter), specific density (Gravimetric method and water replacement), moisture content (Oven-drying of 1 g of pellet sample at 104-110°C for 1 hour), combustion ash, volatile matter, and fixed carbon were analyzed following the ASTM methods.

## 3. Results and discussıons

*3.1. Oil content of spent coffee ground (SCG)*

Extraction of oil from spent coffee grounds was carried out using solvents such as n-hexane, ether and mixture of isopropanol to hexane ratio (50:50) under reflux conditions. At least three extraction tests were run for each solvent and the mean ratio of oil extracted from SCG to dry weight of SCG used was recovered after solvent rotavopor. The solvent extracted crude oil yields are (15.6% w/w hexane, 17.5% w/w ether and 20.6 %w/w mixture of isopropanol to hexane ratio (50:50). The solvent that allowed for the higher oil recovery was a mixture of isopropanol to hexane ratio (50:50) but with a lower solvent recovery. On the other hand, hexane was the solvent that allowed for lower oil recovery, with a longer extraction time. Also hexane was one of the solvents with higher recovery rate while ether showed the opposite result. But ether was also the solvent that took the longest time to finish the extraction and one of the solvents with a lower recovery rate. Considering the cost of the solvents and the energy consumption needed to perform the oil extraction and the solvent recovery by distillation, the mixture 50:50 (Vol/Vol) of hexane and isopropanol was chosen to perform further extractions.

The obtained oil content is relatively higher than that reported in the SCG oil (10-15%) depending on the coffee species as in (Kondamudi et al., 2008, Deligiannis et al., 2011). The oil content of this study is higher than previous studies (19.73 %w/w) using hexne. Higher value (28.3 %w/w) as in (Caetano et al., 2012) for SCG oil was extracted with a mixture of isopropanol and hexane. The variation in the oil yield could be attributed to differences in varieties of coffee, solvent type and cultivation climate. Compared to some other oilseeds such as olive, soybeans and cotton seeds, which have average oil contents of only 17%, 20% and 14%, respectively, this may result in lower operating costs as in (Rossel, 1987).

*3.2. Properties of the extracted coffee oil*

SCG oil was characterized to determine its physiochemical properties to utilize as a feedstock in the production of biodiesel. The obtained results were compared with that of American Standard for Testing Materials (ASTM D 6751). The standard values and the results obtained are summarized in Table 1.

**Table 1**
Characterization of the oil extracted from waste coffee ground.

| Property | Units | Test Methods | Limits | Results |
|---|---|---|---|---|
| Density (15 °C) | g /cm ³ | D1298 | 0.86–0.90 | 0.917 |
| Kinematic Viscosity (40 °C) | mm² /s | D 445 | 1.9– 6.0 | 42.65 |
| Gross calorific value | MJ/kg | D 240 | Report | 38.22 |
| Cloud point | °C | D 97 | Report | 11 |
| Iodine Value | gI2/100g | EN14111 | 120 max | 79 |
| Water and sediment | %volume | D2709 | 0.050 max | 0.03 |
| Acid value | Mg | D974 | 0.8 max | 9.85 |
| Flash point | °C | D93 | 130 min | >200 |
| Saponification value | mgKOH/g | AOCSCd 3-25 | -------- | 167.28 |

The viscosity of the oil is too high to be used in direct combustion engines. The oil was found to be very acidic and the saponification value was also high as in "Table 1" to be directly converted into biodiesel without pre-treatment, which may indicate a higher degree of oxidation and occurrence of hydrolysis reactions (Knothe, 2007). In order to overcome this problem the oil was first esterifed using hydrocloric acid as a catalyst prior to transesterification. The relatively high HHV suggests that this oil can be used for direct combustion. The SCG oil has high density, so cannot be used directly as fuel; since this high density would polymerize and leads to the formation of deposits in the car engine. The high cloud point of SCG oil indicates its unsuitability as biodiesel feedstock. The peroxide value is low, means its rancidity is low and can store for a long time. The flash point is very high, making it better suited for biodiesel production with respect to safety during storage and transportation.

*3.3. Biodiesel yield of SCG oil*

Biodiesel yield estimation was done after the separation and purification of the transesterified product. The 82 %w/w yield of SCG oil synthesized was calculated. The %FFA of the SCG oil was decreased from 4.9% to 0.5% after three consecutive esterification steps. After esterification, the oil pretreatment loss was found to be 6.8 %w/w based on the initial sample of SCG oil.

*3.4. Spent Coffee Ground biodiesel Fuel properties*

Transesterification of the oil to biodiesel was carried out using methanol and KOH. The properties of biodiesel fuel prepared from spent coffee grounds are analyzed by ASTM analysis (Table 2). The analysis of the results shows that biodiesel obtained from spent coffee grounds is a strong candidate as an alternative to diesel (Barnwal and Sharma, 2005).

**Table 2**
Characterization of the oil extracted from spent coffee ground.

| Property | Units | Test Methods | Limits | | SCG oil methyl ester |
|---|---|---|---|---|---|
| | | | ASTM D6751 | EN 14214 | |
| Density (15⁰C) | g /cm ³ | D1298 | -------- | 0.86 – 0.9 | 0.88 |
| Kinematic viscosity (40°C) | mm ²/s | D445 | 1.9 -6.0 | 3.5-5 | 5.4 |
| Gross calorific value | MJ/kg | D240 | -------- | ------- | 39.6 |
| Cloud point | °C | D97 | Report | ------- | 13 |
| Iodine value | gI2/100g | EN14111 | -------- | 120 max | 74 |
| Water and sediment | %volume | D2709 | 0.05 max | 500 mg/kg | 0.01 |
| Ash content | W% | D874 | 0.02 max | 0.02 max | 0.01 |
| Acid value | mg KOH/g | D664 | 0.5 max | 0.50 max | 0.7 |
| Carbon residue | % mass | D189 | 0.05 max | 0.3max | 0.033 |
| Flash point | °C | D93 | 93.0 min | 120 min | 220 |

*3.5. Fatty acid composition of spent coffee ground oil methyl ester*

Investigation of the fatty acid components of SCG biodiesel was carried out using GC. The fatty acid profile of the methyl ester was identified and quantified as shown in figure 1. GC analysis showed the presence of C16-C20 methyl esters of fatty acids. SCG biodiesel consists of both saturated and unsaturated esters (Figure 1), where more than 98.3% of the total composition were methyl esters of linoleic acid (37.3%), palmitic acid (35.8%), oleic (13.9%), stearic acid (8.1%) and arachidic(3.2%).

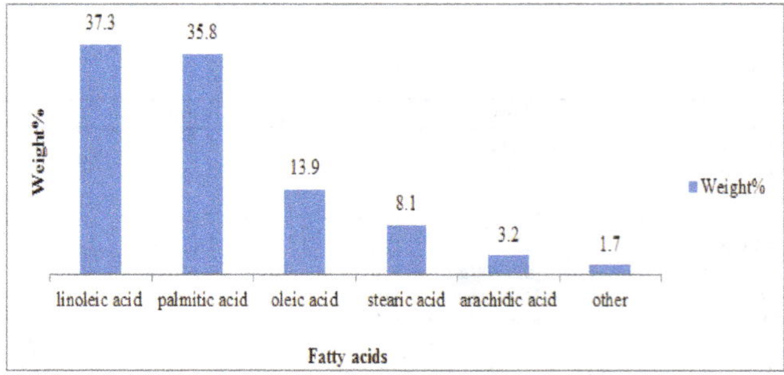


**Fig.**1. SCG methyl ester fatty acid composition.

*3.6. Pelletized fuel quality*

SCGs are composed of 13.8% of cellulose, 36.7% of hemicelluloses, 13.6% of proteins and 33.6% of lignin (Caetano et al.,2012; Mussatto et al., 2011). Cellulose is considered as a major source of volatile content in the biomass, while lignin serves as a natural binder in the biomass (Biswas et al., 2011; Wahyudiono and Goto, 2013). Although the SCG contain relatively high lignin content, the addition of a binder, like glycerin, is expected to enhance combination of biomass particles well. Table 3 presents the bulk density, the proximate analysis results, the heating value and pellet quality demands suggested by EUBIA (EUBIA, 2007).

**Table 3**
Properties of pelletized fuel prepared from SCG after oil extraction and glycerin.

| Glycerin content (%) | Bulk density (kg/m$^3$) | Proximate Analysis (% wt) | | | | Heating value (MJkg-1) |
|---|---|---|---|---|---|---|
| | | Moisture content | Volatile matter | Ash | Fixed carbon | |
| 20 | 996.6 | 5.12 | 73.4 | 5.08 | 16.4 | 19.3 |
| 25 | 995.7 | 4.99 | 76.8 | 4.4 | 13.81 | 19.7 |
| 30 | 996.0 | 4.33 | 78.0 | 3.7 | 13.97 | 19.8 |
| 35 | 997.8 | 3.12 | 79.1 | 3.2 | 14.58 | 20.2 |
| 40 | 995.5 | 5.03 | 77.8 | 2.8 | 14.37 | 21.6 |
| EUBIA [21] | >650 | <10 | - | <0.5 | - | >17 |

All the tested properties of the studied fuel sample pellets were found to comply with the suggested EUBIA values, except for the ash content. An increase of glycerin in the solid waste remaining after oil extraction appears to benefit the reduction of the ash content of the fuel. However, the maximum glycerin of 40% still produced an amount of ash greater than the criterion value of 0.5%. The test results show that the fuel sample can satisfy the fuel pellet quality demands for domestic use as suggested by the European Biomass Industry Association (EUBIA) for pellet characteristics of the bulk density and moisture content (EUBIA, 2007). Since the finished fuel pellet was oven-dried at 105°C for 24 h, it contained relatively low moisture content of 3.12- 5.12 %wt. The heat of combustion of the fuel sample indicates that the pelletized fuel provided a better fuel quality. The pelletized fuel had the heating value greater than the suggested EUBIA value of 17 MJ/kg. An increase of the glycerin content from 20% to 40% can increase the combustion heating value from 19.3 to 21.6 MJ/kg. The use of glycerin as a binder helps to improve the biomass pelletized fuel for its calorific property. This fuel pellet has high calorific value compared to the biomass energy obtained from groundnut (12.60 MJ/Kg) (Musa, 2007), cowpea (14.37 MJ/Kg) and soybeans (12.94 MJ/Kg) reported by (Enweremadu et al., 2004). The energy values and combustion qualities of the fuel produced in this study is sufficient enough to produce the required heat for domestic cooking and also for industrial application especially the energy requirement of the small-scale industries.

## 4. Conclusion

The aim of this study was to evaluate spent coffee grounds as a potential source to produce biodiesel and its by-products as fuel pellets. Oil extraction with a mixture of hexane/isopropanol (50:50 vol/vol) that allowed for the higher oil recovery (21.5 %) at a relatively lower cost was converted to biodiesel with 82% yield. GC analyses indicated that the spent coffee ground biodiesel consisted of both saturated (47.1%) and unsaturated (51.2%) esters. The quality of the obtained biodiesel was evaluated according the ASTM standard, showing that the biodiesel obtained is within the standard limits for all the evaluated parameters except for carbon residue.

The solid waste remaining after oil extraction and glycerin, which are by-products from the biodiesel production, are experimentally proven to be promising raw materials for producing fuel pellets (21.6 MJ/kg).The production of the solid waste after oil extraction and glycerin pellets can also promote a zero waste approach for the biodiesel production process and can eliminate disposal cost of the crude glycerin. This work will give new insight to producing biofuels without growing plants and/or converting food to fuel.

## References

Balat, M., Balat, H.A., 2009. Critical review of biodiesel as vehicular fuel. Energ. Conv. Mgmt. 49, 2727-2741.

Barnwal, B.K., Sharma, M.P., 2005. Prospects of biodiesel production from vegetable oils in India. Renewable Sustainable Energy ReV. 9, 363–378.

Biswas, K., Yang, W., Blasiak,W., 2011. Steam pretreatment of Salix to upgrade biomass fuel for wood pellet production. Fuel Process Technol., vol. 92, pp. 1711-1717.

Bureau of African Affairs, Background Note: Ethiopia. U.S. Department of state (2012).

Caetano, N.S., Silva, V.F.M., Mata, T.M., 2012. Valorization of Coffee Grounds for Biodiesel Production. The Italian Association of Chemical Engineering, VOL. 26.

Campo, P., Zhao, Y., Suidan, M.T., Venosa, A. D., Sorial, G.A., 2007. Biodegradation kinetics and toxicity of vegetable oil triacylglycerols under aerobic conditions. Chemosphere, 68, 2054-2062.

Deligiannis, A., Papazafeiropoulou, A., Anastopoulos, G., Zannikos, F., 2011. Waste coffee grounds as an energy feedstock. Proceedings of the 3rd International CEMEPE & SECOTOX Conference Skiathos, ISBN 978-960-6865-43-5, June 19-24.

Enweremadu, C.C., Ojediran, J.O., Oladeji, J.T., Afolabi, I.O., 2004. Evaluation of Energy Potential of Husks from Soy-beans and Cowpea. Science Focus, 8:18-23.

EUBIA, 2007. Comparison between briquettes and pellets. European Biomass Industry Association. [Online]. Available http://www.eubia.org/197.0.html.

Franca, A.S., Oliveira, L.S., Ferreira, M.E., 2009. Kinetics and equilibrium studies of methylene blue adsorption by spent coffee grounds. Desalination, 249, 267- 272.

Knothe, G., 2007. Some aspects of biodiesel oxidative stability. Fuel Process Technology, 88:669-77.

Kondamudi, N., Mohapatra, S.K., Misra, M., 2008. Spent coffee grounds as a versatile source of green energy. Journal of Agricultural and Food Chemistry. 56; 11757-11760.

Leifa, F., Pandey, A., Soccol, C. R., 2000. Solid state cultivation—an efficient method to use toxic agro-industrial residues. Journal of Basic Microbiology, 40, 187–197.

Mebrahtu, H., Araya, A., Nigist, A., 2013. Investigation of Waste Coffee Ground as a Potential Raw Material for Biodiesel Production: international journal of renewable energy research, vol.3, No.4.

Murugesan, A., Umarani, C., Chinnusamy, T., Krishnan, M., Subramanian,

R., Neduzchezhain, N., 2009. Production and analysis of bio-diesel from non-edible oils–A review. Renew. Sustain. Energy Rev. 13, 825–834.

Musa, N.A., 2007. Comparative Fuel Characterization of Rice Husk and Groundnut Shell Briquettes. NJREDl. 6(4):23-27.

Mussatto, S.I., Carneiro, L.M., Silva, J.P.A., Roberto, I.C., Teixeira, J.A., 2011. A study on chemical constituents and sugars extraction from spent coffee grounds. Carbohydrate Polymers, 83,368-374.

Nebel, B.A., Mittelbach, M., 2006. Biodiesel from extracted fat out of meat and bone meal. Eur. J. Lipid Sci. Technol. 108, 398-403.

Rossel, J.B., 1987. Classical analysis of oils and fats: In: Hamilton RJ, Rossel JB (Eds): Analysis of oils and fats. Elsevier Applied Science, London and New York, pp. 1-90.

Santori, G., Nicola, G.D., Moglie, M., Polonara, F., 2012. A review analyzing the industrial biodiesel production practice starting from vegetable oil refining. Applied Energy, 92, 109-132.

Tokimoto, T., Kawasaki, N., Nakamura, T., Akutagawa, J., Tanada, S., 2005. Removal of lead ions in drinking water by coffee grounds as vegetable biomass. Journal of Colloid and Interface Science, 281, 56-61.

Wahyudiono, S.M., Goto, M., 2013. Utilization of sub and supercritical water reactions in resource recovery of biomass wastes. Engineering Journal, vol. 17, pp. 1-12.

Yanagimoto, K., Ochi, H., Lee, K.G., Shibamoto, T., 2004. Antioxidative Activities of Fractions Obtained from Brewed Coffee. J. Agric. Food Chem. 52, 592-596.

7

# Evaluation of the cultivation conditions of marine microalgae *Chlorella* sp. to be used as feedstock in ultrasound-assisted ethanolysis

Mateus S. Amaral[1], Carla C. Loures[2], Patrícia C.M. Da Rós[1], Sara A. Machado[1], Cristiano E. R. Reis[3], Heizir F. de Castro[1], Messias B. Silva[1,2,]*

[1] *Engineering School of Lorena-University of São Paulo, Lorena, SP, Brazil.*

[2] *Engineering Faculty of Guaratingueta – State University Julio de Mesquita Filho-UNESP, Guaratingueta, SP, Brazil.*

[3]*Department of Bioproducts and Biosystems Engineering, College of Food, Agricultural and Natural Resource Sciences, University of Minnesota, USA.*

## HIGHLIGHTS

➢*Culture conditions of Chorella* sp. *were simultaneously investigated employing a Taguchi design experiment.*

➢ *Highest lipid content of 19.8% was obtained.*

➢ *Main fatty acids were C16 and C18 which are suitable for the production of high-quality biodiesel.*

➢ *Ultrasound-assisted biodiesel synthesis led to 78.4% conversion rate.*

## GRAPHICAL ABSTRACT

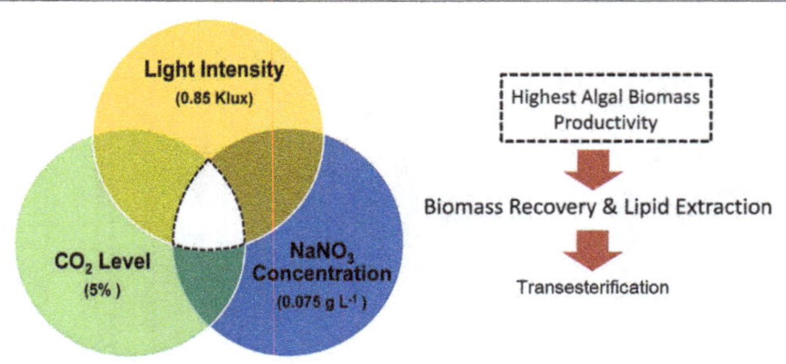


**Keywords:**
*Chlorella* sp.
Experimental Design
Microalgal oil
Transesterification
Fatty acid ethyl esters (FAEE)

## ABSTRACT

A total of 8 assays was conducted to study the influence of different variables namely, light intensity, $CO_2$ level, $NaNO_3$ concentration and aeration rate, on the cultivation of the marine microalgae *Chlorella* sp. to enhance the biomass feedstock availability for biodiesel. The experiments were designed using a Taguchi L8 experimental array set at two levels of operation, having light intensity (0.85 and 14.5 klux), $CO_2$ (5 and 10%), $NaNO_3$ (0.025 and 0.075 g $L^{-1}$) and aeration rate (3:33 and 1.67 vvm) as independent variables and considering biomass productivity and lipid content as response variables. All the experiments were performed in six photobioreactor vessels connected in series with a total volume of 8.4 L and working volumes of 2 L and 4 L, depending on the conditions assessed. The highest biomass productivity was 210.9 mg $L^{-1}day^{-1}$, corresponding to a lipid content of 8.2%. Such results were attained when the culture conditions were set at 0.85 klux light intensity, 5% $CO_2$ and 0.075 g $L^{-1}$ $NaNO_3$. The aeration rate showed no significant influence on the biomass productivity. On the other hand, the highest lipid content was achieved when the cultures were grown using the lowest concentration of $NaNO_3$ (0.025 g $L^{-1}$) and an aeration rate of 1.67 vvm, while the other factors had no statistical significance. Under these conditions, the lipid content obtained was 19.8%, at the expense of reducing the biomass productivity to 85.9 mg $L^{-1}day^{-1}$.The fatty acid profile of the lipid material characterized by gas chromatography identified fourteen fatty acids with carbon chain ranging from C8 to C20 in which most of the fatty acids present were saturated (58.7 %) and monounsaturated (36.1%) fatty acids. Those obtained at higher proportions were the oleic (22.8%), palmitic (20.7%) and lauric (17.7 %) acids, indicating a suitable composition for fatty acid ethyl esters (FAEE) synthesis. This was confirmed by acid catalysis performed under ultrasound irradiations reaching a conversion rate of 78.4% within only 4 h.

* Corresponding author
E-mail address: messias@dequi.eel.usp.br

## 1. Introduction

An increase in greenhouse gas emissions, followed by a simultaneous depreciation on fossil fuel reservoirs have incited many studies to aim at finding environmentally-sustainable alternative feedstocks for biofuel production (Beetul et al., 2014; Girard et al., 2014; Zhu, 2015). In this context, microalgae have been highlighted as promising raw materials for biofuel production owing to its numerous advantages, such as being able to grow on low-value agricultural land, great capacity of capturing atmospheric $CO_2$, as well as lipid accumulation (Koller et al., 2014). As a complimentary source to the limited fossil fuels, lipids extracted from microalgae can be processed into biodiesel by catalyzed transesterification reactions (D'oca et al., 2011a; Teo et al., 2014).

Microalgae are photosynthetic microorganisms (endowed with chlorophyll-a) that can grow quickly and be found throughout the most diverse ecosystems e.g. on the surface of certain types of soils, most abundantly in aquatic environments, or grouped in the form of linked linear segments of cells (Chisti, 2007; Mata et al., 2010).

Many studies have been conducted with numerous species of microalgae to assess their potential for lipid accumulation, such as *Isochrysis* sp. (25-33% lipids), *Nannochloris* sp. (20-35% lipids), *Nannochloropsis* sp. (31-68% lipids), *Neochlorisoleo abundans* (35-54% lipids) and *Tetraselmessueica* (15-23% lipids) (Mata et al, 2010). According to Huntley and Redalje (2007), the *Chlorella* species are considered as a promising candidate for commercial lipid production due to its rapid growth, and also because it is highlighted as one of the most robust species for cultivation in open ponds. However, factors associated with microalgae cultivation, including nutrient requirements (nitrogen, phosphorus, metals, among others), pH, temperature, light intensity and carbon source (glucose, glycerol, lactose, $CO_2$), could influence both the cell growth and the accumulation of lipids (Millán-Oropeza et al., 2015; Nascimento et al., 2015). The conventional optimization method involves the variation of one parameter at a time while keeping the others constant. This is a time-consuming procedure and often does not identify the interaction effect among various parameters as compared to statistical methods. Taguchi method uses orthogonal arrays allowing the investigation of various factors with a reduced number of experiments, while is able to identify the significant parameters of an ideal process with multiple qualitative aspects (Ross, 1995).

This statistical tool involves the calculation of the signal to noise ratio (S/N) throughout experimental replicates, representing a magnitude relation between sensitivity and variability of a particular output measurement, and the analysis of variance (ANOVA) to evaluate the significance of the variables studied on the process (Ross, 1995; Sharma et al., 2005).

The present study was aimed at establishing the cultivation conditions of the marine microalgae *Chlorella* sp. to obtain high cell productivity and lipid content. To achieve this, a Taguchi's L8 array with four independent variables at two levels was performed. The feasibility of using the resulting lipids as feedstock for biodiesel synthesis was also assessed. Owing to the high free fatty acids levels generally found in microbial oils (Da Rós et al, 2013), homogeneous acid catalysis was attempted.

## 2. Materials and methods

### 2.1. Microalgae strain

All experiments were performed with the Chlor-CF strain of the marine microalgae *Chlorella* sp., isolated in Cabo Frio (Rio de Janeiro-Brazil), obtained from the Seaweed Culture Collection of the Oceanographic Institute at the University of São Paulo (São Paulo, SP, Brazil).

### 2.2. Photobioreactors and microalgae cultivation

The experimental set up comprised of six closed photobioreactor vessels (14 × 20 × 30 cm) connected in series with a total volume of 8.4 L, as shown in **Figure 1**. The photobioreactor was run on modified Guillard f/2 medium (Guillard, 1975), supplemented with $NaNO_3$ (0.025 and 0.075 g $L^{-1}$) and $CO_2$ (5 and 10%) and was stirred with sterile air (filtered with 0.22-µm filter). The photobioreactor vessels were inoculated by adjusting the initial cell concentration to 3 × $10^6$ $mL^{-1}$. The reactor vessels were sparged with sterile air at different rates (3.33 and 1.67 vvm) and maintained at 27±1°C under photoperiods of 12:12 h (light: dark) at different light intensities (0.85 and 14.5 Klux) using halogen lamps placed above the photobioreactor vessels.

### 2.3. Experimental design and data treatment

The experimental design was performed using Taguchi L8 orthogonal array being operated according to the factors and levels displayed in **Table 1**. All runs were carried out in triplicates. Biomass productivity and lipid content were taken as response variables. The statistical analysis was performed using the Statistica software (version 12.0) by analyzing the graph effects and tables of the analysis of variance (ANOVA). S/N ratio was calculated for larger-the-better response using the **Equation 1**:

$$S/N = -10 \log 1/n \, \Sigma(1/Y_i^2) \quad \text{(Eq.1)}$$

**Table 1.**
Factors evaluated in the experimental design and their respective coding levels.

| Code | Factor | Level | |
|---|---|---|---|
| | | 1 | 2 |
| A | Light intensity (klux) | 0.85 | 14.50 |
| B | $CO_2$ (%) | 5 | 10 |
| C | $NaNO_3$ (g/L) | 0.025 | 0.075 |
| D | Aeration rate (vvm) | 3.33 | 1.67 |

### 2.4. Harvesting and biomass determination

The cell growth was monitored using a counting Neubauer hemocytometer and by measuring the optical density at 570 nm. The resulting biomass at the end of the batch run was recovered by coagulation using 1 mol.$L^{-1}$ sodium hydroxide. The supernatant was then removed and the cells were washed with a 0.6 mol.$L^{-1}$ ammonium formate solution in order to remove sea salt. The cells were subsequently lyophilized. After drying, the biomass was weighed to determine the biomass productivity (mass of dried biomass per culture volume per day of cultivation, mg $L^{-1}$ day $^{-1}$).

### 2.5. Lipid extraction

Lipids were extracted from the lyophilized biomass according to a modified Bligh's and Dyer's method (Bligh and Dyer, 1959), using a mixture of chloroform, methanol and water as extracting solvent, in the following respective ratio: 1: 2: 0.8 (v/v/v). The lipid extracted was dried in a rotary evaporator to remove solvent residues and was subsequently dried at 60 °C to constant weight. The total lipids were measured gravimetrically, and then lipids contents and yields were calculated (Da Rós et al., 2013).

### 2.6. Lipid feedstock characterization

The AOCS's method (American Oil Chemist's Society, 2004) was used for the determination of total free fatty acids (FFA), which was expressed in terms of free oleic acid (%). The fatty acid composition analysis was performed in a capillary gas chromatography (CGC Agilent 6850 Series GC System) according to the methodology described by Silva et al., (2014).

### 2.7. Biodiesel synthesis

The biodiesel synthesis reaction from the oil extracted from the *Chlorella* sp. biomass with ethanol (molar ratio 30:1 alcohol/lipid material) was performed in a cylindrical reactor (2×10 cm) in the presence of 10% $H_2SO_4$ (98%, Synth-SP, Brazil) as catalyst, using an ultrasonic bath (Unique Model USC 1800) to control the stirring and keep the temperature at 60 °C. The reactions were conducted for 4 h.

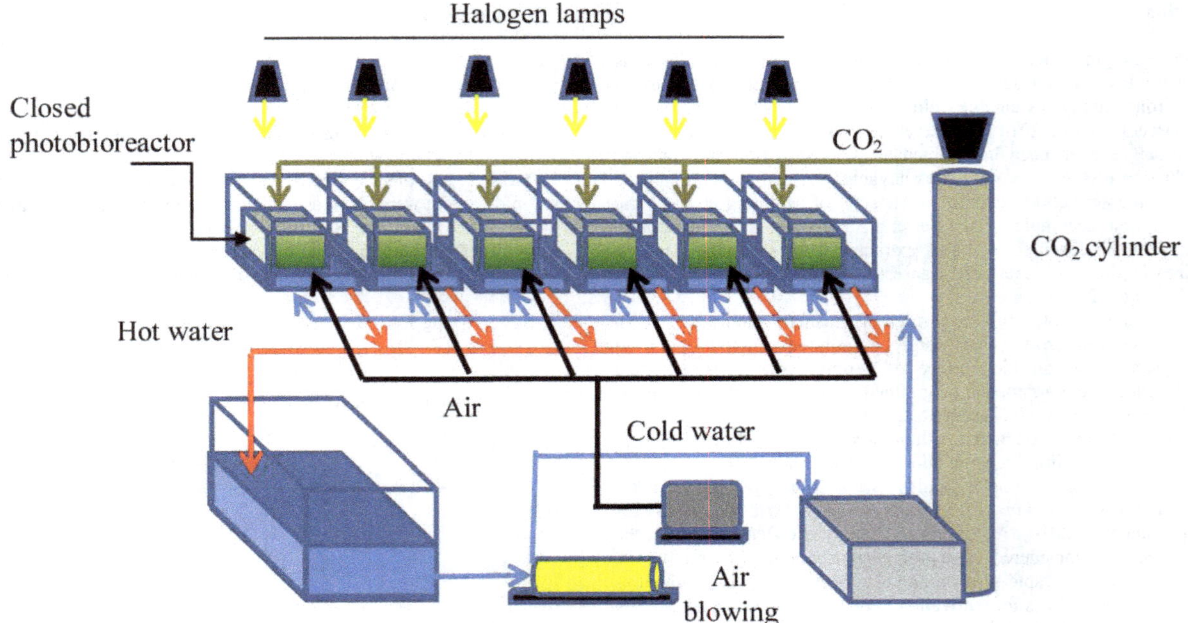


**Fig.1.** Schematic representation of the photobioreactor system.

*2.8. Downstream procedure*

Once the reaction was complete, the reaction medium was subjected to natural cooling for approximately 1 h. The sample containing ethyl esters was purified according to the method described by Da Rós et al., (2013), and was further subjected to analysis.

*2.9. Ester content quantification*

The conversion into ethyl esters was evaluated by proton nuclear magnetic resonance spectrometry ($^1H$ NMR) in a Mercury 300 MHz-Varian spectrometer, with 5 mm glass tubes, using $CDCl_3$ as solvent and 0.3% tetramethylsilane (TMS) as internal standard. The calculations involving the conversion of esters were performed using an equation according to the methodology described by Paiva et al., (2013). It allowed the identification of ester molecules produced during transesterificationthrough the peaks in the region of 4.05 to 4.35 ppm.

**3. Results and discussion**

*3.1. Statistical analysis*

Screening experiments were conducted to identify the factors influencing the biomass concentration and lipid contents during the cultivation of marine *Chorella* sp. and to verify if any changes to their settings should be made to improve the process. The effects of different experimental variables on this bioprocess were simultaneously investigated, employing a Taguchi design experiment. Four variables (A: Light Intensity, B: $CO_2$, C: $NaNO_3$, and D: aeration rate) were taken into consideration. The experimental matrix and the results are shown in **Table 2**. Each line in the table represents an experimental run and each column stands for an independent variable (factor), along with the mean values and S/N ratios of the response variables obtained for each condition.

The biomass productivity values ranged from 54.3 to 210.9 mg $L^{-1}day^{-1}$, while the lipid content varied between 8.2 and 19.8%. The highest biomass productivity was achieved for the experiment 2, in which the employed conditions were: 0.85 klux of light intensity; $CO_2$ at 5%, $NaNO_3$ at 0.075 g $L^{-1}$, and aeration rate at 1.67 vvm. Under these conditions, the lipid content was 8.2%. On the other hand, the experiment 3 yielded higher levels of lipids (19.8%), at the expense of reducing the biomass productivity to 85.9 mg $L^{-1}$ $day^{-1}$. In this experiment the following conditions were set: 0.85 klux of light intensity; $CO_2$ at 10%; $NaNO_3$ at 0.025 g $L^{-1}$, and aeration rate of 1.67 vvm.

In general, the results obtained showed that the most important variable was the $NaNO_3$ concentration, although conditions that maximized the biomass productivity were different from those attained for the lipid contents.

*3.1.1. Effect of factors on biomass productivity*

Among the factors studied (i.e. A: Light Intensity, B: $CO_2$, C : $NaNO_3$, and D: aeration rate), factor C showed less variability and greater significance in the process due to the highest value of F statistics and lowest p-value, followed by factor B. Factor A showed moderate significance, while factor D (aeration rate) was not significant on biomass productivity under the used conditions (Supplementary Data, **Table S1**).

The graphical analysis (**Fig. 2**) in relation with the effect of the factors on biomass productivity showed that factor C at level 2 ($NaNO_3$ at 0.075 g $L^{-1}$) was the most significant factor, followed by factor B at level 1 ($CO_2$ at 5%), and factor A at level 1 (0.85 Klux of light intensity). The adjustment of the variables to optimize the process concerning this response variable, i.e. maximizing biomass productivity, suggest that factors A, B, and C (light intensity, $CO_2$% and $NaNO_3$ concentration, respectively) should be set at levels 1 (0.85 KLux), 1 (5 %) and 2 (0.075 g $L^{-1}$), respectively.

Light intensity, $CO_2$ and nitrogen concentration in the culture medium are reported as important factors by a number of studies (Gushina and Harwood, 2006; Mata et al., 2010; Suali and Sarbatly, 2012). Light intensity is directly related to the photosynthesis process and has different effects on microalgae species. Different species require different levels of light energy to conduct the process; however excessive luminosity can cause photo inhibition and cell death.

$CO_2$ concentration in the culture medium could significantly affect the photosynthesis process, but each species responds differently to different $CO_2$ concentrations (Wang et al., 2013a). Nitrogen is a key component to obtain three classes of structural substances of cells: proteins, nucleic acids and photosynthetic pigments. If nitrogen supply is abundant in the culture, the production rates of proteins and chlorophyll in the cells increase. On the other hand, when the available nitrogen concentration is

**Table 2.**
Experimental matrix and results obtained for biomass productivity and lipid content according to experimental design from growth of the Chorella sp.

| Experiment | Independent variable Factor | | | | Response variable Biomass productivity ($mg.L^{-1}.day^{-1}$) | | | | | Lipids (%) | | | | |
|---|---|---|---|---|---|---|---|---|---|---|---|---|---|---|
| | A | B | C | D | Triplicate | | | Average | S/N | Triplicate | | | Average | S/N |
| 1 | 1 | 1 | 1 | 1 | 64.4 | 50.0 | 68.8 | **61.0** | **35.5** | 13.5 | 13.7 | 13.9 | **13.7** | **22.7** |
| 2 | 1 | 1 | 2 | 2 | 210.0 | 214.7 | 208.2 | **210.9** | **46.5** | 8.6 | 7.8 | 8.2 | **8.2** | **18.3** |
| 3 | 1 | 2 | 1 | 2 | 78.5 | 100.0 | 79.2 | **85.9** | **38.5** | 19.7 | 20.0 | 19.5 | **19.7** | **25.9** |
| 4 | 1 | 2 | 2 | 1 | 158.8 | 66.3 | 91.3 | **105.4** | **39.1** | 9.5 | 8.7 | 7.7 | **8.6** | **18.6** |
| 5 | 2 | 1 | 1 | 2 | 64.3 | 64.3 | 34.3 | **54.3** | **35.5** | 15.9 | 17.6 | 15.2 | **16.2** | **24.2** |
| 6 | 2 | 1 | 2 | 1 | 199.2 | 193.1 | 198.5 | **196.9** | **45.9** | 10.4 | 12.6 | 10.7 | **11.2** | **20.9** |
| 7 | 2 | 2 | 1 | 1 | 98.3 | 38.3 | 65.0 | **67.2** | **34.7** | 10.7 | 13.4 | 12.8 | **12.3** | **21.7** |
| 8 | 2 | 2 | 2 | 2 | 78.6 | 71.4 | 68.6 | **72.9** | **37.2** | 7.3 | 9.9 | 10.2 | **9.1** | **18.9** |

A: Light intensity (klux) C: $NaNO_3$ (g/L)
B: $CO_2$ (%) D: Aeration rate (vvm)

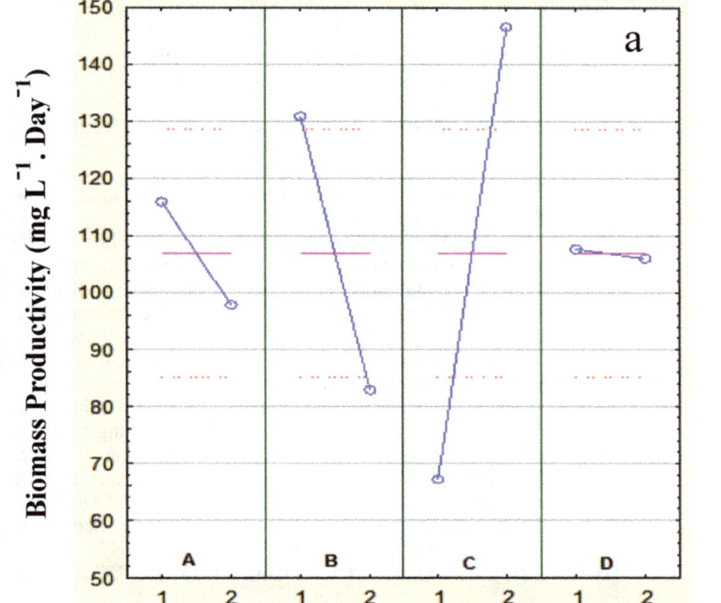


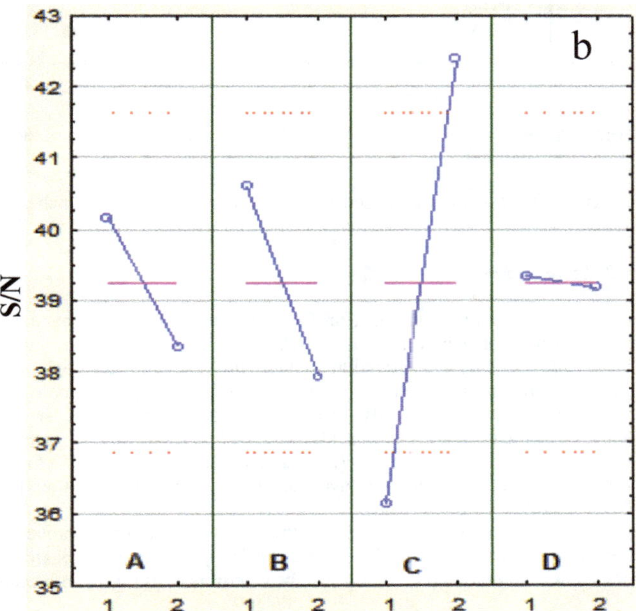


**Fig.2.** Influence of factors on the biomass productivity (a) on the average, (b) on the signal /noise ratio. Factors: A - Light Intensity (klux), B - Carbon Dioxide (%), C - concentration of NaNO3 (g / L), D – Aeration rate (vvm).

low, there is a significant decrease in the biomass productivity (Mata et al., 2010; Praveenkumar et al., 2012).

*3.1.2. Effect of factors on lipid content*

The statistical analysis of the results obtained for the variable related to the lipid content during the cultivation of the marine microalgae *Chlorella* sp. showed that the factors which presented a significant influence were the concentration of $NaNO_3$ (factor C) and aeration rate (factor D) (Supplementary Data, **Table S2**).

Through the analysis of variance (ANOVA) of the mean value and the S/N ratio shown in **Figure 3**, it could be observed that factor C at level 1 ($NaNO_3$ at 0.025 g $L^{-1}$) followed by factor C at level 2 (1.67 vvm) led to the least variability and the greatest significance on the process. This could be ascribed to the fact that these factors at the mentioned levels yielded the highest S/N values and the lowest p-values. The other factors (light intensity and $CO_2$) were not statistically significant for this response variable within the range studied.

Thus, the process adjustment proposal aiming at maximizing the response variable (lipid content) suggests that factors C and D (concentration of $NaNO_3$ and vvm) should be set at levels 1 (0.025 g $L^{-1}$) and 2 (1.67 vvm), respectively.

It is reported that microalgae cultivation performed under stress conditions, such as low homogenization of the culture medium, and nutrient deprivation, e.g. nitrogen, tends to stimulate the synthesis and accumulation of lipids (Mata et al., 2010; Wang et al., 2013b). The nitrogen limitation can generate a physiological stress which increases as the concentration of nutrients decreases. This stress changes the cell's metabolism, directing the metabolic processes to the production of lipid

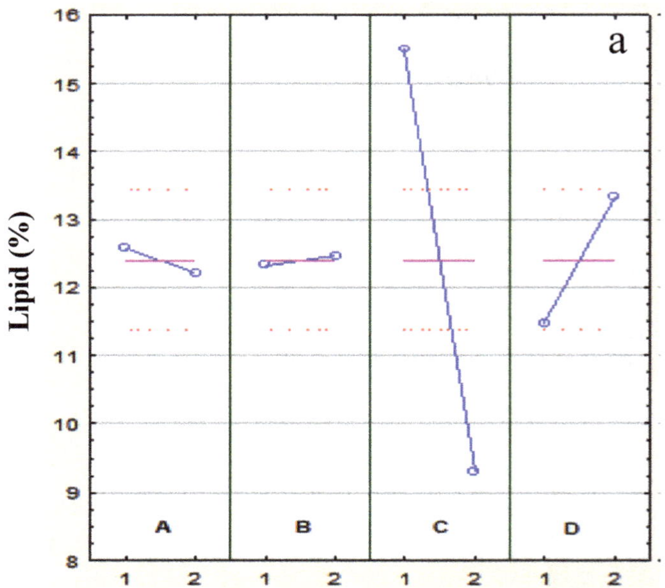


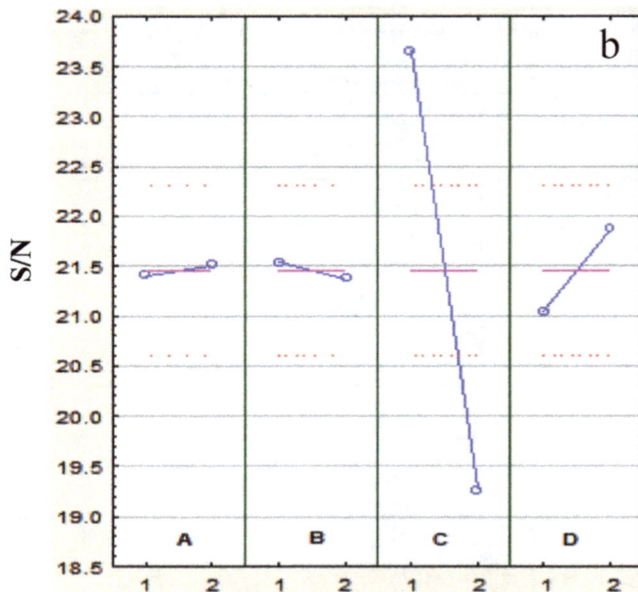


**Fig. 3.** Influence of factors on the lipid levels (a) on the average, (b) on the signal / noise ratio. Factors: A - Light Intensity (klux), B - Carbon Dioxide (%), C - concentration of NaNO3 (g / L), D - Aeration rate (vvm).

reservoirs, thus preparing the cell for a period of deprivation (Sanchez et at., 1996).

*3.2. Fatty acid profile*

**Table 3** shows the fatty acid profile of the lipid material recovered from the cultivation of marine microalgae *Chlorella* sp. along with the profile of other *Chlorella* strains described in the literature. The lipid feedstock from the *Chlorella* sp. strain studied in the present work presented a composition of 57.8% of saturated fatty acids, 31.6% of monounsaturated fatty acids, and 8.3% of polyunsaturated fatty acids. Those obtained at higher proportions were the oleic (27.8%), palmitic (20.7%) and lauric (17.7%) acids. From the composition of fatty acids, the average molecular weight was found to be 743.5 g $mol^{-1}$.

From a comparative point of view, the resulting lipid profile proved to be distinct from other strains of the same species, particularly in regard to the degree of unsaturation (DU) which can be determined by taking into account, the amount of monounsaturated and polyunsaturated fatty acids (wt.%) present in a feedstock (Da Rós et al., 2013).

While the microalgal oils obtained through the cultivation of *Chlorella* strains reported in different studies (D'oca et al., 2011b; Lam and Lee, 2013; Wang et al., 2013a) have DU values higher than 87.8%, the microalgal oil obtained from the *Chorella* sp. in the present study had only 48.1% as shown in **Table 3.**

It is known that the amount of each fatty acid present in the triglyceride molecule used in the transesterification reaction, as well as the chain length and unsaturation number are important factors for determining the physical characteristics of both the feedstock and the resulting biodiesel (Ramos et al., 2009). One of the critical parameters for biodiesel commercialization is its oxidative stability which is closely related to the amount of polyunsaturated fatty acids, especially the linoleic (C18: 2) and linolenic (C18: 3) acids present in the raw materials to be used in the process (Knothe, 2005; Francisco et al., 2010).

Thus, the fatty acid profile of the marine microalgae *Chlorella* sp.

**Table 3.**
Comparison of the fatty acid profile of the lipid material from different strains of the *Chlorella* sp.

| Strain | | | *Chlorella* sp. | *Chlorella vulgaris* | *Chlorella vulgaris* | *Chlorella pyrenoidosa* |
|---|---|---|---|---|---|---|
| **Saturated FA (%wt)** | | | | | | |
| C8:0 | Caprilic | | 8.2 | 0.0 | 0.0 | 0.0 |
| C10:0 | Capric | | 3.9 | 0.0 | 0.0 | 0.0 |
| C12:0 | Lauric | | 17.7 | 0.0 | 0.0 | 0.0 |
| C14:0 | Myristic | | 6.3 | 0.0 | 0.0 | 0.7 |
| C16:0 | Palmitic | | 20.7 | 14.0 | 16.6 | 17.3 |
| C17:0 | Margaric | | 0.0 | 0.3 | 0.0 | 0.0 |
| C18:0 | Stearic | | 1.0 | 2.0 | 4.0 | 1.2 |
| | | **Total** | **57.8** | **16.3** | **20.6** | **19.2** |
| **Monounsaturated FA** | | | | | | |
| C16:1 | Palmitoleic | | 4.2 | 0.0 | 2.4 | 0.8 |
| C18:1 | Oleic | | 27.2 | 70.8 | 53.6 | 3.3 |
| C20:1 | Gadoleic | | 0.3 | 0.0 | 0.0 | 0.0 |
| | | **Total** | **31.6** | **70.8** | **56.0** | **4.1** |
| **Polyunsaturated FA** | | | | | | |
| C16:2 | Hexadecadienoic | | 0.0 | 0.0 | 4.2 | 7.0 |
| C16:3 | Hexadecatrienoic | | 0.0 | 0.0 | 0.0 | 9.3 |
| C18:2 | Linoleic | | 3.9 | 10.5 | 11.8 | 18.5 |
| C18:3 | Linolenic | | 4.3 | 0.8 | 4.1 | 41.8 |
| | | **Total** | **8.3** | **11.3** | **20.1** | **76.6** |
| Others | | | 2.3 | 1.6 | 3.3 | 0.0 |
| **Degree of unsaturation* (DU)** | | | 48.1 | 93.4 | 87.8 | 150.3 |
| **Reference** | | | This work | Wang et al. (2013B) | Lam and Lee, (2013) | D'oca et al. (2011) |

* Calculated according to the equation DU = (wt.% monounsaturated)+2.(wt.% polyunsaturated) as described by Francisco et al. (2010) and Da Rós et al. (2013).

reported in this study shows advantages over the other strains of the same species, favoring its use as lipid feedstock for biodiesel synthesis. In contrast to the other studies that reported, respectively, 11.3%, 20.1% polyunsaturated acids for *Chlorella vulgaris* (Lam and Lee, 2013; Wang et al., 2013a), 76.6% for *Chlorella pyrenoidosa* (D'oca et al., 2011b), the resulting lipid material from the microalgae *Chlorella* sp. investigated herein contained only 8.3% of these acids. It should be noted that polyunsaturated fatty acids present in high amounts in the lipid material may result in a biodiesel which is prone to undergo polymerization when subjected to intense heat, causing engine deposits (Francisco et al., 2010; Ehimen et al., 2010).

### *3.3. Biodiesel synthesis*

The acid value of the microalgal oil was determined to be 32.7 mg KOH/g, with a high FFA content of 18.7%. Lipids with FFA content over 5 wt % are not suitable for alkaline-catalyzed transesterification, as the FFA will tend to consume the catalyst and form soap, leading to serious separation problems. Hence, acid-catalyzed transesterification was employed in the present study but to overcome the slow rate of the reaction, it was conducted under ultrasound irradiations (Meng et al., 2009).

The $^1$H-NMR spectrum of the ethyl esters obtained through the transesterification reaction using lipids from the marine microalgae *Chlorella* sp. and ethanol by acid catalysis showed a quartet signal at 4.1 ppm, referring to the ethylene hydrogens of the alcoholic portion of the ester [$CH_3$-$CH_2$-OC(=O)-R] (Supplementary Data, **Fig. S1**). Based on the yield calculation involved in this technique, an estimated conversion value of 78.4% was achieved. Similar results were achieved by Ehimen et al. (2010) who used an acid catalyzed transesterification of lipid materials from the microalgae *Chlorella* sp. at 90 °C and reported conversion rates of between 70 and 92%.

## 4. Conclusion

The conditions for enhancing *Chlorella* sp biomass yield were successfully established. This work demonstrated that the factors studied in the cultivation of microalgae *Chlorella* sp. could influence each response variable differently (biomass productivity and lipid content). The concentration of $NaNO_3$ played a key role in optimizing the total biomass yield *vs.* lipid content in cultivations designed to harvest lipids for downstream transesterification into biodiesel. For enhancing biomass productivity, concentrations of $CO_2$ and $NaNO_3$were found to be statistically significant and should be kept at 5% and 0.075 g $L^{-1,}$ respectively. The factors that maximized the accumulation of lipids were the concentration of $NaNO_3$ (0.025 $gL^{-1}$) and aeration rate (1.67 vvm). Moreover, the lipid from *Chlorella* sp had suitable composition for biodiesel production, yielding high amounts of C16-C18 fatty acids required for biodiesel synthesis. The transesterification of the lipid material employing acid catalysis led to a conversion rate of 78.4%.

## 5. Acknowledgments

The authors would like to thank the Coordenação de Aperfeiçoamento de Pessoal de Nível Superior (CAPES) and the Conselho Nacional de Desenvolvimento Científico e Tecnológico (CNPq) for financial support and scholarships. We thank Dr. J.C. S. Barboza for $^1$H-NMR analysis.

## References

American Oil Chemists' Society, 2004. Official Methods and Recommended Practices of the AOCS,5th ed, Champaign: AOCS.

Beetul, K., Sadally, S.B., Taleb-Hossenkhan, N., Bhagooli, R., Puchooa, D., 2014. An investigation of biodiesel production from microalgae found in mauritian waters. Biofuel Res. J. 2, 58-64.

Bligh, E.G., Dyer, W.J., 1959. A rapid method of total lipid extraction and purification. Can. J. Biochem. Physiol. 37, 911-917.

Chisti, Y., 2007. Biodiesel from microalgae. Biotechnol. Adv. 25, 294-306.

Da Rós, P.C.M., Silva, C.S.P., Silva-Stenico, M.E., Fiore, M.F., de Castro, H.F., 2013. Assessment of chemical and physico-chemical properties of cyanobacterial lipids for biodiesel production. Mar. Drugs. 11, 2365-2381.

D'oca, M.G.M., Haertel, P.L., Moraes, D.C., Callegaro, F.J.P., Kurz, M.H.S., Primel, E.G., Clementin, R.M., Morón-Villarreyes, J.A., 2011a. Base/acid-catalyzed FAEE production from hydroxylated vegetable oils. Fuel. 99, 912-916.

D'oca, M.G.M., Viêgas, C.V., Lemões, J.S., Miyasaki, E.K., Morón-Villarreyes, J.A., Primel, E.G., Abreu, P.C., 2011b. Production of FAMEs from several microalgal lipidic extracts and direct transesterification of the *Chlorella pyrenoidosa*. Biomass Bioenergy. 35, 1533-1538.

Ehimen, E.A., Sun, Z.F., Carrington, C.G, 2010. Variables affecting the *in situ* transesterification of microalgae lipids. Fuel. 89, 677-684.

Francisco, E.C., Neves, D.B., Jacob-Lopes, E., Franco, T.T., 2010. Microalgae as feedstock for biodiesel production: Carbon dioxide sequestration, lipid production and biofuel quality. J. Chem. Technol. Biotechnol. 85, 395-403.

Girard, J.M., Roy, M.-L., Hafsa, M.B., Gagnon, J., Faucheux, N., Heitz, M., Tremblay, R., Deschênes, J.S., 2014. Mixotrophic cultivation of green microalgae *Scenedes musobliquus* on cheese whey permeate for biodiesel production. Algal Res. 5, 241-248.

Guillard, R.R.L., 1975. Culture of phytoplankton for feeding marine invertebrates, In: Smith W.L., Chanley, M.H., Culture of Marine Invertebrate Animals, New York: Plenum, pp. 29-60.

Gushina, L.A., Harwood, J.L., 2006. Lipids and lipid metabolism in eukaryotic algae. Prog. Lipid Res. 45, 160-185.

Huntley, M.E., Redalje, D.G., 2007. $CO_2$ mitigation and renewable oil from photosynthetic microbes: a new appraisal. Mitig. Adapt. Strat. Gl. 12, 573-608.

Knothe, G., 2005. Dependence of biodiesel fuel properties on the structure of fatty acid alkyl esters. Fuel Process. Technol. 86, 1059-1070.

Koller, M., Muhr, A., Braunegg, G., 2014. Microalgae as versatile cellular factories for valued products. Algal Res. 6, 52-63.

Lam, M.K., Lee, K.T., 2013. Effect of carbon source towards the growth of *Chlorella vulgaris* for $CO_2$ bio-mitigation and biodiesel production. Int. J. Greenh. Gas. Con. 14, 169-176.

Mata, T.M., Martins, A.A., Caetano, N.S., 2010. Microalgae for biodiesel production and other applications: A review. Renew. Sustain. Energy Rev. 14, 217-232.

Meng, X., Yang, J., Xu, X., Zhang, L., Nie, Q., Xian, M., 2009. Biodiesel production from oleaginous microorganisms. Renew. Energy. 34, 1-5.

Millán-Oropeza, A., Torres-Bustillos L.G., Fernández-Linares L., 2015. Simultaneous effect of nitrate ($NO_3^-$) concentration, carbon dioxide ($CO_2$) supply and nitrogen limitation on biomass, lipids, carbohydrates and proteins accumulation in *Nannochloropsis oculata*. Biofuel Res. J. 5, 215-221.

Nascimento, I.A., Cabanelas, I.T.D., dos Santos, J.N., Nascimento, M.A., Souza, L., Sansone, G., 2015. Biodiesel yields and fuel quality as criteria for algal-feedstock selection: Effects of $CO_2$-supplementation and nutrient levels in cultures. Algal Res. 8, 53-60.

Paiva, E.J.M., Silva, M.L.C.P., Barboza, J.C.S., Oliveira, P.C., de Castro, H.F., Giordani, D.S., 2013. Non-edible babassu oil as a new source for energy production-a feasibility transesterification survey assisted by ultrasound. Ultrason. Sonochem. 20, 833-838

Praveenkumar, R., Shameera, K., Mahalakshmi, G., Akbarsha, M.A., Thajuddin, N., 2012. Influence of nutrient deprivations on lipid accumulation in a dominant indigenous microalga *Chlorella* sp., BUM11008. Evaluation for biodiesel production. Biomass Bioenergy. 37, 60-66.

Ramos, M.J., Fernández, C.M., Casas, A., Rodríguez, L., Pérez, Á., 2009. Influence of fatty acid composition of raw materials on biodiesel properties. Bioresour. Technol. 100, 261-268.

Ross, P.J., 1995. Taguchi techniques for quality engineering, **2nd edition**, McGraw-HillProfessional, New York.

Sanchez, J.L.G., Perez, J.A.S., Camacho, F.G., Sevilla, J.M.F., Grima, E.M., 1996. Optimization of light and temperature for growing *Chlorella* sp using response surface methodology. Biotechnol. Tech. 10, 329-334.

Sharma, P., Verma, A., Sidhu, R.K., Pandey, O.P., 2005. Process parameter selection for strontium ferrite sintered magnets using Taguchi L9 orthogonal design. J. Mater. Process. Technol. 168, 147-151.

Silva, C.S.P., Silva-Stenico, M.E., Fiore, M.F., deCastro, H.F., Da Rós,

P.C.M., 2014. Optimization of the cultivation conditions for *Synechococcus* sp. PCC7942 (Cyanobacterium) to be used as feedstock for biodiesel production. Algal Res. 3, 1-7.

Suali, E., Sarbatly, R., 2012. Conversion of microalgae to biofuel. Renew. Sustain. Energy Rev. 16, (2012) 4316-4342.

Teo, C.L., Jamaluddin, H., Zain, N.A.M., Idris, A., 2014. Biodiesel production via lipase catalyzed transesterification of microalgae lipids from *Tetraselmis* sp. Renew. Energy. 68, 1-5.

Wang, C., Yu, X., Li, H., Yang, J., 2013a. Nitrogen and phosphorus removal from municipal wastewater by the green alga *Chlorella* sp. J. Environ. Biol. 34, 421-425.

Wang, W., Zhou, W., Liu, J., Li, Y., Zhang, Y., 2013b. Biodiesel production from hydrolysate of *Cyperus esculentus* waste by *Chlorella vulgaris*. Bioresour. Technol. 136, 24-29.

Zhu, L., 2015. Biorefinery as a promising approach to promote microalgae industry: An innovative framework. Renew. Sustain. Energy Rev. 41, 1376-1384.

**Supplementary Data**

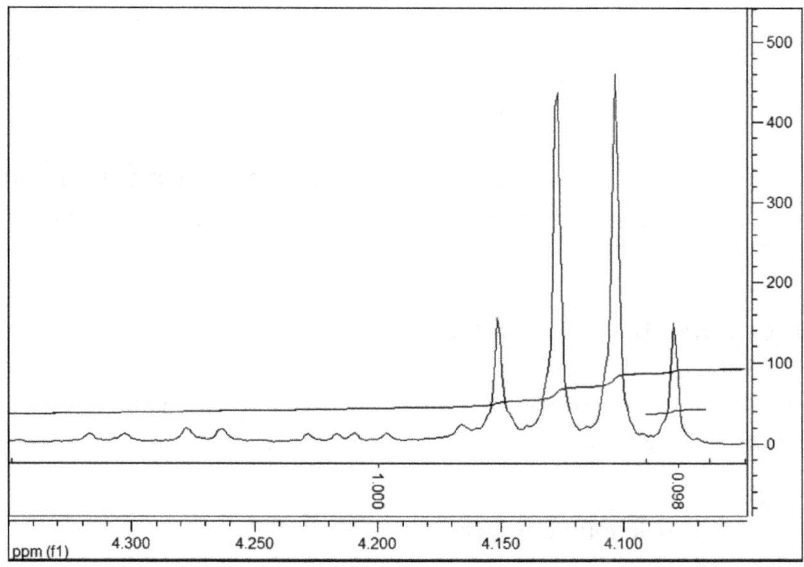


**Fig.S.1.** Nuclear magnetic resonance of protons of the ethyl esters obtained from the lipid feedstock of the microalgae *Chlorella* sp.

**Table S.1.**
ANOVA analysis, mean value, and signal to noise ratio (S/N) for biomass productivity of the microalgae *Chlorella* sp. grown under different conditions.

| | Factor | SS | DF | MS | F | *p* |
|---|---|---|---|---|---|---|
| Mean | **A** | 1949.40 | 1 | 1949.40 | 4.1168 | 0.0594 |
| | **B** | 13800.01 | 1 | 13800.01 | 29.1435 | 0.0001 |
| | **C** | 37865.87 | 1 | 37865.87 | 79.9669 | 0.0000 |
| | **D** | 16.50 | 1 | 16.50 | 0.0349 | 0.8543 |
| | **Residual error** | 7576.31 | 16 | 473.52 | -- | -- |
| S/N | **A** | 19.65 | 1 | 19.65 | 3.4562 | 0.0815 |
| | **B** | 42.77 | 1 | 42.77 | 7.5242 | 0.0144 |
| | **C** | 233.65 | 1 | 233.65 | 41.1025 | 0.0000 |
| | **D** | 0.16 | 1 | 0.16 | 0.0273 | 0.8708 |
| | **Residual error** | 90.95 | 16 | 5.68 | -- | -- |

SS: sum of squares DF: degree of freedom MS: Mean sum of squares F:F-test p: significance value.

**Table S.2.**
ANOVA analysis, mean value, and signal to noise ratio (S/N) for lipid content of the microalgae *Chlorella* sp. grown under different conditions.

| | Factor | SS | DF | MS | F | *p* |
|---|---|---|---|---|---|---|
| Mean | **A** | 0.76 | 1 | 0.76 | 0.7132 | 0.4108 |
| | **B** | 0.10 | 1 | 0.10 | 0.0932 | 0.7641 |
| | **C** | 231.01 | 1 | 231.01 | 217.900 | 0.0000 |
| | **D** | 20.72 | 1 | 20.72 | 19.545 | 0.0004 |
| | **Residual error** | 16.96 | 16 | 1.06 | - | - |
| S/N | **A** | 0.06 | 1 | 0.06 | 0.0893 | 0.7689 |
| | **B** | 0.16 | 1 | 0.16 | 0.2205 | 0.6450 |
| | **C** | 115.63 | 1 | 115.63 | 160.08 | 0.0000 |
| | **D** | 4.12 | 1 | 4.12 | 5.7096 | 0.0295 |
| | **Residual error** | 11.56 | 16 | 0.72 | - | - |

SS: sum of squares DF: degree of freedom MS: Mean sum of squares F: F-test p: significance value

# 8

# Simultaneous effect of nitrate ($NO_3^-$) concentration, carbon dioxide ($CO_2$) supply and nitrogen limitation on biomass, lipids, carbohydrates and proteins accumulation in *Nannochloropsis oculata*

Aarón Millán-Oropeza, Luis G. Torres -Bustillos, Luis Fernández-Linares*

*Departamento de Bioprocesos, Unidad Profesional Interdisciplinaria de Biotecnología, Instituto Politécnico Nacional (UPIBI - IPN), Av. Acueducto s/n Col. Barrio la Laguna Ticomán, 07340, Mexico City, Mexico.*

## HIGHLIGHTS

- Novel approach to simultaneously study the effects of culture conditions on *Nannochloropsis oculata.*
- Comprehensive experimental study on *N. oculata* lipids production.
- Importance of initial $NO_3^-$ concentration, $CO_2$ and N limitation on lipid production.
- Importance of biomass, protein and carbohydrates responses was also simultaneously evaluated.

## GRAPHICAL ABSTRACT

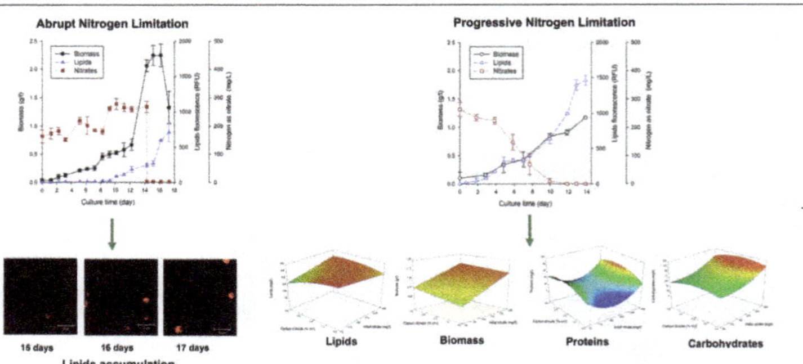


## ABSTRACT

Biodiesel from microalgae is a promising technology. Nutrient limitation and the addition of $CO_2$ are two strategies to increase lipid content in microalgae. There are two different types of nitrogen limitation, progressive and abrupt limitation. In this work, the simultaneous effect of initial nitrate concentration, addition of $CO_2$, and nitrogen limitation on biomass, lipid, protein and carbohydrates accumulation were analyzed. An experimental design was established in which initial nitrogen concentration, culture time and $CO_2$ aeration as independent numerical variables with three levels were considered. Nitrogen limitation was taken into account as a categorical independent variable. For the experimental design, all the experiments were performed with progressive nitrogen limitation. The dependent response variables were biomass, lipid production, carbohydrates and proteins. Subsequently, comparison of both types of limitation i.e. progressive and abrupt limitation, was performed. Nitrogen limitation in a progressive mode exerted a greater effect on lipid accumulation. Culture time, nitrogen limitation and the interaction of initial nitrate concentration with nitrogen limitation had higher influences on lipids and biomass production. The highest lipid production and productivity were at 582 $mgL^{-1}$ (49.7 % lipid, dry weight basis) and 41.5 $mgL^{-1}d^{-1}$, respectively; under the following conditions: 250 $mgL^{-1}$ of initial nitrate concentration, $CO_2$ supply of 4% (v/v), 12 d of culturing and 2 d in state of nitrogen starvation induced by progressive limitation. This work presents a novel way to perform simultaneous analysis of the effect of the initial concentration of nitrate, nitrogen limitation, and $CO_2$ supply on growth and lipid production of *Nannochloropsis oculata,* with the aim to produce potential biofuels feedstock.

**Keywords:**
*Nannochloropsis oculata*
Biofuels
Microalgae
Lipids
Carbohydrates
Protein

* Corresponding author
E-mail address: lfernand36@gmail.com

## 1. Introduction

The potentials of microalgal biomass as feedstock for renewable fuels is widely known. This is due to the several advantages that microalgae offer compared to other renewable sources, i.e., they present higher growth rate and require less water than terrestrial crops (Balat, 2011); microalgae have a very short harvesting cycle (1–10 d) compared to other land-based feedstock (which are harvested once or twice a year)(Harun et al., 2010); and algae cultivation does not cause competition with food over arable lands and could be well achieved over desert and seashore lands (Demirbas and Demirbas, 2011). Finally, their phototrophic growth has a favorable environmental impact, since the $CO_2$ released to the atmosphere during hydrocarbons combustion is recycled by microalgae in photosynthetic processes (100 tons of biomass uptakes about 183 tons of $CO_2$) (Chisti, 2007).

Particularly, biodiesel is one of the most attractive final products derived from microalgae due to their high lipid content; and after the removal of the lipid fraction, the remaining residual biomass (mainly carbohydrates and proteins) can also be used for energy generation or high value by-products (Pienkos and Darzins, 2009).

Phototrophic metabolism of microalgae is mainly influenced by light irradiance, culture temperature and nutrients supply. However, it is almost impossible to control diurnal fluctuations of irradiance and temperature in open large-scale systems (Grima et al., 1999; Borowitzka and Moheimani, 2013). For this reason, the availability of nutrients is the most documented approach to enhance lipid production. Nutrients limitation strategy is based on a two-stage cultivation with the first one focused on cell growth in a rich nutrient medium; while the second stage triggers fatty acids accumulation in cells through nutrients limitation. In this context, there are two different types of nutrients limitation. Progressive limitation occurs due to continuous substrates consumption by cells, which induces the nutrient- limited stage. While abrupt limitation is based on transferring the cells from a rich culture medium to another lacking one or more than one type of nutrients. Certain microalgae species are able to accumulate up to 40-63 % of lipid per dry cell weight (DCW) in media with low nitrogen concentrations (Illman et al., 2000). However, the nitrogen deprivation strategy involves a reduction of cell duplication that results in low biomass productivities during fatty acids accumulation (Rodolfi et al., 2009). One proposed strategy to overcome this problem was to supply inorganic carbon to the nitrogen-limited culture. It was observed that the addition of bicarbonate ($NaHCO_3$) and carbon dioxide ($CO_2$) to microalgal cultures not only enhanced biomass and fatty acid methyl esters (FAME) productivities, but also promoted nitrogen assimilation (Chiu et al., 2009; Lin et al., 2012). Moreover, the specific growth rate of microalgal cultures could increase up to three folds when low concentrations of $CO_2$ were supplied during the aeration (2-5% v/v) (Chiu et al., 2009).

Since microalgal cultures depend on multiple factors, multivariate statistic techniques are required to study several variables simultaneously. The Response Surface Methodology (RSM) is a collection of mathematical and statistical techniques developed to model experimental responses. It is based on the fit of a polynomial equation to the experimental data, which describes the behavior of a data set with the aim to find the optimal response within specified ranges of the factors (Montgomery and Myers, 2003). This methodology includes the interactive effects of all the variables studied. The experimental design is a crucial aspect of the RSM, which is a specific set of experiments defined by a matrix composed by the different level combinations of the variables studied (Bezerra et al., 2008).

In this work, a novel and comprehensive experimental design using the microalgae strain *Nannochloropsis oculata* was performed. By RSM the simultaneous effect of initial nitrate concentration, $CO_2$ supply, and nitrogen limitation on biomass and lipid production, as well as carbohydrates and proteins contents in the residual biomass after oil extraction were analyzed. Furthermore, the effect of progressive and abrupt limitation of nitrogen on biomass and lipids production, as well as carbohydrates and residual proteins was also investigated.

## 2. Material and method

### 2.1. Microalgal cultures

The microalgae strain *N. oculata* used in this study was obtained from the Centro de Investigación Científica y de Educación Superior de Ensenada collection (CICESE, Mexico). The strain was grown in cylindrical glass photobioreactors (1 L), at 25±1 °C, under photoperiods of 12:12 h (light: dark) using cool-white fluorescent light with an intensity of 100 μmol photons $m^{-2}s^{-1}$. Microalgal cultures were aerated continuously (1.6 $Lmin^{-1}$). Cells were first cultured for proliferation until stationary phase; then, algal cultures were inoculated by adjusting the initial cell concentration to 2 x $10^6$ cells $ml^{-1}$ approximately. For abrupt nitrogen limitation experiments, cell cultures were daily fed with nitrate ($NaNO_3$) to ensure nitrate excess until the beginning of the nitrogen limitation phase. At that time, cells were centrifuged for 15 min at 3500 ×g, the supernatant was removed and cells were resuspended in a fresh medium lacking nitrate.

### 2.2. Medium composition

Microalgae cells were cultivated in the modified f/2 medium with artificial sea water using the following composition ($gL^{-1}$): 29.23 NaCl, 1.105 KCl, 1.21 tris-base, 2.45 $MgSO_4 \cdot 7H_2O$, 1.83 $CaCl_2 \cdot 2H_2O$, 0.25 $NaHCO_3$, and 3 mL of trace elements solution. The trace elements solution contained ($gL^{-1}$): 5 $NaH_2PO_4 \cdot H_2O$, 4.1 $Na_2EDTA$, 3.16 $FeCl_3 \cdot 6H_2O$, 0.18 $MnCl_2 \cdot 4H_2O$, 0.01 $CoCl_2 \cdot 6H_2O$, 0.01 $CuSO_4 \cdot 5H_2O$, 0.023 $ZnSO_4 \cdot 7H_2O$, and 0.006 $Na_2MoO_4$. Nitrogen source was $NaNO_3$ and three different initial concentrations of the ion ($NO_3^-$) were tested (250, 200 and 150 mg $L^{-1}$). All chemicals were analytical grade.

### 2.3. Response surface experimental design

An experimental design was established by taking into account initial nitrogen concentration, culture time and $CO_2$ aeration as independent numerical variables, and nitrogen limitation as a categorical independent variable (**Table 1**). All the experiments were performed with progressive nitrogen limitation. The dependent response variables were biomass and lipid production, as well as carbohydrates and proteins contents. The experimental results were analyzed using the software Design Expert 8.0.7.1.

**Table 1.**
Experimental design.

| Parameters / Range | Initial nitrate concentration (mg/l) | $CO_2$ supply (% v/v) | Culture time (days) | Nitrogen limitation* |
|---|---|---|---|---|
| -1 | 150 | 0 | 7 | N/A |
| 0 | 200 | 2 | 10 | With |
| 1 | 250 | 4 | 14 | Without |

* Categorical parameter

Analysis of variance (ANOVA) was used for graphical analyses of the data to obtain interactions between the independent and dependent variables. The quality of the fit polynomial model was expressed by the coefficient of determination ($R^2$) and its statistical significance was checked by the Fisher F-test. Model terms were accepted or rejected based on the *p* value (probability) with a 95% of confidence level. Three dimensional plots were obtained based on the effect of the levels of two factors.

### 2.4. Biomass determination

Biomass was determined by: a) dry weight, i.e., culture samples of 5 ml were filtered in pre-weighed glass-fiber filters (Ahlstrom ©, 0.7 μm), excess of salts were removed with 15 ml of $NH_4HCO_3$ 0.5 N, filters were dried completely after 24 h of oven incubation at 48°C and weighed, b) direct cell count was done by using a counting Neubauer hemocytometer and the software Cell C Counting (Selinummi et al., 2005); and c) optical density was obtained at 750 nm with a GENESYS-10S spectrophotometer.

### 2.5. Lipid extraction

Culture samples of 30 ml each were concentrated by centrifugation (6000 rpm for 15 min), the pellet was washed with 20 ml of deionized water and centrifuged again (6000 rpm for 15 min), and the biomass was dried at 48°C. Dried biomass was milled and resuspended in 4 ml of hexane. Suspensions were sonicated during 30 min (Branson 1510 sonicator, 30 KHz) and then stored for 12 h at 5°C. Afterwards, the lipid extract was transferred to a pre-weighed vial and two rounds of washing were performed by adding 3 ml of hexane to the residual biomass and the solvent was recovered with the initial extract (10 ml of final lipids extract). Afterwards, hexane was removed with nitrogen gas and the vials were dried until reached a constant weight and finally weighed.

*2.6. Neutral lipid analysis by Nile Red (NR) staining and Confocal Laser Scanning Microscopy*

One ml samples of fresh algal suspensions were adjusted to cell concentrations in a range between $1\times10^4$ to $5\times10^5$ cell $ml^{-1}$. Cell suspensions were stained with 2 µl of NR (20 µg NR $L^{-1}$ in acetone). The reaction was allowed to occur during 20 min in darkness at room temperature. The relative fluorescence was measured in a Trilogy™ spectrofluorometer using excitation and emission wavelengths of 530 and 575 nm, respectively and non-stained cells and medium without cells were used as control. For image acquisition, 3 µl of the stained mixture were placed on microscope slides and analyzed in a Carl Zeiss LSM710 Confocal Lasser Scanning Microscope (Germany). Photographs with image size of 512×512 pixels were taken using a Plan-Apochromat 63×/1.4 Oil DIC M27 objective. The variable band pass filter for emission was centered at 488 nm with 2.0% of potency and emission light was detected at 539-628 nm. Transmitted light and fluorescent images were merged and colored using the Zen software.

*2.7. Analysis of FAMEs*

The extracted lipid was transesterified with 750 µl of methanolicchlorhydric acid 0.5 N at 80 °C during 3 h. FAMEs were dissolved in 1 ml of hexane and filtered with a 0.2 µm polytetrafluoroethylene membrane. Two µl samples of FAMEs were analyzed by gas chromatography (GC) on a Clarus ® 500 gas chromatograph (Perkin Elmer). A 30 m column was used for the separation (AT-WAX, Polyethylene glycol (PEG) with internal diameter of 0.25 mm, and film thickness of 0.2 µm). Helium was used as the carrier gas at a flow rate of 0.96 ml $min^{-1}$. The injector temperature was set at 230 °C and the detector (flame ionization detector, FID) was set at 250 °C. The temperature ramp for the oven was set as follows: 5 min at 140 °C, a temperature rate increase of 8 °C $min^{-1}$ to 240 °C, then 15 min at 240 °C. Retention times of FAMEs components were taken from a standard that was used as reference to identify the FAMEs in the samples (Supelco 37 FAME mix, Sigma Aldrich No. 47885-U).

*2.8. Carbohydrates and proteins determination*

After lipids extraction, 5 mg of the exhausted biomass were placed in glass tubes for carbohydrates and proteins quantification. For carbohydrates, 3 ml of HCl 2 N were added and biomass was hydrolyzed at 100°C during 2 h in a Major Science dry bath block (USA). The hydrolyzate was analyzed according to the Dubois method (Dubois et al., 1951). For protein determination, 3 ml of NaOH 1 N were added to the exhausted biomass and hydrolyzed at 100°C during 1 h in a Major Science dry bath block (USA). The hydrolyzate was analyzed using the Bradford method (Bradford, 1976).

*2.9. Nitrate quantification*

Nitrate concentration in the medium was measured according to a modified version of the Keeney Nelson method (Keeney and Nelson, 1982). Briefly, 2 ml of the culture were centrifuged at 17000 ×g for 5 mins, 500 µl of supernatant were transferred into glass tubes and dried in a Major Science dry bath block (USA). Then, 500 µl of 2-4 biphenyl sulfonic acid were added to the dried samples, followed by gentle addition of 2.2 ml of KOH 12 N. The reaction was allowed to occur during 5 min and 100 µl of the resulting supernatant were placed in plastic cuvettes containing 900 µl deionized water. The samples were analyzed at 410 nm in a GENESYS-10S spectrophotometer.

## 3. Results and discussion

*3.1. Response Surface analysis*

Microalgal cultures showed a regular cellular growth behavior. Biomass was accumulated while nutrients were consumed. This progressive consumption induced a nitrogen-limited state at different culture times based on the initial nitrate concentration. Thus, nitrogen limitation was achieved after 6, 8 and 8 d in cultures with initial nitrate concentrations of 150, 200 and 250 mg $L^{-1}$, respectively (**Fig. 1**). Cellular growth was affected by nitrogen starvation, showing a stationary phase after 10 d in cultures with initial nitrate concentrations of 150 and 200 mg $L^{-1}$, whereas stationary cell growth was observed after 13 d in cultures with initial nitrate concentration of 250 mg $L^{-1}$ after 5 d of nitrogen starvation.

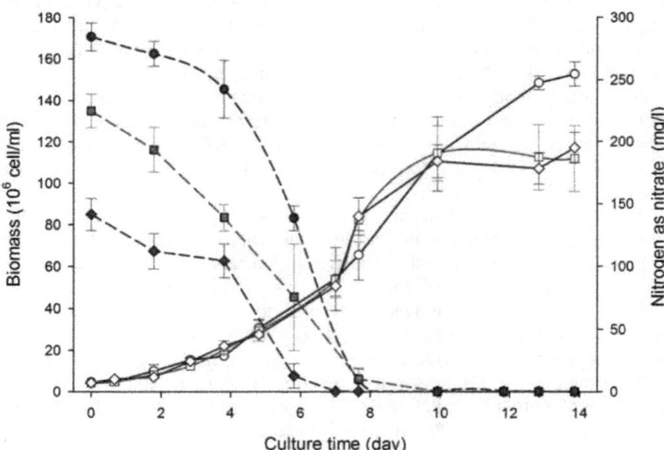


**Fig.1.** Progressive nitrate limitation and biomass production in *N. oculata* cultures at different initial nitrate concentrations. Continuous lines indicate cellular growth and dashed-lines represent initial nitrate ($NO_3^-$) consumption of 150 (diamond), 200 (square), and 250 (circle) mg $L^{-1}$.

A statistically significant difference at the 95% level ($p$ = 0.0024) was observed on final cells concentration between the highest level of initial nitrogen concentration (250 mg $L^{-1}$) and the other two levels (200 and 150 mg $L^{-1}$). Although nitrogen limitation plays an important role in cell growth, the presence of other macronutrients (e.g. Fe, K, and Mg) also affects biomass generation (Converti et al., 2009; Chen et al., 2011), which might be one explanation for the cell growth observed during nitrogen restriction.

An experimental design was performed in *N. oculata* cultures considering initial nitrate concentration, $CO_2$ supply, culture time and state of nitrogen limitation as independent variables (**Table 1**). Lipid content, biomass, carbohydrates and proteins were analyzed as response variables due to their importance as a raw material for food supplement or bioenergetics feedstock.

ANOVA results were assessed with various descriptive statistic tools, such as the p-value, that refers to the probability of the test to be distributed under the null hypothesis; probability of lack of fit (PLOF); and the adequate precision (**Table 2**).

**Table 2.**
Experimental ANOVA and regression analysis of the experimental design.

| Response | PLOF | p | Adequate precision | $R^2$ | Model equation |
|---|---|---|---|---|---|
| Lipids (mg/L) | 0.3348 | <0.0001 | 16.97 | 0.9325 | 286.71 + 198.50 D + 0.63 AB – 9.06 BC – 55.59 BD |
| Biomass (g/L) | 0.3833 | 0.0003 | 8.14 | 0.7436 | 0.93 + 0.263 C + 0.109 D – 0.0954 AD |
| Carbohydrates (mg/L) | 0.1030 | <0.0001 | 18.20 | 0.9349 | 23.03 – 17.9 A – 2.39 B + 9.86 C – 19.95 D - 10.01 AC – 14.79 AD + 15.29 $A^2$ – 14.28 $B^2$ + 27.01$C^2$ |
| Proteins (mg/L) | 0.8835 | <0.0001 | 11.44 | 0.8681 | 72.05 – 12.61 A – 7.68 B + 16.17 C +26.77 D – 67.80 AC – 82.72 AD – 15 BD + 90.35 CD + 36.78 $A^2$ – 17.95 $B^2$ + 56.75 $C^2$ |

PLOF = probability of lack of fit p = probability of error
A = Initial ($NO_3^-$) nitrate concentration (mg/l) B = $CO_2$ supply (% v/v)
C = Culture time (day) D = Nitrogen limitation (categorical factor)

Adequate precision measures the signal to noise ratio and numerical values greater than 4 indicate that the models can be used to navigate the designed space. The terms of the equations were selected based on their significance of the *p*-values <0.05. The fit of the model yielded $R^2$ values up to 0.9325 for lipids and 0.9349 for carbohydrates, but not good values were obtained for proteins and biomass (**Table 2**). The independent factors that showed higher influence on the response variables were listed hierarchically in **Table 3**, based on their parameter estimation.

**Table 3.**
Hierarchical influence of equations terms of the response variables i.e. lipids, biomass, residual carbohydrates and residual proteins.

| Response variables | Equation terms (descendant order of influence) |
|---|---|
| Lipids | D, BD, BC, AB |
| Biomass | C, AD, D |
| Carbohydrates | $C^2$, D, A, $A^2$, AD, $B^2$, AC, C, B |
| Proteins | CD, AD, AC, $C^2$, $A^2$, D, $B^2$, C, BD, A, B |

A = Initial nitrate ($NO_3^-$) concentration (mg/l) B = $CO_2$ suply (% v/v)
C = Culture time (day) D = Nitrogen limitation (categorical factor)

Concerning the enhancement of lipid accumulation in *N. oculata,* the effect of nitrogen limitation on lipid content in microalgal cultures has been previously reported (Pruvost et al., 2009; Rodolfi et al., 2009). Additionally, in this study simultaneous effect of different factors interaction i.e. nitrogen limitation with carbon dioxide, culture time with carbon dioxide, and initial nitrate concentration with carbon dioxide were observed. The highest lipid concentration obtained was at 582 mg/l (**Fig. 2A**), corresponding to 49.7% of lipid DCW. This was achieved under the following conditions: 250 mg $L^{-1}$ of initial nitrate concentration, supply of 4% $CO_2$ (v/v) and 14 d of cultivation. Two days out of these 14 d, the cells were exposed to nitrogen starvation.
The terms with higher influence on biomass production were: culture time, nitrogen limitation and the interaction of initial nitrate concentration with nitrogen limitation (**Table 3**). The highest biomass concentration was recorded at 1.17g/l (**Fig. 2B**) under the same optimal conditions for lipid accumulation. Compared to the work of Chiu et al.(2009), where they observed higher biomass concentration in cultures of *N. oculata* enriched with 2% of $CO_2$ (v/v) and reported growth inhibition at $CO_2$ concentrations higher than 5% (v/v), in this work the addition of $CO_2$ did not have a statistically significant effect ($p$ = 0.2871) in the levels of the experimental design. In general, the higher the initial concentration of nitrate, the larger the concentration of biomass obtained. This emphasizes the more important role of initial nitrate concentration over $CO_2$ supply on biomass generation.

The highest concentration of carbohydrates was measured at 108.9 mg/l, and was obtained using 250 mg $L^{-1}$ of initial nitrate concentration, 4% $CO_2$ supply (v/v) and a period of 14 d cultivation including 2 d nitrogen limitation (**Fig. 2C**). Carbohydrate accumulation is inversely proportional to the lipids production, since the lipid precursor glycerol-3-phosphate is produced by glucose catabolism (Chen et al., 2013). However, microalgal glucose polymers produced via cellulose/starch are the predominant components in the cell walls, and both, starch and most of the cell wall polysaccharides can be converted into fermentable sugars for subsequent bioethanol production via microbial fermentation (Wang et al., 2011).

The highest concentration of proteins was at 84.1 mg/l and it was achieved in cultures with 250 mg $L^{-1}$of initial nitrate, without $CO_2$ supply, and after 10 d of cultivation; however, the final concentration of proteins under the optimal conditions for lipids and biomass production was 81.6 mg/l (**Fig. 2D**). Carbohydrates and proteins models showed two similar terms with interaction of factors: initial nitrate concentration with nitrogen limitation, and initial nitrate concentration with culture time. This implied the more significant effect of nitrate addition over $CO_2$ supply (**Table 3**).

Despite the fact that low concentrations of carbohydrates and proteins were obtained after lipids extraction (9.2 and 6.9% of DCW, respectively), these molecules would play an important role in sustainable microalgae-based bioprocesses at large scale.

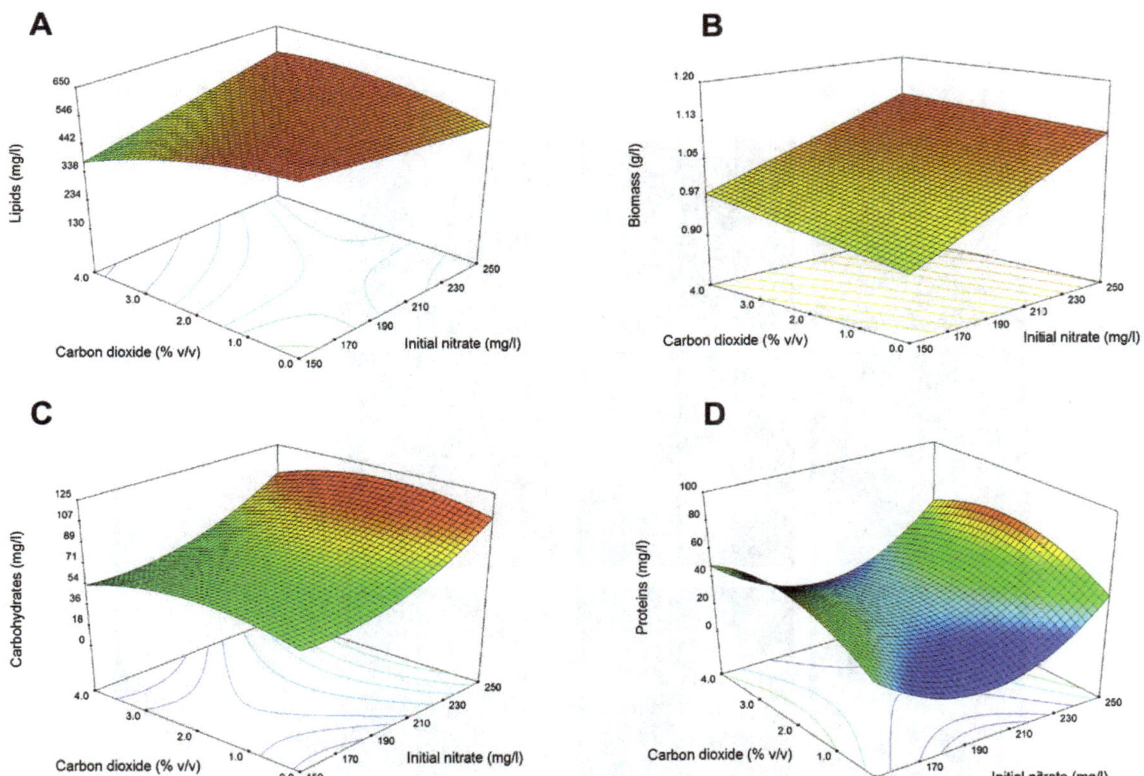


**Fig.2.** Response surface graphs for the effect of initial nitrate concentration and $CO_2$ supply on lipids (A), biomass (B), residual carbohydrates (C) and residual proteins (D) in *N. oculata*.

*3.2. Comparison of abrupt and progressive nitrogen limitations*

Since nitrogen limitation was the common representative term for all the response variables in the experimental design (**Table 3**), it was decided to compare two different types of nitrogen limitation under the optimal conditions led to the highest biomass, lipids and carbohydrates production using the experimental design. Thereby, abrupt nitrogen limitation was performed by maintaining nitrate concentration (roughly 250 mg $L^{-1}$) for 14 d, and after this period of time, the cells were transferred to another culture medium lacking nitrate, while maintaining a continuous $CO_2$ supply (4% v/v).

The cultures that were maintained with sufficient amounts of nitrate reached the highest biomass concentration (i.e., 2.25 g $L^{-1}$) after 14 d (**Fig. 3A**); two folds the highest value obtained under progressive nitrate consumption (i.e., 1.17g $L^{-1}$) (**Fig. 3B**). Three days after the abrupt nitrate limitation, the biomass concentration reduced to 40.9% from the highest value, and a slight reduction in cells concentration was also observed (data not shown). While in cultures with progressive nitrate consumption, the trend of biomass accumulation was to increase despite of nitrogen limitation conditions.

In both abrupt and progressive limitation modes, relative fluorescence of lipids bodies stained with NR dye revealed increased lipid accumulation. The highest lipid concentrations for abrupt and progressive limitations were 284 mg $L^{-1}$ after 16 d and 582 mg $L^{-1}$ after 14 d of cultivation, respectively. The lower lipids content (12–20% DCW) coupled to biomass diminution in cultures under abrupt limitation was also observed in other studies involving *N. oculata* (Van Vooren et al., 2012). In fact, cell growth diminution was a result of photo-oxidative stress due to nitrogen source scarceness. Surprisingly, higher concentrations of residual protein were achieved in the present experiments with abrupt nitrogen limitation reaching 1.31 g $L^{-1}$ (58% DCW) 16 d after cultivation; just 2 d after the abrupt limitation.

With regards to lipids accumulation, qualitative analysis was conducted by Confocal Laser Scanning Microscopy of cells stained with NR dye after 24, 48 and 72 h of abrupt nitrogen limitation (**Fig. 4**). Total lipids quantification showed a consistent correlation with triacylglycerol measurements (relative fluorescence). In better words, higher lipids quantities led to stronger relative fluorescence observed in the NR-stained cells.

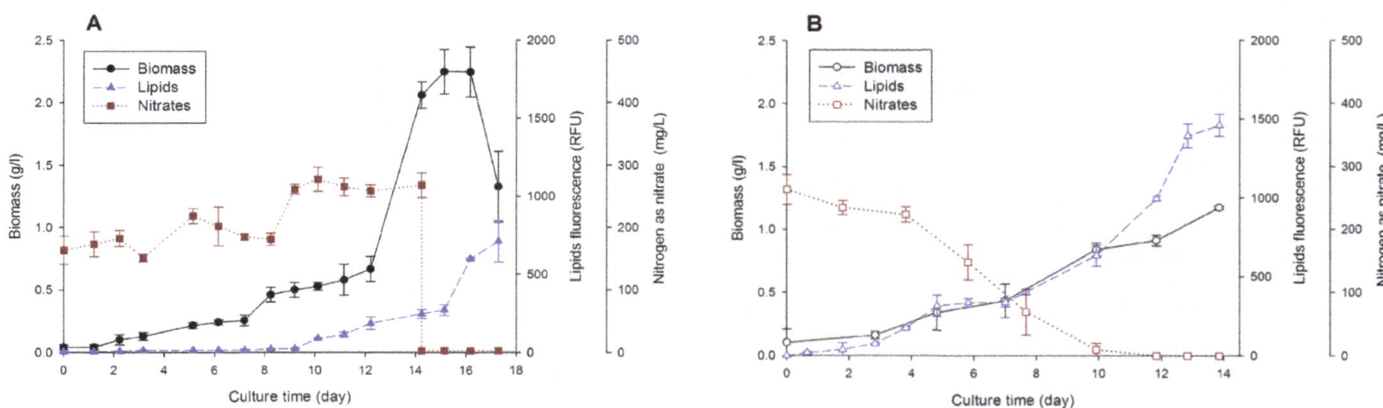


**Fig.3.** Effect of abrupt (A) and progressive (B) nitrate limitation on biomass (g $L^{-1}$) and lipids accumulation (relative fluorescence).

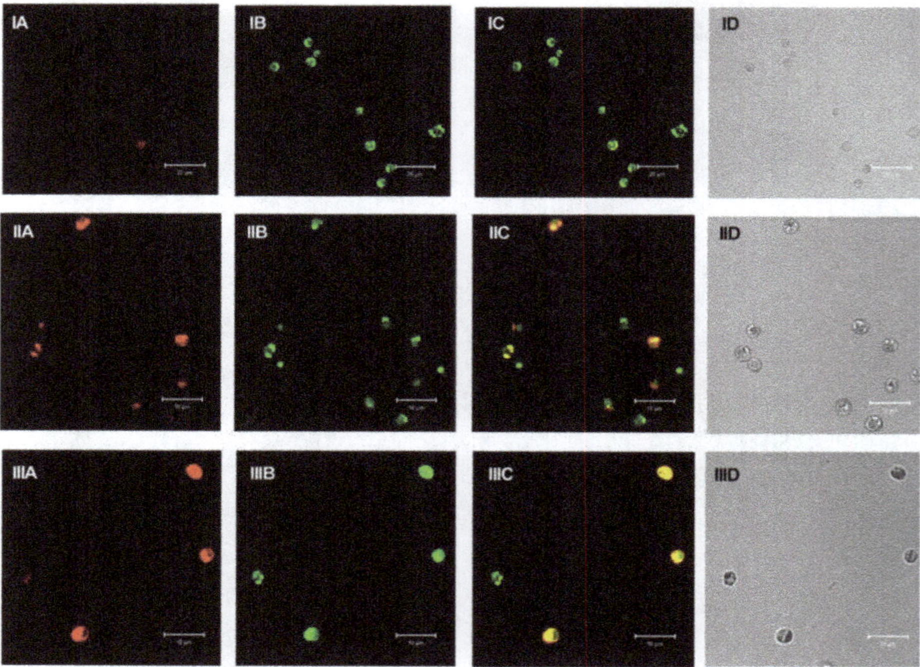


**Fig.4.** Confocal Laser Scanning Microscopy images of *N. oculata* cultures at (I) 24, (II) 48, and (III) 72 h after the abrupt nitrate limitation. In red (A): lipids stained with NR dye, in green (B): chlorophyll auto-fluorescence, in yellow (C): lipids and chlorophyll co-localization, and in gray (D) optical images.

Chlorophyll fluorescence was constant while the cell exhibited higher lipids fluorescence over the process time. The fact that lipid content was duplicated from 24 to 72 h under abrupt nitrogen limitation can be attributed to the simultaneous phenomena of biomass diminution and high lipids expression in nitrate-stressed cells. In general, these results are in agreement with those reported previously (Pruvost et al., 2009; Van Vooren et al., 2012) progressive nitrate consumption. Maintaining nitrate concentration at 250 mg $L^{-1}$ coupled to 4% $CO_2$ supply allowed higher biomass concentrations (2.25 g $L^{-1}$) and higher protein content (1.31 g $L^{-1}$). FAMEs profile did not differ between the abrupt and progressive limitation. In this work, it was shown that the lipid production was mainly influenced by the combination of nitrogen limitation and carbon dioxide concentration, culture time and carbon dioxide concentration, as well as initial nitrate concentration and carbon dioxide concentration.

**Table 4.**
FAMEs profile in the abrupt and progressive nitrogen (N-$NO_3^-$) limitation experiments.

| Type of (N-$NO_3^-$) limitation | FAME (% abundance) | | | | | | | | |
|---|---|---|---|---|---|---|---|---|---|
| | C14:0 | C16:0 | C16:1Δ9 | C18:0 | C18:1 Δ9 | C18:2 Δ9,12 | C18:3 Δ9,12,15 | C20:3 Δ11,14,17 | C20:5 Δ5,8,11,14,17 |
| (A) Progressive limitation | 8.09 ± 0.17 | 41.23 ± 3.09 | 21.14 ± 3.50 | 10.19 ± 3.92 | 5.79 ± 1.46 | 2.01 ± 0.45 | 0.76 ± 0.26 | 3.23 ± 1.03 | 6.08 ± 1.49 |
| (B) Abrupt limitation | 8.28 ± 0.44 | 38.52 ± 3.58 | 21.53 ± 6.86 | 8.26 ± 4.25 | 7.24 ± 0.39 | 2.24 ± 0.44 | 0.95 ± 0.14 | 4.23 ± 1.88 | 7.33±2.81 |

A: 14 d of cultivation
B: 2 d of limitation, 16 d of cultivation

In this work, the highest lipid productivity achieved in cultures of *N. oculata* was recorded at 41.5 mg $L^{-1}d^{-1}$ under progressive nitrogen limitation (N-$NO_3^-$) under the optimal conditions previously described. Compared to the other studies on *Nannochloropsis* species, this value was slightly higher. For instance, Rodolfi et al. (2009) reported lipid productivity of 37.6 mg $L^{-1}d^{-1}$ in their experiments performed in 250 ml culture systems, and the value reported by Gouveia and Oliveira (2009) was 25.8 mg $L^{-1}d^{-1}$ in cultures grown in wall panel systems of 100 l.

Several studies concerning the effect of nitrogen limitation on lipid and biomass have been carried out in diverse microalgae species (Converti et al., 2009; Xin et al., 2010; Chen et al., 2011), pointing out the importance of the lack of nitrogen source for lipids accumulation triggering. In this work, we observed the same effect on *N. oculata* cultures under progressive and abrupt nitrogen limitation. Since nitrogen restriction decreases cellular growth, the addition of inorganic carbon sources has been studied to overcome this issue (Gordillo et al., 1998; Chiu et al., 2009; Chai et al., 2012; Lin et al., 2012). In general, higher biomass concentrations were achieved with the addition of CO to the cultures. However, there still exists the inhibiting effect of high $CO_2$ concentration i.e. 5% (v/v) on cell growth in *N. oculata* (Chiu et al., 2009). In this work, the highest level of $CO_2$ supply (4% v/v) applied in the experimental design for the RSM did not show growth inhibition effects in the cultures. Overall, most of the previous studies on lipids and biomass production in microalgae analyzed only one variable at a time, but in the present investigation, we showed the simultaneous effect of nitrogen restriction, initial nitrate concentration and $CO_2$ supply on lipids accumulation, biomass production, as well as residual carbohydrates and residual proteins content information that would provide a complete outlook on *N. oculata* in the framework of biofuels feedstock production.

### *3.3. FAME composition*

Lipid extracts composition was analyzed by GC-FID in order to identify FAMEs profile for biodiesel production. No significant differences in FAMEs profiles were observed between abrupt and progressive limitation experiments (**Table 4**). The most abundant components were: palmitic acid (C16:0), palmitoleic acid (C16:1), and stearic acid (C18:0) with abundance values of 39.87, 21.34 and 9.23%, respectively. Saturated fatty acids (SFA) content was recorded at 55.1 and 59% for abrupt and progressive limitation, respectively.

## 4. Conclusion

Response surface analysis allowed observing the simultaneous effect of different interactions among important factors promoting lipid accumulation such as the combination of nitrogen limitation and carbon dioxide concentration, culture time and carbon dioxide, as well as initial nitrate concentration and carbon dioxide concentration. The optimal conditions for biomass, lipids and carbohydrates production were: 250 mg $L^{-1}$ of initial nitrate concentration, supply of 4% $CO_2$ (v/v) and 14 d of cultivation under

## Acknowledgments

The authors are grateful for the financial support provided by Consejo Nacional de Ciencia y Tecnologia (CONACYT) and Secretaria de Investigación y Posgrado del Instituto Politécnico Nacional (SIP-IPN) Grants SIP-20130388 and SIP-20144620. Authors also acknowledge the technical assistance of M.J. Perea Flores of the Centro de Nanociencias y Micro y Nanotecnologías del Instituto Politécnico Nacional, where Confocal Laser Scanning Microscopy images were taken and the revision of the manuscript by Claudia Guerreo, PhD.

## References

Balat, M., 2011. Potential alternatives to edible oils for biodiesel production – A review of current work. Energy. Convers. Manag. 52, 1479-1492.

Bezerra, M.A., Santelli, R.E., Oliveira, E.P., Villar, L.S., Escaleira, L.A., 2008. Response surface methodology (RSM) as a tool for optimization in analytical chemistry. Talanta. 76, 965-977.

Borowitzka, M., Moheimani, N., 2013. Open Pond Culture Systems, in: Borowitzka, M.A., Moheimani, N.R. (Eds.), Algae for Biofuels and Energy. Springer Netherlands, pp, 133-152.

Bradford, M.M., 1976. A rapid and sensitive method for the quantitation of microgram quantities of protein utilizing the principle of protein-dye binding. Anal. Biochem. 72, 248-254.

Converti, A., Casazza, A.A., Ortiz, E.Y., Perego, P., Del Borghi, M., 2009. Effect of temperature and nitrogen concentration on the growth and lipid content of *Nannochloropsis oculata* and *Chlorella vulgaris* for biodiesel production. Chem. Eng. Process. 48, 1146-1151.

Chai, X., Zhao, X., Baoying, W., 2012. Biofixation of carbon dioxide by *Chlorococcum sp.* in a photobioreactor with polytetrafluoroethene membrane sparger. Afr. J. Biotechnol. 11, 7445-7453.

Chen, C.Y., Zhao, X.Q., Yen, H.W., Ho, S.H., Cheng, C.L., Lee, D.J., Bai, F.W., Chang, J.S., 2013. Microalgae-based carbohydrates for biofuel production. Biochem. Eng. J. 78, 1-10.

Chen, M., Tang, H., Ma, H., Holland, T.C., Ng, K.Y.S., Salley, S.O., 2011. Effect of nutrients on growth and lipid accumulation in the green algae *Dunaliella tertiolecta*. Bioresour. Technol. 102, 1649-1655.

Chisti, Y., 2007. Biodiesel from microalgae. Biotechnol. Adv. 25, 294-306.

Chiu, S.Y., Kao, C.Y., Tsai, M.T., Ong, S.C., Chen, C.H., Lin, C.S., 2009. Lipid accumulation and $CO_2$ utilization of *Nannochloropsis oculata* in response to $CO_2$ aeration. Bioresour. Technol. 100, 833-838.

Demirbas, A., Demirbas, M.F., 2011. Importance of algae oil as a source of biodiesel. Energy. Convers. Manag. 52, 163-170.

Dubois, M., Gilles, K., Hamilton, J.K., Rebers, P.A., Smith, F., 1951. A Colorimetric Method for the Determination of Sugars. Nature.168,167-167.

Gordillo, F.L., Goutx, M., Figueroa, F., Niell, F.X., 1998. Effects of light intensity, $CO_2$ and nitrogen supply on lipid class composition of *Dunaliella viridis*. J. Appl. Phycol. 10, 135-144.
Gouveia, L., Oliveira, A.C., 2009. Microalgae as a raw material for biofuels production. J. Ind. Microbiol. Biotechnol. 36, 269-274.
Grima, E.M., Ancién Fernández, F.G., García Camacho, F., Chisti, Y., 1999. Photobioreactors: light regime, mass transfer, and scaleup. J. Biotechnol. 70, 231-247.
Harun, R., Danquah, M.K., Forde, G.M., 2010. Microalgal biomass as a fermentation feedstock for bioethanol production. J. Chem. Technol. Biotechnol. 85, 199-203.
Illman, A.M., Scragg, A.H., Shales, S.W., 2000. Increase in *Chlorella* strains calorific values when grown in low nitrogen medium. Enzym. Microb. Technol. 27, 631-635.
Keeney, D.R., Nelson, D.W., 1982. Nitrogen Inorganic Forms, in: Black, C.A. (Ed.), Methods of soil analysis, Part 2: Agronomy. American Society of Agronomy, Madison, Wisconsin. pp, 643-698.
Lin, Q., Gu, N., Li, G., Lin, J., Huang, L., Tan, L., 2012. Effects of inorganic carbon concentration on carbon formation, nitrate utilization, biomass and oil accumulation of *Nannochloropsis oculata* CS 179. Bioresour. Technol. 111, 353-359.
Montgomery, D.C., Myers, R.H., 2003. Response Surface Methodology. Process and Product Optimization Using Designed Experiments. John Wiley & Sons, NJ, USA.
Pienkos, P.T., Darzins, A., 2009. The promise and challenges of microalgal-derived biofuels. Biofuels. Bioprod. Biorefining. 3, 431-440.
Pruvost, J., Van Vooren, G., Cogne, G., Legrand, J., 2009. Investigation of biomass and lipids production with Neochloris oleoabundans in photobioreactor. Bioresour. Technol. 100, 5988-5995.
Rodolfi, L., Chini Zittelli, G., Bassi, N., Padovani, G., Biondi, N., Bonini, G., Tredici, M.R., 2009. Microalgae for oil: Strain selection, induction of lipid synthesis and outdoor mass cultivation in a low-cost photobioreactor. Biotechnol. Bioeng. 102, 100-112.
Selinummi, J., Seppala, J., Yli-Harja, O., Puhakka, J.A., 2005. Software for quantification of labeled bacteria from digital microscope images by automated image analysis. BioTechniques. 39, 859-863.
Van Vooren, G., Le Grand, F., Legrand, J., Cuiné, S., Peltier, G., Pruvost, J., 2012. Investigation of fatty acids accumulation in *Nannochloropsis oculata* for biodiesel application. Bioresour. Technol. 124, 421-432.
Wang, X., Liu, X., Wang, G., 2011. Two-stage hydrolysis of invasive algal feedstock for ethanol fermentation. J. Integr. Plant. Biol. 53, 246-252.
Xin, L., Hu, H.Y., Ke, G., Sun, Y.X., 2010. Effects of different nitrogen and phosphorus concentrations on the growth, nutrient uptake, and lipid accumulation of a freshwater microalga *Scenedesmus sp*. Bioresour. Technol. 101, 5494-5500.

**Supplementary Data**

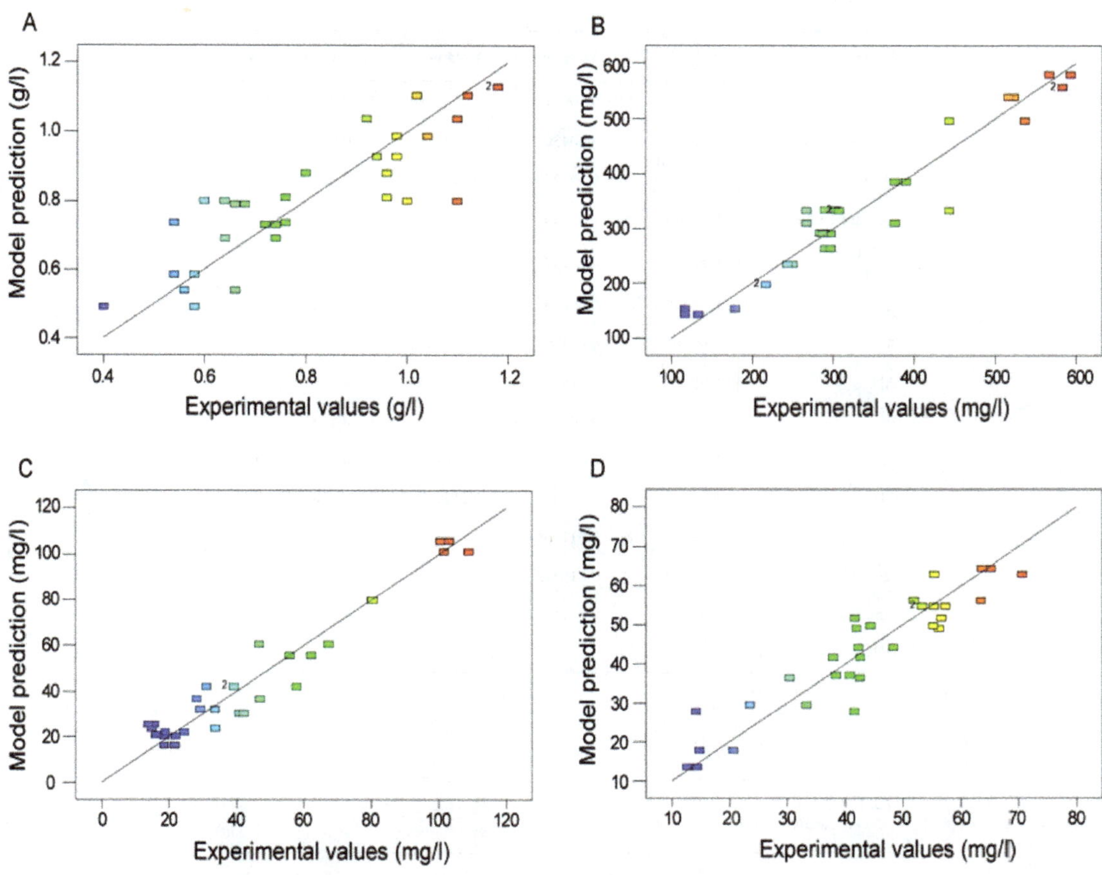


**Fig. S.1.** Experimental values vs. model prediction of: (A) biomass $R^2 = 0.7436$; (B) lipids $R^2 = 0.9325$; (C) residual carbohydrates $R^2 = 0.9349$ and (D) residual proteins $R^2 = 0.8681$

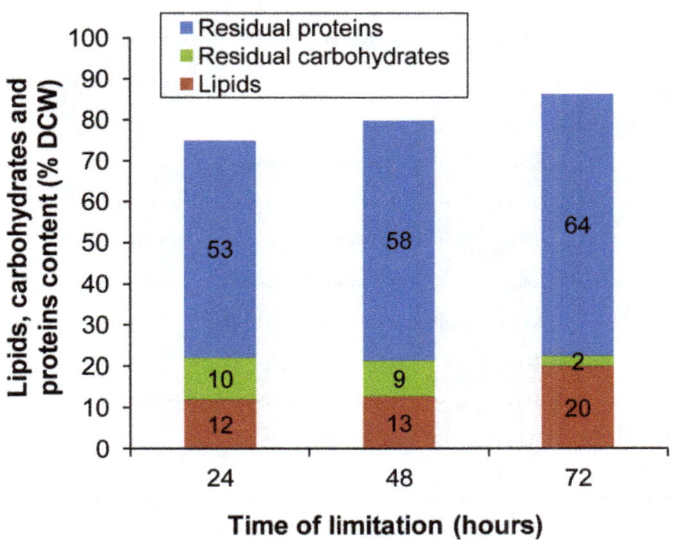


**Fig. S.2.** Lipid, carbohydrates and proteins content after: 24, 48 and 72 h of abrupt nitrogen limitation in cultures of *N. oculata*.

**Table S.1.**
ANOVA analysis of the biomass response.

| Source | Sum of squares | df | Mean square | F Value | p-value Prob > F |
|---|---|---|---|---|---|
| Model | 1.09411743 | 10 | 0.10941174 | 6.09321869 | 0.0003 |
| A-Initial nitrate | 0.00420205 | 1 | 0.00420205 | 0.23401514 | 0.6336 |
| B-Carbon dioxide | 0.00979072 | 1 | 0.00979072 | 0.54525215 | 0.4684 |
| C-Culture time | 0.25746318 | 1 | 0.25746318 | 14.33831 | 0.0011 |
| D-Nitrogen limitation | 0.03052348 | 1 | 0.03052348 | 1.69987454 | 0.2064 |
| AB | 0.00073443 | 1 | 0.00073443 | 0.04090116 | 0.8417 |
| AC | 0.00100207 | 1 | 0.00100207 | 0.05580587 | 0.8155 |
| AD | 0.02579629 | 1 | 0.02579629 | 1.43661394 | 0.2440 |
| BC | 0.00645884 | 1 | 0.00645884 | 0.35969772 | 0.5551 |
| BD | 0.00429467 | 1 | 0.00429467 | 0.23917303 | 0.6299 |
| CD | 0.00635706 | 1 | 0.00635706 | 0.35402943 | 0.5582 |
| Residual | 0.37708257 | 21 | 0.01795631 | — | — |
| Lack of Fit | 0.07818257 | 4 | 0.01954564 | 1.11166257 | 0.3833 |
| Pure Error | 0.2989 | 17 | 0.01758235 | — | — |
| Cor Total | 1.4712 | 31 | — | — | — |

df = degrees of freedom

**Table S.2.**
ANOVA analysis of the lipids response.

| Source | Sum of squares | df | Mean square | F value | p- value Prob > F |
|---|---|---|---|---|---|
| Model | 571798.618 | 10 | 57179.8618 | 29.938123 | < 0.0001 |
| A-Initial nitrate | 7985.20599 | 1 | 7985.20599 | 4.18087892 | 0.0536 |
| B-Carbon dioxide | 2278.60514 | 1 | 2278.60514 | 1.19302773 | 0.2871 |
| C-Culture time | 37.0484046 | 1 | 37.0484046 | 0.01939773 | 0.8906 |
| D-Nitrogen limitation | 16598.679 | 1 | 16598.679 | 8.69070468 | 0.0077 |
| AB | 35303.3898 | 1 | 35303.3898 | 18.4840815 | 0.0003 |
| BC | 34586.9582 | 1 | 34586.9582 | 18.1089736 | 0.0004 |
| BD | 61769.1813 | 1 | 61769.1813 | 32.3409901 | < 0.0001 |
| CD | 4899.79059 | 1 | 4899.79059 | 2.56542301 | 0.1242 |
| B^2 | 5723.86347 | 1 | 5723.86347 | 2.99688952 | 0.0981 |
| C^2 | 3615.14506 | 1 | 3615.14506 | 1.89281075 | 0.1834 |
| Residual | 40108.63 | 21 | 1909.93476 | — | — |
| Lack of Fit | 9007.61191 | 4 | 2251.90298 | 1.23090345 | 0.3348 |
| Pure Error | 31101.0181 | 17 | 1829.47165 | — | — |
| Cor Total | 611907.248 | 31 | — | — | — |

df = degrees of freedom

**Table S.3.**
ANOVA analysis of the residual carbohydrates response.

| Source | Sum of quares | df | Mean square | F value | p-value Prob > F |
|---|---|---|---|---|---|
| Model | 24625.7225 | 9 | 2736.19139 | 35.0887695 | < 0.0001 |
| A-Initial nitrate | 1001.73898 | 1 | 1001.73898 | 12.8462461 | 0.0017 |
| B-Carbon dioxide | 114.37602 | 1 | 114.37602 | 1.46675185 | 0.2387 |
| C-Culture time | 369.020491 | 1 | 369.020491 | 4.73229869 | 0.0406 |
| D-Nitrogen limitation | 1045.27481 | 1 | 1045.27481 | 13.4045473 | 0.0014 |
| AC | 332.876698 | 1 | 332.876698 | 4.26879267 | 0.0508 |
| AD | 336.4311 | 1 | 336.4311 | 4.31437412 | 0.0497 |
| A^2 | 377.596888 | 1 | 377.596888 | 4.84228195 | 0.0386 |
| B^2 | 583.618644 | 1 | 583.618644 | 7.48429374 | 0.0121 |
| C^2 | 2427.92192 | 1 | 2427.92192 | 31.1355386 | < 0.0001 |
| Residual | 1715.54065 | 22 | 77.9791206 | — | — |
| Lack of Fit | 572.792288 | 5 | 134.558458 | 2.19371601 | 0.1030 |
| Pure Error | 1042.74837 | 17 | 61.3381391 | — | — |
| Cor Total | 26341.2631 | 31 | — | — | — |

df = degrees of freedom

**Table S.4.**
ANOVA analysis of the residual proteins response.

| Source | Sum of quares | df | Mean square | F value | p-value Prob > F |
|---|---|---|---|---|---|
| Model | 6955.76521 | 11 | 532.342292 | 11.9642879 | < 0.0001 |
| A-Initial nitrate | 424.019227 | 1 | 424.019227 | 8.02269305 | 0.0103 |
| B-Carbon dioxide | 754.170043 | 1 | 754.170043 | 14.2693406 | 0.0012 |
| C-Culture time | 562.092573 | 1 | 562.092573 | 10.6351219 | 0.0039 |
| D-Nitrogen limitation | 1002.6363 | 1 | 1002.6363 | 18.9704682 | 0.0003 |
| AC | 2297.06542 | 1 | 2297.06542 | 43.4618281 | < 0.0001 |
| AD | 2423.34314 | 1 | 2423.34314 | 45.8510768 | < 0.0001 |
| BD | 2879.08559 | 1 | 2879.08559 | 54.4739919 | < 0.0001 |
| CD | 2729.89239 | 1 | 2729.89239 | 51.6511688 | < 0.0001 |
| A^2 | 1202.18174 | 1 | 1202.18174 | 22.7459853 | 0.0001 |
| B^2 | 644.480059 | 1 | 544.480059 | 12.1939416 | 0.0023 |
| C^2 | 2779.4837 | 1 | 2779.4837 | 52.5894657 | < 0.0001 |
| Residual | 1057.04961 | 20 | 52.8524804 | — | — |
| Lack of Fit | 38.9118876 | 3 | 12.9706292 | 0.21657256 | 0.8835 |
| Pure Error | 1018.13772 | 17 | 59.8904542 | — | — |
| Cor Total | 8012.81482 | 31 | — | — | — |

df = degrees of freedom

**Table S.5.**
Parameter estimation for the lipid model equation.

| Factor | Coefficient Estimate | Standard Error | 95% CI Low | 95% CI High |
|---|---|---|---|---|
| Intercept | 286.706957 | 193.108286 | -114.883708 | 688.297623 |
| A-Initial nitrate concentration | -0.82238544 | 0.40219959 | -1.65880527 | 0.01403438 |
| B-Carbon dioxide | -39.143468 | 35.8371967 | -113.670999 | 35.3840624 |
| C-Culture time | -5.76532963 | 41.3950731 | -91.8510968 | 80.3204376 |
| D-Nitrogen limitation | 198.501716 | 67.334365 | 58.4722381 | 338.531193 |
| AB | 0.63030551 | 0.14660614 | 0.32542135 | 0.93518968 |
| BC | -9.06135733 | 2.1293465 | -13.4895758 | -4.63313886 |
| BD | -55.5930929 | 9.77561706 | -75.9226015 | -35.2635843 |
| CD | -11.8786036 | 7.4162762 | -27.3015942 | 3.5443871 |
| B^2 | -8.52965187 | 4.92715177 | -18.7762249 | 1.71692116 |
| C^2 | 2.78159806 | 2.02181204 | -1.42299024 | 6.98618636 |

**Table S.6.**
Parameter estimation for the biomass model equation.

| Factor | Coefficient Estimate | Standard Error | 95% CI Low | 95% CI High |
|---|---|---|---|---|
| Intercept | 0.93473627 | 0.09382155 | 0.73962367 | 1.12984887 |
| A-Initial nitrate concentration | -0.03023344 | 0.06249792 | -0.16020497 | 0.0997381 |
| B-Carbon dioxide | -0.03496409 | 0.04735039 | -0.13343461 | 0.06350643 |
| C-Culture time | 0.26383107 | 0.069675 | 0.11893398 | 0.40872816 |
| D-Nitrogen limitation | 0.10964432 | 0.08409646 | -0.06524384 | 0.28453249 |
| AB | -0.00909116 | 0.04495226 | -0.10257449 | 0.08439217 |
| AC | 0.01428607 | 0.06047459 | -0.11147772 | 0.14004987 |
| AD | -0.09548049 | 0.07966079 | -0.26114416 | 0.07018319 |
| BC | 0.02662707 | 0.0443971 | -0.06570174 | 0.11895589 |
| BD | -0.02931768 | 0.05994783 | -0.15398602 | 0.09535067 |
| CD | 0.05054952 | 0.08495665 | -0.1261275 | 0.22722654 |

**Table S.7.**
Parameter estimation for the residual carbohydrates model equation.

| Factor | Coefficient Estimate | Standard Error | 95% CI Low | 95% CI High |
|---|---|---|---|---|
| Intercept | 23.0252243 | 4.78909833 | 13.0932422 | 32.9572063 |
| A-Initial nitrate concentration | 17.8977805 | 4.99356905 | 7.54175218 | 28.2538089 |
| B-Carbon dioxide | -2.39140147 | 1.97457743 | -6.48642442 | 1.70362149 |
| C-Culture time | 9.8611089 | 4.53304135 | 0.46015653 | 19.2620613 |
| D-Nitrogen limitation | -19.9469631 | 5.44817073 | -31.2457777 | -8.64814857 |
| AC | -10.0082486 | 4.84401581 | -20.0541225 | 0.03762536 |
| AD | -14.7885063 | 7.11976091 | -29.5539867 | -0.02302591 |
| A^2 | 15.2859475 | 6.94652056 | 0.87974554 | 29.6921494 |
| B^2 | -14.2798791 | 5.21974299 | -25.1049635 | -3.45479473 |
| C^2 | 27.0114394 | 4.84082671 | 16.9721793 | 37.0506996 |

**Table S.8.**
Parameter estimation for the residual proteins model equation.

| Factor | Coefficient Estimate | Standard Error | 95% CI Low | 95% CI High |
|---|---|---|---|---|
| Intercept | 72.0460147 | 6.28644486 | 58.9327205 | 85.1593089 |
| A-Initial nitrate concentration | -12.6098061 | 4.45192994 | -21.8963692 | -3.323243 |
| B-Carbon dioxide | -7.67590611 | 2.03201871 | -11.9146229 | -3.43718936 |
| C-Culture time | 16.1670358 | 4.95745937 | 5.82595676 | 26.5081148 |
| D-Nitrogen limitation | 26.7747073 | 6.1473184 | 13.9516259 | 39.5977888 |
| AC | -67.8034835 | 10.2848512 | -89.2573071 | -46.3496599 |
| AD | -82.7219655 | 12.2164796 | -108.205096 | -57.2388356 |
| BD | -14.9976185 | 2.03201871 | -19.2363353 | -10.7589018 |
| CD | 90.3469274 | 12.5711009 | 64.1240705 | 116.569784 |
| A^2 | 36.7757319 | 7.71096884 | 20.6909328 | 52.8605311 |
| B^2 | -17.9510454 | 5.14064589 | -28.6742448 | -7.22784593 |
| C^2 | 56.7472732 | 7.82520311 | 40.4241856 | 73.0703608 |

9

# Dry anaerobic digestion of lignocellulosic and protein residues

Maryam M. Kabir*, Mohammad J. Taherzadeh, Ilona Sárvári Horváth

*Swedish Centre for Resource Recovery, University of Borås, 501 90, Borås, Sweden.*

## HIGHLIGHTS

- ➢AD digestion/co-digestion of wool and wheat straw were evaluated with TS contents of between 6 to 30%.
- ➢Methane yield of wheat straw was highest at TS of 21% and lowest at TS of 30%.
- ➢Methane yield of wool textile was highest at TS of 13% and lowest at TS of 30%.
- ➢The addition of enzyme on wool and wheat straw led to improvement in methane yield at TS of 13%.
- ➢Synergetic effects were observed when these substrates were co-digested.

## GRAPHICAL ABSTRACT

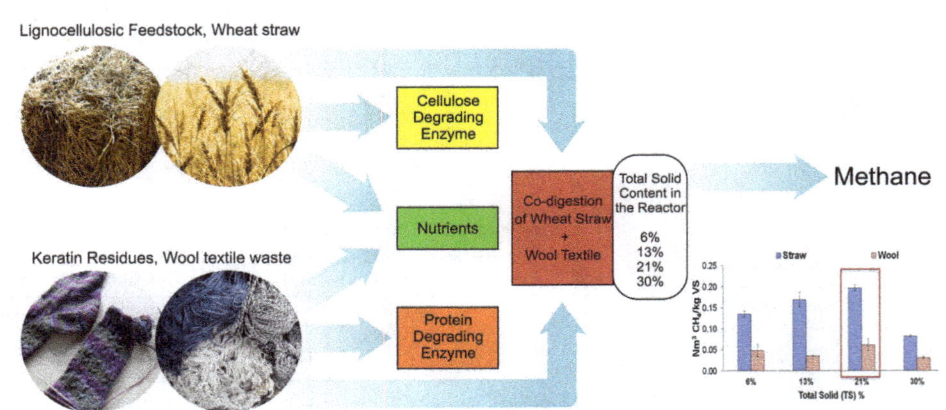


## ABSTRACT

Utilisation of wheat straw and wool textile waste in dry anaerobic digestion (AD) process was investigated. Dry-AD of the individual substrates as well as co-digestion of those were evaluated using different total solid (TS) contents ranging between 6 to 30%. Additionally, the effects of the addition of nutrients and cellulose- or protein-degrading enzymes on the performance of the AD process were also investigated. Dry-AD of the wheat straw resulted in methane yields of 0.081 – 0.200 $Nm^3CH_4/kgVS$ with the lowest and highest values obtained at 30 and 21% TS, respectively. The addition of the cellulolytic enzymes could significantly increase the yield in the reactor containing 13% TS (0.231 $Nm^3CH_4/kg$ VS). Likewise, degradation of wool textile waste was enhanced significantly at TS of 13% with the addition of the protein-degrading enzyme (0.131 $Nm^3CH_4/kg$ VS). Furthermore, the co-digestion of these two substrates showed higher methane yields compared with the methane potentials calculated for the individual fractions at all the investigated TS contents due to synergetic effects and better nutritional balance.

**Keywords:**
Dry anaerobic digestion
Lignocellulosic biomass
Wheat straw
Wool
Keratin
Enzyme addition

* Corresponding author
E-mail address: maryam.kabir@hb.se

## 1. Introduction

Anaerobic digestion (AD) has received tremendous attention recently as a consequence of a growing demand for renewable energies together with increasingly-stringent environmental regulations. This technology has been used in processing of municipal and industrial sludge for almost 100 years (Chandler and Jewell, 1980). The product of the AD, the biogas, is a versatile energy source, which can be utilised for the generation of heat and/or electricity, either separated or combined and in addition, it can be upgraded and used as vehicle fuel (Ghosh et al., 2000; Shafiei et al., 2011; Salehian et al., 2013). The AD process is classified as wet-AD and solid-state/dry-AD, based on the total solids (TS) content of the feedstock in the digester. Generally, TS content in wet-AD digesters is lower than 10%, whereas in dry fermentation, it is higher than 10% (De Baere and Mattheeuws, 2008; Zhu et al., 2010). During the past few years, over 63% of the new AD plant installations in Europe were based on dry-AD (De Baere and Mattheeuws, 2008; Kiran et al., 2014).

The increasing interest in dry-AD based plants can generally be explained by the advantages of the dry-AD processes over the wet-AD processes. More specifically, due to the increase in TS content in the dry-AD, this process requires a smaller reactor volume, which consequently reduces the material cost and the energy required for heating. Moreover, the digestate can easily be subjected to composting and then used as soil conditioner or fertilizer as no or minimal dewatering is needed. Furthermore, problems related to the wet-AD, such as floating of scum layer can be avoided (Chanakya et al., 1993; Nordberg and Edström, 1997).

Lignocellulosic biomass and textile wastes could be ideal substrates for dry-AD, due to their high abundance and their low moisture content. The global production of fibres has shown a constant increase in the past few decades (Akia et al., 2014; Barchyn and Cenkowski, 2014; Pothiraj et al., 2015). According to the Food and Agriculture Organization of the United Nations (FAO), the annual wool production is estimated at approximately 2.1 million tonnes per year (International Wool Textile Organization: IWTO, 2015). These fibres, after recycling and reuse for textile and other applications, would finally end up in waste stations. Consequently, the disposal of this large volume of fibre wastes has evolved into a major concern for the textile industry today.

Wool is mainly composed of a recalcitrant protein, keratin, (≈ 97%) and lipids (≈ 1%) (Heine and Höcker, 1995). Due to its high protein content, it can be used as an alternative renewable biomass source for the production of value-added products, *via* chemical, physicochemical, and microbial processes. One novel approach is the application of biological degradation of wool-based textiles for biogas production.

On the other hand, among the large amounts of different lignocellulosic biomass produced within the EU, straw generation is projected to be approximately 127 million tonnes in the year 2020. This is equal to 49.3 Mtoe (million tonnes oil equivalent) (Kretschmer et al., 2012). This high potential as well as the availability of straw and its high carbon content make it a promising substrate for biofuel production.

Hence, the main aim of this study was to explore the performance of AD of wheat straw and wool textile waste alone as well as in co-digestion using wet, semi dry, and dry-AD processes. The TS contents were therefore varied between 6 – 30% during the investigations. The effect of the addition of cellulose- or protein-degrading enzymes, as well as the effect of the addition of nutrients on the performance of the AD processes was also determined. Furthermore, chemical compositions and structural variations of these substrates during the degradation process were also investigated. To the best of our knowledge, this is the first report on the dry-digestion process of wheat straw and wool textile waste as co-substrates.

## 2. Materials and methods

### *2.1. Preparation of wool and wheat straw substrates*

Wool textile waste, consisting of 70% wool (protein) and 30% polyamide, was obtained from the Woolpower AB (Östersund, Sweden). This textile waste was then ground to particles approximately 2 mm in size using a Retsch GmbH SM 100 comfort miller (Germany) and stored until further investigation. Wheat straw used in this study was supplied by Lantmännen Agroetanol (Norrköping, Sweden). The composition of the wheat straw was determined based on the dry matter of the wheat straw: 35% cellulose, 22% hemicellulose, 18% insoluble lignin, and 16% extractives. The straw was milled into particles of between 2 to 5 mm in size, using a Octagon 200 sieve shaker (UK) and stored prior to analyses and anaerobic digestion.

### *2.2. Anaerobic digestion assays*

The methane potential of different substrate combinations under different conditions were determined using batch anaerobic digestion assays. The milled wheat straw or wool textile waste as well as a mixture of these two were mixed with a predetermined amount of inoculum and deionized water to achieve a substrate to inoculum ratio (S/I) of 2:1 (based on the volatile solids -VS- content) (Liew et al., 2012), and an initial TS contents of 6%, 13%, 21% and 30%. The inoculum was obtained from a farm-scale digester located at Vårgårda-Herrljunga Biogas AB (Sweden), operating at mesophilic temperature (37°C). Because of the low TS content, the inoculum was centrifuged and the supernatant was discharged to achieve the desirable TS for the dry-AD. The remaining sludge was mixed to obtain a homogenous inoculum. The AD of the wheat straw or wool textile waste was investigated even with the addition of enzymes, Cellic® CTec3 enzyme (Novozomes, Denmark) or an alkaline endopeptidase (Savinase® 16 L, Type EX), respectively. The cellulolytic enzyme load was 10 FPU (Filter Paper Unit)/g straw and the endopeptidase load was 10 KNPU (Kilo Novo Protease Unit)/g textile. These enzymes were added to the digesters directly when starting up the AD process.

Furthermore, wool or straw was also digested with the addition of nutrients. The final nutrient concentrations for the basal medium (1g/L substrate, containing inorganic macronutrients) were (mg/L): $NH_4Cl$ (76.4), $KH_2PO_4$ (5.18), $MgSO_4{\cdot}7H_2O$ (0.27), $CaCl_2{\cdot}2H_2O$, (10.00), and trace nutrients, 1 ml/L (Osuna et al., 2003). Finally, the co-digestion of the wheat straw and wool textile waste (ratio 1:1 based on the VS content) was investigated using similar TS concentrations as in the mono-digestion assays.

After the set-ups were prepared, anaerobic conditions were obtained by purging the reactors with nitrogen gas for about 2 min, and the reactors were then incubated in a convection oven under mesophilic conditions (37 ± 1°C). Inoculum without adding any substrate was also evaluated as blank to determine the methane production of the inoculum itself. The accumulated methane production was measured by taking gas samples regularly from the headspace during a 50-d long examination period.

### *2.3. Analytical methods*

The TS and VS of the investigated substrates were determined according to Sluiter et al. (2008a). Total nitrogen content of the wool textile waste before and after digestion was determined by Kjeldahl digestion, using a 2020 Kjeltec Digestor and a 2400 Kjeltec Analyser unit (FOSS analytical A/S Hilleröd, Denmark). The extractive content of the wheat straw was measured according to Sluiter et al. (2005). Furthermore, the structural carbohydrates content was determined using a two-step hydrolysis method described by Sluiter et al. (2008b). The sugars were quantified by high performance liquid chromatography (HPLC) (Waters 2695, Millipore and Milford, USA) equipped with a refractive index (RI) detector (Waters 2414, Millipore and Milford, USA), using a Pb-based ion exchange column (Aminex HPX-87P, Bio-Rad, USA). The eluent was pure water with a flow rate of 0.6 mL/min at 85°C.

Fourier transform infrared spectroscopy (FTIR) (Impact 410, Nicolet Instrument Corp., Madison, WI) was used to measure the cellulosic crystallinity. The spectra were achieved with an average of 64 scans and a resolution of 4 $cm^{-1}$ in the range from 600 to 4000 $cm^{-1}$ and controlled by the Nicolet OMNIC 4.1 analysing software. The crystallinity was analysed by considering the absorbance bands at 1422 and 898 $cm^{-1}$, assigned to cellulose I and cellulose II (amorphous cellulose), respectively (Carrillo et al., 2004). The absorbance ratio of $A_{1422}/A_{898}$ was used to measure the crystallinity index of the wheat straw before and after digestion, and that of the digested straw in the presence of the enzyme.

The protein secondary structure in the wool textile waste was also investigated by FTIR spectrometery. The most sensitive spectral region to the protein secondary structural components is the amide I band, which appears in the absorption bands of 1700 – 1600 $cm^{-1}$ due to the C=O stretch vibrations of the peptide linkage. To analyse the amide I band component, the second derivative spectra was curve fitted and used to identify the composite absorptions attributed to α-Helix, β-Sheet and disordered microstructural components, respectively, before and after 50 days of anaerobic digestion (Kong and Yu, 2007). The percentage of the α-Helix, β-Sheet and disordered microstructures were determined by adding the sum of absorptions for each and stating their sums as a fraction of total amide I band area (Cardamone, 2010).

Methane production was measured using a gas chromatograph (GC) (Auto System, Perkin Elmer, USA) equipped with a packed column (Perkin Elmer, 6' x 1.8" OD, 80/100, Mesh, USA) and a thermal conductivity detector (Perkin Elmer) with an inject temperature of 150°C. Nitrogen was used as carrier gas at 75°C with a flow rate of 20 mL/min. A 250μL pressure-tight syringe (VICI, Precision Sampling Inc., USA) was used for gas sampling. To avoid overpressure built up in the digesters, the excess gas was released through a needle after each gas sampling. The results were presented as $Nm^3$ $CH_4$/kg VS and were calculated as the volume of methane gas produced per kg of VS loaded into each reactor at the start-up and were corrected by subtracting the methane volume obtained from the blank reactor containing the inoculum only. The volumetric productivity of methane expressed as $V_{methane}/V_{working\ volume}$, meaning the $m^3$ of methane gas produced ($V_{methane}$) per $m^3$ of working volume of the reactor ($V_{working\ volume}$), was also determined.

### 2.4. Statistical analyses

All AD assays as well as the compositional and structural analyses were run in triplicates. Statistical analysis was performed using the software package MINITAB®. All error bars and intervals reported represent one standard deviation. ANOVA general linear model was applied to evaluate the methane production and the effect of the varying TS content in the digesters with a significance threshold (p-value) of 0.05. The effects of the addition of nutrients and enzymes on the anaerobic digestion process were evaluated *via* two sample comparative t-test using the MINITAB®.

## 3. Results and discussion

### 3.1. Anaerobic digestion of wheat straw and wool textile waste

The AD of wheat straw and wool textile waste was investigated using four different TS contents (i.e., 6, 13, 21 and 30%) in batch operation mode. The accumulated methane production was determined during a 50-d long digestion period. The S/I ratio (based on the VS content) was set to 2 in each digester, while the TS concentrations varied between 6 – 30%. The obtained accumulated methane yields and volumetric productivities are presented in **Figure 1** and **Table 1**.

#### 3.1.1. Anaerobic digestion of wheat straw

In the AD of wheat straw (without addition of enzyme and nutrients), the methane yields obtained were between 0.081 – 0.200 $Nm^3CH_4$/kg VS (**Table 1** and **Fig.1**). The highest methane yields were determined when the TS concentrations of 13 and 21% were used, with methane productions of 0.170 and 0.200 $Nm^3CH_4$/kg VS, respectively. To the contrary, the methane production rate was significantly reduced when the straw was digested using the TS contents of 6% (0.135 $Nm^3CH_4$/kg VS) or 30% (0.081 $Nm^3CH_4$/kg VS). However, according to the statistical analyses, the methane production rate of the dry-AD using the TS contents of 13 or 21% did not differ significantly ($p>0.05$).

**Table 1.**
Accumulated methane production (Nm3CH4/ kg VS), methane volumetric productivity (Nm3 /m3 Working Volume), and the expected calculated methane yield from the co-digested mixture, at TS contents of 6%, 13%, 21%, and 30%.

| | Feedstock | Accumulated methane yield ( $Nm^3$ $CH_4$/ kg VS) | Volumetric productivity ( $m^3$ CH4/$m^3$ Working Volume) | Expected methane yields of co-digested mixture ( $Nm^3$ $CH_4$/ kg VS) |
|---|---|---|---|---|
| 6% TS | Straw | 0.135 ± 0.007 | 0.011 | – |
| | Straw+ Enzyme | 0.171 ± 0.030 | 0.014 | – |
| | Straw+ Nutrient | 0.141 ± 0.013 | 0.012 | – |
| | Wool | 0.048 ± 0.013 | 0.004 | – |
| | Wool+ Enzyme | 0.108 ± 0.005 | 0.009 | – |
| | Wool+ Nutrient | 0.058 ± 0.009 | 0.005 | – |
| | Wool+ Straw | 0.136 ± 0.009 | 0.014 | 0.091 |
| | Wool+ straw + nutrient | 0.143 ± 0.011 | 0.012 | 0.102 |
| 13% TS | Straw | 0.170 ± 0.017 | 0.021 | – |
| | Straw+ Enzyme | 0.231 ± 0.011 | 0.029 | – |
| | Straw+ Nutrient | 0.204 ± 0.006 | 0.025 | – |
| | Wool | 0.035 ± 0.000 | 0.005 | – |
| | Wool+ Enzyme | 0.131 ± 0.005 | 0.016 | – |
| | Wool+ Nutrient | 0.088 ± 0.002 | 0.011 | – |
| | Wool+ Straw | 0.100 ± 0.010 | 0.012 | 0.102 |
| | Wool+ straw + nutrient | 0.133 ± 0.003 | 0.016 | 0.146 |
| 21% TS | Straw | 0.200 ± 0.007 | 0.033 | – |
| | Straw+ Enzyme | 0.211 ± 0.002 | 0.035 | – |
| | Straw+ Nutrient | 0.225 ± 0.013 | 0.037 | – |
| | Wool | 0.061 ± 0.014 | 0.010 | – |
| | Wool+ Enzyme | 0.040 ± 0.007 | 0.000 | – |
| | Wool+ Nutrient | 0.089 ± 0.009 | 0.015 | – |
| | Wool+ Straw | 0.154 ± 0.001 | 0.026 | 0.130 |
| | Wool+ straw + nutrient | 0.177 ± 0.021 | 0.030 | 0.157 |
| 30% TS | Straw | 0.081 ± 0.002 | 0.016 | – |
| | Straw+ Enzyme | 0.052 ± 0.002 | 0.010 | – |
| | Straw+ Nutrient | 0.070 ± 0.000 | 0.014 | – |
| | Wool | 0.030 ± 0.002 | 0.006 | – |
| | Wool+ Enzyme | 0.033 ± 0.003 | 0.006 | – |
| | Wool+ Nutrient | 0.013 ± 0.008 | 0.003 | – |
| | Wool+ Straw | 0.090 ± 0.005 | 0.020 | 0.055 |
| | Wool+ straw + nutrient | 0.097 ± 0.006 | 0.020 | 0.041 |

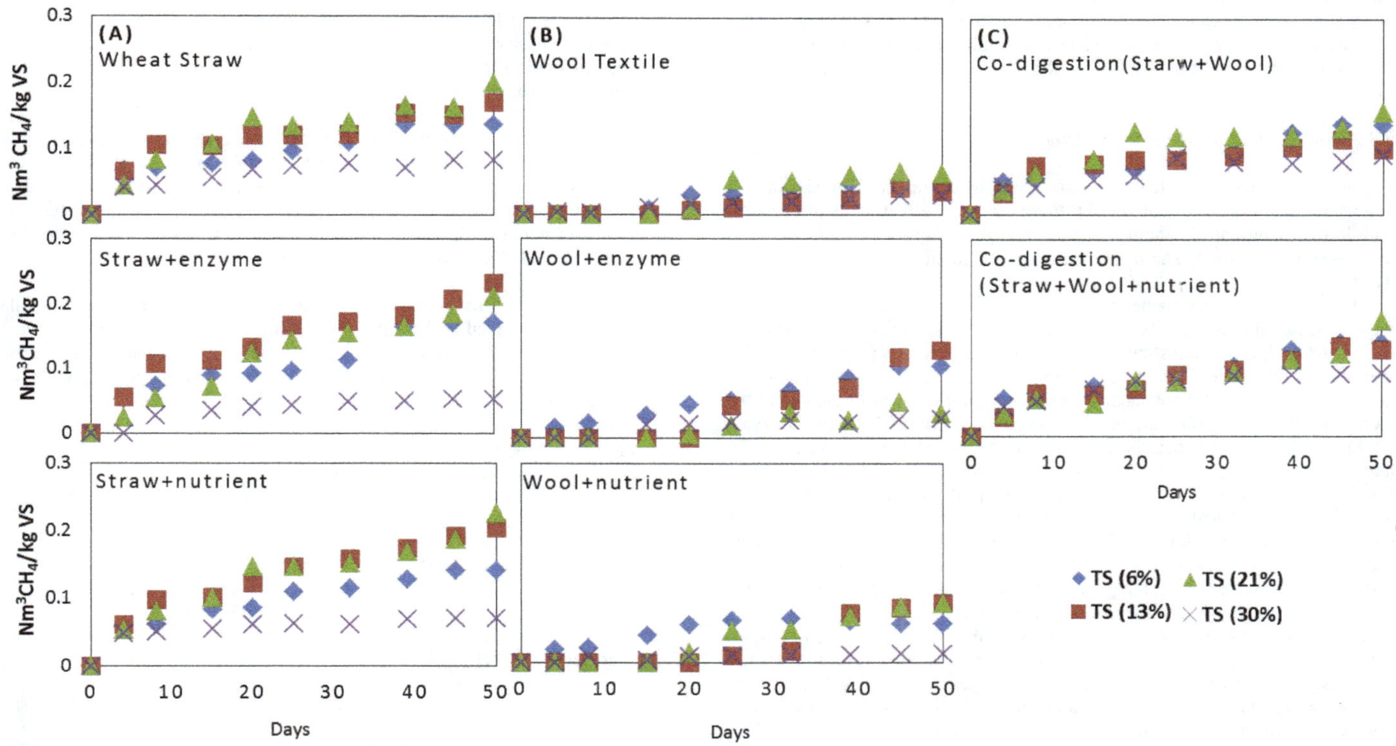


**Fig.1.** Accumulated methane production expressed in $Nm^3\,CH_4/kg$ VS for wheat straw (A), wool textile waste (B), and co-digestion of these two (C) obtained during 50-day AD, at TS contents of 6%, 13% 21% and 30%. The digestion conditions are described in the figure.

The effect of the addition of nutrients in the AD of wheat straw was also examined, and the obtained methane yields were 0.141, 0.204, 0.225 and 0.070 $Nm^3CH_4/kg$ VS, in reactors containing 6, 13, 21 and 30% TS, respectively. The statistical analyses of the results showed that there was no significant difference between the methane yields obtained without and with the nutrients addition.

The enzymes that are responsible for degradation of biomass in the AD system are intrinsically present in digesters as they are excreted by the microorganisms. However, to improve the breakdown process, excess enzyme loading was applied in the present study. The enzymes used were directly added to the vessels at the start-up of the digestion process. AD of the wheat straw supplied with excess cellulolytic enzyme resulted in methane yields of 0.171, 0.213, 0.211 and 0.052 $Nm^3CH_4/kg$ VS, for TS contents of 6, 13, 21 and 30%, respectively.

By comparing the methane potential reported previously for wheat straw and what obtained in batch wise wet-AD studies (ranging between 0.189 – 0.200 $Nm^3CH_4/kg$ VS) at mesophilic conditions (Jurado et al., 2013) with the results of the current work, it can be concluded that almost similar methane yields (0.170 – 0.200 $Nm^3CH_4/kg$ VS) were achieved even in the dry digestion mode, with the advantage of more organic loading in the system. Moreover, the simultaneous addition of cellulose- degrading enzymes at the start up of the digestion process could significantly enhance the methane production at the TS content of 13%, leading to 0.231 $Nm^3CH_4/kg$ VS methane production. However, the statistical analyses showed that the addition of enzyme caused no significant differences ($p>0.05$) at all the other TS contents applied. A possible explanation for that might be the presence of a lignin shield surrounding the cellulose fraction of the straw, which ultimately hindered the enzyme accessibility to the substrate. As reported earlier (Karimi et al., 2013), the binding of lignin to the cellulase enzymes is irreversible, and it can lead to the deactivation of the enzymes, thus reducing their effectiveness.

The volumetric methane productivity (shown in **Table 1**) was also evaluated. The determination of the volumetric methane productivity is an important factor when considering the economy of AD processes. The volumetric productivities determined during this study showed that the highest methane production from straw could be achieved using the TS contents of 13% (i.e., 0.021 – 0.030 $m^3CH_4/\,m^3$working volume) or 21%, (i.e., 0.033 – 0.037 $m^3CH_4/\,m^3$working volume), respectively (**Table 1**).

Generally, the higher methane yields obtained at the TS contents of 13 and 21% might be related to the nature of the straw, which can act as a structural biofilm carrier. In AD systems, when biofilm carriers are present in the reactor, syntrophic interactions between a large variety of microorganisms can occur, which could subsequently lead to higher methane yields. This is in agreement with the findings obtained in previous studies where additional biofilm carriers were applied using inert or naturally-degradable materials, like straw, to which the microbial population could be attached (Kazda et al., 2012; Ward et al., 2008). Due to the cell's natural tendency to form dense granules, additional biofilm carriers would enhance process stability, which will lead to increased methane yields (Ward et al., 2008). Previously, Andersson and Björnsson (2002) also found that the addition of straw in a two-stage anaerobic digestion system resulted in an improvement in the digestion performance with enhanced methane production.

Therefore, one can expect that the AD of straw in a dry system would lead to increased cell tendencies for attachment to the surface of the straw making the degradation more efficient. This could also be easily comprehended from the results of this study, revealing high gas production per unit volume in digesters operating at 2 or 3 times higher TS contents of straw, compared with wet AD systems. During these conditions, highly-efficient degraders can be developed and/or retained in the system due to an increased accessible surface area. Hence, the cost effectiveness of the dry-AD process increases (Jewell et al., 1993). On the other hand, the biomethane potential of straw in reactors with the highest TS concentration of 30% was not high in this study. This can be the consequence of excessive shortage of moisture in the reactor, which finally deteriorates the digestion process. This result is in accordance with those of the previous studies reporting a dramatic decrease in biogas production with TS contents ranging between 30 to 50% (Jewell et al., 1993). Moreover, according to Wujcik and Jewell (1980), high TS contents of 30 to 40% cause inhibition of the AD due to the accumulation of volatile fatty acids (VFAs).

### *3.1.2. Anaerobic digestion of wool textile waste*

The AD of wool textile waste was investigated as well at similar TS concentrations (i.e., 6 – 30%). Furthermore, the effect of simultaneous addition of a protein-degrading enzyme and the addition of nutrients were also taken into account. The results obtained are summarised in **Figure 1** and **Table 1**.

The methane yields obtained without the addition of enzyme and nutrients were low, ranging between 0.030 to 0.061 $Nm^3CH_4/kg$ VS at the TS contents of 6 – 30% (**Fig. 1**). Interestingly, no significant differences in methane yields were observed as the TS increased from 6 to 30% ($p > 0.05$). Similarly, the addition of nutrients did not cause a significant improvement in the AD of the wool textile waste. However, the addition of the protein-degrading enzyme could significantly enhance the methane production at the TS contents of 6 and 13%, resulting in 0.108 and 0.131 $Nm^3CH_4/kg$ VS, respectively. Previously, in a wet-AD system and at thermophilic conditions, only 5% of the expected theoretical yield from the protein fraction of similar textile waste could be achieved (i.e., 0.020 $m^3CH_4/kg$ VS) (Kabir et al., 2013). While in the present study, the degradation of wool textile waste in the dry-AD mode resulted in 3 and 6.5 times higher methane yields, without and with the addition of a protein-degrading enzyme, respectively.

The highest protein conversion of the wool textile waste into biomethane was obtained using the enzyme supplement at the TS content of 13% (0.131 $m^3CH_4/kg$ VS), corresponding to 38% of the expected theoretical yield from this substrate. However, a descending trend in the methane yield was observed when enzyme addition was conducted at the TS contents of 21% and 30% (of 0.040 and 0.032 $m^3CH_4/kg$ VS, respectively). The reason for this low gas production at higher TS concentrations might be the increase in the mass transfer barrier for the diffusion of VFAs (Bollon et al., 2013), which leads to the accumulation of VFAs in the system. Besides, as the addition of enzyme accelerates the degradation rate of proteins (Forgács et al., 2013), the consequent accumulation of ammonium nitrogen might be another reason behind the low methane yield (Forgács et al., 2013).

Furthermore, there was a 20-d long lag period observed in the digestion of the wool textile waste, at all TS concentrations investigated (**Fig. 1**). Nevertheless, the process was initiated, and the methane production started up in the reactors containing 6 – 21% TS while the digestion failed when the TS concentration was increased to 30%. These results are in accordance with those obtained previously by Forgács et al. (2013) who used chicken feather (β-keratin protein) as feedstock in a semi-continuous anaerobic co-digestion process. In their study, after an initial period of 20 d, the ammonium-nitrogen concentration increased to 4,200 mg/L in the reactors where enzyme was added together with feather, resulting in a decrease in the methane production. However, the system became stabilised afterwards together with a slow decrease in ammonium-nitrogen concentration.

**Table 1** presents the volumetric methane productivity determined when the wool textile waste was digested at different TS levels. The results obtained showed that the volumetric productivity was highest when the TS content of 13% was used. However, the volumetric productivity decreased when the TS was further increased to 21 and 30%.

### *3.1.3. Co-digestion of wheat straw and wool textile*

The co-digestion of wool textile wastes and wheat straw (in a ratio of 1:1 based on the VS) was also investigated using similar TS contents with or without the addition of nutrients. The expected methane yields for the co-digested mixture can be calculated from the methane potentials obtained for the individual fractions (**Table 1**). All the co-digestion assays showed higher accumulated methane yields than the calculated expected levels, with an exception of when the TS content of 13% was used. According to Alvarez and Lidén (2008) and Pagés Díaz et al. (2011), the accumulated methane yield in a co-digestion process can be higher than the calculated expected value, due to the synergic effects developed in the digester. Interestingly, the methane production of the co-digested wheat straw and wool textile waste was the highest, showing an increase of 58%, compared with the expected value, when the TS content of 30% together with addition of nutrients was applied. Furthermore, similar TS content without nutrients supplementation led to an increase of 39%, compared with the expected methane yield. Since, higher TS concentrations or higher S/I ratio generally results in a slowdown in the microbial growth (Batstone et al., 2002; Yang and Li, 2014), it is expected that digesters containing 30% TS, more likely suffer from nutritional imbalances, compared with digesters running on lower TS contents. Therefore, co-digestion of the carbon-rich straw with the nitrogen-rich wool textile waste provided a nutrient efficient feedstock with a more balanced C/N ratio even at the TS content of 30%; and this consequently resulted in a noticeable improvement in methane production.

Furthermore, this improvement was even higher when additional nutrients were added to the digestion system. The results of this study are in agreement with those of the study conducted by Kayhanian and Rich (1995), which suggested that mixing of two or three organic wastes could provide a nutrient-sufficient feedstock for a dry-AD process.

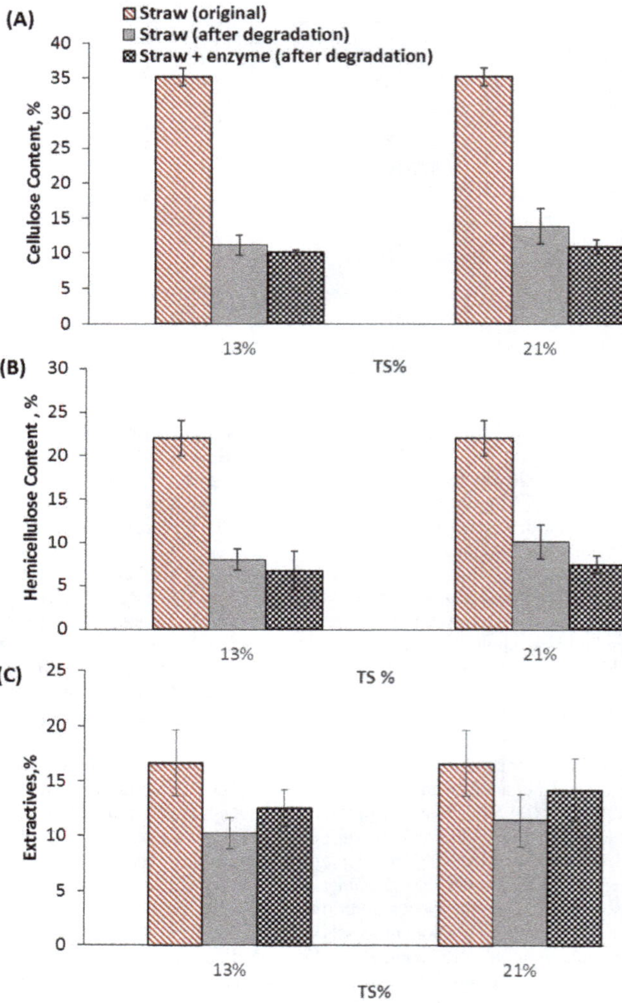


**Fig.2.** Degradation of cellulose % (A), hemicellulose % (B), and extractives % (C) during the 50-d dry-AD. The error bars correspond to one-standard deviation.

### *3.2. Degradation of substrates and compositional changes*

Compositional analyses of the wheat straw and wool textile waste were performed, before and after the 50-day long digestion period at the TS contents of 13% and 21%. When investigating the straw samples, changes in the composition (cellulose, hemicellulose, and extractives) were defined by measuring the corresponding mass reductions between the beginning and end of the 50-day digestion (**Fig. 2**) The compositional analysis of the un-treated wheat straw showed 35.2% cellulose and 22.2% hemicellulose content (**Fig. 2A** and **2B**). The cellulose content of the straw residues obtained after the dry-AD digestion at the 13% TS with and without the addition of enzymes were 10.0% and 11.2%, respectively. The highest levels of cellulose degradation of 69 and 71% were observed when cellulolytic enzymes were added to the digestion process using the TS contents of 13 and 21%, respectively. A similar trend was observed in the hemicellulose removal as well, showing a reduction in hemicelluloses by 69 and 66%, when the digestion process was performed at the TS contents of 13 and 21%, respectively, both with the addition of enzymes. On the other hand, the hemicellulose reduction rate achieved through the dry digestion of the straw, without the addition of enzymes was lower, i.e., 63 and 54% at the TS contents of 13 and 21%, respectively. Hence, the highest cellulose and hemicellulose degradations were achieved after simultaneous enzymatic hydrolysis and dry-digestion. The optimum temperature for the cellulolytic enzymes added is about 38°C, which is very close to the mesophilic operation temperature of the dry-AD conducted in this study (Tengborg et al., 2001). This temperature supported the activity of the enzymes resulting in a more efficient first step of the digestion, i.e., the hydrolysis step.

The extractives' content of the wheat straw was 16.6% of the dry weight, and the degraded wheat straw with and without the addition of cellulolytic enzymes showed 14 – 38% reduction in extractives (**Fig. 2C**). Since the extractives include compounds, such as free sugars, oligomers, and organic acids (Chen et al., 2007; Chen et al., 2010), these compounds can be easily degraded and can potentially contribute to biogas production (Tong et al., 1990). The composition of the extractives was not analysed in this study; however, further research on the degradation of extractive compounds and their contributions to the AD process would be interesting. The lignin content of the wheat straw remained intact before and after digestion, meaning that this fraction was not subjected to degradation during the digestion period applied.

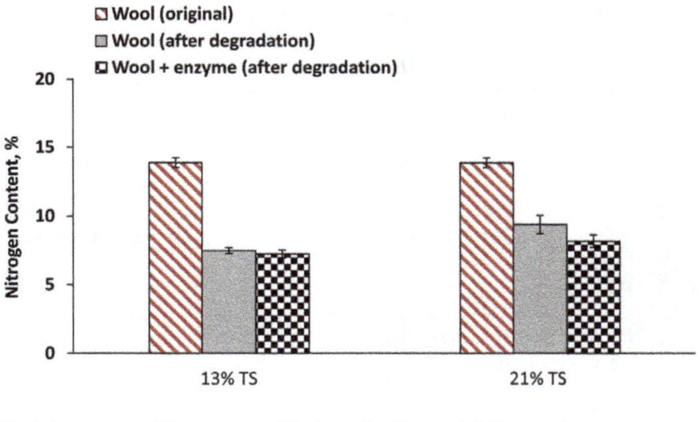

**Fig.3.** Degradation of nitrogen content %, during the 50-d dry-AD. The error bars correspond to one-standard deviation.

The nitrogen content of the wool textile waste was determined by the Kjeldahl method. The N-content of the wool textiles was 14%, and a reduction of nearly 50% in the N-content was observed after the digestion, with or without the addition of a protein-degrading enzyme (**Fig. 3**). The nitrogen content of the wool textile waste after the 50-d degradation period without and with the enzyme addition was 7.5 and 7.24% in reactors containing 13% TS, respectively. While the nitrogen content stood at 9.4 and 8.2% when the 21% TS without and with the enzyme addition was applied, respectively (**Fig. 3**). However, these high degradation rates of the wool proteins did not necessarily lead to higher methane productions (**Fig. 1**). This might be due to the accumulation of ammonium nitrogen as discussed earlier. A lag phase of 20 d was observed when the wool textile waste was digested at the TS content of 13 and 21%, and this lag phase was prolonged. Moreover, the conversion of organic matters for gas production did not start up at all when the TS content was increased to 30%. However, the lag phase was shorter, and methane production rate was faster in the more diluted system, i.e., by using the TS content of 6%. The ammonia levels in the digesters at the end of the digestion period were not measured; however, the obtained decrease in the N-content after the digestion on one hand, and the low accumulated methane production determined on the other hand, can indicate an imbalance in the digestion system. It is worth quoting that nitrogen is an important nutrient for microbial growth. However, high concentrations of nitrogen may lead to a severe disturbance in the AD performance, which can cause a dramatic decrease in microbial activities (Zhang et al., 2011). This in turn results in decreased methane production rates and increased concentrations of intermediate digestion products, like VFAs, which will finally lead to a total termination of methanogenic activity (Sung and Liu, 2003; Calli et al., 2005).

### *3.3. Structural changes in wheat straw and wool obtained after dry-AD*

The cellulosic crystallinity of the wheat straw was investigated by FTIR spectrometery. The crystallinity index of the un-treated wheat straw as well as the digested straw obtained after the dry-AD which was performed at the TS concentrations of 13 and 21% (with and without the addition of enzymes) is shown in **Figure 4**. The crystallinity index of the un-treated straw was 0.62; however, throughout the 50-d long dry-AD process, the crystallinity index was increased to 1.3 – 1.5 when the TS contents of 13 or 21% were applied, without or with the addition of the cellulolytic enzymes (**Fig. 4**). As previously mentioned, the crystallinity index is the absorbance ratio between crystalline cellulose I and the amorphous form of cellulose II ($A_{1422}/A_{898}$). As shown in **Figure 4**, there was a significant increase in the crystallinity index after the 50-d digestion period, which was confirmed by the statistical analyses as well (p= 0.000). Since the cellulolytic enzymes can target the less crystalline regions of cellulose fibres (Taherzadeh and Karimi, 2007), this increase in crystallinity indicates that the amorphous cellulose part was initially subjected to biological degradation, subsequently leaving the crystalline cellulose fraction relatively intact.

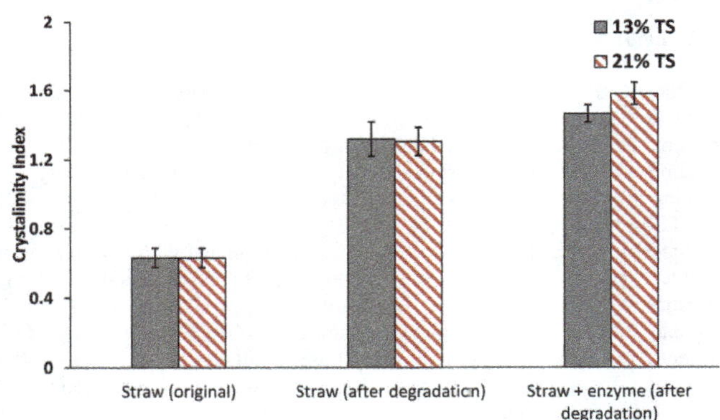


**Fig.4.** Crystallinity index of wheat straw before and after the 50-d dry-AD. The error bars correspond to one-standard deviation.

The structural changes in the protein microstructure of the wool textile waste were also studied by FTIR (**Fig. 5**). The amide I (in the range between 1700 – 1600 cm$^{-1}$) and the amide II (between 1545 – 1400 cm$^{-1}$) bands, as the most predominant vibrational bands of the protein backbone were investigated (data not shown) (Kong and Yu, 2007). These bands are mainly associated with the stretching vibrations of the peptide bonds. The most sensitive spectral region to the protein secondary structure is amide I; therefore, the de-convoluted amide I band was studied in detail to

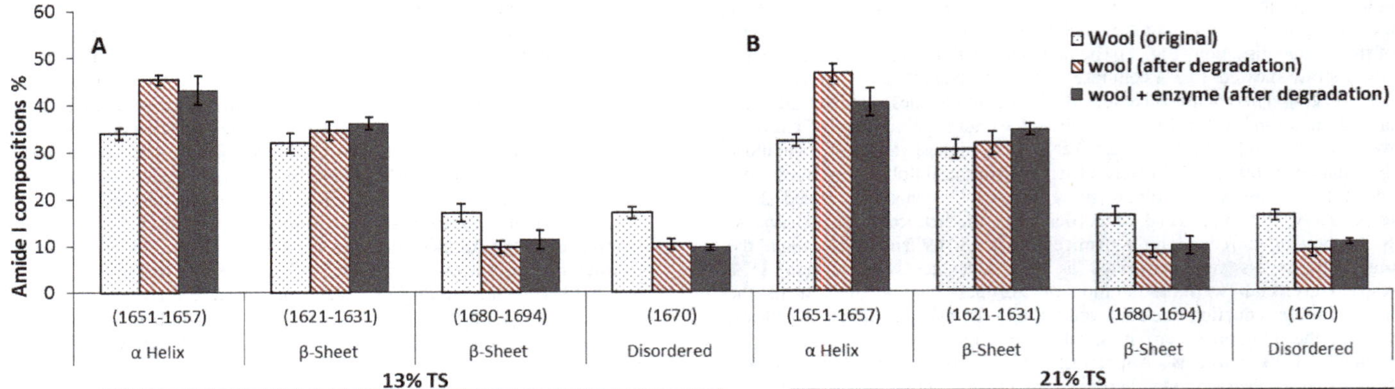


**Fig.5.** The secondary structure of wool textile protein (α-helix, β-sheet, and disordered region), before and after the 50-d dry-AD. The error bars correspond to one-standard deviation.

understand the changes in the secondary structure of the wool textile protein due to the digestion. The absorption regions of 1631 – 1621 $cm^{-1}$ and 1694 – 1680 $cm^{-1}$ show the secondary structures of β-sheet and the absorption regions of 1657 – 1651 $cm^{-1}$, and 1679 – 1670 $cm^{-1}$ represent the secondary structures of α-helix, and disordered regions, respectively. Originally, the wool textile waste contained 29.3% α-helix, 44.4% β-sheet, and 17% disordered regions. There was a decrease (from 17% to 10%) obtained in these disordered regions after the 50-d long digestion period when the TS contents of 13 or 21% were applied. Consequently, the degradation of the wool textile waste led to an increase in the α-helix conformation by 42.5 – 49.1% at both TS conditions, both with and without the addition of the protein-degrading enzyme. Furthermore, there was a slight decrease in the β-sheet conformation after the 50-d dry-AD (**Fig. 5**). Therefore, it can be concluded that as expected the disordered regions were the most affected regions during the digestion process. However, there is certainly a knowledge gap in the AD of keratin-rich substrates and their structural changes during the degradation process, which should be investigated further in the future.

## 4. Conclusion

Overall, the AD of the straw and wool textile waste using dry digestion technology was found as a promising option compared with the wet AD systems due to the higher methane yield and productivity, reduction in reactor volume and thus less energy consumption for heating, and finally more economic maintenance. Nevertheless, the results of this study revealed that the TS content applied must be properly adjusted to avoid overloading the system. Moreover, it was also shown that the problems with overloading due to the high TS content could be solved by the co-digestion of different substrates. More specifically, the co-digestion of a carbon-rich substrate (wheat straw) and a nitrogen-rich substrate (wool textile waste) can lead to higher methane production rates than those calculated based on the methane potentials for the single substrates. Moreover, due to the synergetic effects and a better nutritional balance, the TS could be increased to 30% during the co-digestion without causing serious problems to the digestion system. These findings can open up a new path for cost-effective utilisation of complex biomass-oriented substrates by further optimisation of their co-digestion through dry-AD systems.

## References


[1] Akia, M., Yazdani, F., Motaee, E., Han, D., Arandiyan, H., 2014. A review on conversion of biomass to biofuel by nanocatalysts. Biofuel Res. J. 1(1), 16-25.

[2] Alvarez, R., Lidén, G., 2008. Semi-continuous co-digestion of solid slaughterhouse waste, manure, and fruit and vegetable waste. Renew. Energy. 33(4), 726-734.

[3] Andersson, J., Björnsson, L., 2002. Evaluation of straw as a biofilm carrier in the methanogenic stage of two-stage anaerobic digestion of crop residues. Bioresour. Technol. 85(1), 51-56.

[4] Barchyn, D., Cenkowski, S., 2014. Process analysis of superheated steam pre-treatment of wheat straw and its relative effect on ethanol selling price. Biofuel Res. J. 1(4), 123-128.

[5] Batstone, D.J., Keller, J., Angelidaki, I., Kalyuzhnyi, S.V., Pavlostathis, S.G., Rozzi, A., Sanders, W.T.M., Siegrist, H., Vavilin, V.A., 2002. The IWA Anaerobic Digestion Model No 1 (ADM 1). Water Sci. Technol. 45(10), 65-73.

[6] Bollon, J., Benbelkacem, H., Gourdon, R., Buffière, P., 2013. Measurement of diffusion coefficients in dry anaerobic digestion media. Chem. Eng. Sci. 89, 115-119.

[7] Calli, B., Mertoglu, B., Inanc, B., Yenigun, O., 2005. Effects of high free ammonia concentrations on the performances of anaerobic bioreactors. Process Biochem. 40(3), 1285-1292.

[8] Cardamone, J.M., 2010. Investigating the microstructure of keratin extracted from wool: Peptide sequence (MALDI-TOF/TOF) and protein conformation (FTIR). J. Mol. Struct. 969(1), 97-105.

[9] Carrillo, F., Colom, X., Sunol, J.J., Saurina, J., 2004. Structural FTIR analysis and thermal characterisation of lyocell and viscose-type fibres. Eur. Polym. J. 40(9), 2229-2234.

[10] Chanakya, H.N., Borgaonkar, S., Meena, G., Jagadish, K.S., 1993. Solid phase fermentation of untreated leaf biomass to biogas. Biomass Bioenergy. 5(5), 369-377.

[11] Chandler, J.A., Jewell, W.J., 1980. Predicting methane fermentation biodegradability. Solar Energy Res. Inst. Golden, Colorado.

[12] Chen, S.F., Mowery, R.A., Scarlata, C.J., Chambliss, C.K., 2007. Compositional analysis of water-soluble materials in corn stover. J. Agric. Food. Chem. 55(15), 5912-5918.

[13] Chen, S.F., Mowery, R.A., Sevcik, R.S., Scarlata, C.J., Chambliss, C.K. 2010. Compositional analysis of water-soluble materials in switchgrass. J. Agric. Food. Chem. 58(6), 3251-3258.

[14] De Baere, L., Mattheeuws, B., 2008. State-of-the-art 2008-anaerobic digestion of solid waste. Waste management world. 9 (5), 1-8.

[15] Forgács, G., Lundin, M., Taherzadeh, M., Sárvári Horváth, I., 2013. Pretreatment of chicken feather waste for improved biogas production. Appl. Biochem. Biotechnol. 169(7), 2016-2028.

[16] Ghosh, S., Henry, M., Sajjad, A., Mensinger, M., Arora, J., 2000. Pilot-scale gasification of municipal solid wastes by high-rate and two-phase anaerobic digestion (TPAD). Water Sci. Technol. 41(3), 101-110.

[17] Heine E, Höcker H., 1995. Enzyme treatments for wool and cotton. Rev. Prog. Color. Relat. Top. 25(1), 57-70.

[18] International Wool Textile Organization (IWTO), 2015. Wool The Natural Fiber, IWTO.

[19] Jewell, W.J., Cummings, R.J., Richards, B.K., 1993. Methane fermentation of energy crops: maximum conversion kinetics and in situ biogas purification. Biomass Bioenergy. 5(3), 261-278.

[20] Jurado, E., Gavala, H.N., Skiadas, I.V., 2013. Enhancement of methane yield from wheat straw, miscanthus and willow using aqueous ammonia soaking. Environ. Technol. 34(13-14), 2069-2075.

[21] Kabir, M.M., Forgács, G., Sárvári Horváth, I., 2013. Enhanced methane production from wool textile residues by thermal and enzymatic pretreatment. Process Biochem. 48(4), 575-580.

[22] Karimi, K., Shafiei, M., Kumar, R., 2013. Progress in physical and chemical pretreatment of lignocellulosic biomass. in: Biofuel Technologies, Springer. pp. 53-96.

[23] Kayhanian, M., Rich, D., 1995. Pilot-scale high solids thermophilic digestion of municipal solid waste with an emphasis on nutrient requirements. Biomass Bioenergy. 8(6-8), 433-444.

[24] Kazda, M., Zak, M., Bengelsdorf, F.R., 2012. Effects of additional biofilms carrierson anaerobic digestion of food wates: results from laboratory experiments and a full-scale application. Anaerobic Digestion of Solid Biomass and Biowaste, Berlin.

[25] Kiran, E.U., Trzcinski, A.P., Liu, Y., 2014. Glucoamylase production from food waste by solid state fermentation and its evaluation in the hydrolysis of domestic food waste. Biofuel Res. J. 1(3), 98-105.

[26] Kong, J., Yu, S., 2007. Fourier transform infrared spectroscopic analysis of protein secondary structures. Acta Biochim. Biophys. Sin. 39(8), 549-559.

[27] Kretschmer, B., Allen, B., Hart, K., 2012. Mobilising Cereal Straw in the EU to feed Advanced biofuel production. Institute for European Environmental Policy.

[28] Liew, L.N., Shi, J., Li, Y., 2012. Methane production from solid-state anaerobic digestion of lignocellulosic biomass. Biomass Bioenergy. 46, 125-132.

[29] Nordberg, Å., Edström, M., 1997. Co-digestion of ley crop silage, straw and manure. in: The Future of Biogas in Europe. Bio Press, Herning, Denmark.

[30] Osuna, M.B., Zandvoort, M.H., Iza, J.M., Lettinga, G., Lens, P.N.L., 2003. Effects of trace element addition on volatile fatty acid conversions in anaerobic granular sludge reactors. Environ . Technol. 24(5), 573-587.

[31] Pagés Díaz, J., Pereda Reyes, I., Lundin, M., Sárvári Horváth, I., 2011. Co-digestion of different waste mixtures from agro-industrial activities: Kinetic evaluation and synergetic effects. Bioresour. Technol. 102(23), 10834-10840.

[32] Pothiraj, C., Arun, A., Eyini, M., 2015. Simultaneous saccharification and fermentation of cassava waste for ethanol production. Biofuel Res. J. 2(1), 196-202.

[33] Salehian, P., Karimi, K., Zilouei, H., Jeihanipour, A., 2013. Improvement of biogas production from pine wood by alkali pretreatment. Fuel. 106(0), 484-489.

[34] Shafiei, M., Karimi, K., Taherzadeh, M.J., 2011. Techno-economical study of ethanol and biogas from spruce wood by NMMO-pretreatment and rapid fermentation and digestion. Bioresour. Technol. 102(17), 7879-7886.

[35] Sluiter, A., Hames, B., Ruiz, R., Scarlata, C., Sluiter, J., Templeton, D., 2008a. Determination of ash in biomass. NREL/TP-510-42622 NREL Laboratory Analytical Procedure. National Renewable Energy Laboratory.

[36] Sluiter, A., Hames, B., Ruiz, R., Scarlata, C., Sluiter, J., Templeton, D., Crocker, D., 2008b. Determination of structural carbohydrates and lignin in biomass. NREL/TP-510-42618 NREL Laboratory Analytical Procedure. National Renewable Energy Laboratory.

[37] Sluiter, A., Ruiz, R., Scarlata, C., Sluiter, J., Templeton, D., 2005. Determination of extractives in biomass. Laboratory Analytical Procedure (LAP), 1617.

[38] Sung, S., Liu, T., 2003. Ammonia inhibition on thermophilic anaerobic digestion. Chemosphere. 53(1), 43-52.

[39] Taherzadeh, M.J., Karimi, K., 2007. Enzymatic-based hydrolysis processes for Ethanol. BioResources. 2(4), 707-738.

[40] Tengborg, C., Galbe, M., Zacchi, G., 2001. Influence of enzyme loading and physical parameters on the enzymatic hydrolysis of steam-pretreated softwood. Biotechnol. Progr. 17(1), 110-117.

[41] Tong, X., Smith, L.H., McCarty, P.L., 1990. Methane fermentation of selected lignocellulosic materials. Biomass. 21(4), 239-255.

[42] Ward, A.J., Hobbs, P.J., Holliman, P.J., Jones, D.L., 2008. Optimisation of the anaerobic digestion of agricultural resources. Bioresour. Technol. 99(17), 7928-7940.

[43] Wujcik, W.J., Jewell, W.J., 1980. Dry anaerobic fermentation. Biotechnol. Bioeng. Symp.;(United States). Cornell Univ., NY.

[44] Yang, L., Li, Y., 2014. Anaerobic digestion of giant reed for methane production. Bioresour. Technol. 171, 233-239.

[45] Zhang, Y., Cañas, E.M.Z., Zhu, Z., Linville, J.L., Chen, S., He, Q., 2011. Robustness of archaeal populations in anaerobic co-digestion of dairy and poultry wastes. Bioresour. Technol. 102(2), 779-785.

[46] Zhu, J., Wan, C., Li, Y., 2010. Enhanced solid-state anaerobic digestion of corn stover by alkaline pretreatment. Bioresour. Technol. 101(19), 7523-7528.

# 10

# Production and characterization of biodiesel using palm kernel oil; fresh and recovered from spent bleaching earth

Abiodun Aladetuyi[1], Gabriel A. Olatunji[1], David S. Ogunniyi[2], Temitope E. Odetoye[2*], Stephen O. Oguntoye[1]

[1]Department of Chemistry, University of Ilorin, P.M.B 1515, Ilorin, Nigeria.

[2]Department of Chemical Engineering, University of Ilorin, P.M.B 1515, Ilorin, Nigeria.

## HIGHLIGHTS

- *Utilization of agricultural residue; cocoa pod ash, in place of KOH catalyst for biodiesel production.*
- *Palm kernel oil was recovered from spent bleaching earth.*
- *Cocoa pod ash as catalyst led to higher yields of methyl esters from palm kernel oil and spent bleaching earth oil.*

## GRAPHICAL ABSTRACT

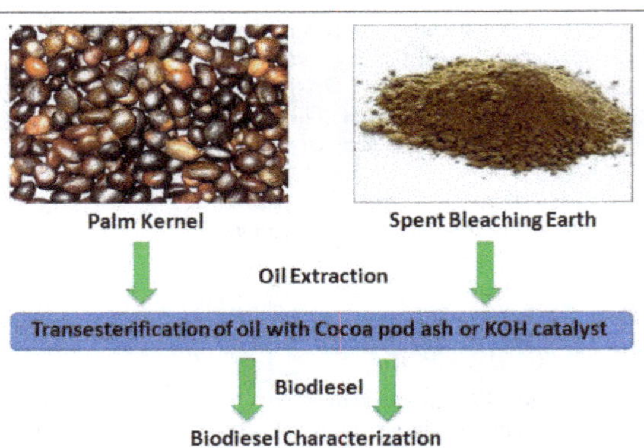


## ABSTRACT

Palm kernel oil (PKO) was recovered from spent bleaching earth with a yield of 16 %, using n-hexane while the fresh oil was extracted from palm kernel with n-hexane and a yield of 40.23% was obtained. These oils were trans-esterified with methanol under the same reaction conditions: 100 °C, 2 h reaction time, and oil-methanol ratio of 5:1 (w/v). The cocoa pod ash (CPA) was compared with potassium hydroxide (KOH) as catalyst. The percentage yields of biodiesel obtained from PKO catalysed by CPA and KOH were 94 and 90%, respectively. While the yields achieved using the recovered oil catalysed by CPA and KOH were measured at 86 and 81.20 %. The physico-chemical properties of the biodiesel produced showed that the flash point, viscosity, density, ash content, percentage carbon content, specific gravity and the acid value fell within American Society for Testing and Materials (ASTM) specifications for biodiesel. The findings of this study suggest that agricultural residues such as CPA used in this study could be explored as alternatives for KOH catalyst for biodiesel production.

**Keywords:**
Palm kernel oil
Spent bleaching earth
Agricultural residues
Cocoa pod ash
Novel basic catalyst
Biodiesel

## 1. Introduction

Fossil fuel depletion, concern for the environment and unstable crude oil prices have led to intensified search for alternative non-fossils fuels. Oil recovered from spent bleaching earth (SBE) and palm kernel (PK) has been found to be an alternative source of energy (Knothe et al, 2005). These oils cannot be used directly in internal combustion engine due to two main reasons: low volatility and high viscosity (Knothe et al, 2005; Atabani et al.,

* Corresponding author
E-mail address: todetoye@yahoo.com

2012). To overcome these problems, vegetable oils or animals fats are reacted with simple alcohol (methanol, ethanol, propanol) to produce fatty acid methyl esters (FAME), fatty acid ethyl esters (FAEE) and fatty acid propyl esters (FAPE) which are also known as biodiesel. The methods to achieve such conversion include pyrolysis, micro-emulsion and trans-esterification (Knothe et al, 2005). Among them, trans-esterification was found to be the best route with minimal engine complication (Knothe et al, 2005; Gupta *et al,* 2007; Shahid and Jamal, 2011).

Biodiesel is found to be the best substitute for petro-diesel, not only because of its comparable calorific value but also for its several other advantages such as: low toxic emissions, biodegradability, high flash point, excellent lubricity and environmental compatibility (Balat and Balat, 2010; Knothe et al, 2005). Biodiesel has been promoted and reported as a promising long term renewable energy source which has the potential to address net emission of carbon dioxide ($CO_2$) to the atmosphere, security concerns and the fluctuating prices of fossil fuels (Alamu et al., 2007a, 2007b ; Balat and Balat, 2010).

Triglycerides used for the production of biodiesel come from various sources: edible oils, non-edible oils, waste/used oils, animal fats and from microorganisms (Meng, et al; 2009). However, there is an economic sense in the use of waste oil feedstocks such as recovered oil from SBE or PK. Bleaching earth is used in vegetable oil refinery to remove colouring matters, soap, gums, metal (iron, nickel), oxidized compounds and polymers. Bleaching earth retains 20 – 40 % of oils and importantly, the adsorbed oil represents a considerable part of the bleached oil (Ong, 1983; Nursulihatimarsyila et al., 2010). It is estimated that about 600,000 metric tons or more of bleaching earth are utilized worldwide in the refining process based on the worldwide production of more than 60 million tons of oils (Kaimal et al., 2002).

SBE is usually disposed as waste by dumping in landfills without any attempts to recover the oil. Its disposal by incineration, inclusion in animal feeds and concrete manufacturing are sometimes practiced. Overall, SBE due to the substantial oil content in the earth, poses a serious threat to the environment by causing fire and releasing pollution hazards. Hence, it has been suggested that the oil in SBE be recovered and re-used as an alternative energy source which could also reduce the cost associated with refining processes (Nursulihatimarsyila et al., 2010).

Palm kernel oil (PKO) in its fresh form or recoverd from SBE can be used as biodiesel. Various studies have been carried out on the conversion of PKO and oil recovered from SBE to biodiesel using conventional homogenous catalysts (i.e. sodium hydroxide or potassium hydroxide) or heterogeneous catalysts e.g. calcium oxide through the trans-esterification reaction (Alamu et al., 2007a; Boey et al., 2009; Huang and Chang, 2010). In the present work, we report the use of cocoa pod ash (CPA) as catalyst in the trans-esterification of PKO and oil recovered from SBE to obtain biodiesel.

## 2. Materials and methods

### *2.1. Materials*

The two types of oil used included oil recovered from SBE and neat PKO. The SBE was obtained from Unilever Nigeria Plc., Agbara, Ogun State, Nigeria while the palm kernel seeds used were purchased from the Ajebamidele market at Odo-Oro Ekiti in Ikole local government area, Ekiti State, Nigeria. All solvents including n-hexane, methanol and ethanol were of analytical grade and purchased from Sigma Aldrich (England). The KOH catalyst used for the conversion of the oils to biodiesel was a commercial product obtained from BDH Chemical Limited, Poole, England. Cocoa pod was sourced for in a local cocoa farm at Odo-Oro Ekiti, Ekiti State, Nigeria and was subsequently converted to CPA.

### *2.2. Methods*

#### *2.2.1. Cocoa pod ash (CPA) preparation*

The cocoa pods were washed with distilled water, air dried and burnt. The burnt cocoa pods were converted into ash in a muffle furnace at a temperature of 550 °C for 50 min. The ash obtained was sieved to obtain a fine powder. The ash was analysed for its mineral content using an atomic absorption spectrometer (AAS) at International Institute of Tropical Agriculture (IITA) laboratory, Ibadan, Nigeria.

#### *2.2.2. Oil extraction*

The oil was recovered from the bleaching earth by cold-extraction with n-hexane solvent. One kg of fresh SBE was extracted with 2 L of n-hexane in a 4-L extracting jar, i.e. 2 L of n-hexane was poured into a 4-L jar containing 1kg of fresh SBE. The jar was covered and allowed to stay at room condition for a period of 72 h. After which the resulting solution was decanted off the jar and filtered using Whatman filter paper. The filtrate (oil + n-hexane) was then concentrated by distillation with the aid of a vacuum pump at room temperature to recover the SBE oil. 160 g of oil was recovered from the 1kg SBE giving an oil yield of 16%. Fresh n-hexane solvent was used for the extraction. The solvent recovered could be re-used, but in this work it was not re-used since the extraction was a one-batch experiment.

In the same way, dried PK was mechanically ground to fine granules. Then, 1.2 kg of the ground sample was cold-extracted with 2 L of n- hexane for 3 d; this was then filtered and concentrated in-vacuo to give 482.67 g palm PKO i.e. 40.23 % extraction yield.

#### *2.2.3. Determination of total alkali in the ash*

Determination of the total alkali in the CPA was achieved by titration with acid solution. 0.1M HCl acid solution was firstly standardized using 0.1M $Na_2CO_3$. The standardized 0.1M HCl acid solution was also titrated against 0.1M KOH as a reference basic solution. The standardized 0.1M HCl acid solution was then titrated against the CPA solution. Hence, the molar concentration of equivalent alkali in the CPA which corresponds to that of 0.5 g potassium hydroxide (KOH) used for the conversion catalyst was determined by stoichiometry.

#### *2.2.4. Transesterification procedure*

The method described by Alamu et al. (2007a) was adopted with some modifications. Fifty g of the oil recovered from SBE was heated for 20 min at 45 °C in a 100 ml, two-neck quick-fit round bottom flask on a water bath. Then, 0.5 g KOH, dissolved in 14 mL methanol was added through the second neck to the heated oil. The reaction mixture (oil-methanol ratio of 5.62:1 mol/mol (5:1 w/v)), was refluxed and stirred at 100 °C for 2 h. Preliminary experiments indicated poor yields at lower temperatures (especially below 60 °C)(data not shown). Therefore, trans-esterification was carried out under reflux at 100 °C that was higher than methanol boiling point. The reaction product was transferred into a separating funnel and allowed to stand for 12 h. After glycerol separation, the obtained FAMEs were washed three times with water and then dried over anhydrous $MgSO_4$.

While using CPA as catalyst, 1.09 g of CPA (equivalent amount of alkali) was leached in 14 mL of methanol, filtered with Whatman filter paper and also added through the second neck of the flask to the heated oil to produce biodiesel from SBE as described above. Similarly, the same procedure was followed as of PKO. The experiments were carried out in triplicates and the average methyl ester yield was determined. Biodiesel yield was estimated using the following equation (Eq. 1):

$$Y = Vp/Vs \times 100 \qquad \text{Eq. 1}$$

Where, Y is the yield (%), Vp represents the volume of product obtained (ml) and Vs is the volume of starting material (ml).

#### *2.2.5. Biodiesel fuel characterization*

The fuel-related properties of the extracted oil from the SBE, PKO and their corresponding biodiesel products, obtained with KOH and CPA catalysts, were determined according to ASTM standard methods. Comparisons were also made considering the European biodiesel standard EN14214, commercial petroleum diesel, and the ASTM specification for pure biodiesel (B100). The viscosity and specific gravity measurements were made using the Ostwald viscometer thermostatted at 40 °C and thermal-Hydrometer apparatus following the ASTM standards D445 and D1298, respectively. All measurements were performed in triplicates and the mean values were reported.

The flash points and percentage carbon contents of the oils and their corresponding biodiesels were determined at LUBCON Nigeria Ltd, Ilorin, using flash point apparatus and by pyrolysis following the ASTM standard D93 and D4530, respectively. The density measurements were obtained according to the ASTM standard test method D4052. The oxidative ash contents of the oils and biodiesel samples were determined through sulphated acid test in a muffle furnace at 700 °C until oxidation of carbon was complete in accordance with the ASTM standard test method D874. Acid value measurements were carried out using a modified method of the ASTM D664. The refractive index was determined using a refractometer at the Laboratory of Chemistry Department, University of Ilorin, Nigeria. The FAMEs compositions of the biodiesel were determined using a QP2010 PLUS, GC interfaced with a BG mode analytical mass spectrometer. Helium gas was used as carrier gas at 1.2 ml/mm. The MS operating conditions were: ionization voltage 70 ev, and ion source 230 °C. The expected and actual compound on each line of a given retention time was identified by choosing the hit number that corresponded with the mass spectra of the authentic samples and the most frequent ret-index number.

## 3. Results and discussion

The percentage yield of the residual oil recovered from EBE with n-hexane was 16%. The oil had a light yellow colour. Also, the percentage yield of the golden yellow oil extracted from PK was 40.34%. These results are comparable to those reported by other researchers (Boey *et al.*, 2009 ; Huang and Chang, 2010). Table 1 shows the concentrations of the metals present in the CPA compared to the values reported in the literature (Osinowo and Taiwo, 2001; Ayeni, 2011).

**Table 1.**
Metal contents present in CPA and the values reported in the literature.

| Metals | AAS result obtained (%) | Literature value (%)* |
|---|---|---|
| Potassium (K) | 11.33 | 16.07 |
| Sodium (Na) | 0.09 | 0.2 |
| Magnesium (Mg) | 0.57 | 1.00 |
| Calcium (Ca) | 0.04 | 5.40 |
| Iron (Fe) | Trace | 1.22 |

* (Osinowo and Taiwo, 2001; Ayeni, 2011)

The relatively high concentration of potassium (11.3 %) might be due to the use of CPA as potassium based fertilizer in the soil. The metal content values of the CPA obtained were compared with literature values (Osinowo and Taiwo, 2004; Ayeni, 2011). The difference in metal contents may be due to the difference in samples origin. Generally, the study revealed that the use of CPA as a catalyst gave the highest methyl ester yield of 94% compared to the 90% yield obtained through the KOH-catalysed reaction (Table 2).

**Table 2.**
Percentage yield of methyl ester obtained using KOH and CPA.

| Oil type | Catalyst | Biodiesel Sample | Yield (%) |
|---|---|---|---|
| PKO | KOH | PKO-KOH | 90.0 |
| | Cocoa pod ash | PKO-CPA | 94.0 |
| SBE Oil | KOH | SBE-KOH | 81.2 |
| | Cocoa pod ash | SBE-CPA | 86.0 |

The relatively high yield of biodiesel from the use of CPA catalyst is attributed predominantly to high content of potassium (K) as given by the AAS result analysis (Table 1). The oxide/hydroxide of potassium has the ability to co-catalyse the trans-esterification process. Oil recovered from SBE resulted in a lower yield of biodiesel with the use of both catalysts. This might be as a result of impurities introduced to the extracted oil from the SBE for the earth was initially used for refining purpose. Despite this hindrance, CPA catalyst led to a better biodiesel yield of 86% compared to 81.20 % obtained from KOH-catalysed process. The results obtained demonstrated that CPA is better catalyst than KOH for the production of biodiesel from SBE and PKO. The yield value of 90% for KOH-catalysed PKO biodiesel was quite comparable to the yield of 90 – 92% achieved by Alamu et al. (2007a) from the same oil feedstock.

### *3.1. Fuel-related properties*

#### *3.1.1. PKO and SBE oil*

The fuel-related properties of PKO and oil recovered from SBE are indicated in Table 3. These properties such as viscosity, density, specific gravity, ash content and refractive index values were higher for PKO compared to those of the oil recovered from SBE. However, the percentage carbon residue, flash point and the acid value of PKO were lower than those obtained for SBE oil (Table 3).

**Table 3.**
Fuel-related properties of PKO, SBE oil, and their corresponding biodiesels obtained using KOH and CPA catalysts. The ASTM specifications for B100 are also presented.

| Properties | PKO | SBE OIL | SBE KOH | SBE CPA | PKO KOH | PKO CPA | Petro-diesel | ASTM method | Biodiesl standard limits |
|---|---|---|---|---|---|---|---|---|---|
| Kinematic Viscosity @ 40°C (cSt) | 29.60 | 27.45 | 3.97 | 4.03 | 4.76 | 5.48 | 2.85 | D445 | 1.9 -6.0 |
| Flash point (°C) | - | 90 | 70 | 130 | 100 | 170 | 64 | D93 | 100 min |
| Density (g/cm$^3$) | 0.9720 | 0.7064 | 0.7064 | 0.7367 | 0.7679 | 0.8624 | 0.850 | D4052 | - |
| Ash content (w%) | 0.1160 | 0.0154 | 0.0156 | 0.0172 | 0.0040 | 0.0137 | - | D4530 | 0.80max |
| Refractive index | 1.447 | 1.338 | 1.327 | 1.326 | 1.443 | 1.443 | - | - | - |
| Carbon content (%) | 27.69 | 41.08 | 41.08 | 33.11 | 53.03 | 55.23 | - | D874 | 0.05max |
| Acid value (mgKOH/g) | 3.37 | 8.90 | 6.21 | 4.88 | 0.80 | 0.81 | - | D664 | 0.80max |

#### *3.1.2. PKO and PKO-biodiesel*

Table 3 tabulates the fuel properties of the extracted PKO and its biodiesel products. The results show that all the investigated fuel properties for the PKO had higher values than the corresponding biodiesel product. This is an indication of the conversion of the oil to methyl esters. The PKO methyl esters fuel properties were comparable to those of petroleum diesel and agreed with the ASTM standard values for B100, thereby making the biodiesel product suitable as a substitute for petroleum diesel. The viscosity of PKO (29.60 cSt) reduced to 4.76 cSt for PKO biodiesel catalysed with KOH and 5.48 cSt for CPA-catalysed biodiesel. These reductions correspond to percentage decrease of 83.92% and 81.49%, respectively which will also enhance the biodiesel fluidity in the engine. The flash point (170 °C) of CPA-catalysed PKO biodiesel was greater than that of PKO biodiesel catalysed with KOH. Both these values were greater than the PKO flash point and were in agreement with the ASTM flash point specification for B100.

Ash contents of KOH-catalysed PKO and CPA-catalysed PKO biodiesel products were lower than that of PKO but fell within the ASTM specification for B100. The densities of the PKO methyl esters catalysed with CPA and KOH were lower than the density of PKO. Only PKO CPA-catalysed methyl ester had a density of 0.8624gcm$^{-3}$, greater than that of petroleum diesel with (0.850gcm$^{-3}$). Also, the acid value of PKO methyl ester catalysed with KOH (0.80 mgKOH/g) was lower than that of the corresponding CPA-catalysed biodiesel (0.81 mgKOH/g), but acid value of PKO (3.37 mgKOH/g) was higher than the values obtained for the corresponding biodiesels. High percentage carbon residue was obtained for both CPA-catalysed PKO biodiesel and KOH-catalysed biodiesel.

### 3.1.3. SBE oil and SBE oil-biodiesel

All the examined fuel properties for the oil recovered from SBE had higher values than their corresponding biodiesel products. The viscosity of SBE oil (27.45 cSt) was reduced to 3.97 cSt and 4.03 cSt for KOH and CPA-catalysed biodiesels, respectively. These correspond to a percentage decrease of 85.54% for KOH-catalysed biodiesel and 85.32% for CPA-catalysed biodiesel. High specific gravity value for the SBE oil was obtained compared to its biodiesel product. Flash point of SBE oil (90 °C) was greater than its biodiesel product catalysed with KOH (70 °C) but was less than 130 °C obtained for CPA-catalysed methyl ester. Only SBE oil biodiesel catalyzed with CPA fell within the ASTM specification for B100. The ash content of the SBE oil (0.0154 w%) was less than those of the KOH and CPA-catalysed biodiesels at 0.0156 w% and 0.0172 w%, respectively. SBE oil and KOH-catalysed SBE oil biodiesel possessed the same density value of 0.7064 $gcm^{-3}$. This value was lower than that obtained for CPA-catalysed biodiesel. The density values of SBE oil biodiesels (0.7064 and 0.7367 $gcm^{-3}$) were measured lower than that of petroleum diesel (0.850 $gcm^{-3}$).

### 3.1.4. PKO biodiesel and SBE oil biodiesel

From Table 3, higher viscosity values were obtained for PKO biodiesel with the use of both KOH and CPA catalysts compared to SBE oil biodiesels. Also, high flash point values were obtained for PKO biodiesel with the use of both catalysts compared to SBE oil biodiesel catalysed with KOH. CPA-catalysed SBE oil biodiesel possessed a flash point greater than that of PKO biodiesel obtained by using KOH catalyst. Nevertheless, the flash point value of the latter also fell within ASTM specification for B100 and was higher than the flash point of petroleum diesel. The densities of PKO biodiesels obtained using KOH and CPA catalysts were higher than SBE oil biodiesels. The specific gravity of PKO methyl ester obtained using CPA catalyst was measured higher than that of the SBE oil biodiesels catalysed by KOH and CPA. All specific gravity values obtained fell within the ASTM (D1298) specification for B100.

With the use of both catalysts, the ash contents of the PKO biodiesels were lower than that of the SBE oil biodiesels and were in line with the ASTM (D4530) specification for B100, thereby marking the biodiesel product a suitable substitute for petroleum diesel. The percentage carbon contents of the PKO biodiesels were higher than those of SBE oil biodiesels. Relatively higher acid values of SBE oil methyl ester compared to the values obtained for the PKO biodiesels catalysed with KOH and CPA could be attributed to the high free fatty acids content of the SBE oil and probably due to the impurities picked up by the oil in the course of refining process.

From the economic point of view, the biodiesel raw materials used in this study i.e. CPA and SBE oil could make the biodiesel production more affordable and economical for CPA is an agricultural residue and SBE is an industrial waste. Utilization of these wastes in the production of biodiesel could also have a double waste management effect on the environment while the biodiesel produced is preferred over its petroleum counterpart owing to its environmental advantages.

Notwithstanding, more effort needs to be directed towards optimizing the extraction process of SBE oil to make it more economically viable including recovery and re-use of the solvent used in extraction.

### 3.2. GC-MS analyses

Figures SI to S3 shows the gas chromatogram of the SBE oil biodiesels obtained using CPA. The GC-MS results also confirmed that higher conversion was achieved in biodiesel production from the PKO compared to biodiesel production from SBE oil when the CPA catalyst was used (Tables 4 and 5).

**Table 4.**
Chemical compositions (%) of the transesterified PKO using CPA catalyst.

| Compounds | Composition (%) | Ret index |
|---|---|---|
| Methyl octanoate | 2.52 | 1083 |
| Methyl nonanoate | 0.34 | 1183 |
| Methyl decanoate | 3.10 | 1282 |
| Methyl decanoate | 0.44 | 1282 |
| Methyl dodecanoate | 19.49 | 1481 |
| Methyl tetradecanoate | 9.97 | 1680 |
| 11-Hexadecenal | 1.67 | 1808 |
| 14-methyl pentadecanoate | 10.40 | 1814 |
| Hexadecanoic acid | 4.93 | 1968 |
| Methyloctanoate | 0.60 | 2077 |
| 12-Methyl octadecanoate | 17.90 | 2085 |
| Octadecanoic acid | 9.74 | 2167 |
| 6-Octadecenoic acid | 14.69 | 2175 |
| 2-hydroxyl-l-Methylhexadecanoate | 2.18 | 2498 |
| 3-hydroxyl-9-Methyl octadecenoate | 2.62 | 2527 |

**Table 5.**
Chemical composition (%) of the transesterified SBE oil using CPA catalyst.

| Compounds | Composition (%) | Ret index |
|---|---|---|
| Hexanoic acid | 0.78 | 974 |
| Heptanoic acid | 1.13 | 1073 |
| Octanoic acid | 2.06 | 1173 |
| Nonanoic acid | 3.68 | 1272 |
| Decanoic acid | 0.80 | 1380 |
| Decenoic acid | 0.29 | 1380 |
| Heptylhexanoate | 5.41 | 1481 |
| Dodecanoic acid | 0.82 | 1570 |
| Dodecenoic acid | 0.30 | 1578 |
| Methyltetradecanoate | 0.02 | 1680 |
| Tetradecanoic acid | 1.75 | 1769 |
| Pentadecanoic acid | 0.27 | 1869 |
| Methylhexadecanoate | 2.00 | 1878 |
| Hexadecanoic acid | 15.63 | 1968 |
| 12-methyloctadecenoate | 5.50 | 2085 |
| 9-octadecenoic acid | 30.40 | 2175 |
| 2, 3-dihydroxylhexanoate | 4.05 | 2482 |
| Tetradecylhexadecanoate | 4.74 | 3171 |
| 2-hydroxyl-1,3-propanoate | 12.22 | 3997 |
| 9-methyloctadecenoate | 8.17 | 6149 |

## 4. Conclusion

This study shows that, CPA as trans-esterification catalyst led to higher yields of methyl esters with both PKO and SBE oil as feedstock. This demonstrates the alkaline strength of the potassium contained in the ash. Hence, biodiesel produced from PKO and SBE oil using CPA could be a potential substitute for petroleum diesel and may be blended with petroleum diesel. All fuel properties examined fell within the ASTM specification for B100. This investigation also suggests that agricultural residues could be explored and utilized in place of conventional catalysts e.g. NaOH/KOH which in turn could improve the economic features of biodiesel production processes.

**References**


Alamu, O.J., Waheed, M.A., Jekayinfa, S.O., 2007a. Biodiesel Production from Nigerian Palm Kernel Oil: Effect of KOH concentration on yield. Energy Sustain. Dev. 11(3), 77-82.

Alamu, O.J., Waheed, M.A., Jekayinfa, S.O., 2007b. Alkali-catalysed Laboratory Production and testing of biodiesel fuel from Nigeria Palm Kernel Oil. Agricultural Engineering International. CIGR J. 9, 1–11.

Atabani, A.E., Silitonga, A.S., Badruddin, I.A., Mahlia, T.M.I., Masjuki, H.H., Mekhilef, S., 2012. A comprehensive review on biodiesel as an alternative energy resource and its characteristics. Renew. Sustain. Energy Rev. 16, 2070.

Ayeni, L.S., 2011. Effect of Soil and Combined Cocoa pod ash, Poultry Manure and N.P.K 20:10:10 Fertilizer on Soil Organic Carbon Available P and forms of Nitrogen on Alfisols in South Western Nigeria. Int. Res. J. Agric. Sci. 1(2), 77.

Balat, M., Balat, H., 2010. Progress in biodiesel processing. Appl. Energy. 87(6), 1815.

Boey, P.L., Gaanty, P.M., Shafida, A.H., 2009. Biodiesel from Adsorbed Waste Oil on Spent Bleaching Clay using CaO as Heterogeneous Catalyst. Eur. J. Sci. Res. 33, 347.

Gupta, P.K., Kumar, R., Panesar, B.S., Thapar, V.K., 2007. Parametric Studies on Biodiesel prepared from Rice Bran Oil. Agric. Eng. Int.: CIGR EJ 2007; IX(April) [Manuscript EE 06 007].

Huang, Y., Chang, J.I., 2010. Biodiesel production from residual oils recovered from spent bleaching earth. Renew. Energy. 35(1), 269.

Kaimal, I.N.B., Vigayalakshmi, P., Laximi, A.A., Ramakinga, B., 2002. Process for Simultaneous Conversion of Adsorbed Oil to Alkyl Esters and Regeneration of Commercial Spent Bleaching Earth for Reuse. US Patent, 0115875Al.

Knothe, G., Gerpen, J.V., Krahl, A., 2005. Introduction to Biodiesel Handbook. American Oil Chemists Society Press, Urbana. II: 1-3.

Meng, X., Yang, J., Xu, Z., Zhang, L., Nie, Q., Xian, M., 2009. Biodiesel Production from Oleaginous Micro-organisms. Renew. Energy. 34, 1.

Nursulihatimarsyila, A.W., Cheah, K.Y., Chuah, T.G., Siew, W.L., Choong, T.S.Y., 2010. Deoiling and Regeneration Efficiencies of Spent Bleaching Clay. Am. J. Appl. Sci. 7 (3), 434.

Ong, J.T.L., 1983. Oil Recovery from Spent Bleaching Earth and Disposal of the extracted Material. J. Am. Oil Chem. Soc. 60, 314.

Osinowo, F.A., Taiwo, E.O., 2001. Evaluation of Various Agro-wastes for Traditional Black Soap Production. Biores. Technol. 79, (1) 95.

Shahid, E.M., Jamal, J., 2011. Production of biodiesel: a technical review. Renew. Sustain. Energy Rev. 15(9), 4732.

**Supplementary Data**

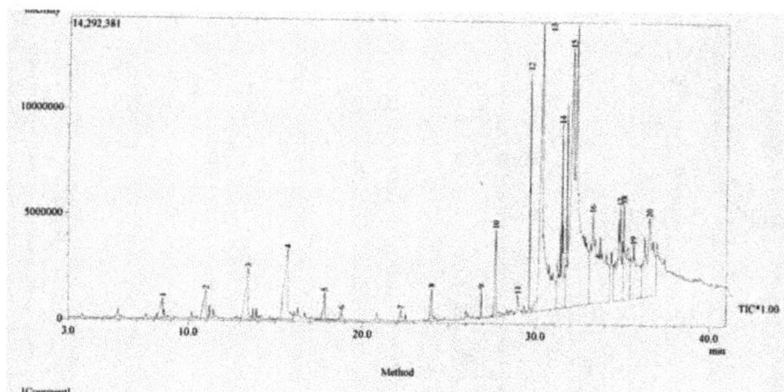

Fig. S1. Gas chromatogram of SBE oil biodiesels obtained using CPA catalyst.

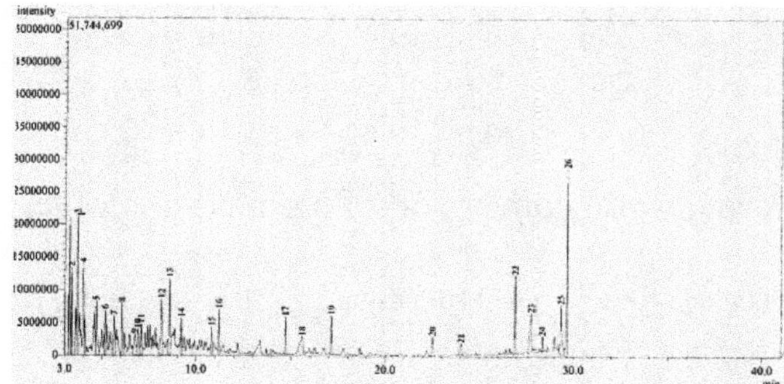

Fig. S2. Gas chromatogram of SBE biodiesels obtained using KOH catalyst.

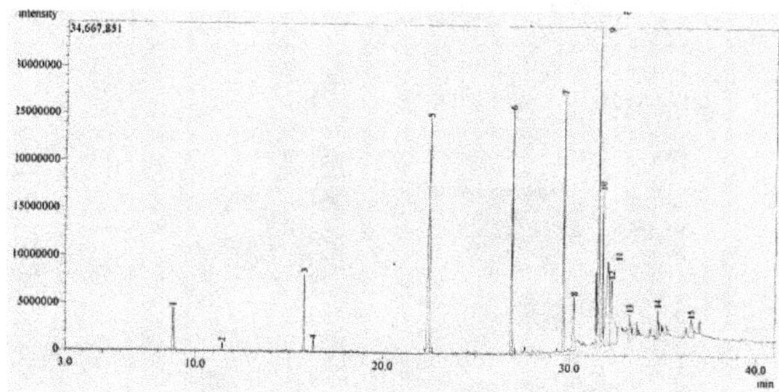

Fig. S3. Gas chromatogram of PKO biodiesels obtained using CPA catalyst.

# 11

# A review on conversion of biomass to biofuel by nanocatalysts

Mandana Akia[1, 2*], Farshad Yazdani[1, 2], Elahe Motaee[1, 2], Dezhi Han[3], Hamidreza Arandiyan[4]

[1]*Chemistry & Chemical Engineering Research Center of Iran (CCERCI), P.O. Box 4335-186, Tehran, Iran*

[2]*Nano/catalysts Research Group, Biofuel Research Team (BRTeam), Karaj, Iran*

[3]*Key Laboratory of Biofuels, Qingdao Institute of Bioenergy and Bioprocess Technology, Chines Academy of Sciences, Qingdao 266101, PR China*

[4]*State Key Joint Laboratory of Environment Simulation and Pollution Control (SKLESPC), School of Environment, Tsinghua University, Beijing 100084, China*

**HIGHLIGHTS**

➢Significant improvements in biomass conversion using nanocatalysts.
➢Feasibility of utilization milder operating conditions by using nanocatalysts compared to the bulk catalysts.
➢The role of nanocatalysts to overcome some challenges in biomass conversion, improving the products quality.

**GRAPHICAL ABSTRACT**

Products from nanocatalytic conversion of biomass

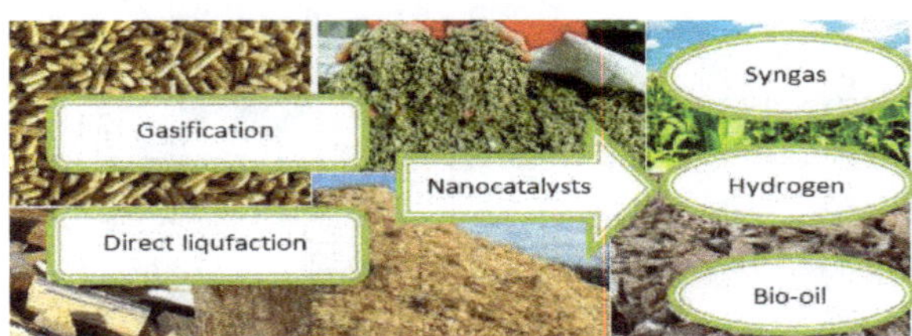


**ABSTRACT**

The world's increasing demand for energy has led to an increase in fossil fuel consumption. However this source of energy is limited and is accompanied with pollution problems. The availability and wide diversity of biomass resources have made them an attractive and promising source of energy. The conversion of biomass to biofuel has resulted in the production of liquid and gaseous fuels that can be used for different means methods such as thermochemical and biological processes. Thermochemical processes as a major conversion route which include gasification and direct liquefaction are applied to convert biomass to more useful biofuel. Catalytic processes are increasingly applied in biofuel development. Nanocatalysts play an important role in improving product quality and achieving optimal operating conditions. Nanocatalysts with a high specific surface area and high catalytic activity may solve the most common problems of heterogeneous catalysts such as mass transfer resistance, time consumption, fast deactivation and inefficiency. In this regard attempts to develop new types of nanocatalysts have been increased. Among the different biofuels produced from biomass, biodiesel has attained a great deal of attention. Nanocatalytic conversion of biomass to biodiesel has been reported using different edible and nonedible feedstock. In most research studies, the application of nanocatalysts improves yield efficiency at relatively milder operating conditions compared to the bulk catalysts.

**Keywords:**
Biomass
Biofuel
Nanocatalysts
Gasification
Liquefaction

## 1. Introduction

The demand for energy is increasing in the world due to the rapidly growing global population and urbanization. Throughout history, mankind has used wood as a source of producing energy. After the industrial revolution, the main source of energy shifted to fossil fuels. The accurate amounts of the world's total fossil fuel reserves are not known. However, it has been forecasted by the International Energy Agency that the global peak

* Corresponding author
E-mail address: akia.mandana@gmail.com (M. Akia).

in oil production will be between 1996 and 2035 (Kumar et al., 2009; Hansen et al., 2010; Liu et al., 2012; Demirbas, 2008; Sinag, 2012; Shirazi et al., 2013; Hamze et al., 2013a, 2014b). Furthermore, the increase in crude oil prices and the pollution caused by petroleum-based energy sources have created serious environmental problems, e.g. global warming. Such concerns about fossil fuels have led to the utilization of alternative energy sources (Dominik and Janssen, 2007; Demirbas, 2008). The primary alternative sources of energy systems that can replace fossil fuels are water, wind, solar energy, and biomass. Currently biomass is gaining a great deal of attention in terms of supplying the world's energy demands. Due to its availability and environmentally friendly nature such as causing no net increase in carbon dioxide levels and producing very low amounts of sulfur, biomass energy is believed to contribute one half of the total energy demand in industrial countries by 2050. Annually, approximately 27 billion tons of $CO_2$ is emitted from the burning of fossil fuels, and this is predicted to increase about 60% by 2030. Therefore, using bioderived fuels is crucial to reducing the carbon footprint (Demirbas, 2008).

Biomass energy is supplying about 10-15% (or 45± 10 EJ) of today's demand. Biomass feedstock include a broad range of organic material such as wood, wood-based energy crops, corn stover, grass, algae, wheat straw, rice straw, corn, miscanthus, nonedible oils, green and wood landfill waste, animal fats, waste frying oils, agricultural residues, municipal wastes, forest product wastes, paper, cardboard and food waste. Most biomass is produced through photosynthesis. The photosynthesis yield of approximately 720 billion tons per year, make the largest organic raw material cellulose resource in the world (Demirbas, 2008; Luo and Zhou, 2012). In general all kinds of biomass can be used as feedstock including starchy, triglyceride and lignocellulosic feedstock. Biomass can be converted to biofuels and biopower via thermochemical and biochemical processes. Thermochemical conversion is a significant route for producing products such as bio-methanol, biodiesel, bio-oil, bio-syngas and biohydrogen. Biochemical conversion routes produce liquid or gaseous fuels through fermentation or anaerobic respiration. The production of biofuels through thermochemical conversion processes with a broad range of technologies has drawn the most attention in the world. The main advantage of thermochemical processes for biomass conversion in comparison to other methods such as biochemical technologies is the feedstock used (Mitrovi et al., 2012; Kumar and Tyagi, 2013; Kumar et al., 2009). There are three main routes for biomass' thermochemical conversion including combustion, gasification and pyrolysis. Combustion is the most direct and technically easiest process which converts organic matter to heat, carbon dioxide and water using an oxidant. Gasification of biomass is a heating process within the presence of an oxidant produces a mixture of carbon monoxide and hydrogen referred to as synthesis gas (syngas) by partial oxidation. Gasification has many advantages over combustion. It can use low-value feedstock and convert them into electricity and also vehicle fuels. Within the forthcoming years, it will serve as a major technology for complementing the energy demands of the world (Alonso et al., 2010; Sinag, 2012).

Pyrolysis is a thermal heating of materials in the absence of oxygen, which produces three forms of products including gases, pyrolytic oil and char. Pyrolytic oil, also known as "tar or bio-oil", which is viscous, corrosive, relatively unstable and chemically very complex, cannot be used as transportation fuel directly due to its high oxygen value (40-50 wt%), high water content (15-30 wt%) and also low H/C ratios. Biomass gasification/pyrolysis is one of the promising technologies used for converting biomass to bioenergy (Hansen, 2010; Liu et al., 2010; Aradi et al., 2010a).

One of the most important challenges of the 21st century is the use of nanoscience in the development of sustainable and renewable energy production schemes. In the first decade of the current century, the formation of new fields in catalyst science, named nanocatalysis, has attracted everyone's attention. In general having a large surface to volume ratio of nanoparticles compared to bulk materials, makes them excellent potential catalysts (Aravind and Jong, 2012;Malik and Sangwan, 2012; Kim et al., 2013).The effect of changes in acidic properties, type of metal content and porosity of catalysts are widely known in biomass conversion. However to improve the products' quality, new types of catalysts have been developed. The catalyst utilization, blending of biomass with coal before firing to improve the quality of products and optimization of the experimental conditions are some important attempts. Also in such applications, increased surface area nanoparticles are extremely attractive candidates (Liu et al., 2010; Wilcoxon, 2012).

There are several methods to prepare nanosized materials. These materials may be used directly or in the form of supported nanoparticles on solids such as oxides, carbon or zeolites. Some usual methods for nanocatalysts preparation are impregnation, precipitation, chemical vapor deposition, and electrochemical deposition. Precipitation and impregnation methods are simple, cheap, and well-studied but it is difficult to control the size of particles. Chemical vapor deposition is widely used in the electronics industries but it is an expensive method. Electrochemical deposition is an inexpensive method that does not need high temperature and concentration. This method would allow a good control on size and chemical properties of the deposited nanomaterials but usually forms one dimensional nanomaterials (Liu et al., 2010; Aradi et al., 2010a; Wilcoxon, 2012). In most research, impregnation and precipitation methods have been used for biomass catalysts preparation. Meanwhile, nanoparticles properties usually adjust by changing synthesis parameters such as pH of the solution, concentration, the reducing agent, and calcination temperature (Pleisson et al., 2012).

Recent catalyst developments have led to the upgrade of biomass gasification processes to increase the syngas production and reduce the tar formation. In catalytic biomass liquefaction the main aims are to increase liquid yield and quality of products. Furthermore, nanocatalyst characteristics, such as high catalytic activities and high specific surface area have helped overcome some limitations on heterogeneous catalysts for their application in biodiesel production from biomass. In this paper the latest progress in nanocatalytic conversion of biomass has been reviewed. However, research studies have not been extended in all biomass conversion fields.

## 2. Conventional technologies for biomass to biofuel conversion

The process of refining lignocellulosic feedstock to hydrocarbon biofuels can be divided into two general parts. Whole biomass is deconstructed to provide upgradeable gaseous or liquid platforms. This step is usually applied through thermochemical methods to produce synthesis gas (by gasification) or bio-oils (by pyrolysis or liquefaction), or through the hydrolysis route to provide sugar monomers that then deoxygenated to form upgraded intermediates. Thermochemical conversion process is a major method of biomass upgrading for biofuels production, offering a wide range of potential technologies (Hansen et al., 2010). In the following section gasification and direct liquefaction processes have been described as two main thermochemical conversion methods.

### *2.1. Gasification*

Gasification processes provide a competitive route for converting various, highly distributed and low-value lignocellulosic biomass to synthetic gas for generation of a broad sort of outputs: electricity, heat and power, liquid fuels, synthetic chemicals as well as hydrogen production ($H_2$). The importance of gasification is that it is not constrained to a particular plant-based feedstock, and thus any lignocellulosic biomass can be considered appropriate (Luo and Zhou, 2012; Demirbas, 2008).Gasification of biomass is classified into two different ways, Low-temperature gasification (LTG) and high-temperature gasification (HTG).Production of hydrogen and synthesis of gas conducted through the LTG process is an attractive method, especially for low calorific value biomass, such as livestock manure compost and waste activated sludge. The advantages of LTG process is its easiness and efficient operation while avoiding ash problems such as sintering, agglomeration, deposition, erosion and corrosion. Moreover the tar components in LTG include lighter hydrocarbons which are different from the ones used in the HTG process (Ozaki et al., 2012). Advanced gasification processes are currently being investigated for co-production of liquid fuels in research activities and also in pilot plant scales.

The gasification process includes a sequence of interconnected reactions as it can be seen in Figure 1. The first step includes a quick drying process. Fast pyrolysis takes place in the second step which is a thermal conversion to char and gas products. The final step is gasification, namely, partial oxidation reaction between pyrolysis production and oxygenant. In this process the char is oxidized with an oxygen source (Luo and Zhou, 2012; Bridgwater, 2003;

Nordgreen, 2011).The most frequently used oxidants for gasification process are pure $O_2$, air, $CO_2$, steam or a mixture of these components (Nordgreen, 2011).

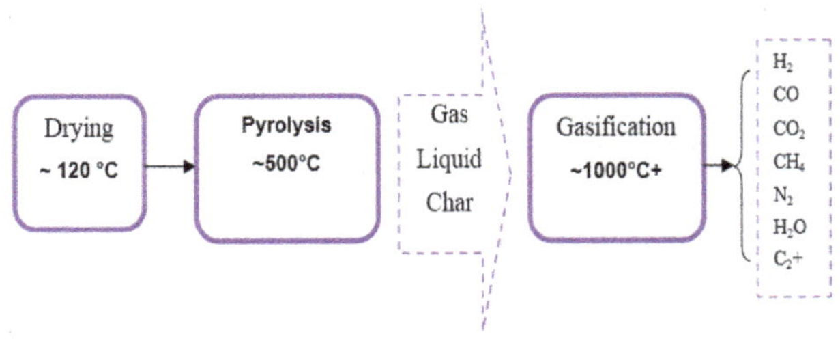


**Fig.1.** Gasification steps.

The gasification products mainly consist of carbon monoxide (CO), carbon dioxide ($CO_2$), hydrogen ($H_2$), methane ($CH_4$), and water ($H_2O$). During the gasification process, part of the biomass is converted to char particles, tars, and ash instead of syngas. Reduction and conversion of char and tar is one of the major issues in biomass gasification which also increase syngas yield and overall conversion efficiency. The produced syngas quality depends on many factors such as raw material composition, water content, temperature, heating rate, type of gasifier, and oxidation of pyrolysis products. Predicting the exact composition of gasifier products is difficult. One of the methods for theoretically gas composition determination is obtained by using the water-gas equilibrium concept at a given temperature (Kumar et al., 2009; Sinag, 2012; Luo and Zhou, 2012; Asadullah, 2014; Nordgreen, 2011).

Some of the major gasification reactions are as follows:

**Exothermic Reactions:** (Eq. 1-4)

Oxidation:

$$C+O_2 \rightarrow CO_2 \quad \text{(Eq. 1)}$$

Partial oxidation:

$$2C+O_2 \rightarrow 2CO \quad \text{(Eq. 2)}$$

Water gas-shift (WGS):

$$CO+H_2O\,(g) \rightarrow CO_2+H_2 \quad \text{(Eq. 3)}$$

Methanation:

$$C+2H_2 \rightarrow CH_4 \quad \text{(Eq. 4)}$$

**Endothermic Reactions:** (Eq. 5 and 6)

Water gas reaction:

$$C+H_2O\,(g) \rightarrow CO+H_2,\; C+2H_2O\,(g) \rightarrow CO_2+2H_2 \quad \text{(Eq. 5)}$$

Boudouard reaction:

$$C+CO_2 \rightarrow 2CO \quad \text{(Eq. 6)}$$

Heat can be provided directly or indirectly to satisfy the necessities of the endothermic reactions. Pyrolysis and gasification reactions will be conducted in a single vessel in directly heated gasification. The required heat for the endothermic reactions can be provided by using air or oxygen as an oxidant, which combusts a portion of the biomass (Eqs. 1, 2).High purity syngas (i.e. low quantities of inerts like $N_2$) is extremely beneficial for fuels and chemicals synthesis since it will reduce the size and cost of downstream equipment (Ciferno and Marano, 2002).

The transformation of biomass into hydrogen rich gas provides a competitive means for producing energy and chemicals from renewable resources. Hydrogen production is very important for solving two major energy problems including reducing dependence on petroleum and reducing pollution and greenhouse gas emissions (Rapagna et al., 1998). Tar elimination and CO conversion to levels less than one volume percent by means of the WGS reaction is important for hydrogen production. Producing hydrogen from biomass consists of multiple reaction steps. In the production of high purity hydrogen, the reforming of fuels is followed by two WGS reaction steps, and the final steps are carbon monoxide purification and carbon dioxide removal. Steam gasification processes appear to be optimum candidates which provide an economical, reliable and convenient process for extracting hydrogen from the produced gas. During the steam gasification process, the chemical energy of biomass can be converted into enriched hydrogen syngas containing up to about 50% by volume of hydrogen on a dry basis. Steam can be achieved from the dehydration reactions of crop residues or from an external source. Steam can react with carbon monoxide to produce hydrogen and carbon dioxide. WGS reaction is the principle reaction in the steam gasification system (Chang et al., 2011; Rapagna et al., 2002; Asadullah et al., 2002). Hydrogen can also be produced from biomass by pyrolysis. Biological technologies for hydrogen generation (biohydrogen) provide a wide range of routes for producing hydrogen. Moreover biological processes are considered to be more environmentally friendly and less energy intensive. In general, biomass gasification and the subsequent production of synthetic fuels (diesel fuel, methanol, dimethylether, etc.) requires complex and more expensive technologies compared to conventional petrochemical processes or current direct liquefaction. Therefore it is essential to ensure about the economic viability of biomass gasification plants (Demirbas, 2008; Hansen et al., 2010).

### *2.2. Direct Liquefaction*

Biomass can be converted to liquefied products through combined physical and chemical reactions, the technology being called direct liquefaction. In these processes the biomass macromolecules are decomposed to small molecules through heating and sometimes in the presence of a catalyst. Direct liquefaction may be divided into pyrolysis and liquefaction methods. Although both are thermochemical conversion methods, their operating conditions are different. The operation temperature in the liquefaction method is lower (250-325°C) but the operation pressure (5-20MPa) is higher than that of pyrolysis. In pyrolysis, unlike liquefaction, it is necessary to dry biomass before feeding (Xu et al., 2011).

#### *2.2.1. Liquefaction*

Two main routes can be considered industrially for the liquefaction of biomass to bio-oils and these include, hydrothermal liquefaction, and catalytic liquefaction (Vertes et al., 2010).Hydrothermal liquefaction is based on the superior properties of water at higher temperatures and pressures. The reactivity of biomass is considerable in water especially under hydrothermal conditions. Biomass consists of components with polar bonds which are attacked by the polar molecules of water. At elevated temperatures and pressures these attacks are more sever. As a result, hemicellulose and cellulose are hydrolyzed very quickly at these conditions (Kruse and Dinjus, 2007). Hydrothermal liquefaction has another important advantage. Usually all biomass sources are wet and it is possible that their water content be at a range of up to 95wt. %. In most biomass upgrading methods it is necessary to dry feeds before processing. In hydrothermal liquefaction conversions of biomass perform with its high water content. The water content of the biomass not only is not a disadvantage but it is also useful by reducing the process' required fresh water. Using water as both reactant and solvent in the liquefaction has some other benefits as well. The degradable products of the process are completely soluble in water under elevated temperatures and pressures, which prevent any polymerization. In addition, no solid products, such as coke and char, are formed because water acts as both a reactant and solvent in hydrothermal liquefaction. Water in liquefaction has another important advantage. It is an economic and environmentally friendly solvent, because it will not produce pollution. Mixed solvents (ethanol-water) (Liu et al., 2013) and also other pure solvents (ethanol) (Zheng et al., 2013) have been used in thermochemical conversion of biomass. However, these solvents increase the operating costs of the biomass conversion.

Catalytic liquefaction is similar to hydrothermal liquefaction. However, using a catalyst brings some advantages to the biomass process. Catalyst reduces the residence time and operating temperature and pressure. Catalyst has useful effects on hydrothermal products, it increases liquid products and reduces gaseous one. Also a catalyst can improve liquid products quality (Vertes et al., 2010). As an instance, hydrothermal liquefaction of a microalgae with heterogeneous catalysts showed improvements in products quality (Duan and Savage, 2011). Different heterogeneous catalysts (Pd/C, Pt/C, Ru/C, Ni/$SiO_2$-$Al_2O_3$, CoMo/$\gamma$-$Al_2O_3$ (sulfide), and zeolite) have been used in hydrothermal liquefaction of the microalgae nannochloropsis sp., at 350°C. It was seen that the crude bio-oils produced from liquefaction with these catalysts, except zeolite, flowed easily and were much less viscous than the biocrudes of noncatalyzed liquefaction.

*2.2.2. Pyrolysis*

Thermal, anaerobic decomposition of biomass at temperatures of 377-527°C is called pyrolysis. A temperature of at least 400°C in pyrolysis process is needed to completely decompose the organic structure of the biomass into monomers and oligomers fragments. The noncondensable portion of pyrolysis products rise by increasing temperature to above 600°C. Pyrolysis operations are based on the size of biomass feeds and are divided into two main processes, slow pyrolysis and fast pyrolysis. The slow pyrolysis can disport to conventional charcoal production and intermediate pyrolysis (Vertes et al., 2010).In conventional charcoal production, large pieces of wood are slowly heated to 400°C for a long time (up to 18 hr). The sole product of such process is charcoal when wood was used as a raw material, in the conventional kilns. However, in large retorts with capacities of $100m^3$ and more, which is used in conventional industrial charcoal production, non-solid products are also achieved. Refining facilities are combined with pyrolysis units to collect and condense gas products. Nevertheless, more than 65% of pyrolysis products are solid (charcoal) and less than 20% are liquid, in conventional charcoal production (Luo and Zhou, 2012; Henrich, 2005).

Intermediate pyrolysis differs from conventional charcoal production in terms of biomass residence time. Wooden feeds are entered into the screw tubular kiln and moved forward by screw rotation. The pyrolysis temperature of 380-400°C is initiated in the kiln by transferring heat from the wall of the kiln, shaft of the screw and also heat carriers. The carriers are balls of various materials with small size. With such heat transfer arrangements, the biomass heated faster than in conventional charcoal production but not as quickly as fast pyrolysis, which is why this process is called intermediate. Another difference between these two pyrolysis methods is related to the products quality. The solid portion of intermediate pyrolysis reduced to about 35%, while the liquid products increased to more than 45% (Schnitzer et al., 2007).

In fast or flash pyrolysis grained biomass with less than 3mm diameter is converted to a combustible liquid fuel in one simple step. The dry feed (less than 20% moisture content) quickly mixes with grainy heat carrier of sand, steel shot, or etc. at approximately500°C. More than 70% of pyrolysis products are condensed to liquid due to their fast heating and vapor condensation of less than two seconds. The pyrolysis condensates show some hydrophilic behavior caused by their high oxygen content, up to 45%. This behavior makes some problems when these condensates mix with usual hydrocarbon fuels or form two phase liquids. However, this may be solved by mixing the condensates with methanol or ethanol. Furthermore, more oxygen content means lower energy content that is not desirable. One method of reducing oxygen in organic tar is increasing water in the reactor which means entering excess hydrogen in reaction. But the laboratory experiments did not provide efficient results. Another promising method is catalytic deoxygenation of hot stream of biomass pyrolysis products (Xu et al., 2011; Vertes et al., 2010).

In a research work (Malik and Sangwan, 2012) fast pyrolysis products have been used to generate electricity. It is shown that generation of electricity by pyrolysis products is more beneficial than that of any other biomass conversion method in the long term and has a lower cost. In a recent work macroalgae powder has been converted to bio-oil by fast pyrolysis method in a free fall reactor. Macroalgae (Enteromorphaprolifera) has been converted into bio-oil by this process at a temperature range of 100-750°C. Characteristics of the resulted bio-oil were investigated and seen that the average heat value and oxygen content were 25.33 MJ $kg^{-1}$ and 30.27wt. % respectively (Zhao et al., 2013).

## 3. Application of nanocatalysts in biofuel production

*3.1. Nanocatalysts for biomass gasification*

In biomass gasification, preventing tar and char formation is an important issue. Tar is a complex mixture of condensable hydrocarbons including aromatic compounds of single ring to 5-ring along with other oxygen containing hydrocarbons and complex polycyclic aromatic hydrocarbons (PAHs). The boiling temperature of tar is high and it condenses at temperatures below 350°C which creates major problems such as corrosion or failure of engines as well as blockage of pipes and filters. Tars may also act as poison for catalysts. Biomass chars are highly disordered carbonaceous materials with a short-range polycrystalline structure which consist of small aromatic structural units (Asadullah et al., 2002; Li et al., 2008; Luo and Zhou, 2012; Duman et al., 2013a, 2014b; Nordgreen et al., 2011). Two main approaches employed for controlling the production of tar are, including treatments inside the gasifier (primary methods) and hot gas cleaning after the gasifier (secondary methods). Although the secondary methods are effective, primary methods are also gaining much attention because of economic benefits. The most important parameters in the primary methods are including temperature, gasifying agent, equivalence ratio, residence time and catalysts which have significant effects on tar formation and decomposition. The primary methods have not been applied commercially because they are not still fully understood (Balat, 2009; Luo and Zhou, 2012).

The effect of catalysts on gasification products is very important. Catalysts not only reduce the tar content; but also improve the quality of gas products and the conversion efficiency. The presence of a catalyst decreased the char yield during the final step of the gasification process while it increased the char formation during the volatilization stage (Balat, 2009; Aradi et al., 2010a, 2011b).The successful gasification catalysts have some criteria including being effective at removing tars, being resistant to deactivation as a result of carbon fouling or sintering, can easily be regenerated and are inexpensive (Sutton et al., 2001; Wilcoxon, 2012; Aravind and Jong, 2012).

Char formation during the pyrolysis step of gasification can be expressed by the following equation (Eq. 7):

$$\text{Biomass} \rightarrow \text{Volatiles} + \text{char} \qquad \text{(Eq. 7)}$$

Depends on using steam, oxygen or $CO_2$ as gasification agent, conversion of the residual chars can be presented sequentially as follows (Duman et al., 2014)(Eq. 8-10):

$$\text{Char } C + H_2O \rightarrow CO + H_2 \qquad \text{(Eq. 8)}$$

$$2C + 3/2\ O_2 \rightarrow CO + CO_2 \qquad \text{(Eq. 9)}$$

$$C + CO_2 \rightarrow 2CO \qquad \text{(Eq. 10)}$$

Potassium, sodium and calcium have been found to be the most effective catalysts for promoting char gasification in steam or carbon dioxide media. There may be other metallic species beneficial for biomass conversion, although some elements present in waste biomass may prevent char gasification by poisoning the catalysts (Nzihou et al., 2013).

Catalytic tar cleaning is potentially attractive because no additional input energy is necessary. The important reactions during tar reduction include steam reforming, dry reforming, thermal cracking, hydrocracking, hydro reforming and WGS reactions (Anis and Zainal, 2011; Han and Kim, 2008). The proposed reactions are as follows (Nordgreen, 2011) (Eq 11-17).

In general catalysts for tar conversion are classified into mineral or synthetic. Mineral includes calcined rocks, olivine, clay minerals and ferrous metal oxides. Transition metals, activated alumina, alkali metal carbonates, FCC catalysts and chars are the main synthetic catalysts.

| | | |
|---|---|---|
| tar→ $C^* + C_nH_m$ + gas | tar cracking (C* carbon on the catalyst surface) | (Eq. 11) |
| $C^* + H_2O \rightarrow CO + H_2$ | carbon-steam reaction | (Eq. 12) |
| $CH_4 + H_2O \leftrightarrow CO + 3H_2$ | $H_2$ methane reforming | (Eq. 13) |
| $C_nH_m + H_2O \rightarrow CO + H_2$ | light hydrocarbon reforming | (Eq. 14) |
| $CO + H_2O \leftrightarrow CO_2 + H_2$ | WGS reaction | (Eq. 15) |
| $C_nH_{2n} + H_2 \leftrightarrow C_nH_{2n}+2$ | hydrogenation | (Eq. 16) |
| tar + $H_2O \rightarrow CO + H_2$ | tar reforming | (Eq. 17) |

Tar conversion by using dolomite, nickel-based and other catalysts such as alkali metals at elevated temperatures of typically 800-900°C achieved near 99%. $MgCO_3CaCO_3$ (Dolomite) is a magnesium ore widely used in biomass gasification. The tar content of the produced gases during the biomass conversion process is significantly reduced in the presence of Dolomite (Sutton et el., 2001; Han and Kim, 2008; Nzihou et al., 2013; Asadullah, 2014).However, dolomite catalysts are efficient in tar cracking; they have some disadvantages such as sensitivity to elevated pressure and thermal instability which leads to loss of surface area due to sintering (Nordgreen, 2011).

Nickel-Based Catalysts are very effective for the catalytic hot gas cleanup during biomass gasification. Elimination of tar is also achieved by Ni-based catalysts with a high rate. Moreover Ni-based catalysts have been used for the production of hydrogen-rich product gas (Balat et al., 2009; Sinag, 2012). Anis and Zainal (2011) reported that among all catalysts for converting tar into fuel gas, nickel catalysts are the most efficient ones. The stability of nickel catalysts increased with co-impregnation of nickel on mineral catalysts (olivine, dolomite, and zeolite). Even if nickel catalysts have a remarkable effect on tar conversion, it may not be recommended for applications in atmospheric biomass gasification due to its high costs and severe risk for deactivation via sulphur chemisorption and carbon deposition (Nordgreen, 2011).

Alkali metals such as lithium, sodium, potassium, rubidium, and cesium can be used directly as catalysts in the form of alkali metal carbonates or supported on other materials such as alumina and silica. Addition of alkali metals to biomass can also be achieved by impregnation. These metals are highly reactive. Alkali metal catalysts lead to an enhancement in the biomass gasification reactions, especially for char formation reactions. The presence of $Na_2CO_3$, $K_2CO_3$ or $CsCO_3$ as catalyst in biomass steam gasification decreased the carbon conversion degree to gas with an increase in the rate and total amount of produced gas (Sutton et al., 2001; Han and Kim, 2008; Basker et al., 2012; Nzihou et al., 2013; Asadullah, 2014).

In recent years, nanomaterials have obtained extensive interests for their unique properties in various fields in comparison with their bulk counter parts. Among the nanocatalysts for biomass gasification, nano-sized NiO (nano-NiO) particles have received a great deal of attention for their catalytic properties. In specific, supported catalyst can be prepared by loading nano particles of NiO on the surface of distinct carriers (such as alumina) which can be more economic (Li et al., 2008a). Li et al. (2008b) investigated the effect of nano-NiO particles and micro-NiO particles as catalysts on biomass pyrolysis. They obtained char yield results for both catalysts and compared the results with the pyrolysis process without using catalysts. Based on the presented data in Table 1, the decomposition of cellulose in the presence of micro-NiO was 10°C lower than that of the pure cellulose, while the decomposition of cellulose with nano-NiO started at 294°C, which was 19°C lower than that of the pure cellulose. The final char yield (5.64 wt. %) was further decreased compared to when micro-NiO particles were applied. They proved the effectiveness of nano-NiO catalysts in pyrolyzing of biomass at a relatively lower temperature.

The results of using nano-Ni catalyst (NiO supported on gamma alumina) in direct gasification of sawdust demonstrated that this catalyst can considerably improve the quality of the produced gas while significantly eliminating tar production (Li et al., 2008a, b).

Aradi et al. (2010) examined the organometallic nanocatalysts of Ni compound and $Ni_3Cu\ (SiO_2)_6$ nanoalloy catalyst for biomass gasification. The results showed a significant increase in $H_2$ production which is well suited for further processing such as Fischer-Tropsch. Their findings revealed that the nanoalloy catalyst increased biomass conversion efficiency at relatively low gasification temperatures (Aradi et al., 2010a). Other nanocatalysts that have been used in biomass gasification are nano-ZnO and nano-$SnO_2$ structures. Sinag et al. (2011) have shown that nano-ZnO is an effective catalyst for low temperature WGS reaction, while nano-$SnO_2$ is an effective catalyst for high-temperature WGS reaction during the cellulose gasification in hot compressed water. As it can be seen in Table 2, results showed a remarkable effect for nano-ZnO on cellulose conversion at 300°C while nano-$SnO_2$was an effective catalyst for the cellulose conversion at 400-500°C. The data presented in Table 2 has been obtained based on the information reported by Sinag et al. (2011).

**Table 1**
Comparison of the char formation in cellulose pyrolysis process using nano-NiO, micro-NiO catalysts and without using catalyst (Li et al., 2008b).

| Catalyst | Initial decomposition temperature (°C) | Char yield |
|---|---|---|
| Without catalyst | 313 | 6.14 |
| Micro-NiO | 303 | 6.09 |
| Nano-NiO | 294 | 5.64 |

**Table 2**
Comparison the effects of bulk and nanocatalysts of ZnO and $SnO_2$ for cellulose conversion at different temperatures.

| Temperature (°C) | Conversion% | | | |
|---|---|---|---|---|
| | Nano-ZnO | Bulk ZnO | Nano-$SnO_2$ | Bulk $SnO_2$ |
| 300 | 92.4 | 83.0 | 71.0 | 64.0 |
| 400 | 83.2 | 83.0 | 88.2 | 75.2 |
| 500 | 89.4 | 83.0 | 88.4 | 76.6 |
| 600 | 86.8 | 75.0 | 84.2 | 78.8 |

The gaseous species obtained at 300°C in the presence of bulk and nano-ZnO mainly consisted of $CO_2$ and $H_2$, which revealed the progress of WGS at lower temperatures. The rate of WGS in the presence of nano-ZnO is faster in comparison with the nano-$SnO_2$. They found that larger surface areas of nano-ZnO enhanced the WGS reaction. Based on a research that used nano zinc-based oxides as catalyst for conversion of glucose into $H_2$ in supercritical water (SCW), it was found that the existence of both $H_2O_2$ and ZnO catalysts in the reactor enhanced hydrogen production (Sinag et al., 2011).Hao et al. (2005) investigated tar removal efficiency using $CeO_2$ particles, nano-$CeO_2$ and nano-(CeZr) $xO_2$ catalysts during the cellulose and sawdust gasification process. The experimental results showed a higher activity for the nano-(CeZr)$xO_2$ catalyst compared to the bulk and nano-$CeO_2$(Han and Kim, 2008).The same results for efficient performance of nanoalloy catalysts were obtained by Aradi et al. (2010). The available researches on nanocatalysts in biomass gasification are very limited.

### *3.2. Nanocatalysts for biomass liquefaction*

Alkaline salts, $Na_2CO_3$, KOH and so on, are commonly used as homogenous catalysts in liquefaction processes (Duan and Savage, 2011). The effects of some other catalysts on the liquefaction of biomass have also been investigated, such as $NaHCO_3$ (Sun et al., 2010), Ca $(OH)_2$, Ba $(OH)_2$, $FeSO_4$ (Xu and Lad, 2007). The heterogeneous catalysts have been used in catalytic conversion of biomass. Different heterogeneous catalysts Pd/C, Pt/C, Ru/C, Ni/$SiO_2$-$Al_2O_3$, CoMo/$\gamma$-$Al_2O_3$ (sulfided), zeolite (Duan and Savage, 2011), and Fe (Sun et al., 2010) have been studied in conversion of the biomass. In catalytic hydro conversion of biomass, liquid catalysts have the advantage of being mono dispersed in reaction mixtures. In other words, solid catalysts have the superiority of higher catalytic activity in addition to being easily separated from the products. Acid-functionalized paramagnetic nanoparticles are promising materials for use in catalytic hydro conversion of biomass. These functionalized nanoparticles can easily separate and be recycled in the catalytic hydro conversion process. Nanocatalysts have some other advantages which make them attractive for use in biomass to liquid (BTL) processes. Having the fluid solution characteristics, mono dispersed nanocatalysts have excellent accessibility to the oxygen atoms of the cellulose ether linkage (Guo et al., 2012). Conversion of biomass to liquid compounds such as paraffinic, naphtenic and aromatic hydrocarbons can supplement part of the worldwide petrochemical demand. Although use of nanocatalysts in the catalytic conversion of biomass to liquid chemicals has had several advantages; most of the research attention has been paid to conversion of biomass to biodiesel. Having relatively high prices, diesel fuel is in large demand in today's world. So an individual subsection has been assigned to nanocatalytic conversion of biomass to biodiesel. There are very limited studies on the conversion of biomass to other bio- oils. Nanoparticles of Co were used as catalysts in the conversion of spent tea to biochemical (Mahmood and Hussain, 2010). It is claimed that in this pyrolysis process, Co nanoparticles reduce reaction temperature by up to 650◦C. The liquid products yield of the reaction at 300◦C and atmospheric pressure was about 60%.

Hydrothermal conversions of cellulose in the presence of two metal oxides ($SnO_2$ and ZnO) have been studied (Sinag et al., 2011). It has been found that using bulk ZnO as the catalyst increased the amount of glycolic acid by five times compared to when bulk $SnO_2$ was used in the hydrothermal conversion of cellulose at 300°C. Interestingly, when nano sized particles of these catalysts were used, the produced glycolic acid with ZnO catalyst was 12 times higher than that of nano-$SnO_2$. This result illustrates the excellent catalytic properties of the nanosized catalyst. The production of biogasoline and organic liquid products (OLP) were also studied in a fixed bed reactor with nanocrystalline zeolite as the catalyst and waste cooking palm oil as the biofeed (Taufiqurrahmi et al., 2011). Results showed that under different operating conditions, the conversion of 87.5-92.9 wt% of the feed is attainable. In such a condition, a gasoline fraction yield of 33.61- 37.05 wt% and an OLP fraction yield of 46.1-53.4wt% can be obtained. For zeolite Y with pore sizes of 0.67 nm as the catalyst, the optimum conditions of 458°C and an oil: catalyst ratio of 6 have been reported. The NiW-nano-hydroxyapatite (NiW-nHA) composite was used as the catalyst in hydrocracking of Jatropha oil. In the operating conditions of 360°C and 3 MPa about 92% of the feed was converted. The yield of $C_{15}$-$C_{18}$ alkanes in the product was up to 83.5wt%. By increasing operating temperatures it is possible to obtain 100% conversion of Jatropha oil in this process (Zhou et al., 2012).

### 3.2.1. *Nanocatalysts for biodiesel production*

In the biodiesel production method, transesterification is the chemical reaction between triglycerides and alcohol within the presence of a catalyst for producing monoesters. The triglyceride molecules are transformed to monoesters and glycerol. The transesterification method incorporates a sequence of three reversible reactions. The conversions of triglycerides to diglycerides, diglycerides to monoglycerides and glycerides into glycerol yield one ester molecule in each stage. The general transesterification reaction can be represented by Figure 2 (Gerpen, 2005).

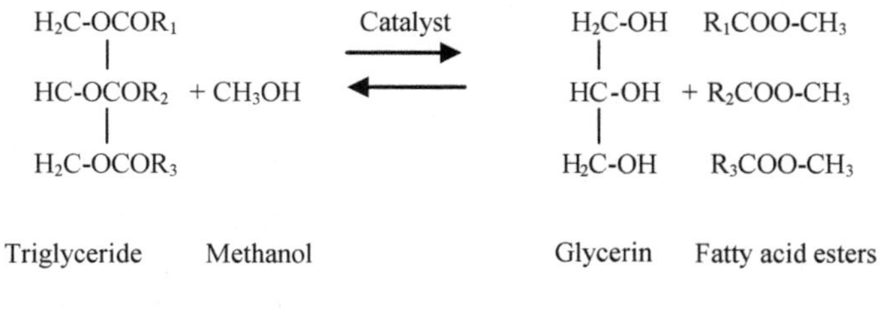

| | | | | |
|---|---|---|---|---|
| $H_2C$-$OCOR_1$ | | Catalyst | $H_2C$-OH | $R_1COO$-$CH_3$ |
| HC-$OCOR_2$ | + $CH_3OH$ | ⇄ | HC-OH | + $R_2COO$-$CH_3$ |
| $H_2C$-$OCOR_3$ | | | $H_2C$-OH | $R_3COO$-$CH_3$ |
| Triglyceride | Methanol | | Glycerin | Fatty acid esters |

**Fig.2.** Transesterification of triglycerides with methanol.

The transesterification reaction of oil and alcohol with a homogeneous catalyst is the general method for the preparation of biodiesel. However, the homogeneous catalysts have many shortcomings, such as requiring large amounts of water, difficulties in product isolation, and environmental pollution caused by the liquid wastes. The use of "a green" method based on heterogeneous catalysts is a new trend in the preparation of biodiesel. Biodiesel synthesis using solid catalysts instead of homogeneous ones could potentially lead to cheaper production costs by enabling reuse of the catalyst and opportunities to operate in a fixed bed continuous process. Heterogeneous catalytic methods are usually mass transfer resistant, time consuming and inefficient. Despite the solid phase, catalytic methods are intensively studied, the industrial applications are limited. This fact suggests that further research is necessary to solve current problems (Narasimharao et al. 2007). Nanocatalysts that have high specific surface and high catalysis activities may solve the above problems. A number of researchers have studied the preparation of nanosized heterogeneous catalysts to increase the catalytic activity. It is evident that the large surface area, which is characteristic of nanosized material, resulted in a rise within the amount of the catalytically basic and acidic sites. The nanocatalysts used for biomass to biodiesel conversion have been presented in Table 3.

Feyzi et al. (2013) used the magnetic Cs/Al/$Fe_3O_4$ as a nanocatalyst for transesterification reaction of sunflower oil, the optimal catalyst showed high catalytic activity for biodiesel production and the biodiesel yield reached 94.8%.For the transesterification of Pongamia oil with methanol, Obadiah et al. (2012), used calcined Mg-Al hydrotalcite as a solid base catalyst. The reaction conditions of the system were optimized to maximize the methyl esters conversion (about 90.8%).

Mguni et al. (2012) studied the transesterification of sunflower oil with nano-MgO precipitated and deposited on $TiO_2$ support as catalyst. Conversions of 84, 91 and 95% were measured at 225°C compared to 15, 35 and 42% at 150°C respectively for 10, 20 and 30wt. %MgO catalyst. Verziu et al. (2007) obtained biodiesel from rapeseed oil and sunflower oil using different nanocrystalline MgO catalysts in nanosheets form, which were prepared by conventional and aerogel method. Working under microwave conditions with these systems led to higher conversions and selectivity when preparing methyl esters, as compared to autoclave or ultrasound conditions. MgO can be used effectively as a heterogeneous catalyst for biodiesel transesterification. The exposed facet of the MgO has an important influence on activity and selectivity. A new nanocatalyst with potassium bitartrate as an active component on zirconia support was synthesized by Qiu et al. (2011). The transesterification reaction of soybean oil and methanol was catalyzed heterogeneously. The highest biodiesel yield of about 98.03%was obtained at methanol to oil molar ratio of 16:1, reaction time of 2 hr, a reaction temperature of 60°C and a catalyst amount of 6.0 wt%.

Lithium ion impregnated calcium oxide as a nanocatalyst for the biodiesel production from karanja and jatropha oils was studied by Kaur and Ali (2011). They reached a yield of 99% by using Li-CaOnanocatalyst for karanja oil transestrification after 1hr and jatropha oil after 2 hr.

Chang et al. (2010), reported preparation of CaO / $Fe_3O_4$ the nanometer magnetic solid base catalyst for production of biodiesel. They found that the conversion yield of transesterification reaction catalyzed by Ca $(OH)_2$ ($Ca^{+2}$: $Fe_3O_4$=7) can reach 95% in 80 min, and the conversion of 99% obtained after 4 hr. Nanometer magnetic base solid catalyst was proposed as an efficient catalyst for biodiesel production because of its easy separation which led to a reduction in operating costs . Wen et al. (2010) reported the preparation of KF/CaO solid basic nanocatalyst for Chinese tallow seed oil. Optimal conditions for obtaining a 96.8% yield was a 12:1 molar ratio of alcohol to oil, 4wt. %of catalyst, reaction temperature of 65°C and reaction time of 2.5 hr. Deng et al. (2011) observed that hydrotalcite-derived particles with Mg/Al molar ratio of 3:1could be an effective method for the production of biodiesel from Jatropha oil with a 95.2% yield after 1.5 hrs at ultrasonic conditions. Reddy et al. (2006) achieved a 99% biodiesel yield from soybean oil in room temperature by using nanocrystalline calcium oxides as a catalyst. They studied various forms of nano-CaO such as powder, pellets, and granules. They found that high yields of transesterification of soybeans oil are due to the higher surface area associated with small crystallite size and defects. Wang and Yang (2007) investigated nano magnesium oxides as a heterogeneous catalyst and its effect on biodiesel synthesis from soybeans oil. Nano-MgO showed higher catalytic activity in supercritical/ subcritical temperatures. It was evidently superior to that of non-catalysts. The apparent

**Table 3**
Nanocatalysts used for biodiesel production along with their operating conditions.

| No. | Nano Catalyst | Size (nm) | Feedstock | Operation Condition: Temp. (°C) | Alcohol:oil ratio | Catalyst loading (wt. %) | Reaction Time (min) | Biodiesel yield (%) | Ref. |
|---|---|---|---|---|---|---|---|---|---|
| 1 | Cs/Al/$Fe_3O_4$ | 30-35 | Sunflower oil | 58 | 14:1 | 4 | 120 | 94.80 | (Feyzi et al. 2013) |
| 2 | Hydrotalcite (Mg-Al) | 4.66-21.1 | Pongamia oil | 65 | 6:1 | 1.5 | 240 | 90.8 | (Obadiah et al. 2012) |
| 3 | MgO Supported on Titania | - | Soybean oil | 150-225 | 18:1 | 0.1-7 | 60 | 95 | (Mguni et al. 2012) |
| 4 | MgO | 50-200 | Sunflower oil / Rapeseed oil | 70-310 | 4:1 | - | 40 -120 | 98 | (Verziu et al. 2007) |
| 5 | $ZrO_2$ loaded with $C_4H_4O_6HK$ | 10-40 | Soybean oil | 60 | 16:1 | 6 | 120 | 98.03 | (Qiu et al. 2011) |
| 6 | Lithium impregnated calcium oxide (Li-Cao) | 40 | Karanja oil / Jatropha oil | 65 | 12:1 | 5 | 60 / 120 | 99 | (Kaur and Ali 2011) |
| 7 | Magnetic solid base catalysts CaO / $Fe_3O_4$ | 49 | Jatropha oil | 70 | 15:1 | 2 | 80 | 95 | (Chang et al. 2010) |
| 8 | KF/CaO | 30-100 | Chinese tallow seed oil | 65 | 12:1 | 4 | 150 | 96 | (Wen et al. 2010) |
| 9 | Hydrotalcite-derived particles with Mg/Al molar ratio of 3:1 | 7.3 | Jatropha oil | 45 | 4:1 | 1 | 90 | 95.2 | (Deng et al. 2011) |
| 10 | Cao | 20 | Soybean oil | 23-25 | 27:1 | - | 720 | 99 | (Reddy et al. 2006) |
| 11 | Mgo | 60 | Soybean oil | 200- 260 | 6:1 | 0.5-3 | 12 | 99.04 | (Wangand Yang 2007) |
| 12 | KF/CaO–$Fe_3O_4$ | 50 | Stillingia oil | 65 | 36:1 | 4 | 180 | 95 | (Hu et al. (2011) |
| 13 | $TiO_2$-ZnO / ZnO | 34.2 / 28.4 | Palm oil | 60 | 12:1 | - | 300 | 92.2 / 83.2 | (Madhuvilakku and Piraman, 2013) |
| 14 | KF/ $Al_2O_3$ | 50 | Canola oil | 65 | 6:1 | 3 | 480 | 97.7 | (Boz et al. 2009) |
| 15 | ZnOnanorods | - | Olive oil | 150 | 15:1 | 1 | 480 | 94.8 | (Molina 2012) |
| 16 | CaO/MgO | - | Jatropha oil | 64.5 | 18:1 | 2 | 210 | 92 | (Chang et al. 2010) |
| 17 | Ca $(OH)_2$-$Fe_3O_4$($Ca^{+2}$: $Fe_3O_4$=7) | - | Jatropha oil | 70 | 15:1 | 2 | 240 | 99 | (Chang et al. 2010) |

activation energy with nano-MgO was lower than that without MgO. They found that Nano-MgO can catalyze the transesterification reactions, but its catalytic ability was quite weak under normal temperature. At a temperature of 60°C, the methyl ester yield was only about 3% in 3 hr when 3 wt. % of nano- MgO was added. Thus, it is desirable to find a more efficient methodfor transesterification of triglycerides by using methanol with a higher reaction rate under more moderate temperature and pressure.observed with a 99.04% yield in 12 min. Hu et al. (2011) investigated nano-magnetic solid base catalyst KF/CaO-$Fe_3O_4$ for transtrifaction of stillingia oil, extracted from the seeds of Chinese tallow (*Sapium sebiferum*). The best activity obtained with nano-magnetic solid base catalysts with 25 wt. % KF loading and 5 wt. % $Fe_3O_4$, calcined at 600°C for 3 hr.

Madhuvilakku and Piraman(2013) used both ZnO and $TiO_2$-ZnO nanocatalysts for production of biodiesel from palm oil. The substitution of Ti ions on the Zinc lattice ed to the creation of defects, responsible for stable catalytic activity. A 92.2% yield was attained with 5 hr. at lower catalyst loading of 200 mg of $TiO_2$-ZnOnanocatalystat 6:1 methanol to oil molar ratio and 60°C. The $TiO_2$-ZnO mixed oxide nanocatalysts illustrated a significantly improved performance which could be a potential catalyst in the large-scale biodiesel production compared to the ZnO nanocatalyst.

Boz et al. (2009) reported that the optimum loading amount of KF on nano-$\gamma$-$Al_2O_3$ was 15 wt. %. The conversion of triglycerides to biodiesel reached values which were as high as 97.7 ± 2.14%. Such high biodiesel yields reflect the benefits of reaching relatively high basicity and the use of nanosized catalyst particles for Canola oil transestrification. Molina (2012) studied on ZnO nanorods as catalyst for biodiesel production from olive oil. The catalytic performance of the ZnO nanorods was slightly better than that of the conventional ZnO. The reported yield of Olive oil to biodiesel was 94.8 % by using ZnO nanorods compared to 91.4% by commercial ZnO.

Based on different studies on nanocatalyst application for biodiesel production, it is evident that the large porous catalytic surface increased the contact between alcohol and oil, leading to an increase in nanocatalytic effectiveness. Utilization of different edible and nonedible oils for transestrification reactions by using both acid and alkali nanocatalysts show the important influence of these catalysts regarding activity and selectivity. The presented results reveal that the high specific surface area of nanostructure materials in comparison with bulk catalysts is favorable for contact between catalyst and substrates, which effectively improve the yield of products.

## 4. Conclusion

Without doubt, it is necessary to replace fossil energy resources with new safe sources. Among the existing choices, biomass seems to be the best option. The energy released from biomass is renewable and environmentally friendly, so it is strongly recommended to be applied. It is obvious that many methods are available for converting biomass to biofuel. However, some of the key challenges in biomass conversion provide new research potential for improving quality of products and solving its related environmental problems. Introducing nanotechnology research to biomass conversion has witnessed rapid growth, which is mainly related to unique property of possessing high specific surface area. In this paper a review of thermochemical nanocatalytic processes as a major technology for biomass conversion has been provided. Thermochemical biomass gasification converts biomass to a combustible gas mixture through partial oxidation at relatively high temperatures. The products mainly include carbon monoxide and hydrogen (syngas). In biomass gasification the nanocatalystswhich have mostly been used to reduce tar formation are NiO, $CeO_2$, ZnO,$SnO_2$. Moreover, the application of nanoalloys such as (CeZr) $xO_2$ and $Ni_3Cu$ $(SiO_2)_6$ provide higher performances at relatively lower gasification temperatures. In biomass liquefaction nanocatalysts have been successfully used to increase the liquids yields and also enhance the value-added products. The higher temperature in the liquefaction process increases gaseous products. Nanocatalysts successfully reduce reaction temperatures causing an increase in the liquids products which means an improvement in the liquefaction operation. In pyrolysis of spent tea Co nanoparticles reduce the operating temperature to 300°C and increase the liquid product yield to 60%. Use of ZnO and $SnO_2$ nanoparticles in hydrothermal conversion of cellulose shows better liquid product yield in comparison with using these catalysts in bulk dimension. Nanocrystalline zeolite was used in catalytic conversion of cooking palm oil which has attained about 93% conversion at optimum temperature of 458°C. The nanocomposite catalyst of NiW-hydroxyapatite may convert 100% of Jatropha oil in catalytic the hydrocracking process. Nanocatalysts for biodiesel production significantly improve the yield of products. The main nanometal oxides that have been used for biodiesel production are Zn, Ca, Mg, Zr. These have either been used individually or supported on different materials. However some other catalysts such as Li, Cs, KF have been utilized for edible and nonedible feedstock. In addition, magnetic nanoparticles functionalized with different catalysts have been implied in biodiesel production, facilitate the catalyst recovery. The results of using KF/CaO and nanomagnetic KF/CaO-$Fe_3O_4$ catalysts for biodiesel preparation show better performance of KF/CaO catalyst with a higher surface area of about 109 $m^2g^{-1}$ at the same operating conditions compared to KF/CaO-$Fe_3O_4$ catalyst with a surface area of 20.8 $m^2g^{-1}$. Loading KF on nano $Al_2O_3$ support could obtain a higher yield of about 97.7% compared to 96.8% for KF/CaO catalyst. But it must be noted thatthe operating conditions of using KF/ $Al_2O_3$ catalyst were relatively higher than that of KF/CaO catalyst. The alcohol: oil ratio, catalyst loading and reaction time were 15:1, 3 wt. % and 480min in comparison with 12:1, 4wt. % and 150min respectively for KF/ $Al_2O_3$ and KF/CaO catalysts. Despite using different feedstocks, the better performance of KF/CaO catalyst may be related to its higher surface area (109 $m^2g^{-1}$) compared to the value of 41.7 $m^2g^{-1}$ for KF/ $Al_2O_3$ catalyst. The highest biodiesel yield was obtained using nanocatalysts Li-CaO and CaO at reaction time of 120 and 720min, temperature of 65 and 25°C and methanol to oil molar ratio of 12 and 27, respectively. Similar yield has been obtained using a nano-MgO catalyst but at a higher temperature of about 200-260°C. In general to achieve high performances at relatively mild operating conditions, it is necessary to increase the reaction time while at ordinary reaction times, it is necessary to apply severe operating conditions. Comparing the results of using different supported and unsupported MgO catalysts revealed that decreasing the reaction time led to an increase in reaction temperatures for achieving higher performances. As a whole using milder operating conditions led to a reduction in energy consumption requirements of the process which could be feasible with using nanocatalysts.

## Reference


Alonso, D. M., Bond, J. Q., Dumesic, J. A., 2010. Catalytic conversion of biomass to biofuels. Green Chem., 12, 1493–1513.

Anis, S., Zainal, Z.A., 2011. Tar reduction in biomass producer gas via mechanical, catalytic and thermal methods: A review. Renew. Sust. Energ. Rev. 15, 2355–2377.

Aradi, A., Roos, J., Jao, T.C., 2010. Nanoparticle catalyst compounds and/or volatile organometallic compounds and method of using the same for biomass gasification. US 20100299990A1.

Aradi, A., Roos, J., Jao, T.C., 2011. Nanoparticle catalysts and method of using the same for biomass gasification. US20110315931A1.

Aravind, P.V., Wiebren, J. D., 2012. Evaluation of high temperature gas cleaning options for biomass gasification product gas for Solid Oxide Fuel Cells. Prog. Energ. Combust. 38, 737-764.

Asadullah, M., 2014. Barriers of commercial power generation using biomass gasification gas: Areview. Renew. Sust. Energ. Rev. 29, 201–215.

Asadullah, M., Ito, S. I., Kunimori, K., Yamada, M., Tomishige, K., 2002. Biomass Gasification to Hydrogen and Syngas at Low Temperature: Novel Catalytic System Using Fluidized-Bed Reactor. J.Catal.208, 255–259.

Balat, M., Balat, M., Kirtay, E., Balat, H., 2009. Main routes for the thermo-conversion of biomass into fuels and chemicals. Part 2: Gasification systems. Energ. Convers. Manage. 50, 3158–3168.

Basker, C., Basker, S., Dhillon, R.S., 2012. Biomass Conversion: The Interface of Biotechnology, Chemistry and Materials Science, Springer, E-book.

Boz, N., Degirmenbasi, N., M. Kalyon, D., 2009. Conversion of biomass to fuel: Transesterification of vegetable oil to biodiesel using KF loaded nano-γ-Al2O3 as catalyst. Appl. Catal. B.89, 590–596.

Bridgwater, A.V., 2003. Renewable fuels and chemicals by thermal processing of biomass, Chem. Eng. J. 91, 87–102.

Bridgwater, A.V., 2012. Review of fast pyrolysis of biomass and product upgrading. Biomass. Bioenerg. 38 pp, 68-94.

Chang, A.C., Louh, R.F., Wong, D., Tseng, J., Lee, Y.S., 2011. Hydrogen production by aqueous-phase biomass reforming over carbon textile supported Pt-Ru bimetallic catalysts. Int. J. Hydrogen Energ. 36, 8794-8799.

Ciferno, J.P., Marano, J.J., 2002. Benchmarking Biomass Gasification Technologies for Fuels, Chemicals and Hydrogen Production, National Energy Technology Laboratory of the U.S. Department of Energy.

Demirbas, A., 2008. Biofuels sources, biofuel policy, biofuel economy and global biofuel projections. Energ. Convers. Manage. 49, 2106–2116.

Deng, X., Fang, Zh., Liu, Y.H., Liu Yu, Ch., 2011. Production of biodiesel from Jatropha oil catalyzed by nanosized solid basic catalyst. Energy. 36, 777-784.

Duan, P., Savage, P.E., 2011. Hydrothermal liquefaction of a Microalga with heterogeneous catalyst. Ind. Eng. Chem. Res.50, 52–61.

Duman, G., Uddin, M. A., Yanik, J., 2014. The effect of char properties on gasification reactivity. Fuel. Process. Technol.118, 75–81.

Feyzi, M., Hassankhani,A., Rafiee, H., 2013. Preparation and characterization of $CsAlFe_3O_4$ nanocatalysts for biodiesel production. Energ. Convers. Manage. 71, 62-68.

Gerpen, J.V., 2005. Biodiesel processing and production. Fuel. Process. Technol. 86, 1097-1107.

Guo, F., Fang, Z., Xu, C.C., Smith Jr., R. L., 2012. Solid acid mediated hydrolysis of biomass for producing biofuels. Prog. Energ. Combust. 38, 672-690.

Hamze, H., Akia, M., Yazdani, F., 2013. Comparison of one step and two stepprocess for biodiesel production from waste cooking oil using TGA analysis, presented at 3nd National Fuel, Energy and Environment National Congress, Tehran, Iran.

Hamze, H., Akia, M., Yazdani, F., 2014. Viscosity variations of biodiesel produced from waste cooking oil under different process conditions, accepted at the 8th International Chemical Engineering Congress & Exhibition (IChEC), Kish, Iran.

Han, J., Kim, H., 2008. The reduction and control technology of tar during biomass gasification/pyrolysis: An overview. Renew. Sust. Energ. Rev. 12, 397–416.

Hansen, A.C., Kyritsis, D.C., Lee, C.F., 2010. Characteristics of biofuels and renewable fuel standards, in: Vertes, A.A., Qureshi, N., Blaschek, H.P., Yukawa, H. (Eds.), Biomass to Biofuels: Strategies for Global Industries. John Wiley & Sons, Ltd., pp.1-25. ISBN: 978-0-47 0-51312-5

Henrich, E., Clean syngas from biomass by pressurized entrained flow gasification of slurries from fast pyrolysis. Presented at SYNBIOS conference, 18–20 May 2005, Stockholm, Sweden.

Hu, Sh., Guan, Y., Wang, Y., Han, H., 2011. Nano-magnetic catalyst KF/CaO–Fe3O4 for biodiesel production. Appl.Energ. 88, 2685–2690.

K. Narasimharao, K., Lee, A., Wilson, K., 2007. Catalysts in Production of Biodiesel: A Review. JBMB. 1, 19–30.

Kaur,M., Ali, A., 2011. Lithium ion impregnated calcium oxide as nano catalyst for the biodiesel production from karanja and jatropha oils. Renew. Energ. 36, 2866-2871.

Kim, K., Kim, Y., Yang, C., Moon, J., et al., 2013. Long-term operation of biomass-to-liquid systems coupled to gasification and Fischer–Tropsch processes for biofuel production. Bioresour. Technol. 127, 391–399.

Kruse, A., Dinjus, E., 2007. Hot compressed water as reaction medium and reactant: Properties and synthesis reactions. J. Supercrit. Fluid.39, 362-380.

Kumar, A., Jones, D.D., Hanna, M.A., 2009. Thermochemical biomass gasification: A review of the current status of the technology. Energies 2, 556-581.

Li, J., Yan, R., Xiao, B., et al., 2008. Development of Nano-NiO/$Al_2O_3$ Catalyst to be used for Tar Removal in Biomass Gasification. Environ. Sci. Technol. 42, 6224–6229.

Li, J., Yan, R., Xiao, B., et al., 2008. Preparation of Nano-NiO Particles and Evaluation of TheirCatalytic Activity in Pyrolyzing Biomass Components. Energ. Fuel. 22, 16–23.

Liu, C.J., Burghaus, U., Besenbacher, F., Wang, Z. L., 2010. Preparation and Characterization of Nanomaterials for Sustainable Energy Production. 240th ACS National Meeting in Boston, MA. 10, 5517-5526.

Liu, Ch., Lv, P., Yuan, Zh., Yan, F., Luo, W., 2010. The nanometer magnetic solid base catalyst for production of biodiesel. Renew. Energ. 35, 1531–1536.

Liu, H., Su, L., Shao, Y., Zou, L., 2012. Biodiesel production catalyzed by cinder supported CaO/KF particle catalyst. Fuel. 97, 651–657.

Liu, Y., Yuan, X., Huang, H., Wang, X., Wang, H., Zeng, G., 2013. Thermochemical liquefaction of rice husk for bio-oil production in mixed solvent (ethanol–water). Fuel Process. Technol.112, 93–99.

Luo, Z., Zhou, J., 2012. Thermal conversion of biomass, in: Chen, W.Y., Seiner, J., Suzuki, T., Lackner M. (Eds.), Handbook of Climate Change Mitigation, Springer Science Business Media, LLC. pp. 1002-1037.

Madhuvilakku, R. and Piraman, K., 2013. Biodiesel synthesis by TiO2–ZnO mixed oxide nanocatalyst catalyzed palm oil transesterification process. Bioresour. Technol. 150, 55–59.

Mahmood, T., Hussain, S.T., 2010. Nanobiotechnology for the production of biofuels from spent tea. Afr. J. Biotech.9, 858-868.

Malik,P., Sangwan, A., 2012. Nanotechnology: A tool for improving efficiency of bio-energy. J. Eng. Appl. Sci. 1, 37-49.

Mettam, G.R., Adams, L.B., 2009. How to prepare an electronic version of your article, in: Jones, B.S., Smith, R.Z. (Eds.), Introduction to the Electronic Age. E-Publishing Inc., New York, pp. 281–304.

Mguni, L.L., Meijboom, R., Jalama, K., 2012. Biodiesel Production over nano-MgO Supported on Titania. World Acad. Sci., Eng. Technol. 64.

Mitrovi, D. M., Janevski, J. N., Lakovic, M.S., et al., 2012. Primary energy saving using heat storage for biomass heating systems. Therm. Sci. 16, 423-431.

Molina, C.M.M., 2013. ZnoNanorods as Catalyts for Biodiesel Production from Olive Oil. Ms.C. Thesis, University of Louisville.

Nordgreen, T., 2011. Iron-based materials as tar cracking catalyst in waste gasification, Ph.D. Thesis, KTH-Royal Institute of Technology Department of Chemical Engineering and Technology Chemical Technology SE-100 44 Stockholm, Sweden.

Nzihou, A., Stanmore, B., Sharrock, P., 2013. A review of catalysts for the gasification of biomass char, with some reference to coal. Energy. 58, 305-317.

Obadiah, A., Kannan, R., Ravichandran, P., Ramasubbu, A., Vasanth Kumar, S., 2012. Nano HydrotalciteAs A Novel Catalyst for Biodiesel Conversion. Digest J. Nanomat. Biostruc. 7, 321 – 327.

Ozaki, J., Takei, M., Takakusagi, K., Takahashi, N., 2012. Carbon deposition on a Ni/Al2O3 catalyst in low-temperature gasification using C6-hydrocarbons as surrogate biomass tar, Fuel Process. Technol. 102, 30–34.

Pelisson, C. H., Vono, L.L. R., Hubert, C., Nowicki, A.D., et al., 2012. Moving from surfactant-stabilized aqueous rhodium colloidal suspension to heterogeneous magnetite-supported rhodium nanocatalysts: Synthesis, characterization and catalytic performance in hydrogenation reactions. Catal. Today 183, 124– 129.

Qiu, F., Li, Y., Yang, D., Li, X., Sun, P., 2011. Heterogeneous solid base nanocatalyst: Preparation, characterization and application in biodiesel production. Bioresour. Technol. 102, 4150–4156.

Rapagna, S., Jand, N., Foscolo, P.U., 1998. Catalytic gasification of biomass to produce hydrogen rich gas. Int. J. Hydrogen Energ. 23, 551-557.

Rapagna, S., Provendier, H., Petit, C., Kiennemann, A.,Foscolo, P.U., 2002. Development of catalysts suitable for hydrogen or syn-gas production from biomass gasification. Biomass Bioenerg.22, 377 – 388.

Rathore, N.S., Kothari, S., Verma, A., 2013. Thermo-Chemical Conversion, in: Kumar, S., Taraghi, S. K., (Eds.), Recent Advances in Bioenergy Research. Sardar Swaran Singh National Institute of Renewable Energy pp. 347-355.

Reddy, C.R.V., Oshel, R., Verkade, J.G., 2006. Room-Temperature Conversion of Soybean Oil and Poultry Fat to Biodiesel Catalyzed by Nanocrystalline Calcium Oxides. Energ. Fuel. 20, 1310-1314.

Rutz, D., Janssen, R., 2007. Biofuel Technology Handbook, pp. 106-129.

Schnitzer, M.I., Monreal, C.M., Facey, G.A., Fransham, P.B., 2007. The conversion of chicken manure to biooil by fast pyrolysis I. Analyses of chicken manure, biooils and char by 13C and 1H NMR and FTIR spectrophotometry. J. Environ. Sci. Heal. B 42, pp. 71-77.

Shirazi, M.M.A., Kargari, A., Tabatabaei, M., Akia, M., Barkhi, M., Shirazi, MJA., 2013. Acceleration of biodiesel-glycerol decantation through NaCl-assisted gravitational settling: A strategy to economize biodiesel production. Bioresour. Technol. 134, 401-406.

Sinag, A., 2012. Catalysts in thermochemical biomass conversion, in: C. Baskar et al. (Eds.), Biomass Conversion,Springer-Verlag Berlin Heidelberg, pp. 187-197.

Sinag, A., Yumak, T., Balci, V., Kruse, A., 2011. Catalytic hydrothermal conversion of cellulose over SnO2 and ZnO nanoparticle catalysts. J. Supercrit. Fluid.56, 179–185.

Sun, P., Heng, M., Sun, S., Chen, J., 2010. Direct liquefaction of Paulownia in hot compressed water: Influence of catalysts. Energy 35, 5421-5429.

Sutton, D., Kelleher, B., Ross, J.R.H., 2001. Review of literature on catalysts for biomass gasification. Fuel. Process. Technol.73, 155–173.

Taufiqurrahmi, N., Mohamed, A.R., Bhatia,S., 2011. Production of biofuel from waste cooking palm oil using nanocrystalline zeolite as catalyst: Process optimization studies. Bioresour. Technol.102, 10686–10694.

Vertes, A.A., Qureshi, N., Blaschek, H.P., Yukawa, H., 2010. Biomass to Biofuels: Strategies for Global Industries. John Wiley & Sons Ltd, United Kingdom.

Verziu, M., Cojocaru,B., Hu, J., Richards, R.,Ciuculescu, C., Filip, P., Parvulescu, V.I., 2008. Sunflower and rapeseed oil transesterification to biodiesel over different nanocrystalline MgO catalysts. Green Chem. 10, 373-378.

Wang, L. and Yang, J., 2007. Transesterification of soybean oil with nano-MgO or not in supercritical and subcritical methanol. Fuel. 86, 328–333.

Wang, L., Yang, J., 2007. Transesterification of soybean oil with nano-MgO or not in supercritical and subcritical methanol. Fuel. 86, 328–333.

Wen, L., Wang, Y., Lu, D., Hu, Sh., Han, H., 2010. Preparation of KF/CaO nanocatalyst and its application in biodiesel production from Chinese tallow seed oil. Fuel. 89, 2267–2271.

Wilcoxon, J.P., 2012. Nanoparticles Preparation, Characterization and Physical Properties. Frontiers of Nanoscience. Vol. 3. DOI: 10.1016/B978-0-08-096357-0.00005-4 .pp. 43-127.

Xu, C., Lad, N., 2007. Production of heavy oils with high caloric values by direct liquefaction of woody biomass in sub/near-critical water. Energ. Fuel. 22, 635-642.

Xu, Y., Hu, X., Li, W., Shi, Y., 2011. Preparation and characterization of bio-oil, in: Shaukat, S. (Eds.), Biomass, Progress in Biomass and Bioenergy Production. InTech, pp.197-222.

Zhao, H., Yan, H.-X., Liu, M., Sun, B.-B., Zhang, Y , Dong, S.-S., Qi, L.-B., Qin, S., 2013. Production of bio-oil from fast pyrolysis of macroalgae *Enteromorpha prolifera* powder in a free-fall reactor. Energ. Source. Part A, 35, 859–867.

Zheng,C., Tao,H., Xie, X., 2013. Distribution and characterization of liquefaction of celluloses in sub- and super-critical ethanol. Bioresour. 8, 648-662.

Zhou, G., Hou, Y., Liu, L., Liu, H., et al., 2012. Preparation and characterization of NiW–nHA composite catalyst for hydrocracking. Nanoscale 4, 7698-7703.

Zuberbühler, U., Specht, M., Bandi, A., 2005. Gasification of biomass - An overview on available technologies. 1st European Summer School on Renewable Motor Fuels Birkenfeld, Germany, 29 – 31 Aug.

# 12

# Mass-energy balance analysis for estimation of light energy conversion in an integrated system of biological $H_2$ production

A.I. Gavrisheva[1], B.F. Belokopytov[2], V.I. Semina[1,3], E.S. Shastik[1], T.V. Laurinavichene[1], A.A. Tsygankov[1,*]

[1] *Institute of Basic Biological Problems, Russian Academy of Sciences, Pushchino, Moscow Region, 142290, Russia.*

[2] *Institute of Physiology and Biochemistry of Microorganisms, Russian Academy of Sciences, Pushchino, Moscow Region, 142290, Russia.*

[3] *LLC "Ecoproject", Mytishchi, 141014, Moscow Region, Russia.*

## HIGHLIGHTS

- The conversion of light energy into $H_2$ was examined in an integrated three-stage scheme.
- Mass-energy balance regularities were applied to estimate energy conversion efficiencies at different stages.
- This three-stage scheme was found counterproductive for light energy bioconversion to $H_2$.

## GRAPHICAL ABSTRACT

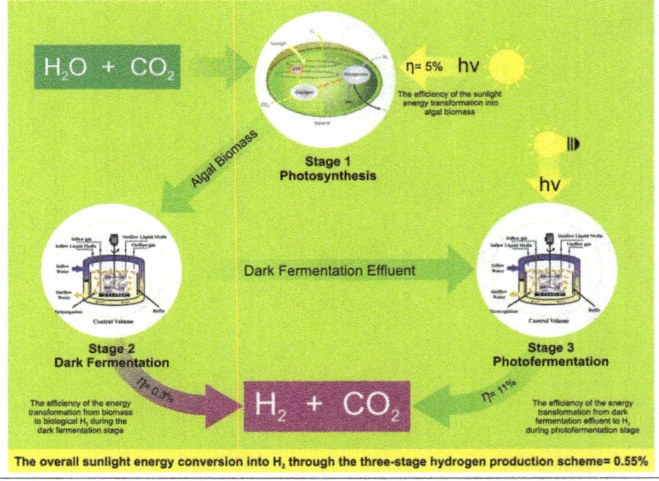


## ABSTRACT

The present study investigated an integrated system of biological $H_2$ production, which includes the accumulation of biomass of autotrophic microalgae, dark fermentation of biomass, and photofermentation of the dark fermentation effluent. Particular emphasis was placed on the estimation of the conversion efficiency of light into hydrogen energy at each stage of this system. For this purpose, the mass and energy balance regularities were applied. The efficiency of the energy transformation from light into the microalgal biomass did not exceed 5%. The efficiency of the energy transformation from biomass to biological $H_2$ during the dark fermentation stage stood at about 0.3%. The photofermentation stage using the model fermentation effluent could improve this estimation to 11%, resulting in an overall efficiency of 0.55%. Evidently, this scheme is counterproductive for light energy bioconversion due to numerous intermediate steps even if the best published data would be taken into account.

**Keywords:**
Microalgae
Energy conversion efficiency
Hydrogen production
Fermentation
Purple bacteria
Mass-energy balance

* Corresponding author
E-mail address: ttt-00@mail.ru

**Abbreviations**

| | |
|---|---|
| Bchl | Bacteriochlorophyll |
| Chl | Chlorophyll |
| FE | Fermentation effluent |
| DW | Dry weight |
| HS | High salt |
| PhBR | Photobioreactor |
| TAP | Tris-Acetate-Phosphate |

## 1. Introduction

Our dependence on fossil fuels correlates with the increasing level of carbon dioxide concentration in the atmosphere. To avoid this problem, new alternative energy sources should be introduced into the practice.

Microalgae are unicellular organisms capable of converting light into chemical energy. They grow faster than plants and do not compete with plants for the land. That is why microalgal biomass is considered as a valuable alternative energy source. Different approaches of microalgal biomass usage as an energy source are under investigations including direct digestion into methane (González-Fernández et al., 2012), biodiesel (Sheehan et al., 1998; Verma et al., 2010) and ethanol (Miranda et al., 2012). Complex utilization of microalgal biomass is also being pursued (Rizwan et al., 2015).

Another possibility of microalgal biomass usage is a three-stage integrated system for $H_2$ production in which the first stage is microalgal biomass production followed by the dark fermentation of the algal biomass as the second stage, and finally the photofermentation of the dark fermentation products (**Fig. 1**).

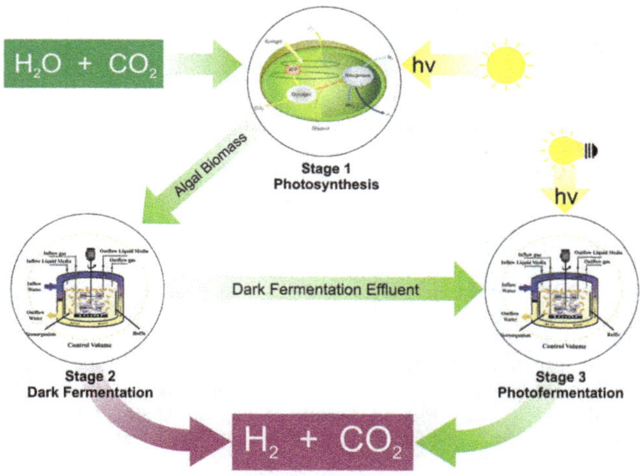


**Fig.1.** An integrated scheme for the light energy accumulation by microalgae, dark fermentation of algae biomass and photofermentation of the dark fermentation effluent.

This three-stage $H_2$ production has been studied by several research teams (Ike et al., 1998; Ike et al., 2001; Kim et al., 2006). The dark fermentation of microalgae biomass with $H_2$ production has been most successful if biomass is rich in starch. Kim et al. (2006) claimed that the highest $H_2$ yield achieved at this stage was 2.58 mol $mol^{-1}$ glucose, calculated on the basis of accumulated starch. They also reported that during the photofermentation of the dark fermentation effluent (FE), the $H_2$ yield increased to 5.7 mol $mol^{-1}$ glucose, and the total $H_2$ yield amounted to 8.3 mol $mol^{-1}$ glucose (Kim et al., 2006). However, it is worth noting that by taking into account the fact that microalgal biomass contains other organics besides starch, the $H_2$ yields may be over-estimated.

In a different study, Kawaguchi et al. (2001) argued that they successfully produced hydrogen from the starch fraction of the biomass using the mixed culture of *Lactobacillus amylovorus* and *Rhodobium marinum* A-501. In their proposed system, *L. amylovorus*, which possesses amylase activity, utilized algal starch for lactic acid production, and *R. marinum* A-501 produced hydrogen in the presence of light using lactic acid as an electron donor. In the described experiments, the biomass of *Dunaliella* and *Chlamydomonas* was freezed-thawed (Kawaguchi et al., 2001). This procedure disrupts most of the cells, simplifying biomass processing in the dark fermentation reactor. In some cases, the processing of microalgal biomass was not described. Though these results are promising in terms of high $H_2$ yield and with respect to algal starch, but the efficiency of the light energy transformation in the system was not considered. In better words, the particular elements of these systems have been explored thoroughly but there is no estimation of the total efficiency of the system, i.e., the ratio of the energy of the $H_2$ obtained to the energy of the incident light. It is worth mentioning that these systems theoretically require only sunlight and are neutral in terms of $CO_2$ production and are therefore, considered very promising.

On such basis, the aim of the present work was to realize the three-step integrated system for $H_2$ production using microalgae, dark fermentative, and purple bacteria, and to estimate the overall efficiency of energy conversion using mass-energy balance regularities.

## 2. Materials and methods

### *2.1. Microalgae and purple bacteria*

Stock cultures of *Chlamydomonas reinhardtii*, strain Dang cc124, and *Chlorella pyrenoidosa 82T* were maintained on agar plates with a standard Tris-Acetate-Phosphate (TAP) medium (pH 6.9) at 28 °C under illumination (36 µE $m^{-2}s^{-1}$). Single colonies were transferred into 10 ml of TAP medium and incubated for 2 d under the same conditions. Then, the cultures were grown autotrophycally on the High-Salt (HS) medium (Sueoka et al., 1967) in 500 ml Erlenmeyer flasks which were bubbled with 2% $CO_2$ in air, filtered through 0.2 mm pore-size membrane filters (Acro 37 TF, Gelman Sciences, Inc., Ann Arbor, MI). The $CO_2$ content in the airflow was analyzed with a DX6100-01 gas analyzer (RMT Ltd., Russia) and maintained using a TRM1 microprocessor system (Oven, Russia).

A microbial consortium (with *Clostridia* predominated) obtained from silo pit liquid (Belokopytov et al., 2009) was used during the dark fermentation stage. The inoculum was grown anaerobically using the medium recommended for biogas-producing microbial communities (Tzavkelova et al., 2012), and was then adapted for starch hydrolysis at 37 °C (Belokopytov et al., 2009).

The purple bacterium *Rhodobacter sphaeroides* N7 (Khusnutdinova et al., 2012) was used at the photofermentative stage. The inoculum was grown 4-5 d on the Ormerod medium (Ormerod et al., 1961) with 10 mM $(NH_4)_2SO_4$ and 20 mM lactate at 28 °C, 60 W $m^{-2}$.

### *2.2. Microalgae cultivation*

To study the efficiency of light energy conversion and to produce microalgae biomass, *C. reinhardtii* was cultivated on the HS medium in a 1.5 L photobioreactor (PhBR) consisting of glass coaxial cylinders (Tsygankov et al., 1994). The thickness of the culture layer was 13 mm. The computerized system was designed to maintain the turbidostat mode (OD 0.05, pH 7.0, and 28.0±0.2 °C). The inoculum was added at 5-10%. The culture was bubbled with 3% $CO_2$ in air at 137 ml $min^{-1}$. The PhBR was illuminated with cool-white fluorescent lamps with the light intensity varying from 36.7 to 256.6 µE $m^{-2}$ $s^{-1}$ PAR. The illuminated area of the culture was 0.083 $m^2$. The outflow culture was collected in 10 L vessels during 7 d. Then, the biomass was harvested by centrifugation (4500 rpm, 15 min) and stored at -10 °C. The microalgae *Chl. pyrenoidosa* was grown in a similar manner in the same medium.

### *2.3. Hydrolysis of C. reinhardtii biomass*

Our preliminary experiments showed that microbial consortium from silo pit liquid did not digest intact cells of *C. reinhardtii* even during 1 month. That is why the pretreatment of biomass appeared to be necessary. The biomass of *C. reinhardtii* was hydrolyzed as follows: 187.5 g of the thawed biomass (corresponding to 15 g of dry weight, $DW_{C.r.}$) was incubated in 112.5 mL of 3.5N $H_2SO_4$ at 120 °C, 1.2 atm, 30 min.

*2.4. Anaerobic dark fermentation of microalgae biomass*

Before fermentation, the pH of the hydrolysate was adjusted to 6.8. Moreover, the hydrolysate was supplied with Mg, Ca, microelements, and phosphates according to the medium composition described by Tzavkelova et al. (2012). Fermentation was performed anaerobically at 37 °C in 500 mL vessels using 4 mL of the microbial consortium as inoculum. At the end of the fermentation, the culture (450 mL) was neutralized to pH 7.0 and harvested by centrifugation (4500 rpm, 15 min). The supernatant was autoclaved, centrifuged again, and was used as fermentation effluent (FE).

*2.5. Utilization of FE for cultivation of purple bacteria (photofermentation)*

To cultivate the purple bacteria, i.e., *R. sphaeroides* N7, the FE was used in non-diluted or diluted form with distilled water, as specified. When indicated some nutrients were added (mg $L^{-1}$): YE, 100; $FeSO_4.7H_2O$, 10; EDTA, 20. Experiments were made in Hungate tubes (16 mL) with 8 mL of the medium with 2% of inoculum, under Ar. Tubes were incubated at 28 °C under illumination; 60 W $m^{-2}$ (incandescent lamps).

*2.6. Other methods*

Chlorophyll (Chl) *a+b* content was assayed spectrophotometrically in 95% ethanol extract (Harris, 1989). Bacteriochlorophyll (Bchl) *a* concentration was measured spectrophotometrically at 772 nm after extraction in 7:2 (v/v) acetone:methanol (Clayton, 1966). Gas production was measured manometrically, and the $H_2$ percentage was analyzed by gas chromatography. The concentration of acetate was determined by gas chromatography as described earlier (Belokopytov et al., 2009). Lactate concentration was assayed by enzymatic method and monitored as NAD reduction at 340 nm (Asatiani, 1969). Glucose concentration was measured by using the Glucose GOD FS kit (DiaSys, Germany). Starch accumulated in the cells was determined as glucose (see above) after enzymatic hydrolysis, according to the method described by Gfeller and Gibbs (1984). The total content of soluble monosacharides and polysaccharides (which could be hydrolyzed by sulfuric acid) was assayed using anthrone reagent and expressed as glucose equivalents (Hanson and Phillips, 1984). Protein concentration was estimated according to the classical Lowry method. The ammonium content was analyzed by the microdiffusion method (Lyubimov et al., 1968). Light intensity was measured in the 400-900 nm region using quantometer (Quantum Meter QMSW-SS) and pyranometer (CM3; Kipp&Zonen, Delft, The Netherlands). During measurements, infrared light with a wavelength more than 850 nm was cut off by filter SZS24. Carbon, hydrogen, and nitrogen (CHN) content in biomass was measured using a CHN analyzer.

## 3. Calculations of energetic efficiency

Calculations were made using the mass and energy balance regularities (Erickson et al., 1978). The energy content of dry algae biomass ($Q_b$) was calculated using the **Equation 1**:

$$Q_b \text{ (kJ g}^{-1}) = 112.8 \times \gamma/M_b \quad (1)$$

where 112.8 is the heat released during the combustion of biomass, which contains 1 g-atom of carbon, with the degree of reduction γ. $M_b$ is the calculated molecular mass of biomass, equal to 25.564 and 25.532 g $mol^{-1}$ for *C. reinhardtii* and *Chl. pyrenoidosa*, correspondingly. The biomass elemental composition is given below.

To calculate γ, the CHN content was measured. The O content was calculated assuming that biomass contains 95% CHNO. According to our measurements, the empirical elemental composition of *C. reinhardtii* and *Chl. pyrenoidosa* was $CH_{0.128}N_{0.178}O_{0.684}$ and $CH_{0.132}N_{0.192}O_{0.666}$, correspondingly. The biomass degree of reduction (γ) was calculated based on the **Equation 2**:

$$\gamma = 4+x+3y-2z \quad (2)$$

where x, y, z represent the numbers of H, N, O atoms, correspondingly, based on biomass composition. Consequently, γ is 3.294 and 3.376, and energy content of biomass is 14.5 and 14.9 kJ $g^{-1}$ for *C. reinhardtii* and *Chl. pyrenoidosa*, respectively.

The efficiency of light energy conversion to energy accumulated in biomass (η) was calculated as shown in the **Equation 3**:

$$\eta \text{ (\%)} = 100E_b/E_{il} \quad (3)$$

where $E_b$ is the energy of heat combustion of biomass produced by the PhBR during 1 h, $E_{il}$ is the incident light energy to the PhBR during 1 h.

The energy of heat combustion of biomass produced by the PhBR during 1 h ($E_b$) was calculated as follows (**Eq. 4**):

$$E_b \text{ (kJ h}^{-1}) = 112.8 \times \mu \times C \times \gamma \times V \quad (4)$$

where μ is specific growth rate ($h^{-1}$), C is the steady-state biomass concentration in the PhBR measured as the number of moles of carbon in biomass per 1 L of culture (mol $L^{-1}$), V is the volume of the culture and is equal to 1.125 and 1.5 L for *C. reinhardtii* and *Chl. pyrenoidosa*, correspondingly.

The incident light energy per 1 h was calculated as follows (**Eq. 5**):

$$E_{il} \text{ (kJ h}^{-1}) = 3600 \times I_o \times S/1000 \quad (5)$$

where $I_o$ (W $m^{-2}$) is the incident light intensity, S ($m^2$) is the illuminated surface of the culture equaling 0.083 and 0.095 $m^2$ for *C. reinhardtii* and *Chl. pyrenoidosa*, respectively.

The specific energy of substrate combustion ($Q_s$) was calculated for acetate, lactate, and glucose based on their CHO formula:

$$Q_s \text{ (kJ g}^{-1}) = 112.8 \times \gamma/M_s \quad (6)$$

where γ is 4 and $M_s$ is 30 (formula of a common type $CH_2O$). Thus, the specific energy of acetate, lactate, and glucose was similar and amounted to 15.04 kJ $g^{-1}$.

The specific energy of hydrogen combustion is 143.1 kJ $g^{-1}$.

## 4. Results and discussion

*4.1. Production of microalgae biomass and efficiency of light energy conversion*

Production of microalgae biomass was studied using *C. reinhardtii* and *Chl. pyrenoidosa* under turbidostat cultivation. The growth rate of *C. reinhardtii* increased with the increase of the light intensity and saturated at 38.5 W $m^{-2}$ reaching the maximal value of 0.115 $h^{-1}$ (**Table 1**).

**Table 1.**
Influence of incident light intensity on the growth parameters and efficiency of light energy conversion in *C. reinhardtii*.

| $I_o$ (W $m^{-2}$)* | $E_{il}$ (kJ $h^{-1}$) | Growth rate, μ ($h^{-1}$) | Biomass concentration ($DW_{C.r.}$) | | $E_b$ (kJ $h^{-1}$) | η (%) |
|---|---|---|---|---|---|---|
| | | | g $L^{-1}$ | mol $L^{-1}$ | | |
| 9.7 | 2.9 | 0.011 | 0.22 | 0.009 | 0.04 | 1.4 |
| 14.1 | 4.2 | 0.046 | 0.27 | 0.011 | 0.20 | 4.8 |
| 18.7 | 5.6 | 0.057 | 0.27 | 0.011 | 0.26 | 4.6 |
| 29.5 | 8.8 | 0.079 | 0.24 | 0.009 | 0.30 | 3.4 |
| 38.5 | 11.5 | 0.110 | 0.24 | 0.009 | 0.41 | 3.6 |
| 64.1 | 19.2 | 0.115 | 0.27 | 0.011 | 0.53 | 2.8 |

* The incident light intensity 9.7 - 64.1 (W $m^{-2}$) corresponded to 36.7 – 256.6 μE $m^{-2}$ $s^{-1}$

The steady-state biomass concentration was about 0.25± 0.02 g $L^{-1}$. Similarly, the energy of heat combustion of biomass produced by the PhBR during 1 h ($E_b$) (Section 3, **Eq. 4**) increased with light intensity up to 0.49 kJ $h^{-1}$. The incident light energy per 1 h ($E_{il}$) increased in proportion to incident light intensity (Section 3, **Eq. 5**). However, the efficiency of light energy conversion to energy accumulated in biomass (ŋ; Section 3, **Eq. 3**) was maximal (4.8 – 4.6%) at rather low light intensities of 14.1 - 18.7 W $m^{-2}$.

Thus, the optimal conditions for the growth rate (and biomass production) distinctly differed from those for efficiency of the light energy conversion. The most efficient light energy conversion took place when the growth rate was only 40% of the maximum value.

Similar results were obtained using the turbidostat culture of *Chl. pyrenoidosa*, while somewhat higher biomass concentration and lower growth rate were obtained (**Table 2**). The highest light energy conversion efficiency of 5.5% was observed at low light intensity (18.7 W $m^{-2}$) when the growth rate was 58% of the maximum value. Since the regularities were the same, *C. reinhardtii* was used in the subsequent experiments.

**Table 2.**
Influence of incident light intensity on the growth parameters and efficiency of light energy conversion in *Chl. pyrenoidosa*.

| $I_o$ (W $m^{-2}$) | $E_{il}$ (kJ $h^{-1}$) | Growth rate, μ ($h^{-1}$) | Biomass concentration ($DW_{Chl.pyr.}$) | | $E_b$ (kJ $h^{-1}$) | ŋ (%) |
|---|---|---|---|---|---|---|
| | | | g $L^{-1}$ | mol $L^{-1}$ | | |
| 9.7 | 3.3 | 0.016 | 0.26 | 0.010 | 0.09 | 2.7 |
| 18.7 | 6.4 | 0.056 | 0.29 | 0.011 | 0.35 | 5.5 |
| 38.5 | 13.2 | 0.093 | 0.28 | 0.011 | 0.58 | 4.4 |
| 64.1 | 21.9 | 0.097 | 0.35 | 0.014 | 0.78 | 3.6 |

These findings were in agreement with the published data on the efficiency of the light energy conversion of 0.2-5.0% (Klass, 1998). It should be noted that this parameter was often calculated in relation to the absorbed light energy bearing in mind that non-absorbed (transmitted) light may be further utilized in some other light-dependent processes. However, it should be emphasized that the intensity of transmitted light is much lower as compared to incident light (10-12 % in the present study). Furthermore, transmitted light is diffused light with modified spectral composition, hence, its utilization is counterproductive. Therefore, the energy of the transmitted light was neglected herein and calculations were made based on the incident light energy.

### *4.2. Pretreatment of the raw microalgae biomass*

Our preliminary fermentation experiments using raw algae biomass (after freeze-thawing) failed and only insignificant production of methane-containing gas was observed (data not shown). Therefore, to improve the fermentation and the availability of carbohydrates to the microorganisms some disruption methods were necessary. Various pretreatment methods have been investigated to produce fermentable sugars from algae biomass ranging from simple heating or freezing-thawing to thermo-acidic or thermo-alkaline hydrolysis (Yang et al., 2011; Liu et al., 2012), ultrasonic disintegration (Jeon et al., 2013; Yun et al., 2013), osmotic shock (Lee et al., 2010), enzymatic pretreatment (Choi et al., 2010). Moreover, different combinations of grinding, enzymatic hydrolysis, hydrogenogens domestication, ultrasonication, microwave-assisted acid heating have also been tested (Cheng et al., 2012). In fact, pretreatment methods are chosen depending on the particular properties of a certain microalgae, especially of their cell wall. Efremenko et al. (2012) applied thermo-acidic pretreatment method for various microalgae and achieved extremely different $H_2$ production rates.

In the present study, the thermo-acidic pretreatment method was used (Section 2.3). This resulted in an increase in the total carbohydrates (measured with anthrone) in the supernatant fraction from 44.8 to 229.7 mg $g^{-1}$ $DW_{C.r.}$ Glucose concentration (measured with glucose oxidase) also increased from 0.09 to 4.3 mg $g^{-1}$ $DW_{C.r.}$

### *4.3. Dark anaerobic fermentation of microalgae hydrolysate*

The utilization of algae biomass for $H_2$ production by using various microorganisms e.g., *C. butyricum* (Kim et al., 2006; Liu et al., 2012), immobilized *Clostridium acetobutylicum* (Efremenko et al., 2012), and anaerobic sewage sludge microflora (Park et al., 2009), through dark fermentation has been reported. For instance, high $H_2$ production of 81-92 mL $g^{-1}$ DW was demonstrated for *Arthrospira platensis* and *Chl. vulgaris* ESP6 hydrolysate obtained through ultrasonic-acid or thermo-acidic pretreatment, correspondingly (Cheng et al., 2012; Liu et al., 2012).

The advantages of monocultures and mixed culture (consortia) have also been widely discussed. Application of pure cultures appears to be very useful approach for experimental estimation of $H_2$ production rates and yields as well as organic acids production. However, in practice, pure culture is not applicable due to high cost of waste sterilization and inability to use wide spectrum of organics (Tekucheva and Tsygankov, 2012).

Even though the dark fermentation of different wastes was studied intensively, this process, however, was not considered in the context of overall efficiency of energy conversion. Hence, the dark anaerobic fermentation of *C. reinhardtii* hydrolysate was performed using a *Clostridia*-predominated consortium as described in Section 2.4. Carbohydrates (glucose) were consumed while acetate (20 mM) and lactate (35 mM) were produced. Gas production started after 24 d and continued during the 56 d-experiment. The total $H_2$ production amounted to 5.8 mM per 1 L FE (**Table 3**), thus, the $H_2$ yield was as low as 0.15 mol $mol^{-1}$ glucose. Evidently, the heterolactic acid fermentation took place, but the low $H_2$ production signified probably that $H_2$ consumers were available in this microbial consortium. Other researchers also reported that low quantities of $H_2$ were produced from *Chl. vulgaris* and *Dunaliella tertiolecta* biomass fermented by anaerobic enrichment cultures derived from digester sludge, and that $H_2$ was subsequently consumed (Lakaniemi et al., 2011).

**Table 3.**
The products of dark anaerobic fermentation of *C. reinhardtii* hydrolysate.

| Substrate/ Products | Concentration of substrate/ products | | Total energetic value** | |
|---|---|---|---|---|
| | mmol L of FE* | mg $g^{-1}$ of $DW_{C.r.}$ | kJ $g^{-1}$ of $DW_{C.r.}$ | kJ 100 $kJ^{-1}$ of biomass |
| Carbohydrates (glucose): | | | | |
| Initial | 42.5 | 230.00 | 3.46 | 23.86 |
| Final | 3.1 | 18.00 | 0.27 | 1.86 |
| $H_2$ | 5.8 | 0.34 | 0.05 | 0.34 |
| $CO_2$ | 4.5 | 14.00 | - | - |
| Acetate | 20.0 | 36.00 | 0.54 | 3.72 |
| Lactate | 35.0 | 99.00 | 1.49 | 10.28 |
| Total | - | - | 2.35 | 16.21 |

* 1 L of FE corresponded to initial 33.3 g of $DW_{C.r.}$
** The specific energy of substrate combustion is given in Section 3.

The carbon recovery in the products was close to 65.3%, which probably means that some additional fermentation products were not detected in the hydrolysate or, alternatively, a significant part of microbial biomass was not digested.

Energy content of the glucose consumed as well as the content of the dark fermentation products were calculated based on **Equation 6** (Section 3). The results obtained showed that the energy content in the glucose available in the hydrolysate was 3.46 kJ $g^{-1}$ (**Table 3**). Furthermore, the energy content of consumed glucose during the fermentation was 3.19 kJ $g^{-1}$. On the other hand, the total energy content of all the measured fermentation products (acetate, lactate, hydrogen, and residual glucose) stood at 2.35 kJ $g^{-1}$. Thus, the energy conversion efficiency of the consumed glucose to all of the measured products during the dark fermentation was 68% (probably underestimated because some products were not measured). While the energy conversion efficiency of the consumed glucose to $H_2$ was only 1.6%.

Moreover, by taking into consideration the initial energy content of dry algae biomass amounted to 14.5 kJ $g^{-1}$ of $DW_{C.r.}$ (Section 3, **Eqs. 1** and **2**), the energy conversion efficiency of the consumed glucose during the dark fermentation into products and $H_2$ was recalculated at 16.2 and 0.3%, respectively.

*4.4. Cultivation of purple bacteria using the FE after dark fermentation (photofermentation)*

Theoretically, the VFAs available in the FE could be used for photofermentation by purple bacteria and $H_2$ could be produced according to the **Equations 7-9**:

$$C_2H_4O_2 + 2H_2O \rightarrow 4H_2 + 2CO_2 \quad (7)$$
$$C_3H_6O_3 + 3H_2O \rightarrow 6H_2 + 3CO_2 \quad (8)$$
$$C_6H_{12}O_6 + 6H_2O \rightarrow 12H_2 + 6CO_2 \quad (9)$$

Thus, using 1 L FE of a known composition (**Table 3**), one could obtain 40 mmol of hydrogen from acetate, 105 mmol from lactate and 18.6 mmol from glucose, in total 327.2 mmol $H_2$ per 1 L of FE. Nevertheless, the cultivation of *R. sphaeroides* N7 on non-diluted and diluted FE (25-50%) did not result in $H_2$ production (**Table 4**). This absence was evidently due to the high level of ammonium content (22.4 mM) in the FE, which was detrimental to the nitrogenase-mediating $H_2$ photoproduction. In fact, this is a common problem for photofermentation stage when using FEs or different wastes with inappropriate C/N ratios. Different ways have been suggested to overcome this problem including chemical methods of ammonium removal (Cheng et al., 2012) and application of ammonium insensitive mutants (Heiniger et al., 2012; Ryu et al., 2014).

**Table 4.**
Final characteristics of *R. sphaeroides* N7 culture grown on the FE (diluted with distilled water) after the dark fermentation.

| FE | Bchl, (mg $L^{-1}$) | Protein (mg $mL^{-1}$) | $NH_4^+$ (mM) | Lactate (mM) | Acetate (mM) | Glucose (mM) | $H_2$ |
|---|---|---|---|---|---|---|---|
| 50% | 61.0±1.4 | 2.1±0.1 | 14.1±1.2 | 3.0±0.3 | 0 | 1.9±0.04 | 0 |
| 50%* | 66.0±2.7 | 1.9±0.2 | 16.9±1.4 | 2.4±0.2 | 0 | 2.8±0.50 | 0 |

* In this case, YE, $FeSO_4.7H_2O$, and EDTA were added (Section 2.5).

The results obtained by using *R. sphaeroides* N7 cultivation on 50% FE are presented in **Table 4**. No significant differences were observed in response to dilution of the FE with distilled water or to the addition of some nutrients. It should be noted that the consumption of acetate was about 100% and that of lactate was above 80%, while the low glucose content did not change. The final concentration of bacterial cells (Bchl)

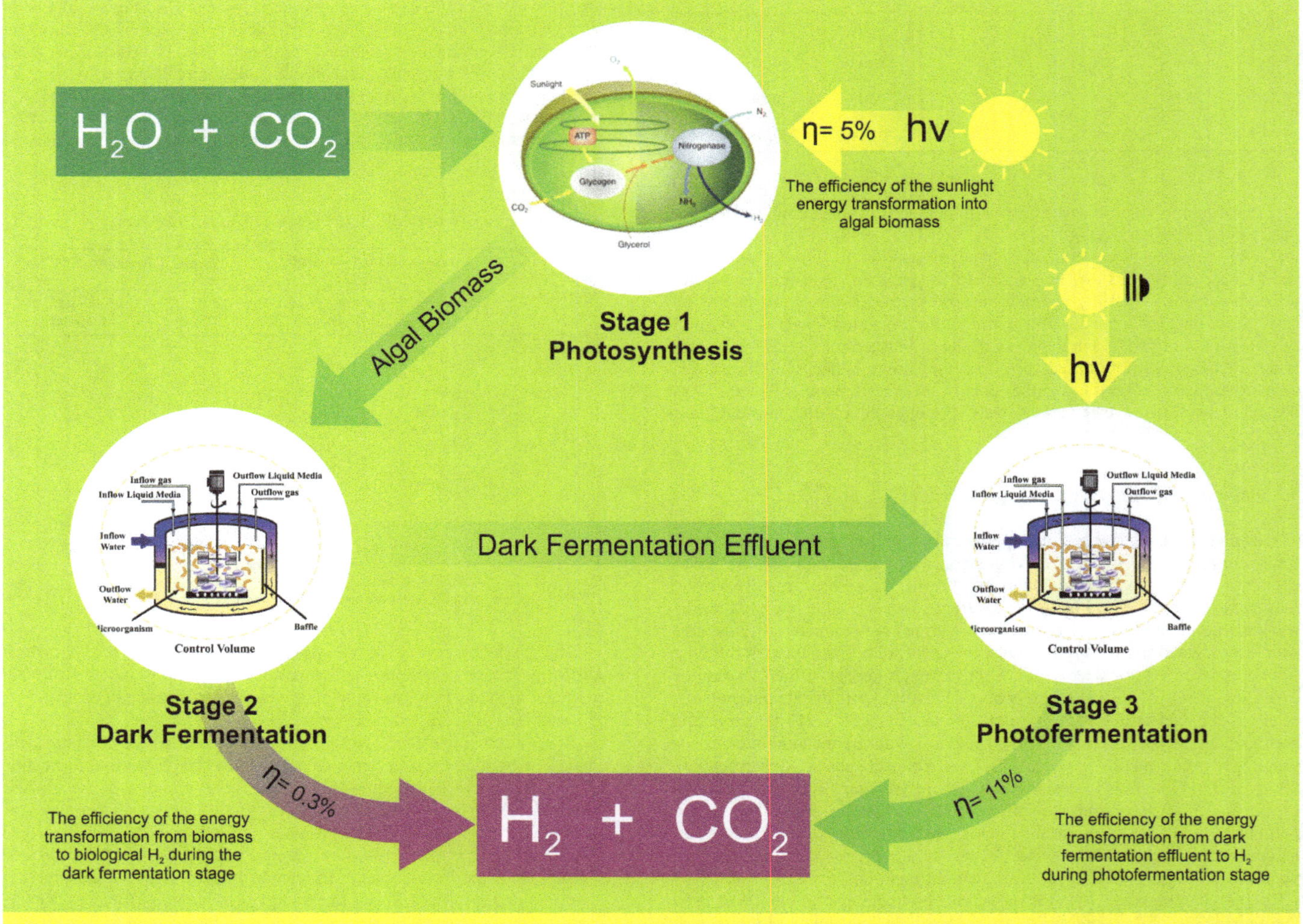


**Fig.2.** The counter-productiveness of the three-stage hydrogen production scheme.

was rather high. Thus, it could be concluded that the medium used was suitable for the growth of the purple bacteria even though the $H_2$ production did not occur.

Nevertheless, to estimate the potential $H_2$ production from the substrates available in the FE (**Table 3**) at non-inhibiting ammonium concentration, the synthetic 50% FE was used where the acetate and lactate concentration were identical to those in the 50% FE but the ammonium content was limited to 3.3 mM. In this case, 187.6 mmol $H_2$ per 1 L FE, i.e., 51.5% of theoretical value was obtained. In another word, the overall $H_2$ production stood at 142.3 mL $g^{-1}$ DW (i.e., 138 mL $g^{-1}$ DW during the photofermentation and 4.3 mL $g^{-1}$ DW during the dark fermentation stage).

The energy of the produced hydrogen was estimated as 1.6 kJ $g^{-1}$ $DW_{C.r.}$ The energy content of the algae biomass was 14.5 kJ $g^{-1}$ $DW_{C.r.}$ (Section 3). Thus, the energy conversion efficiency of the algae biomass to hydrogen was approximately 11%. As demonstrated in Section 4.1, the efficiency of the light energy conversion into the energy of algae biomass was about 5%. This indicates that the overall efficiency of the light energy conversion to the $H_2$ energy (through biomass synthesis and fermentation stages) was not more than 0.55% (**Fig. 2**). Moreover, it should be mentioned that we did not take into account all the additional energy expenditures at various stages (even the light energy consumed by the purple bacteria), which would decrease the efficiency dramatically.

To estimate the potential efficiency of the light energy conversion into $H_2$ using a three-stage scheme, the results obtained in the present study concerning the 1st stage, i.e., microalgae biomass production, could be combined with the most promising results reported previously on the subsequent stages, i.e., dark fermentation and photofermentation. On such basis, the maximal $H_2$ yield reported during dark fermentation and photofermentation was as high as 337 mL $g^{-1}$ $DW_{Ar}$ using *Arthrospira* biomass (Cheng et al., 2012). In their study, the *Arthrospira* biomass was treated by microwave-assisted acid heating, enzymatic hydrolysis, and zeolite to remove ammonium (energy expenditure equivalent of 3.9 kJ $g^{-1}$ $DW_{Ar}$). Assuming that the energy content of *Arthrospira* and *C. reinhardtii* biomass was the same, i.e., 14.5 kJ $g^{-1}$ and that the efficiency of light energy conversion into the energy of algae biomass was about 5%, the efficiency of biomass energy conversion into $H_2$ energy will be 27% and the total energy conversion efficiency of initial light energy into the $H_2$ energy will not exceed 1.4%. Admittedly, the input of the light energy during the 3rd stage was not taken into account, and therefore, the accurate results will be much lower.

## 5. Conclusions

The application of mass and energy balance regularities appeared to be useful for the estimation of the efficiency of light energy conversion into the hydrogen energy at each stage of the three-stage integrated system. Accordingly, the three-stage system was found to possess rather low efficiency of light energy bioconversion even by taking into account the best results available in the literature. Therefore, it could be concluded that this scheme is unproductive for light energy bioconversion due to the numerous intermediate steps. Alternatively, direct light-dependent production of biofuels (ethanol, lipids, or hydrogen) by microalgae as elaborated by Sarsekeyeva et al. (2015) and Tsygankov and Abdullatypov (2015) might be more profitable.

## Acknowledgments

This work was supported by the Russian Science Foundation No 15-14-30007.

## References

[1] Asatiani, V., 1969. Enzymatic methods of analysis (Russ). Nauka, Moscow.

[2] Cheng, J., Xia, A.X., Liu, Y., Lin, R., Zhou, J., Cen, K., 2012. Combination of dark- and photo-fermentation to improve hydrogen production from Arthrospira platensis wet biomass with ammonium removal by zeolite. Int. J. Hydrogen Energy. 37, 13330-13337.

[3] Choi, S.P., Nguyen, M.T., Sim, S.J., 2010. Enzymatic pretreatment of Chlamydomonas reinhardtii biomass for ethanol production. Bioresour. Technol. 101, 5330-5336.

[4] Clayton, R.K., 1966. Spectroscopic analysis of bacteriochlorophylls in vitro and in vivo. Photochem. Photobiol. 5, 669-677.

[5] Efremenko, E.N., Nikolskaya, A.B., Lyagin, I.V., Senko, O.V., Makhlis, T.A., Stepanov, N.A. et al, 2012. Production of biofuels from pretreated microalgae biomass by anaerobic fermentation with immobilized Clostridium acetobutylicum cells. Bioresour. Technol. 114, 342-348.

[6] Erickson, L.E., Minkevich, I.G., Eroshin, V.K., 1978. Application of mass and energy balance regularities in fermentation. Biotechnol. Bioeng. 20, 1595-1621.

[7] Gfeller, R.P., Gibbs, M., 1984. Fermentative metabolism of Chlamydomonas reinhardtii. I: analysis of fermentative products from starch in dark and light. Plant. Physiol. 75, 212-218.

[8] González-Fernández, C., Sialve, B., Bernet, N., Steyer, J.P., 2012. Impact of microalgae characteristics on their conversion to biofuel. Part II: Focus on biomethane production. Biofuels, Bioprod. Biorefin. 6, 205-218.

[9] Hanson, R., Phillips, G., 1984. Chemical composition of bacterial cell, in: Gerhardt, P. et al., (Eds.), Manual of Methods for General Bacteriology (Russian translation). Mir, Moscow, pp. 283-375.

[10] Harris, E.H., 1989. The Chlamydomonas Sourcebook: A comprehensive guide to biology and laboratory use. Academic Press, San Diego, pp. 780.

[11] Heiniger, E.K., Oda, Y., Samanta, S.K., Harwood, C.S., 2012. How posttranslational modification of nitrogenase is circumvented in Rhodopseudomonas palustris strains that produce hydrogen gas constitutively. Appl. Environ. Microbiol. 78, 1023-1032.

[12] Ike, A., Toda, N., Murakawa, T., Hirata, K., Miyamoto, K., 1998. Hydrogen photoproduction from starch in CO2-fixing microalgal biomass by a halotolerant bacterial community, in: Zaborsky, O.R. (Ed.), Biohydrogen. Plenum Press, NY, pp. 311-318.

[13] Ike, A., Kawaguchi, H., Hirata, K., Miyamoto K., 2001. Hydrogen photoproduction from starch in algal biomass, in: Miyake, J., Matsunaga, T., San Pietro, A. (Eds.), Biohydrogen II: an approach to environmentally acceptable technology. Pergamon, Amsterdam, pp. 53-61.

[14] Jeon, B.H., Choi, J.A., Kim, H.C., Hwang, J.H., Abou-Shanab, R.A.I,, Dempsy, B.A. et al, 2013. Ultrasonic disintegration of microalgal biomass and consequent improvement of bioaccessibility/bioavailability in microbial fermentation. Biotechnol. Biofuels. 6, 37-45.

[15] Kawaguchi, H., Hashimoto, K., Hirata, K., Miyamoto, K., 2001. H2 production from algal biomass by mixed culture of Rhodobium marinum A-501 and Lactobacillus amylovorus. J. Biosci. Bioeng. 91, 277-282.

[16] Kim, M.S., Baek, J.S., Yun, Y.S., Sim, S.J., Park, S., Kim, S.-C., 2006. Hydrogen production from Chlamydomonas reinhardtii biomass using a two-step conversion process: Anaerobic conversion and photosynthetic fermentation. Int. J. Hydrogen Energy. 31, 812-316.

[17] Khusnutdinova, A.N., Hristova, A.P., Ovchenkova, E.P., Laurinavichene, T.V., Shastic, E.S., Liu, J. et al, 2012. New tolerant strains of purple nonsulfur bacteria for hydrogen production in a two-stage integrated system. Int. J. Hydrogen Energy. 37, 8820-8827.

[18] Klass, D. L., 1998. Biomass for Renewable Energy, Fuels, and Chemicals. Academic Press, San Diego, CA.

[19] Lakaniemi, A.M., Hulatt, C.H., Thomas, D.N., Tuovinen, O.H., Jaakko A Puhakka J.A., 2011. Biogenic hydrogen and methane production from Chlorella vulgaris and Dunaliella tertiolecta biomass. Biotechnol. Biofuels. 4, 34-46.

[20] Lee, J.; Yoo, C.; Jun, S.; Ahn, C., Oh, H., 2010. Comparison of several methods for effective lipid extraction from microalgae. Bioresour. Technol., 101, 575-577.

[21] Liu, C.H., Chang, C.Y., Cheng, C.L., Lee, D.J., Chang, J.S., 2012. Fermentative hydrogen production by Clostridium butyricum CGS5 using carbohydrate-rich microalgal biomass as feedstock. Int. J. Hydrogen Energy. 37, 15458-15464.

[22] Lyubimov, V.I., L`vov, N.P., Kirshteine, B.E., 1968. Modification of the microdiffusion method for ammonium determination. Prikl. Biochim. Microbiol. 4, 120-121.

[23] Miranda, J.R., Passarinho, P.C., Gouveia, L., 2012. Bioethanol production from Scenedesmus obliquus sugars: the influence of photobioreactors and culture conditions on biomass production. Appl. Microbiol. Biotechnol. 96, 555-564.

[24] Ormerod, J.G., Ormerod, S.K., Gest, H., 1961. Light-dependent utilization of organic compounds and photoproduction of hydrogen by photosynthetic bacteria. Arch. Biochem. Biophys. 64, 449-463.

[25] Park, J-I., Lee, J., Sim, S.J., Lee, J-H., 2009. Production of hydrogen from marine macro-algae biomass using anaerobic sewage sludge microflora. Biotechnol. Bioprocess Eng. 14, 307-315.

[26] Rizwan, M., Lee, J.H., Gani, R., 2015. Optimal design of microalgae-based biorefinery: Economics, opportunities and challenges. Appl. Energy. 150, 69-79.

[27] Ryu, M.H., Hull, N.C., Gomelsky, M., 2014. Metabolic engineering of Rhodobacter sphaeroides or improved hydrogen production. Int. J. Hydrogen Energy. 39, 6384-6390.

[28] Sarsekeyeva, F., Bolatkhan, K., Usserbaeva, Z.A., Bedbenov, V.S., Sinetova, M.A., Los, D.A., 2015. Cyanofuels: biofuels from cyanobacteria. Reality and perspectives. Photosynth. Res. 125, 329-340.

[29] Sheehan, J., Dunahay, T., Benemann, J., Roessler, P., 1998. A look back at t e U.S. Department of Energy's aquatic species program: biodiesel from algae. National Renewable Energy Laboratory, Report NREL/TP-580-24190.

[30] Sueoka, N., Chiang, K.S., Kates, J.R., 1967. Deoxyribonucleic acid replication in meiosis of *Chlamydomonas reinhardtii*. I. Isotopic transfer experiments with a strain producing eight zoospores. J. Mol. Biol. 25, 47-66.

[31] Tekucheva, D.N., Tsygankov, A.A., 2012. Coupled Biological Hydrogen_Producing Systems: A Review. Prikl. Biochim. Microbiol. 4, 357-375.

[32] Tsygankov, A.A., Abdullatypov, A., 2015. Hydrogen Metabolism in Microalgae, in: Allakhverdiev, S.I. (Ed.), Photosynthesis: New Approaches to the Molecular, Cellular, and Organismal Levels, Wiley-Scrivener, Beverly, pp. 133-162.

[33] Tsygankov, A.A., Laurinavichene, T.V., Gogotov, I.N., 1994. Laboratory scale photobioreactor. Biotechnol. Techn. 8, 575-578.

[34] Tsavkelova, E.A., Egorova, M.A., Petrova, E.V., Netrusov, A.I. 2012. Biogas production by microbial communities via decomposition of cellulose and food waste. Appl. Biochem. Microbio. 48(4), 377-384.

[35] Verma, N.M., Mehrotra, S., Shukla, A., Mishra, B.N., 2010. Prospective of biodiesel production utilizing microalgae as the cell factories: A comprehensive discussion. Afr. J. Biotechnol. 9, 1402-1411.

[36] Yang, Z., Guo, R., Xu, X., Fan, X., Li, X., 2011. Thermo-alkaline pretreatment of lipid-extracted microalgal biomass residues enhances hydrogen production. J. Chem. Technol. Biotechnol. 86, 454-460.

[37] Yun, Y.M., Jung, K.W., Kim, D.H., Oh, Y.K., Cho, S.K., Shin, H.S., 2013. Optimization of dark fermentative H2 production from microalgal biomass by combined (acid + ultrasonic) pretreatment. Bioresour. Technol. 141, 220-226.

# 13

# A study of production and characterization of Manketti (*Ricinodendron rautonemii*) methyl ester and its blends as a potential biodiesel feedstock

A.E. Atabani[1,2,3]*, M. Mofijur[1], H.H. Masjuki[1], Irfan A. Badruddin[1], W.T. Chong[1], S.F. Cheng[4], S.W. Gouk[4]

[1] *Department of Mechanical Engineering, Faculty of Engineering, University of Malaya, 50603 Kuala Lumpur, Malaysia.*

[2] *Department of Mechanical Engineering, Erciyes University, 38039 Kayseri, Turkey.*

[3] *Erciyes Teknopark A.Ş, Yeni Mahalle Aşıkveysel Bulvarı Erciyes Teknopark Tekno3 Binası 2. KatNo: 28, 38039 Kayseri, Turkey.*

[4] *Unit of Research on Lipids (URL), Department of Chemistry, Faculty of Science, University of Malaya, 50603 Kuala Lumpur, Malaysia.*

## HIGHLIGHTS

- *Production and characterization of Manketti methyl ester and its blends with diesel.*
- *At all engine speeds and compared to B0, B5 produced lower brake power by 1.18% and higher BSFC by 2.26%.*
- *B5 increased the CO and HC emissions by 32.27% and 37.5%, respectively, but decreased NO emission by 5.26% compared to the neat diesel.*

## GRAPHICAL ABSTRACT

## ABSTRACT

Globally, more than 350 oil-bearing crops are known as potential biodiesel feedstocks. This study reports on production and characterization of Manketti (*Ricinodendron rautonemii*) methyl ester and its blends with diesel. The effect of Manketti biodiesel (B5) on engine and emissions performance was also investigated. The cloud, pour and cold filter plugging points of the produced biodiesel were measured at 1, 3 and 5 °C, respectively. However, the kinematic viscosity of the biodiesel generated was found to be 8.34 $mm^2/s$ which was higher than the limit described by ASTM D6751 and EN 14214. This can be attributed to the high kinematic viscosity of the parent oil (132.75 $mm^2/s$). Nevertheless, blending with diesel improved this attribute. Moreover, it is observed that at all engine speeds, B5 produced lower brake power (1.18%) and higher brake specific fuel consumption (2.26%) compared to B0 (neat diesel). B5 increased the CO and HC emissions by 32.27% and 37.5%, respectively, compared to B0. However, B0 produced 5.26% higher NO emissions than B5.

**Keywords:**
Biodiesel feedstocks
Manketti oil
Transesterification
Physico-chemical properties
Engine performance

* Corresponding author
E-mail address: a_atabani2@msn.com

**Nomenclature**

| | |
|---|---|
| ASTM | American society for testing and materials |
| B0 | (100% diesel and 0% biodiesel) |
| B5 | (95% diesel and 5% biodiesel) |
| BP | Brake Power |
| BSFC | Brake specific fuel consumption |
| CCIO | Crude *Calophyllum inophyllum* oil |
| CCO | Crude coconut oil |
| CIME | *Calophyllum inophyllum* methyl ester |
| CMO | Crude Manketti oil |
| COME | Coconut oil methyl ester |
| CO | Carbon monoxide |
| CO2 | Carbon dioxide |
| CPEO | Crude *Pangium edule* oil |
| GC | Gas chromatography |
| HC | Hydrocarbons |
| MME | Manketti methyl ester |
| NO | Nitrogen oxides |
| PEME | *Pangium edule* methyl ester |
| rpm | Revolution per minute |

## 1. Introduction

The research on alternative fuels is currently gaining worldwide attention due to the increasing energy demands and depleting fossil reserves. Moreover, the increasing global warming and other environmental hazards have forced almost all countries to reduce theie dependence on fossil fuels (Ramaraju and Kumar, 2011). Biodiesel (Greek, bio, life + diesel from Rudolf Diesel) is currently the most common alternative fuel being developed and used as a replacement for petroleum-based diesel. It can be defined as a mixture of mono-alkyl esters of long chain fatty acids (FA) derived from renewable lipids such as vegetable oils and animal fats when reacted with an alcohol (methanol or ethanol) in presence or absence of a catalyst (Demirbas, 2009; Atabani et al., 2012; Kafuku and Mbarawa, 2013). Biodiesel is biodegradable, non-toxic, renewable, and has low emissions of CO, $SO_2$, particulates and hydrocarbons (HC) as compared to conventional diesel (Kafuku and Mbarawa, 2013). The use of vegetable oils (edible and non-edible) plays an important role in biodiesel production. However, availability of these raw materials varies. This necessitates the search for new low-cost agricultural crops (Ibeto et al., 2012; Atabani et al., 2013c). Globally, there are more than 350 oil-bearing crops identified for biodiesel production (Atabani et al., 2012).

Currently one of the main concerns of biodiesel research area is to identify, characterize and perform engine performance analyses on biodiesels derived from many new feedstocks as well as their blends with diesel. For instance, the potential of *Stauntonia chinensis*, Alperujo, Baobab *(Adansonia digitata L.)*, Milkweed *(Calotropis gigantea)*, *Raphanus sativus (oilseed radish)*, *Syagrus romanzoffiana* and *fodder radish* as new sources for biodiesel production has been studied by researchers (de Andrade Ávilaa and Sodré, 2012; Chammoun et al., 2013; Moreira et al., 2013; Hernándeza et al., 2014; Modiba et al., 2014; Phoo et al., 2014; Wang et al., 2014). This indicates the importance of exploring and testing non-conventional biodiesel feesdtocks. Manketti (*Ricinodendron rautonemii*) oil is one of the possible alternative oil crops for biodiesel production. Most previous reports on this plant have been focused on its use in the herbal cosmetic industry (Kafuku and Mbarawa, 2013), and only recently Manketti was explored for biodiesel production (Ruttoa and Enweremadu, 2011; Kafuku and Mbarawa, 2013).

Manketti (*Ricinodendron rautonemii*) belongs to the family *Euphorbiaceae*, locally known as mongongo and feather weight tree. It is a large (7 to 20 m high and 60 cm in diameter), and deciduous tree which grows in the wild on sandy soils between the latitudes of 15 and 21 °S in many parts of Botswana, South Africa, Zambia, Angola, Namibia and up to Mozambique. It is commonly used as a street tree in Victoria Falls, where the Zambesi river falls off the arid southern Zambian plains at North western Zimbabwe (Naturalhub, 2013; Nerd et al., 2013). Some countries such as Mauritania, Guinea, Liberia, Ethiopia, Sudan, Chad, Mali, Niger and Uganda are also suitable for the cultivation of Manketti tree (Beau, 2003)(Figure1). Moreover, the tree was introduced to Australia in the late 1980's (Naturalhub, 2013). This tree prefers hot and dry climates with low amounts of rain (Beau, 2003). Manketti is considered as an important source of food to many rural communities in southern Africa as well (European Commision, 1998).

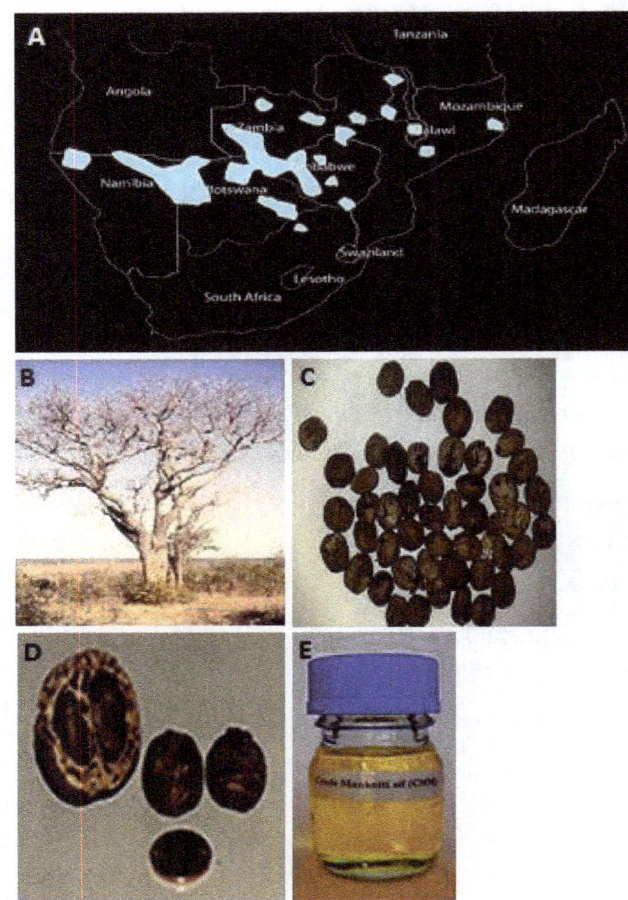


**Fig.1.** Distribution map, tree, fruit, seeds and crude oil of Manketti (*Ricinodendron rautonemii*)
(Beau, 2003; Savaneskin, 2013)

It is worth mentioning that manketti is considered a rapid growing tree and has been designated a protected tree in Namibia since 1952. The tree has a large, straight trunk with stubby and contorted branches and a large spreading crown (Beau, 2003). The branches are stubby and contorted. The trees take around 25 years to commence fruiting. The tree flowers depending on local climatic variations such as temperature and rainfalls, which is around October to December. The small whitish-yellow flowers become a somewhat oval, vaguely plum-like fruit of about 3.5 cm long and 2.5 cm wide. The young fruit is at first covered in fine small hairs on its thin but tough outer skin; under the skin is a narrow spongy layer, at first green, then turning whitish brown with maturity (Naturalhub, 2013). Fruits are egg-shaped, velvety and contain a thin layer of edible flesh around a thick, hard pitted nutshell that encloses the edible oil-bearing kernel (Kafuku and Mbarawa, 2013). The nuts yield a high quality yellow transparent edible oil of which about 60% is used for food and cosmetics. The protein content of the nut is nearly 30%. The shell of the nuts are used as fuel while the leaves are sued as fodder (European Commision, 1998). Recently, the tree has been explored for biodiesel production (Ruttoa and Enweremadu, 2011). Figure1 shows the distribution map, tree, fruit, seeds and crude oil of Manketti (*R. rautonemii*) (Beau, 2003; Savaneskin, 2013).

The purpose of this work was to produce biodiesel from Manketti (*R. rautonemii*) oil using alkaline KOH catalyst followed by a detailed study of physical and chemical properties of the produced biodiesel (MME) and its blends with diesel. The important fuel properties such as density, kinematic viscosity, oxidation stability, viscosity index, calorific value, as well as cold flow properties i.e. cloud point, pour point and cold filter plugging point were measured and compared with ASTM D6751 and EN 14214 standards. Moreover, the current study aimed to investigate the effect of Manketti biodiesel blends (B5) on engine and emissions performance. The success of this study could yield promising and massive new raw material for biodiesel production on a large scale.

## 2. Material and method

### 2.1. *Materials and chemicals*

Crude Manketti oil (CMO) was supplied by Universiti Sains Malaysia (USM), Malaysia, through a personal communication. Reagent grade methanol, potassium hydroxide, sodium sulphate anhydrous and qualitative filter paper (Filtres Fioroni, France) of 150 mm size were used as received.

### 2.2. *Biodiesel production (Alkaline-catalysed Transesterification process)*

In this study, a small scale laboratory reactor consisting of 1L batch reactor (Brand: Favorit), condenser to recover methanol, overhead stirrer (IKA EUROSTAR digital), water bath (WiseCircu® Fuzzy Control System), thermometer and sampling outlet was used to produce biodiesel from CMO. Figure 2 shows the adopted methodology to transesterify CMO to obtain MME. It can be seen that a single step of Alkaline-catalysed transesterification process was adopted as the FFA% of CMO was 1.04% (Table 3) (Atabani et al., 2012). The yield of biodiesel was measured at >90%.

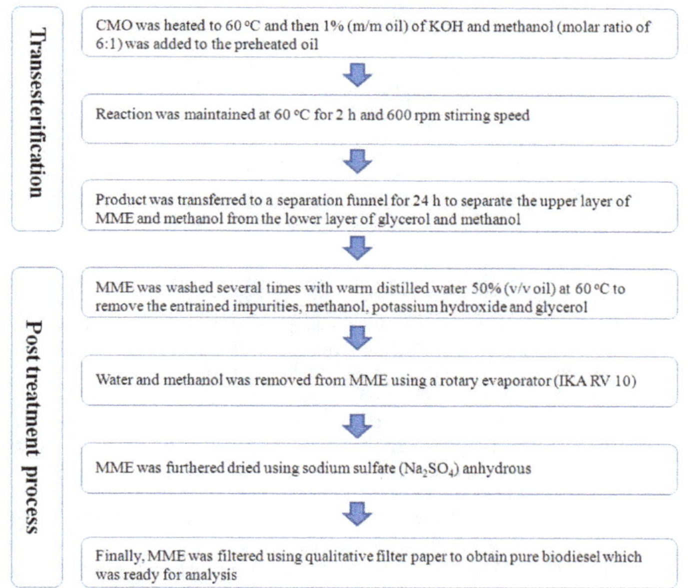


**Fig.2.** Flow chart of transesterification process of CMO to obtain MME.

### 2.3. *Determination of fatty acid compositions of MME using GC*

To determine the FA compositions of MME, a biodiesel sample (1 μL) was injected into a gas chromatography (Shidmadzu, GC-2010A series) equipped with a flame ionization detector and a BPX70 capillary column of 30 m × 0.25 μm × 0.32 mm (Column head pressure: 56.9 kPa). An initial temperature of 140 °C was maintained for 2 min, which was then increased at 8 °C per min to 165 °C, 3 °C per min to reach 192 °C and finally 8 °C per min to reach 220 °C. The column was maintained at the final temperature for another 5 min. The oven, injector and the detector ports were set at 140, 240 and 260 °C, respectively. The carrier gas was helium with the column flow rate at 1.10 mL/min at a 50:1 split ratio (Linear velocity 24.2 cm/sec).

### 2.4. *Analysis of fuel properties*

In this study, the physical and chemical properties of CMO, MME and Manketti-diesel blends (0-20% by volume) were determined. Moreover, Table 1 shows a summary of the equipment and test methods used in this study to analyse properties according to the ASTM D 6751 standard.

**Table 1.**
Equipment list.

| Property | Equipment | ASTM D6751 |
|---|---|---|
| Kinematic viscosity | SVM 3000[a] | ASTM D7042/D445 |
| Flash Point | Pensky-martens flash point-automatic NPM440[b] | D 93 |
| Oxidation stability | 873 Rancimat[c] | D 675 |
| Cloud & Pour point | Cloud and Pour point tester-automatic NTE450[b] | D 2500 & D 97 |
| CFPP | Cold filter plugging point-automatic NTL450[b] | D 6371 |
| Density | SVM 3000[a] | D 1298 |
| Dynamic viscosity | SVM 3000[a] | N/S* |
| Viscosity Index (VI) | SVM 3000[a] | N/S |
| Caloric value | C2000 basic calorimeter[d] | N/S |
| Refractive Index | RM 40 Refractometer[e] | N/S |
| Transmission | Spekol 1500[f] | N/S |
| Absorbance | Spekol 1500[f] | N/S |

Manufacturer: a: Anton Paar (UK), b: Normalab (France), c: Metrohm (Switzerland), d: IKA (UK), e: Mettler Toledo (Switzerland) and, f: Analytical Jena (Germany).
N/S ≡ not specified in ASTM D6751 test methods.

### 2.5. *Engine tests*

A Mitsubishi Pajero (model 4D56T) multi-cylinder diesel engine was used. The experimental investigation was carried out using diesel fuel (B0) and B5 (95% diesel and 5% Manketti methyl esters). Firstly, the engine was run with diesel for 15 min to warm up before running with B5. The engine was run by diesel before the engine was shut down. This was a very important procedure to ensure that the engine was free from biodiesel blends before proceeding to the next test. The engine was run at various speeds ranging from 1000 to 4000 rpm at full load condition. Engine test conditions were monitored by REO-DCA controller connected through a desktop computer to the engine test bed. A BOSCH exhaust gas analyser (model BEA-350) was used to measure the exhaust emission i.e. NO, HC, $CO_2$ and CO. In order to get the average values, all tests were repeated three times. Figure 3 shows the test rig of the engine. The details of the engine specification are tabulated as Table 2.

**Fig.3.** Engine test bed set-up.

**Table 2.**
Detailed specification of the engine.

| Model | | Mitsubishi Pajero (4D56T) |
|---|---|---|
| Engine type | | 4 cylinder inline (Natural aspiration) |
| Displacement | L | 2.5 |
| Cylinder bore x stroke | mm | 91.1 x 95 |
| Valve mechanism | | Single overhead camshaft |
| Rocker arm | | Roller flow type |
| Compression ratio | | 21:1 |
| Maximum engine speed | rpm | 4200 |
| Maximum power | kW | 55 |
| Fuel system | | Distributor type injection pump |
| Lubrication System | | Pressure feed , full flow filtration |
| Oil pump | | Trochoid type |
| Combustion chamber | | Swirl type |
| Cooling system | | Radiator cooling |
| Water pumptype | | Centrifugal impeller type |
| A/F ratio | | 26 |

*2.6. Error analysis*

Errors and uncertainties in the experiments could be associated with instrument selection, condition, calibration, environment, observation, reading and test planning. Uncertainty analysis was needed to prove the accuracy of the experiments. The accuracy of the various parameters were as follows: BP ±0.07 kW, BSFC ±5 g/kWh, CO ± 0.001 vol.%, HC ± 1 ppm, NO ± 1ppm and $CO_2$ ± 0.01 vol.%. An uncertainly analysis was performed using the following formula described by Shahabuddin et al. (2012a) (Eq. 1):

Total uncertainty% = Square root of {(uncertainty of BP)$^2$ + (uncertainty of BSFC)$^2$ + (uncertainty of CO)$^2$ + (HC)$^2$ + (uncertainty of NO)$^2$ + (uncertainty of $CO_2$)$^2$} Eq. 1

For example, the total uncertainty% of this experiment while using B0 was calculated at 2.27% as presented below:

Total uncertainty% = square root of $\{(0.18)^2 + (1.44)^2 + (0.00012)^2 + (0.08)^2 + (0.38)^2 + (0.123)^2\} = 2.27\%$

## 3. Results and discussion

*3.1. Properties of CMO*

Table 3 shows the main properties of CMO in comparison with those of the crude *Pangium edule* oil (CPEO), crude *Calophyllum inophyllum* oil (CCIO) and crude coconut oil (CCO). As seen, CMO was shown to have a very high kinematic viscosity of 132.65 mm$^2$/s compared to 55.677 mm$^2$/s for CCIO, 27.64 mm$^2$/s for CCO and 27.175 mm$^2$/s for CPEO. The acid value of CMO was found favourable for biodiesel production, at only 2.08 mg KOH/g oil compared to 41.74 mg KOH/g oil for CCIO and 19.62 mg KOH/g oil for CPEO. The oxidation stability of CME was only 0.16 h compared to 0.23 h for CCIO, and 6.93 h for CCO.

**Table 3.**
Properties of CMO in comparison with those of CPEO, CCIO and CCO.

| Property | CMO | CCIO[a] | CCO[a] | CPEO[b] |
|---|---|---|---|---|
| Kinematic viscosity *at 40 °C (mm²/s)* | 132.75 | 55.68 | 27.64 | 27.18 |
| Kinematic viscosity *at 100 °C (mm²/s)* | 20.62 | 9.56 | 5.94 | 6.64 |
| Dynamic viscosity *at 40 °C (mpa.s)* | 122.81 | 51.31 | 25.12 | 24.39 |
| Viscosity Index (VI) | 179.9 | 165.4 | 168.5 | 216.2 |
| Flash Point (°C) | 192.5 | 236.5 | 264.5 | N/D |
| Cloud point (°C) | N/D | N/D | N/D | -6 |
| Pour point (°C) | -9 | N/D | N/D | -10 |
| Density *at 40 °C (kg/m³)* | 925.1 | N/D | 908.9 | 897.6 |
| Density *at 15 °C (kg/m³)* | 943.0 | 951.0 | N/D | N/D |
| Specific gravity at 15 °C | 0.9443 | 0.952 | N/D | N/D |
| Acid Value (mg KOH/g oil) | 2.08 | 41.74 | N/D | 19.62 |
| Free fatty acid (FFA) | 1.04 | N/D | N/D | N/D |
| Caloric value (kJ/kg) | 38,682 | 38,511 | 37,806 | 39,523 |
| Oxidation stability (h at 110 °C) | 0.16 | 0.23 | 6.93 | 0.08 |
| Refractive Index | 1.487 | 1.4784 | 1.4545 | 1.4683 |
| Transmission (%T) | 89.3 | 34.7 | 91.2 | 86.1 |
| Absorbance (Abs) | 0.049 | 0.46 | 0.04 | 0.064 |

[a] Atabani et al., 2013b
[b] Atabani et al., 2014

*3.2. Fatty acid compositions of MME*

Table 4 shows the results of FA compositions analysis of MME in comparison with the other biodiesel samples i.e. *Pangium edule* methyl ester (PEME), *Calophyllum inophyllum* methyl ester (CIME) and coconut oil methyl ester (COME). Based on the GC analysis, it was found that MME contained 18.3% saturated FAs and 81.7% unsaturated FAs compared to the values reported previously i.e. 17.2% saturated FAs and 82.9% unsaturated FAs by Kafuku and Mbarawa (2013) and 27.4% saturated FAs and 72.8% unsaturated FAs by Ruttoa and Enweremadu (2011). It can be concluded that MME, PEME and CIME are mainly dominated by unsaturated FAs while COME is dominated by saturated FAs.

**Table 4.**
Fatty acid composition of MME and comparison with CIME, COME and PEME

| Fatty acid composition (as % methyl esters) | MME | MME[a] | MME[b] | CIME[c] | COME[c] | PEME[d] |
|---|---|---|---|---|---|---|
| C8:0 | N/D | N/D | N/D | N/D | 8.2 | N/D |
| C10:0 | N/D | N/D | N/D | N/D | 6.6 | N/D |
| C12:0 | N/D | N/D | N/D | N/D | 48.3 | N/D |
| C14:0 | N/D | N/D | N/D | N/D | 16.4 | 0.1 |
| C16:0 | 10.5 | 9.8 | 15.4 | 14.4 | 9.3 | 8.3 |
| C16:1 | N/D | N/D | 1.6 | 0.3 | N/D | 0.1 |
| C18:0 | 7.8 | 7.2 | 11.5 | 15.2 | 2.4 | 4.0 |
| C18:1 | 18.2 | 16.4 | 23.4 | 41.9 | 7 | 45.2 |
| C18:2 | 43.9 | 43.4 | 45.1 | 26.6 | 1.7 | 39.3 |
| C18:3 | 19.3 | N/D | 1.7 | 0.2 | N/D | 2.5 |
| C20:0 | N/D | 0.2 | 0.5 | 0.8 | N/D | 0.2 |
| C20:1 | 0.3 | 0.3 | 1.0 | 0.2 | N/D | 0.3 |
| C20:2 | N/D | 0.2 | N/D | N/D | N/D | N/D |
| C20:5 | N/D | 22.6 | N/D | N/D | N/D | N/D |
| C22:1 | N/D | N/D | N/D | 0.5 | N/D | N/D |
| Total Saturated FA (%) | 18.3 | 17.2 | 27.4 | 30.4 | 91.3 | 12.6 |
| Total Monounsaturated FA (%) | 18.5 | 16.7 | 26.0 | 42.8 | 7 | 45.6 |
| Total Polyunsaturated FA(%) | 63.2 | 66.2 | 46.8 | 26.8 | 1.7 | 41.8 |

N/D ≡ Not detected [a] Kafuku and Mbarawa, 2013 [b] Ruttoa and Enweremadu, 2011
[c] Atabani et al., 2013a [d] Atabani et al., 2014

*3.3. Physico-chemical properties of MME*

Table 5 shows the physico-chemical properties of the produced MME in comparison with those of the CIME, COME and PEME (Ruttoa and Enweremadu, 2011; Atabani et al., 2013a; Kafuku and Mbarawa, 2013; Atabani et al., 2014). Having considered the data presented in Tables 3 and 4 for CMO and MME, it can be seen that the kinematic viscosity was successfully reduced from 132.75 to 8.03425 mm$^2$/s, refractive index from 1.487 to 1.4698, and density at 40 °C from 925.1 to 887.8 kg/m$^3$. However, the kinematic viscosity of MME was still higher than the limit specified by both the ASTM D 6751 and EN 14214 standards. This might be attributed to the high kinematic viscosity of the parent oil (132.75 mm$^2$/s). The same was observed by Sanford et al. (2009) for castor methyl ester as the kinematic viscosity of castor methyl ester was 15.25 mm$^2$/s compared to 251.2 mm$^2$/s for the crude oil.

The cloud, pour and cold filter plugging points of MME were measured at 1, 3 and 5 °C which were considerably lower than the values previously reported for CIME i.e. 10, 11 and 9 °C, respectively (Atabani et al., 2013a), but were higher than those of COME i.e. 0, -4 and -4 °C (Atabani et al., 2013a) and PEME i.e. -6, -4 and -8 °C (Atabani et al., 2014). Moreover, the oxidation stability of MME (0.52 h) also failed to meet both the ASTM D6751 and EN 14214 standards of 3 and 6 h, respectively. Therefore, it is necessary to supplement the MME with conventional antioxidants such as propyl gallate (PrG) to make up for its low oxidation stability (Hajari et al., 2014).

**Table 5.**
Physico-chemical properties of MME in comparison with those of CIME, COME, and PEME.

| Property | MME | CIME[a] | COME[a] | PEME[b] | ASTM D6751 | EN 14214 |
|---|---|---|---|---|---|---|
| Kinematic viscosity *at 40 °C (mm²/s)* | 8.34 | 5.75 | 4.06 | 5.23 | 1.9-6.0 | 3.5-5.0 |
| Kinematic viscosity *at 100 °C (mm²/s)* | 2.67 | 2.03 | 1.57 | 1.97 | N/A | N/A |
| Dynamic viscosity *at 40 °C (mpa.s)* | 7.41 | 5.04 | 3.52 | 4.56 | N/A | N/A |
| Viscosity Index (VI) | 176 | 174.9 | 180.7 | 211.8 | N/A | N/A |
| CP (°C) | 1 | 10 | 0 | -6 | Report | Report |
| PP (°C) | 3 | 11 | -4 | -4 | Report | Report |
| CFPP (°C) | 5 | 9 | -4 | -8 | Report | Report |
| Density *at 40 °C (kg/m³)* | 887.8 | 877.4 | 866.4 | 871.0 | N/A | N/A |
| Density *at 15 °C (kg/m³)* | 906.0 | N/D | N/D | 890.0 | N/A | 860-900 |
| Specific gravity at 15 °C | 0.9068 | N/D | N/D | 0.8910 | N/A | N/A |
| Caloric value (kJ/kg) | 39,070 | 39,273 | 38,006 | 39,625 | N/A | N/A |
| Refractive Index | 1.4698 | N/D | N/D | 1.4551 | N/A | N/A |
| Transmission (%T) | 71.5 | N/D | N/D | 69.2 | N/A | N/A |
| Absorbance (Abs) | 0.146 | N/D | N/D | 0.160 | N/A | N/A |
| Oxidation stability (h at 110 °C) | 0.52 | 0.09 | 5.12 | 0.57 | 3h (min) | 6h (min) |

[a] Atabani et al., 2013a [b] Atabani et al., 2014

### 3.4. *Physico-chemical properties of MME and its blends with diesel*

Table 6 shows the physico-chemical properties of MME-diesel blends of 5%, 10%, 15% and 20%, as well as theASTM D975 specifications for diesel fuel and blends up to 5% by volume (B5) and ASTM 7467 for blends (B6-B20). Cinsiderable improvements in kinematic viscosity, calorific value and density of MME when blended with diesel were observed (Table 6). However, viscosity index and flash point values dropped as the percentage of diesel increased in the blends.

**Table 6.**
Physico-chemical properties of diesel and MME-diesel blends (5-20%).

| Property | Diesel | MME (5%) | ASTM D975 | MME (10%) | MME (15%) | MME (20%) | ASTM D7467 |
|---|---|---|---|---|---|---|---|
| Calorific value (kJ/kg) | 45,272 | 44,888 | N/A | 44,614 | 44,352 | 43,962 | N/A |
| CP (°C) | 8 | 6 | Report | 7 | 6 | 7 | Report |
| PP (°C) | 0 | 2 | Report | 4 | 1 | 1 | Report |
| CFPP (°C) | 8 | 8 | Report | 8 | 8 | 7 | Report |
| K Viscosity *at 40 °C (mm²/s)* | 3.23 | 3.31 | 1.9-4.1 | 3.47 | 3.62 | 3.79 | 1.9-4.1 |
| D Viscosity *at 40 °C (mm²/s)* | 2.70 | 2.78 | N/A | 2.92 | 3.05 | 3.21 | N/A |
| Density *at 40 °C (kg/m³)* | 827.2 | 838.7 | N/A | 841.2 | 843.5 | 845.8 | N/A |
| K Viscosity *at 100 °C (mm²/s)* | 1.24 | 1.31 | N/A | 1.36 | 1.40 | 1.46 | N/A |
| Viscosity Index | 90 | 138.7 | N/A | 140 | 140.8 | 143.5 | N/A |
| Flash point (°C) | 68.5 | 73.5 | 52 (min) | 74.5 | 74.5 | 75.5 | 52 (min) |

Figure 4 presents the effects of blending MME and diesel on kinematic viscosity, density, viscosity index, and calorific value.

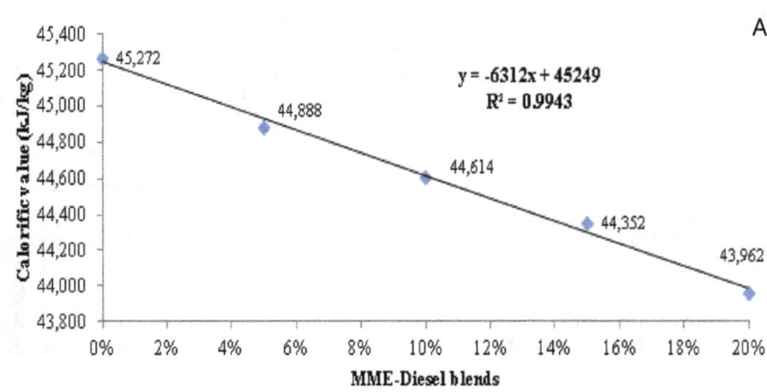


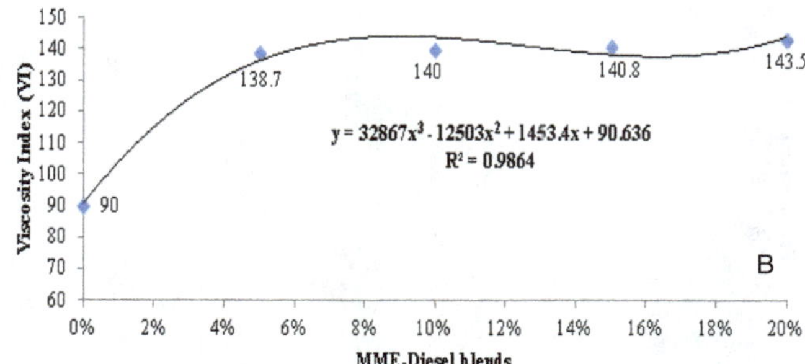


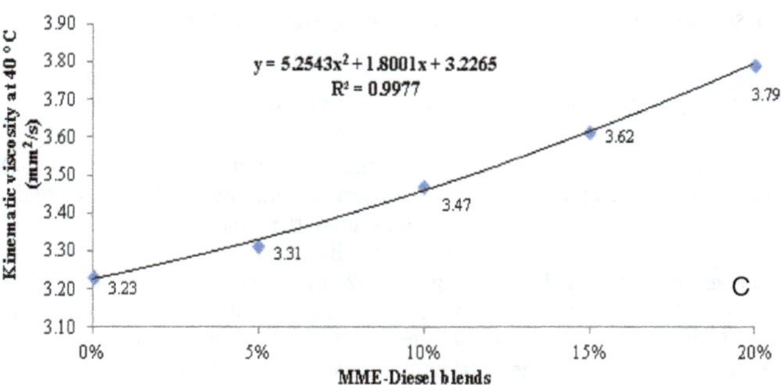


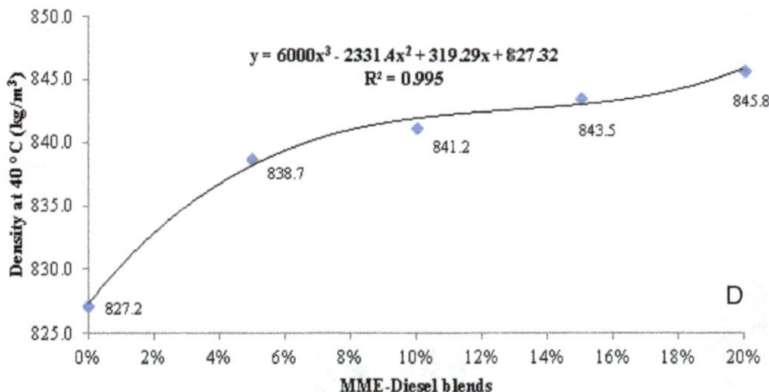


**Fig.4.** Effects of blending MME and diesel on, A) calorific value (kJ/kg), B) viscosity index (VI), C) kinematic viscosity at 40 °C (mm²/s) and, D) density at 40 °C (kg/m³).

### 3.5. *Engine performance*

In this study, engine performance was evaluated in terms of the brake power (BP) and brake specific fuel consumption (BSFC). The details of this evaluation are discussed as follows:

#### 3.5.1. *Brake Power (BP)*

Figure 5 shows the BP output of diesel (B0) and Manketti biodiesel blend (B5) at different engine speeds. For all tested fuels, the BP increased steadily with the engine speed. At all the test speeds, the average BP values of the B0 and B5 fuels were 39.85 and 39.38 kW, respectively. Compared to diesel fuel, the B5 fuel produced lower BP by 1.18% due to its lower calorific value and higher viscosity (Table 6), which influenced the combustion characteristics. Fuels with high kinematic viscosity tend to form larger droplets during injection which can consequently lead to poor combustion. Therefore, the uneven combustion characteristics of the produced biodiesel fuel reduced the engine BP herein (Muralidharan et al., 2011).

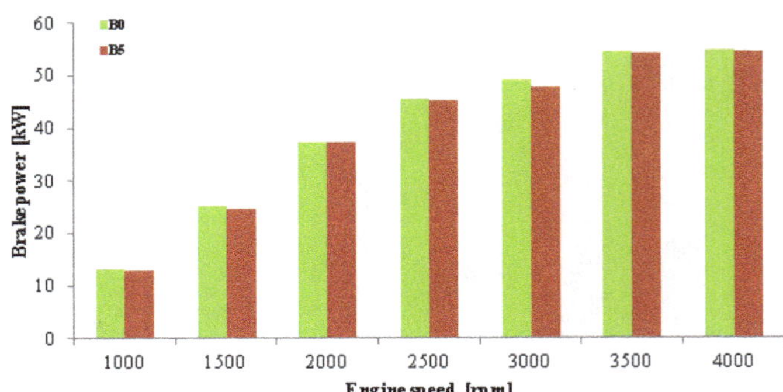


**Fig.5.** Variation of brake power at different engine speeds and full load condition.

### *3.5.2. Brake Specific Fuel Consumption (BSFC)*

Figure 6 illustrates the variation of the BSFC values for the tested fuels at different engine speeds. It can be observed that the biodiesel-blended fuel (B5) resulted in higher BSFC values than the neat diesel fuel (B0). This observation was consistent with the literature (Chauhan et al., 2012; Shahabuddin et al., 2012b; Wang et al., 2013). Factors such as the volumetric fuel injection system, density, kinematic viscosity and the lower heating value could affect the BSFC of diesel engines (Qi et al., 2010). At all speeds, the average BSFCs for the B0 and B5 were 345.18 and 352.98 g/kWh, respectively. Compared to the neat diesel fuel, the BSFC of B5 was 2.26% higher. This could be ascribed to the higher density and kinematic viscosity and lower energy content of the B5 compared to diesel (Mofijur et al., 2013). Fuel is injected into the engine on volume basis.

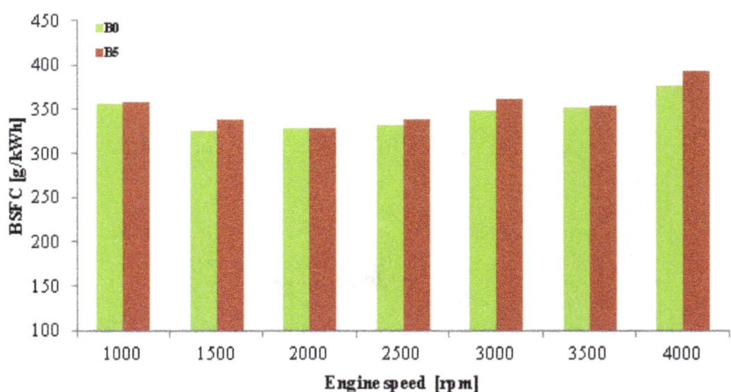


**Fig.6.** Variation of brake specific fuel consumption at different engine speeds and full load condition.

### *3.5.3. Emissions analysis*

In general, factors such as the atomization, volatility, air-fuel ratio, engine speed, injection timing and pressure and fuel type influence CO emissions (Gumus et al., 2012). The variation of CO and HC emissions with diesel and the MME-diesel blend are shown in Figure 7 and 8, respectively. At all speeds, the average CO emissions for the B0 and B5 were 0.12 and 0.15 vol%, and HC emissions for the B0 and B5 were 4 and 5 ppm, respectively. Over the entire range of engine speeds, the B5 increased the CO emissions by 32.27% and HC emission by 37.5% relative to B0. This result was in line with those previously reported elsewhere (Banapurmath et al., 2008; Sahoo et al., 2009). This increase in CO and HC emissions could be attributed to the poor atomization, lower volatility of the investigated biodiesel fuel. In fact, the high kinematic viscosity of the biodiesel led to a less homogenous mixture.

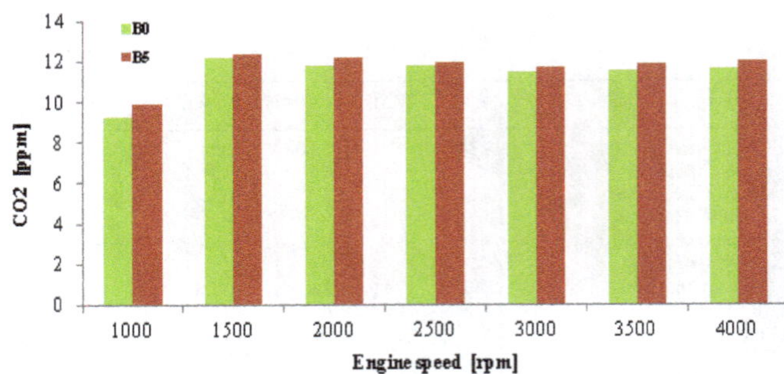


**Fig.7.** Variation of CO emissions at different engine speeds and full load condition.

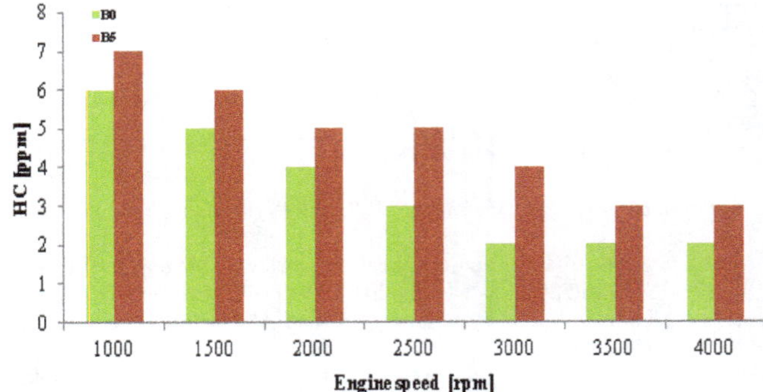


**Fig.8.** Variation of HC emissions with respect to engine speed at full load condition.

The variation of the NO emissions for diesel and biodiesel-blended fuel is presented in Figure 9.

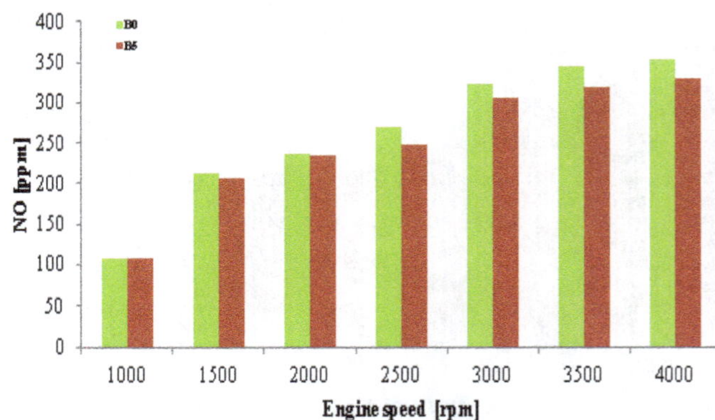


**Fig.9.** Variation of NO emissions at different engine speeds and full load condition.

The NO values were measured higher for diesel than biodiesel-blend fuel. Similar observations were reported by other researchers (Utlu et al., 2008; Qi et al., 2009). In general, the fuel spray properties depend on droplet size, droplet momentum, degree of mixing with air and penetration rate, evaporation rate, and radiant heat transfer rate. A change in any of those properties may change the NOx production.On an average, the B0 generated 5.26% higher NO emissions than B5 fuel over the entire range of speeds. The reason may be less reaction time and temperature in case of biodiesel blend.

The variation of $CO_2$ emissions for the fuel samples at various speeds are shown in Figure 10. B0 and B5 resulted in almost similar amount of $CO_2$ emissions at 11.397 and 11.71ppm, respectively. However, the $CO_2$ emitted through the combustion of biofuel e.g. biodiesel is of biological origin (plant or animal) and therefore, does not increase the atmospheric $CO_2$ carbon level as is the case for fossil fuels (Ramadhas et al., 2005).

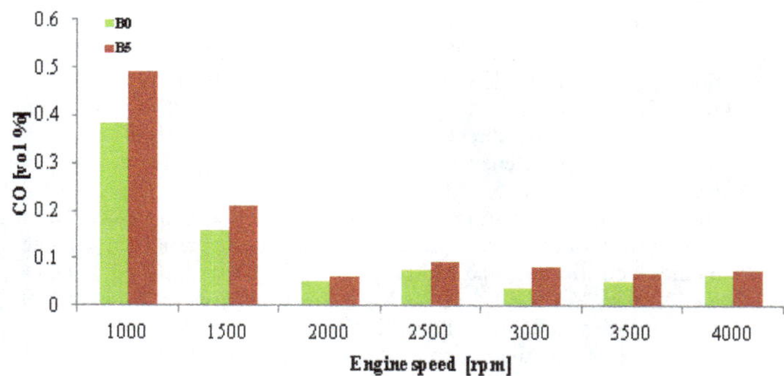


**Fig.10.** Variation of $CO_2$ emissions at different engine speeds and full load condition.

## 4. Conclusion

In this study, MME was produced and tested for various properties. MME was found to have a good potential to be used as a future energy source due to its significant calorific value and comparatively better cold flow properties than some other methyl esters. Blending of MME with diesel resulted in significant improvements in the kinematic viscosity, calorific value and density of MME. Though the B5 blend produced lower BP, higher BSFC, higher CO and HC emission but it significantly reduced NO emission compared to diesel fuel.

## 5. Acknowledgments

The authors would like to acknowledge the Ministry of Higher Education of Malaysia and University of Malaya, Kuala Lumpur, Malaysia for the financial support under the grant number UM.C/HIR/MOHE/ENG/06 (D000006-16001).

## References

Atabani, A.E., Silitonga, A.S., Badruddin, I.A., Mahlia, T.M.I., Masjuki, H.H., Mekhilef, S., 2012. A comprehensive review on biodiesel as an alternative energy resource and its characteristics. Renew.Sustain. Energy Rev. 16(4), 2070-2093.

Atabani, A.E., Badruddin, I.A., Mahlia, T.M.I., Masjuki, H.H., Mofijur, M., Lee, K.T., Chong, W.T., 2013a. Fuel Properties of Croton megalocarpus, Calophyllum inophyllum, and Cocos nucifera (coconut) Methyl Esters and their Performance in a Multicylinder Diesel Engine. Energ. Technol. 1(11), 685-694.

Atabani, A.E., Mahlia, T.M.I., Badruddin, I.A., Masjuki, H.H., Chong, W.T., Lee, K.T., 2013b. Investigation of physical and chemical properties of potential edible and non-edible feedstocks for biodiesel production, a comparative analysis. Renew. Sustain. Energy Rev. 21, 749-755.

Atabani, A.E., Mahlia, T.M.I., Masjuki, H.H., Badruddin, I.A., Yussof, H.W., Chong, W.T., Lee, K.T., 2013c. A comparative evaluation of physical and chemical properties of biodiesel synthesized from edible and non-edible oils and study on the effect of biodiesel blending. Energy. 58, 296-304.

Atabani, A.E., Irfan Anjum Badruddin., H.H. Masjuki, H.H., W.T. Chong, W.T., Lee, K.T., 2014. *Pangium edule* Reinw: a promising non-edible oil feedstock for biodiesel production. Arab. J. Sci. Eng. DOI 10.1007/s13369-014-1452-5.

Beau, M., 2003. Manketti Tree. Available at http://www.blueplanetbiomes.org/manketti.htm. (accessed on 3 July 2014).

Banapurmath, N.R., Tewari, P.G., Hosmath, R.S., 2008. Performance and emission characteristics of a DI compression ignition engine operated on Honge, Jatropha and sesame oil methyl esters. Renewable Energy. 33(9), 1982-1088.

Chammoun, N., Geller, D.P., Das, K.C., 2013. Fuel properties, performance testing and economic feasibility of Raphanus sativus (oilseed radish) biodiesel. Ind. Crops Prod.45, 155-159.

Chauhan, B.S., Kumar, N., Cho, H.M., 2012. A study on the performance and emission of a diesel engine fueled with Jatropha biodiesel oil and its blends. Energy. 37(1), 616-622.

Ávilaa, R.N.D.A., Sodré, J.R., 2012. Physical-chemical properties and thermal behavior of fodder radish crude oil and biodiesel. Ind. Crops Prod. 38, 54-57.

Demirbas, A., 2009. Progress and recent trends in biodiesel fuels. Energy Convers. Manage. 50(1), 14-34.

European Commision, Data Collection and Analysis for Sustainable Forest Management in ACP Countries - Linking National and International Efforts, 1998. Available at ftp://ftp.fao.org/docrep/fao/003/X6694E/X6694E00.pdf. (accessed on 21 October 2014).

Gumus, M., Sayin, C., Canakci, M., 2012. The impact of fuel injection pressure on the exhaust emissions of a direct injection diesel engine fueled with biodiesel-diesel fuel blends. Fuel. 95, 486-494.

Hajjari, M., Ardjmand, M., Tabatabaei, M., 2014. Experimental investigation of the effect of cerium oxide nanoparticles as a combustion-improving additive on biodiesel oxidative stability: mechanism. RSC Adv. 4(28), 14352-14356.

Hernándeza, D., Astudillo, L., Gutiérrez, M., Tenreiro, C., Retamal, C., Rojas, C., 2014. Biodiesel production from an industrial residue: Alperujo. Ind. Crops Prod. 52, 495-498.

Ibeto, C.N., Okoye, C.O.B., Ofoefule, A.U., 2012. Comparative Study of the Physicochemical Characterization of Some Oils as Potential Feedstock for Biodiesel Production. International Scholarly Research Notices: ISRN Renewable Energy, 1-5.

Kafuku, G., Mbarawa, M., 2013. Influence of Fatty Acid Profiles during Supercritical Transesterification of Conventional and Non-Conventional Feedstocks: A Review. Am. J. Anal. Chem. 4, 469-475.

Modiba, E., Osifo, P., Rutto, H., 2014. Biodiesel production from baobab (Adansonia digitata L.) seed kernel oil and its fuel properties. Ind. Crops Prod. 59, 50-54.

Mofijur, M., Masjuki, H.H., Kalam, M.A., Atabani, A.E., 2013. Evaluation of biodiesel blending, engine performance and emissions characteristics of Jatropha curcas methyl ester: Malaysian perspective. Energy. 55(15),879-887.

Moreira, M.A.C., Arrúa, M.E.P., Antunes, A.C., Fiuza, T.E.R., Costa, B.J., Neto, P.H.W., Antunes, S.R.M., 2013. Characterization of Syagrus romanzoffiana oil aiming at biodiesel Production. Ind. Crops . Prod. 48, 57-60.

Muralidharan, K., Vasudevan, D., Sheeba, K.N., 2011. Performance, emission and combustion characteristics of biodiesel fuelled variable compression ratio engine. Energy. 36(8), 5385-5393.

Naturalhub, The Mongongo/Manketti nut, 2013. Available at http://www.naturalhub.com/natural_food_guide_nuts_uncommon_ricinodendron_rautanenii.htm. (accessed on 16 June 2014).

Nerd, A., Aronson, J.A., Mizrahi, Y., 2013. Introduction and domestication of rare and wild fruit and nut trees for desert areas, Available at http://pdf.usaid.gov/pdf_docs/PNABN167.pdf. (accessed on 19 June 2014).

Phoo, Z.W.M.M., Razon, L.F., Knothe, G., Ilham, Z., Goembira, F., Madrazo, C.F., Roces, S.A., Saka, S., 2014. Evaluation of Indian milkweed (Calotropis gigantea) seed oil as alternative feedstock for biodiesel. Ind. Crops Prod. 54, 226-232.

Qi, D.H., Geng, L.M., Chen, H., Bian, Y.Z., Liu, J., Ren, X.C., 2009. Combustion and performance evaluation of a diesel engine fueled with biodiesel produced from soybean crude oil. Renewable Energy. 34(12), 2706-2713.

Qi, D.H., Chen, H., Geng, L,M., Bian, Y.Z., 2005. Experimental studies on the combustion characteristics and performance of a direct injection engine fueled with biodiesel/diesel blends. Energy Convers.Manage. 51(21), 2985-92.

Ramadhas, A.S., Muraleedharan, C., Jayaraj, S., 2005. Performance and emission evaluation of a diesel engine fueled with methyl esters of rubber seed oil. Renewable Energy. 30(12), 1789-1800.

Ramaraju, A., Kumar, A.T.V., 2011. Biodiesel development from high free fatty acid Punnakka oil. ARPN J. Eng. Appl. Scie. 2011. 6(4), 1-6.

Ruttoa, H.L., Enweremadu, C.C, 2011. Optimization of Production Variables of Biodiesel from Manketti Using Response Surface Methodology. Int. J. Green Energy. 8(7), 768-779.

Sahoo, P.K., Das, L.M., Babu, M.K.G., Arora, P., Singh, V.P., Kumar, N.R., Varyani, T.S., 2009. Comparative evaluation of performance and emission characteristics of jatropha, karanja and polanga based biodiesel as fuel in a tractor engine. Fuel. 88(9), 1698-1707.

Sanford, S.D., White, J.M., Shah, P.S., Wee, C., Valverde, M.A., Meier, G.R., 2009. Feedstock and biodiesel characteristics report. Available at http://www.biodiesel.org/reports/20091117_gen-398.pdf. (accessed on 21 October 2014).

Savaneskin, 2013. Manketti oil. Available at http://www.savaneskin.co.za/?ingredients=manketti-oil. (accessed on 25 June 2014).

Shahabuddin, M., Kalam, M.A., Masjuki, H.H., Bhuiya, M.M.K., Mofijur, M., 2012a. An experimental investigation into biodiesel stability by means of oxidation and property determination. Energy. 44(1), 616-622.

Shahabuddin, M., Masjuki, H.H., Kalam, M.A., Mofijur, M., Hazrat, M.A., Liaquat, A.M., 2012b. Effect of Additive on Performance of C.I. Engine Fuelled with Bio Diesel. Energy Procedia. 14, 1624-1629.

Utlu, Z., Koçak, M.S., 2008. The effect of biodiesel fuel obtained from waste frying oil on direct injection diesel engine performance and exhaust emissions. Renewable Energy. 33(8), 1936-1941.

Wang, R., Sun, L., Xie, X., Ma, L., Liu, Z., Liu, X., Ji, N., Xi, G., 2014. Biodiesel production from Stauntonia chinensis seed oil (waste from food processing): Heterogeneous catalysis by modified calcite, biodiesel purification, and fuel properties. Ind. Crops Prod. 62, 8-13.

Wang, X., Ge, Y., Yu, L., Feng, X., 2013. Comparison of combustion characteristics and brake thermal efficiency of a heavy-duty diesel engine fueled with diesel and biodiesel at high altitude. Fuel. 107, 852-858.

14

# Manipulation of carbon flux into fatty acid biosynthesis pathway in *Dunaliella salina* using *AccD* and *ME* genes to enhance lipid content and to improve produced biodiesel quality

Ahmad Farhad Talebi[1,2]*, Masoud Tohidfar[2,3], Abdolreza Bagheri[4], Stephen R. Lyon[5], Kourosh Salehi-Ashtiani[6], Meisam Tabatabaei[2,3]*

[1] *Semnan university, Semnan, Iran.*

[2] *Energy Crops Genetic Engineering Group, Biofuel Research Team (BRTeam), Karaj, Iran.*

[3] *Microbial Biotechnology and Biosafety Dept, Agricultural Biotechnology Research Institute of Iran (ABRII), Karaj, Iran.*

[4] *Biotechnology and Plant Breeding Dept., College of Agriculture, Ferdowsi University of Mashhad, Mashhad, Iran.*

[5] *AlgaXperts, LLC, Milwaukee, Wisconsin, USA.*

[6] *Division of Science and Math, and Center for Genomics and Systems Biology (CGSB), New York University Abu Dhabi, P.O. Box 129188, Abu Dhabi, UAE.*

## HIGHLIGHTS

- *Construction of a vector harboring ME and AccD genes act polycistronically.*
- *Stable integration of the cassette in transcriptionally silent region of chloroplast Zgenome.*
- *12% increase in total lipid content in transgenic microalgae.*
- *Improvement in prospective biodiesel quality especially less oxidation susceptibility.*

## GRAPHICAL ABSTRACT

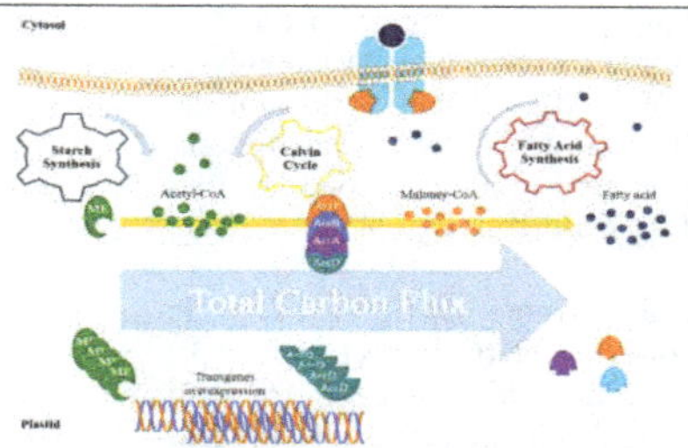


## ABSTRACT

Advanced generations of biofuels basically revolve around non-agricultural energy crops. Among those, microalgae owing to its unique characteristics i.e. natural tolerance to waste and saline water, sustainable biomass production and high lipid content (LC), is regarded by many as the ultimate choice for the production of various biofuels such as biodiesel. In the present study, manipulation of carbon flux into fatty acid biosynthesis pathway in *Dunaliella salina* was achieved using pGH plasmid harboring *AccD* and *ME* genes to enhance lipid content and to improve produced biodiesel quality. The stability of transformation was confirmed by PCR after several passages. Southern hybridization of *AccD* probe with genomic DNA revealed stable integration of the cassette in the specific positions in the chloroplast genome with no read through transcription by endogenous promoters. Comparison of the LC and fatty acid profile of the transformed algal cell line and the control revealed the over-expression of the *ME*/*AccD* genes in the transformants leading to 12% increase in total LC and significant improvements in biodiesel properties especially by increasing algal oil oxidation stability. The whole process successfully implemented herein for transforming algal cells by genes involved in lipid production pathway could be helpful for large scale biodiesel production from microalgae.

**Keywords:**
Carbon flux manipulation
Microalgae
Biodiesel
Genetic engineering
Chloroplast

* Corresponding authors
E-mail addresses: meisam_tab@yahoo.com (M. Tabatabaei); ahmad_farhad64@yahoo.com (A.F. Talebi)

## 1. Introduction

Unicellular microalgae have been at the center of attention of research efforts aimed at developing technologies for the renewable production of biofuels e.g. biodiesel. This is ascribed to the ability of algal cells to survive or proliferate over a wide range of environmental conditions, their remarkable diversity, and the ability to modify heir lipid metabolism efficiently through changing environmental conditions and genetic/metabolic engineering approaches. On the other hand and despite the remarkable attractions of algal fuels, economic and technical challenges are still faced and need to be addressed before algae could efficiently compete with the other biofuels feedstocks e.g. oil crops (Chisti Y, 2008).

Courchesne *et al.* reviewed three possible strategies/approaches for enhanced algal lipid production namely (Courchesne et al., 2009), biochemical engineering (BE), genetic engineering (GE), and transcription factor engineering (TFE). Through the GE strategies, generally one or more key genes such as Acetyl-CoA carboxylase (ACCase) and diacylglycerol acyl-transferase (DGAT) could be up-regulated (Klaus et al., 2004; Bouvier-Nave et al., 2000), while phosphoenolpyruvate carboxylase (PEPC) could be down-regulated (Song et al., 2008; Zhao et al., 2005) . However, GE of the lipid production pathway in the context of the whole cell rather than at a single step, has recently provoked widespread studies. These studies are aimed to channel carbon flux into lipid biosynthesis using multi-genes approach (Blatti et al., 2013) or to use the regulatory factors such as transcription factors (TFs) in order to control the abundance or activity of multiple enzymes relevant to the lipid production process ( Huang et al., 2013; Huanget al., 2014).

ACCaseas is a key enzyme responsible of regulating the rate of de novo fatty acid (FA) biosynthesis in plant's plastid. The modulation of carbon flux is started by the formation of malonyl-CoA through ACCase`s activity. This is the first rate-limiting step in the FA biosynthesis pathway. Plastidic ACCase is composed of 4 subunits i.e. *BC*, *BCCP*, *AccA* and *AccD* (Sasaki et al., 2004). The first evidence on the positive effect of ACCase over-expression on FA synthesis was observed in transformed *Brassica napus* harboring a homomeric ACCase which showed a 5% increase in seed oil content (Roesler et al., 1997). Reports in some of the major peer-reviewed journals published reports suggest that *BC* and *BCCP* subunits are not the limiting factors to the ACCase activity (Shintani et al., 1997; Thelen et al., 2002) , while over-expression of the *AccD* subunit in a study conducted by Madoka *et al.* was shown to have boosted ACCase activity (Madoka et al., 2002).

Apart from the ACCase, GE of Malic enzyme (*ME*); an enzyme responsible for the conversion of malate to pyruvate (Chang et al., 2003), could also further promote carbon flux into lipid biosynthesis. *ME* also reduces $NADP^+$ to NADPH simultaneously and this byproduct is also utilized by the enzymes involved in FA and TAG syntheses which could potentially lead to enhanced lipid production. It was reported that *ME* over-expression led to the cytosolic NADPH increase and this extra reducing power could be used by lipogenic enzymes such as ACCase (Courchesne et al., 2009).

The present study was aimed to enhance lipid biosynthesis pathway both in quantity and quality in a strain of microalgae *Dunaliella salina* for biodiesel production. To achieve that, the FAS pathway was manipulated by transferring the pGH vector harboring the *ME* and *AccD* genes into the chloroplast genome of *D. salina* through the particle bombardment method. Then, lipid content, FA profile and the quality parameters of the resultant biodiesel were investigated in the transformed algal line.

## 2. Material and method

### *2.1. Strain cultivation*

*D. salina* 19/18 was purchased from the Culture Collection of Algae and Protozoa (CCAP)(Sams Research, Scotland) and was cultivated in Johnson Medium (Johnson et al., 1968). The cells were kept at 20 °C and under a constant (24:0) 3klux photon flux of white and red LED lamps. Solid culture was prepared by implementation of 10% agar in the same medium.

### *2.2. Chloramphenicol Inhibition Test of D. salina*

The cell culture of *D. salina* at the exponential phase was exposed to various concentrations of chloramphenicol (Sigma Aldrich, USA), ranging from 10 to 200 µg $mL^{-1}$ liquid Johnson medium, to determine the optimum antibiotic concentration to screen the transformants. The cell concentration in the liquid medium was observed for 30 days. The chloramphenicol inhibition test was performed in triplicates for each concentration and $LC_{50}$ value was determined based on $OD_{620}$ and the cell count data were obtained using the Probit value Method (Ashton, 1972)

### *2.3. Genetic transformation of D. salina*

#### *2.3.1. Sample preparation*

The exponential phase ($OD_{620}$ = 0.5) of *D. salina* cultures was used as the time point for cell bombardment. Briefly, 20 mL of the *D. salina* culture was concentrated using a centrifuge at 1000×g for 10 minutes. The harvested biomass was dissolved in 1 ml Johnson medium and was streaked on a plate containing the same medium with agar. Before bombardment, the plates were kept under semi-dark condition in a phytotron for 24 h.

#### *2.3.2. Plasmid Construction*

The pGH-ME-AccD construct (10.3 kb) harboring 2 types of inducible and constitutive promoters in order to express the ME-AccD gene cassette was synthesized and used to enhance lipid production in *D. salina.* To facilitate cloning of the regulatory elements, all oligonucleotides were synthesized and cloned on the pGH vector by Genestar Co. (Shanghai, China). The sequence for 16S promoter (constitutive promoter) used in the current study was selected as suggested by Rasala *et al.* (2011). As for the inducible promoter (Nit), a partial sequence (636–1206) of the NIT1 gene (AF203033.1) coding nitrate reductase (EC1.7.1.1) in *Chlamydomonas reinhardtii* was used.

In order to regulate the expression of the ME-AccD gene cassette polycistronically, the 118-nucleotide upstream sequence of the rbcl gene was used as the 5' untranslated region (UTR) and the 280-nucleotide downstream sequence of the same gene was used as the 3'UTR.

To optimize the codon usage of *C. reinhardtii ME* nuclear gene (XM_001692632.1), the coding region of this gene was synthesized de novo (Shanghai Genestar Co., Ltd. China) preferringly with A or U at the third position of codons to enhance expression in green microalgae chloroplast. *AccD* was amplified by specific primers of the *AccD* gene of *Brassica napus*. Forward (5'TTTCATGTAAATAGAGGCCAGAAGC3') and reverse (5' CTGTTTTATTTGATTTCATTTTTGTTC3') primers were synthesized by CinnaGen Inc. (Tehran, Iran). The PCR product was directly cloned into pGEM vector using T easy vector systems (promega). *AccD* gene was subcloned from pGEM-AccD vector into pGH-ME vector by *SalI*-*NcoI* digestion. In fact, the resultant pGH-ME-AccD vector was constructed allowing polycistronic expression of these two genes (Fig.1).

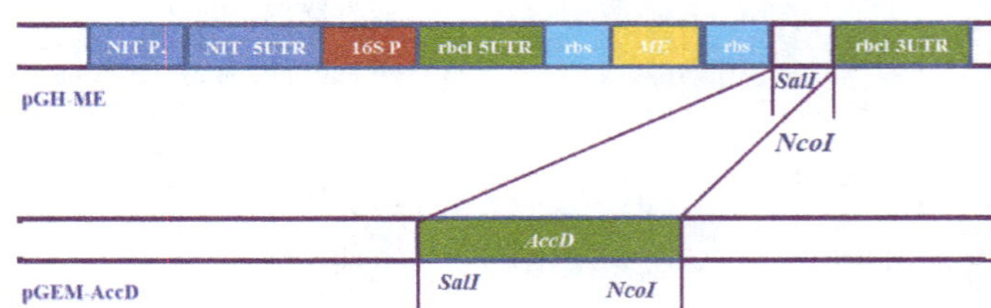


**Fig.1.** Construction of pGH-ME-AccD vector. Polycistronic expression of acetyl CoA carboxylase (*AccD*) and malic enzyme (*ME*) genes achieved by subcloning of *AccD* from pGEM-AccD vector into pGH-ME vector.

The *CAT* gene responsible for antibiotic resistance was isolated from an invitrogene gateway vector (AB752383.1) through PCR. The primers were designed with *BamHI* and *XbaI* restriction site at 3' ends. PCR fragments were digested with *BamHI* and *XbaI* and directly inserted into pGH-AccD-ME vector, downstream of atpB promoter as shown in Figure 2.

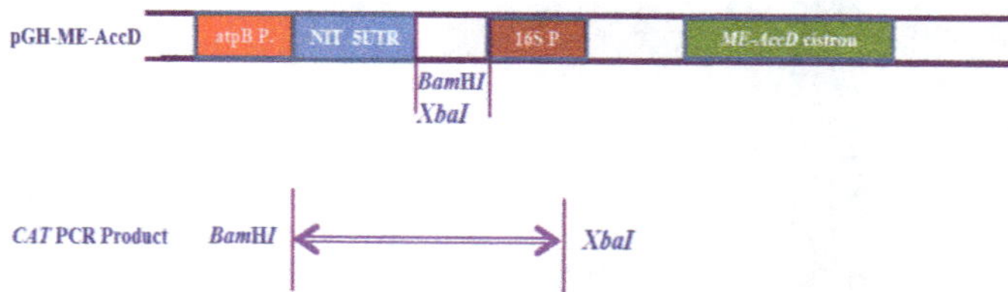


**Fig.2.** Addition of chloramphenicol resistance gene (*CAT*) into the pGH-AccD-ME vector. *CAT* PCR fragments were digested in the flanking restriction sites and were directly inserted into the pGH-AccD

A flanking region was included on both sides of the ME-AccD genes as well as the *CAT* gene. The 2000 bp transcriptionally - silent flanking region was synthesized de novo based on a unique intergenic region (i.e. rrnS-chlB) from *D. salina* strain CCAP 19/18 chloroplast genome (GQ250046.1) (Fig. 3). As a matter of fact, the lack of read through transcription in this intergenic region makes the pGH vector an ideal choice for studying transgene promoter activity.

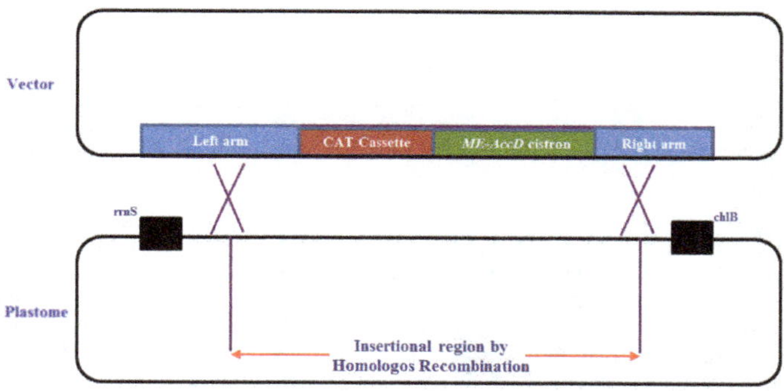


**Fig.3.** Homologues recombination. Two flanking regions (left and right arms) making possible the insertion of the ME-AccD gene cassette into a unique intergenic region (i.e. rrnS-chlB) in *D. salina* strain chloroplast genome.

Finally, *Escherichia coli* strain DH5α carrying the pGH-ME-AccD plasmids was cultured in LB medium containing 50 mg $L^{-1}$ ampicillin and 30 mg $mL^{-1}$ chloramphenicol and the plasmids were isolated using high pure plasmid isolation kit (Roche, Germany). The quantity and purity of the plasmids were determined by nanodrop (Thermo Scientific, Wilmington, DE) and the extracted plasmids were used for the cell bombardment experiment.

### *2.3.3. Micro particle bombardment*

To coat the 0.6 µm diameter gold particles (Bio-Rad Laboratories, USA), with the pGH-ME-AccD vector, the method described by Talebi *et al.* (2013b) was used. Briefly, 60 mg of gold particles was weighed and washed twice by 100% ethanol and then the gold pellet was dissolved in 1 ml of dd water. Then, 60 µL of the resultant gold particles solution (60 mg $mL^{-1}$) was mixed with 5 µL of a pGH-ME-AccD vector (6 µg $\mu L^{-1}$), 50 µL of 2.5 M $CaCl_2$, and 20 µL of 0.1M spermidine. After that, the coated particles were washed in 60 µL of 100% ethanol twice. Finally, 10 µL of the vector-coated particles was layered on a macrocarrier and allowed to air dry at ambient temperature for bombardment.

Then, *D. salina* cells at their exponential phase were bombarded under vacuum with the vector-coated gold particle using a Bio-Rad PDS-1000/He Biolistic Particle Delivery System (Bio-Rad Laboratories, USA) at gas (helium) pressure of 900 psi at distances of 6 cm and 9 cm according to the method described by Jiang (Jiang et al., 2002). Each plate was bombarded for the second time after it was turned for 90 degrees. Prior to bombardment, screens were sterilized by autoclaving and the macrocarriers as well as the rupture disks were sterilized by washing in isopropanol for 15 minutes.

### *2.3.4. Selection of putative clones using chloramphenicol*

After bombardment, the targeted cells were kept under semi dark condition for 24 h at 25 °C. Then, the bombarded and non-bombarded (control) *D. salina* cells were cultured in Johnson liquid medium enriched by 250 µg $mL^{-1}$ of chloramphenicol. The recovered cells were streaked on selective agar plates containing the same medium plus 100 µg $mL^{-1}$ of chloramphenicol. After 2 weeks of incubation at 25 °C, green-colored colonies, representing transformed cells, were visible on the plates. Single colonies were picked using a pipette tip and were inoculated in liquid Johnson medium containing 80 µg $mL^{-1}$ of chloramphenicol. Both bombarded and non-bombarded cells were left to grow for two weeks in antibiotic-contained and antibiotic-free media, respectively. Harvested biomass was used for genetic and expression analyses. The cultures were maintained for further stability and expression analyses.

### *2.4. DNA extraction analysis of D. salina clones survived on the antibiotic enriched medium*

### *2.4.1. DNA extraction*

One hundred mg of *D. salina* cells (OD620 = 1) were harvested from the 20 d old cultures obtained from the single clones grown in the presence of chloramphenicol. This was achieved by centrifugation (10,000×g, 10 min, and ambient temperature) of about 50 ml of cell suspensions.

The total DNA was extracted by DNeasy Plant Mini Kit (QIAGEN, Germany). The quantity and purity of the genomic DNA were determined by a nanodrap at $OD_{260}$ and $OD_{280}$. The quality and integrity of the DNA sample were also verified with 1.0 % (w/v) agarose gel electrophoresis in 1× TAE buffer at 90V for 45 min. The genomic bands were viewed and photographed using AlphaImager TM 2200 (Alpha Innotech Corporation, USA).

### *2.4.2. Polymerase chain reaction (PCR)*

The PCR analysis for the detection of the integrated region of the vector into the *D. salina* chloroplast genome was conducted using specific primers of the *AccD* gene. PCR amplification of *AccD* gene was done in a BIO-RAD DNAEngine® thermocycler through 35 cycles of: 94 °C (1 min), 59°C (1 min) and 72 °C (1.50 min).

### *2.4.3. Southern blot analysis*

To determine the integration of the gene cassette in the chloroplast genome of the transformed lines, 10 µg of the isolated DNA (transformed line and control) and 0.1 µg of plasmid, were digested with *EcoRI* restriction enzyme as recommended by the manufacturer (Fermentas, Vietnam). The digestion pattern was monitored by electrophoresis along with size markers on a 0.8% agarose gel in a 1X TBE buffer and then transferred to nitrocellulose membrane based on the procedure described by Maniatis (Maniatis et al., 1982). A 1467 bp fragment from plasmid pGH containing the entire coding region of *AccD* gene was amplified and labeled by incorporation of DIG-11-dUTP into DNA using PCR (DIG labeling Kit, Roche, Germany). The probe hybridization was carried out at 43 °C for 16 h and washing step was conducted at moderate stringency in 0.5 X SSC, 0.1 SDS. The probed blots were detected with an enzyme-linked immunoassay.

### *2.5. Physiologic studies*

### *2.5.1. Total lipid content and free fatty acid profiling*

By the time the cell growth phase reached the lag stationary phase, physiologic parameters were studied in the transformed and control cell lines. Total lipids content (Lc) and free FA profile were obtained in triplicates for the studied strains. Data comparison was then carried out using the ANOVA test. LC reported as percentage of the total biomass (%dwt), was determined based on the Bligh and Dyer method (Bligh et al., 1995). FA profile was investigated by using Gas Chromatography (GC) analysis based on the procedure described by Talebi *et al.* (2013a). The intracellular neutral lipid distribution in microalgal cells was examined by staining the cells by Nile Red fluorescent dye. Fluorescence-based quantification of the accumulated lipids was achieved following the protocol reported by Chen *et al.* (2009).

The excitation and emission wavelengths of 522 and 628 nm, respectively, were selected based on a previous report by Talebi *et al.* (2014b).

*2.5.2. Bioprospecting biodiesel quality parameters*

The quality characteristics of produced biodiesel from any oil stocks could be predicted by investigating the FA profile of the oil used (Bigelow et al., 2011). A number of studies have previously reported empirical equations based on which all most of the biodiesel quality parameters such as Cetane number (CN), Cloud Point (CP) and the oxidation stability (OS), could be predicted (Ramos et al., 2009; Ramírez-Verduzco et al., 2012). Bioprospecting of biodiesel quality parameters for the oil samples obtained from the control and transformed algal cells was achieved using the BiodiselAnalyser ver. 1.1 software (available on http://www.brteam.ir/biodieselanalyzer) (Talebi et al., 2014a). The data obtained were used to evaluate the effect of transgenes overexpression on the biodiesel quality parameters in the transformed lines in comparison with the control.

**3. Results and discussion**

*3.1. DNA analysis of D. salina transformed lines*

The presence of the 1467 bp *AccD* gene in the algal cells bombarded with the pGH-ME-AccD vector was verified through PCR using specific primers (Fig. 4). The negative controls *i.e.* the non-transformed *D. salina* DNA and the reaction mixture containing no DNA template produced no bands.

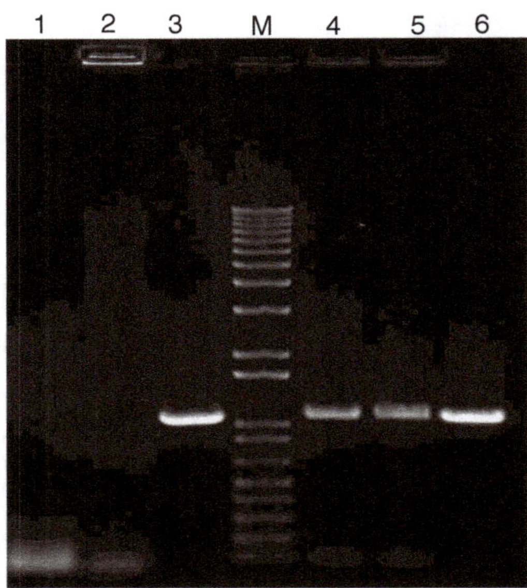


**Fig.4**. Agarose gel electrophoresis verification of the *AccD* PCR products (1467 bp). Lane 1: negative control without any DNA. Lane 2: negative control with untransformed DNA, lanes 3: positive control (pGH-ME-AccD vector), line 4-6: transformed *D. salina*, and M represents the 1 Kb DNA marker (Fermentas).

The insertion stability of the *AccD* gene in the transformed clones was also proved in the repetitive subcultures/generations via PCR analysis. However, the transformed cells lost their resistance to chloramphenicol after the 5[th] subculture (day 100). In a similar observation, Chow (Chow et al., 1999), reported shorter transient maintenance of plasmid in the algal cells, and that the trasformed lines gradually lost their hygromycin resistance.

Southern blot analysis performed with the genomic DNA and the labeled *AccD* probe showed no background formation and the specific hybridization of probe with the transformed DNA revealed stable integration of the flanking regions of the pHG-ME-AccD vector into the chloroplast genome (Fig. 5).

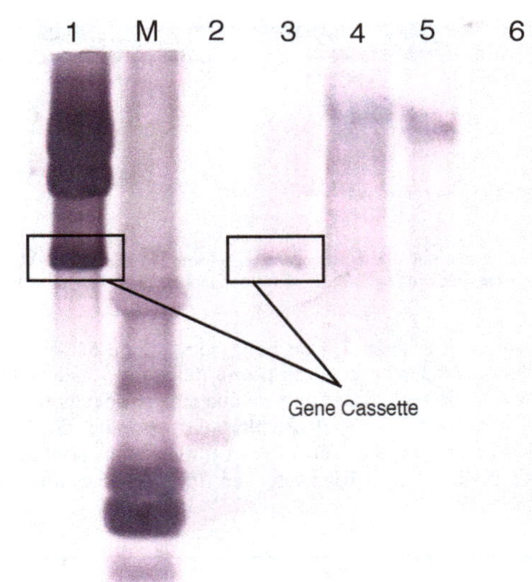


**Fig.5**. Integration of pGH-ME-AccD vector into the chloroplast genome of transformed *D. salina* by Southern analysis. Lane 1: positive control (pGH-ME-AccD vector); lanes 2 and 3: the transformed cells (digested DNA); lanes 4 and 5: the transformed cells (undigested DNA); lane 6: the untransformed cells; and lane M refers to the size marker.

More specifically, a similar band aligned in the patterns observed for the control and the transformed line (Figure 5, lanes 1 and 3, respectively), confirmed the successful insertion of the gene cassette in the targeted positions in the circular genome. Among the optimal transformation procedures reported for *D. salina* by Feng *et al.* (2009) *i.e.* glass beads, particle bombardment and electroporation, PDS1000/He micro-particle bombardment system has been frequently used to introduce different construct into the nuclear genome of *D. salina* (Liu et al., 2010; Tan et al., 2005). The Finding of the present study also further enforced the acceptable efficiency of the bombardment method to introduce a transgene into the *Dunaliella* cells.

*3.2. Effect of transgenes activity on total lipid production*

The effect of the presence of the *AccD-ME* transgenes in lipid metabolism was surveyed by LC measurement in the 35-day old transformed and the control cell lines. The mean value of LC for the transformed cell line showed a slight increase to 25% dwt. which was 12% higher than that of the control. Moreover, high-throughput fluorescence-based technique using a liposoluble fluorescence probe (i.e. Nile Red), was used to quantify neutral lipids in the cells. As shown in Table 1, the fluorescence intensity obtained showed 23% increase in the transformed cell lines in comparison with the control. The lipid measurement results obtained using the fluorometry technique was in line with the LC increase measured by the gravimetrical method.

**Table 1.**
Fluorescence intensity emitted by microalgae stained by Nile red.

| Sample | Fluorescence intensity |
|---|---|
| Water | 3011±188[a] |
| Non stained control cell | 3566±242 |
| Stained control cell | 15933±3155 |
| Transformed cell | 19639±2668 |

In previous attempts, Dunahay (Dunahay et al., 1996) and Wang et al. (Wang et al., 2009) also used genetic manipulation of microalgae through overexpression of TAG biosynthesis pathways genes such as ACCase to improve algal lipid content but failed to achieve any significant increases in lipid accumulation. To the best of our knowledge, there is no report on overexpression of other enzymes involved in lipid biosynthesis like *ME*, nor on blocking competing pathways such as β-oxidation in microalgae.

Plastidic ACCase is responsible for a major part of unsaturated FAs production (Sasaki et al., 2004). Having considered the impermeable nature of the plastid's membrane to the Acetyl-COA ( Ke et al., 2000), it could be concluded that the plastidic ACCase plays an independent role in free FAs synthesis and is not related to the existing ACCase in the cytoplasm. The heteromeric plastidic ACCase is composed of four nuclear- and plastid-encoded subunits. The *BC* and *BCCP* subunits have been shown as non-limiting factors to the ACCase activity (Shintani et al., 1997; Thelen et al., 2002) , while over-expression of the *AccD* subunit has been proved to boost ACCase activity ( Madoka et al., 2002). As Madoka *et al.* (2002) reported that the overexpression of *AccD* caused twice FA production per transformed tobacco in comparison with the wild-type plant. On the other hand, *ME* catalyzes the oxidative decarboxylation of L-malate, producing pyruvate, $CO_2$, and NAD(P)H (Chang et al., 2003) and could play a stimulating role in lipid biosynthesis by providing carbon skeletons and also reducing power i.e. NAD(P)H. This hypothesis was confirmed through a study by Zhang and coworkers (Zhang et al., 2007), in which the recombinant strains of *Mucor circinelloides* harboring *ME* showed an increase in biosynthesis of FAs and formation of unsaturated FAs. They argued that the unique role of *ME* in elimination of the rate-limiting step of FA biosynthesis through supplying NADPH was the cause.

Therefore in the present study, simultaneous overexpression of both *AccD* and *ME* was put to test which resulted in 12% increase in total lipid accumulation in the algal cells. In more details, overexpression of *AccD* and *ME* made LC reach to 25% in the transformed cell line in comparison with 22% in the control line. The introduction and overexpression of *AccD* in chloroplast genome of *D. salina* along with the *ME* gene into the operation site (Fig. 6), must have accelerated the carbon flux into the free FA synthesis leading to increased lipid production.

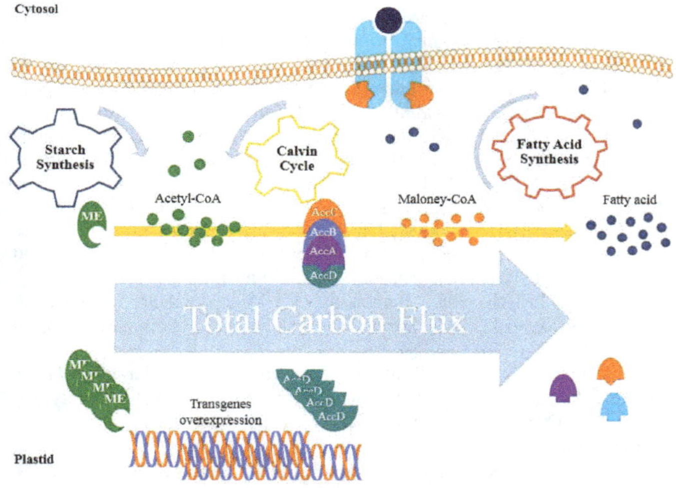


**Fig.6.** Over-expression of *AccD* and *ME* in chloroplast genome of *D. salina.* The carbon flux was switched from starch synthesis to FAs synthesis.

Concerning the regulatory elements used which are responsible for successful overexpression of *AccD* and *ME*, sequence analysis of NIT1 inducible promoter sequence from *C. reinhardtii* revealed a putative CCAAT box at -836, a TFII-I binding site at -664, a TFII-A binding site at -509, and a CCAAT box at -245, relative to the translational start site. The responsive sequence to nitrogen deprivation in the growth media which was used in this study was located -284 nucleotides upstream the start codon. 287 nucleotide of the beginning CDS was inserted to act as a 5' UTR. In other studies, transformation of some microalgae strains such as *Chlamydomonas*, *Chlorella*, *Dunaliella* with the Nit1 promoter is accomplished by electroporation (Champagne et al., 2009). Nit promoter previously was introduced into the *D. salina* genome as a inducible promoter to control expression of the bar and EGFP genes using the inexpensive inducer nitrate and repressor ammonium (Li et al., 2007; Li et al., 2008). On the other hand, the designed construct in the present study harbors 16S rRNA gene promoter as a constitutive promoter. This promoter resembles the bacterial type with typical -10 and -35 elements (Klein et al., 1992).

In result, the combination of nuclear gene promoter, NIT1, and chloroplastic gene promoter, rrnS, guaranteed the high expression of *AccD-ME* transgenes due to the co-presence of nuclear encoded RNA polymerase (NEP) and chloroplast encoded RNA polymerase (PEP) in the chloroplast (Sasaki et al., 2004) . This strategy was a mimetic behavior of the rRNA operon which also benefit two types of promoters in order to express in the chloroplast abundantly (Vera et al., 1995).

### *3.3. Study of transgenes activity on oil quality*

#### *3.3.1. Fatty acid profile*

Fatty acid methyl ester (FAME) profiles of the transformed and control line of *D. salina* are summarized in Table 2. It has been frequently reported that 16-18 carbon chain FAs are dominant in FAME profile of microalgal species and that the FAs have major impact on oil quality as well (Talebi et al., 2013a).

**Table 2.**
Fatty acid composition and properties of the produced algal oils.

| Strain | Fatty acid (%) | | | | | | | SFA/ USFA | Oil content (%) |
|---|---|---|---|---|---|---|---|---|---|
| | 16:0 | 16:1 | 18:0 | 18:1 | 18:2 | 18:3 | 20:1 | | |
| *D. salina* (transformed) | 22.60±0.4 | 1.60±0.2 | 18.60±0.8 | 16.36±0.6 | 5.86±0.5 | 30.80±1.4 | 2.66±0.2 | 72 | 25±2.8[B] |
| *D. salina* (control) | 16.33±0.4 | 1.06±0.3 | 8.60±0.7 | 19.57±0.4 | 6.76±0.8 | 33.60±0.9 | 3.61±0.5 | 35 | 22±1.1[A] |

As a whole, the results of this study showed significant differences between the FAME profile of the studied cell lines. In more details, overexpression of *AccD* and *ME* in the transformed cell line caused an obvious accumulation in the saturated FAs (SFA) like plametic acid (C16:0) and stearic acid (C18:0). On the contrary, this manipulation led to decreased percentages of mono unsaturated FAs (MUFA) i.e. palmitoleic acid (C16:1) and oleic acid (C18:1) and also poly unsaturated FAs (PUFA) i.e. Linolenic acid (C18:3). As a result, SFA/USFA parameter showed over 106% increase in the transformed line. This phenomenon could be explained by the fact that the beginning of the desaturation pathways involving delta-9 desaturase is a highly energy-consuming process and when cells are encouraged to produce and accumulate more lipids in their cytoplasm, they face energy and time scarcity to evolve new-born FAs to desaturated ones. Similar observations have been reported when lipid production is enhanced through biochemical engineering approaches. For instance, Talebi *et al.* (2014b) achieved a 50% increase in total lipid accumulation through myo-inositol supplementation, as a lipid inducer treatment in *D. salina.* while also reported 54% deacrease in USFA/FAME. Zhila *et al.* (2005) while tried to enhance lipid accumulation using N-starvation treatment on *Botryococcus braunii*, observed an increase in the SFAs content (up to 76.8%) and a decrease in the PUFA content (up to 6.8%).

#### *3.3.2. Estimation of biodiesel properties*

FAME profile of algal strains could be used as a tool to predict the characteristics of produced biodiesel (Talebi et al., 2013a). The degree of unsaturation (DU) was decreased by 20% in the transformed line which had major impacts on the BAPE and APE values as well as the oxidation stability (OS) of the produced biodiesel (Table 3). In fact, the APE and BAPE values were reduced by 19 and 20 % in the transformed line, respectively. The OS of the transformed and control lines were estimated at 5.81 and 5.19, respectively. This would mean that the obtained oil from the transformed algal cells was less susceptible to oxidation/rancidity at high temperatures over long-time storage periods. Slim improvements in the estimated CP values were also anticipated. Overall, it was shown that the overexpression of *AccD* and *ME* led to improved fuel properties (Table 3).

**Table3.**
Comparison of the estimated properties of algal biodiesel from the transformed and control cells of *D. salina*.

| | Biodiesel properties | | | | | | | | | |
|---|---|---|---|---|---|---|---|---|---|---|
| **Strains** | CN | CP | BAPE | APE | DU | CFPP | OS | HHV | V | p |
| *D. salina* (transfomed) | 48.05 | 6.90 | 67.46 | 89.62 | 93.88 | 19.84 | 5.81 | 38.78 | 1.28 | 0.86 |
| *D. salina* (control) | 43.15 | 3.58 | 83.90 | 110.20 | 115.46 | 2.15 | 5.19 | 37.42 | 1.18 | 0.83 |

## 4. Conclusion

The feasibility of biodiesel production from genetically modified micro-algae can be expedited if large-scale production facilities can be integrated with other processes, such as wastewater treatment and utilization of carbon dioxide from power plants. Screening of genetic variability between algal isolates could lighten the different potential of variant strains and the result could be used in prone strain selection for gene transformation. 26 years since the first stable introduction of a heterogene into the *Chlamydomonas* chloroplast by the Boynton and Gillham laboratory, we have shown the potential of chloroplast as a realistic platform for GE approaches towards the goal of establishing the microalgae in biofuel production worldwide. We have shown that the introduced transgenes at the insertion site are stable, and that there is no readthrough transcription from outside promoters. In this research we have gained of two promoters combination to reach the highest reported record of enhanced lipid production in a transformed microalgae. It was also demonstrated that GE approaches such as overproduction of multiple involved enzymes has the potential to solve two major obstacles in microalgal biodiesel production namely low lipid content and unfavorable fatty acid profile.

## 5. Acknowledgments

The authors would like to thank Dr. Motahareh Mohsen pour for her kind assistance during the course of this study. We also would like to thank Agricultural Biotechnology Research Institute of Iran (ABRII) and Biofuel Research Team (BRTeam) for financially supporting this study.

## References

Ashton, W.D., 1972. The logit transformation withspecial reference to its uses in bioassay. New York Hafner Publication Corporation.

Bigelow, N.W., Hardin, W.R., Barker, J.P., Ryken, S.A., MacRae, A.C., Cattolico, R.A., 2011. A comprehensive GC–MS sub-microscale assay for fatty acids and its applications. J. Am. Oil Chem. Soc. 88(9), 1329-1338.

Blatti, J.L., Michaud, J., Burkart, M.D., 2013. Engineering fatty acid biosynthesis in microalgae for sustainable biodiesel. Curr. Opin. Chem. Biol. 17(3), 496-505.

Bligh, E.G., Dyer, W.J., 1995. A rapid method for total lipid extraction and purification. Can. J. Biochem. Physiol. 37, 911-917.

Bouvier-Nave, P., Benveniste, P., Oelkers, P., Sturley, S.L., Schaller, H., 2000. Expression in yeast and tobacco of plant cDNAs encoding acyl CoA:diacylglycerol acyltransferase. Eur. J. Biochem. 267, 85-96.

Champagne, M.M., Kuehnle, A.R., 2009. Nuclear based expression of genes for production of biofuels and process co-products in algae. US Patent Application. 12/415,904.

Chang, G.G., Tong, L., 2003. Structure and function of malic enzymes, a new class of oxidative decarboxylases. Biochemistry. 42, 12721-12733.

Chen, W., Zhang, C.H., Song, L., Sommerfeld, M., Qiang, H., 2009. A high throughput Nile red method for quantitative measurement of neutral lipids in microalgae. J. Microbiol. Methods. 77, 41-47.

Chow, K.C., Tung, W.L., 1999. Electrotransformation of *Chlorella vulgaris*. Plant Cell Rep. 18, 778-780.

Chisti, Y., 2008. Biodiesel from microalgae beats bioethanol. Trends Biotechnol. 26(3),126-31.

Courchesne, N.M.D., Parisien, A., Wang, B., Lan, C.Q., 2009. Enhancement of lipid production using biochemical, genetic and transcription factor engineering approaches. J. Biotechnol.141, 31-41.

Dunahay, T.G., Jarvis, E.E., Dais, S.S., Roessler, P.G., 1996. Manipulation of microalgal lipid production using genetic engineering. Appl. Biochem. Biotechnol. - Part A Enzyme Engineering and Biotechnology. 223-231.

Feng, S., Xue, L., Liu, H., Lu, P., 2009. Improvement of efficiency of genetic transformation for *Dunaliella salina* by glass beads method. Mol. Biol. Rep. 36, 1433-1439.

Huang, Y.J., Wang, L., Zheng, M.G., Zheng, L., Tong, Y.L., Li, Y., 2013. Overexpression of NgAUREO1, the gene coding for aurechrome 1 from *Nannochloropsis gaditana*, into *Saccharomyces cerevisiae* leads to a 1.6-fold increase in lipid accumulation. Biotechnol. Lett.1-5.

Huang, Y.J., Wang, L., Zheng, M.G., Zheng, L., Tong, Y.L., Li, Y., 2014. Overexpression of NgAUREO1, the gene coding for aurechrome 1 from Nannochloropsis gaditana, into Saccharomyces cerevisiae leads to a 1.6-fold increase in lipid accumulation. Biotechnol. Lett. 36(3), 575-579.

Jiang, P., Qin, S., Tseng, C.K., 2002. Expression of hepatitis B surface antigen gene (HBsAg) in Laminaria japonica (Laminariales, Phaeophyta). Chin. Sci. Bull. 47, 1438-1440.

Johnson, M.K., Johnson, E.J., Macelroy, R.D., Speer, H.L., Nruff, B.S., 1968. Effect of salts on the halophilic alga Dunaliella viridis. J. Bacteriol. 95, 1461-1468.

Ke, J., Wen, T.N., Nikolau, B.J., Wurtele, E.S., 2000. Coordinate regulation of the nuclear and plastidic genes coding for the subunits of the heteromeric acetyl-coenzyme A carboxylase. Plant physiol. 4, 1057-1072.

Klaus, D., Ohlrogge, J.B., Neuhaus, H.E., Dormann, P., 2004. Increased fatty acid production in potato by engineering of acetyl-CoA carboxylase. Planta. 219, 389-396.

Klein, U., De Cam,p J.D., Bogorad, L., 1992. Two types of chloroplast gene promoters in *Chlamydomonas reinhardtii*. Proc. Nail. Acad. Sci. 89, 3453-3457.

Li, J., Xue, L.X., Yan, H., Wang, L., Liu, L.L., Lu, Y., Xie, H., 2007.The nitrate reductase gene-switch: a system for regulated expression in transformed cells of *Dunaliella salina*. Gene. 403, 132-142.

Li, J., Xue, L.X., Yan, H.X., Wang, L.L., Liu, H.T., Liang, J.Y., 2008. Inducible EGFP expression under the control of the nitrate reductase gene promoter in transgenic *Dunaliella salina*. J. Appl. Phycol. 20, 137-145.

Liu, J., Lu, Y., Xue, L., Xie, H., 2010. A structurally novel salt-regulated promoter of duplicated carbonic anhydrase gene 1 from *Dunaliella salina*. Mol. Biol. Rep. 37, 1143-1154.

Madoka, Y., Tomizawa, K.I., Mizoi, J., Nishida, I., Nagano, Y., Sasaki, Y., 2002. Chloroplast Transformation with Modified accD Operon Increases Acetyl-CoA Carboxylase and Causes Extension of Leaf Longevity and Increase in Seed Yield in Tobacco. Plant Cell Physiol. 12,1518-1525.

Maniatis, T., Fritsch, E.F., Sambrook, J., 1982. Olecular Cloning, A Laboratory Manual. Cold Spring Harbor Laboratory Press, Cold Spring Harbor, NY.

Ramírez-Verduzco, L.F., Rodríguez-Rodríguez, J.E., Jaramillo-Jacob, A.D.R., 2012. Predicting cetane number, kinematic viscosity, density and higher heating value of biodiesel from its fatty acid methyl ester composition. Fuel. 91, 102-111.

Ramos, M.J., Fernandez, C.M., Casas, A., Rodriguez, L., Perez, A.,2009. Influence of fatty acid composition of raw materials on biodiesel properties. Bioresour. Technol. 100(1), 261-268.

Rasala, B.A., Muto, M., Sullivan, J., Mayfield, S.P., 2011. Improved heterologous protein expression in the chloroplast of Chlamydomonas reinhardtii through promoter and 5′ untranslated region optimization. Plant biotechnol. J. 9(6), 674-683.

Roesler, K., Shintani, D., Savage, L., Boddupalli, S., Ohlrogge, J., 1997. Targeting of the Arabidopsis homomeric acetyl-coenzyme A carboxylase to plastids of rapeseeds. *Plant Physiol.* 1, 75-81.

Sasaki, Y., Nagano, Y., 2004. Plant acetyl-CoA carboxylase: structure, biosynthesis, regulation, and gene manipulation for plant breeding. Biosci., Biotechnol., Biochem.. 6, 1175-1184.

Shintani, D., Roesler, K., Shorrosh, B., Savage, L., Ohlrogge, J., 1997. Antisense expression and overexpression of biotin carboxylase in tobacco leaves. *Plant Physiol.* 3, 881-886.

Song, D., Fu J., Shi, D., 2008. Exploitation of Oil-bearing Microalgae for Biodiesel. Chin. J. Biotechnol. 24(3), 341-348.

Talebi, A.F., Mohtashami, S.K., Tabatabaei, M., Tohidfar, M., Zeinalabedini, M., Hadavand, H., 2013a. Fatty Acids Profiling; a Selective Criterion for Screening Microalgae Strains for Biodiesel Production. Algal Res. 2(3), 258-267.

Talebi, A.F., Tohidfar, M., Tabatabaei, M., Bagheri, A., Mohsenpor, M., Mohtashami, S.K., 2013b. Genetic manipulation, a feasible tool to enhance unique characteristic of *Chlorella vulgaris* as a feedstock for biodiesel production. Mol. Biol. Rep. 40(7), 4421-4428.

Talebi, A.F., Tabatabaei, M., Chisti, Y., 2014a. BiodieselAnalyzer: a user-friendly software for predicting the properties of prospective biodiesel. Biofuel Res. J. 2, 55-57.

Talebi, A.F., Tabatabaei, M., Tohidfar, M., Bagheri, A., 2014b. Biochemical modulation of lipid productivity and quality in cultures of local Dunaliella salina strain for biodiesel production. Algal Res. (in press).

Tan, C., Qin, S., Zhang, Q., Jiang, P., Zhao, F., 2005. Establishment of a Micro-Particle Bombardment Transformation System for *Dunaliella salina*. J. Microbiol. 43, 361-365.

Thelen, J.J., Ohlrogge, J.B., 2002. Both antisense and sense expression of biotin carboxyl carrier protein isoform 2 inactivates the plastid acetyl-coenzyme A carboxylase in Arabidopsis thaliana. Plant J. . 4, 419-431.

Vera, A., Sugiura, M., 1995. Chloroplast rRNA transcription from structuallt different tandem promoters: an additional novel-type promoter. Curr. Genet. 27, 280-284.

Wang, Z.T., Ullrich, N., Joo, S., Waffenschmidt, S., Goodenough U., 2009. Algal lipid bodies: stress induction, purification, and biochemical characterization in wild-type and starchless *Chlamydomonas reinhardtii*. Eukaryot. Cell. 8, 1856- 1868.

Zhang, Y., Adams, I.P., Ratledge, C., 2007. Malic enzyme: the controlling activity for lipid production? Overexpression of malic enzyme in Mucor circinelloides leads to a 2.5-fold increase in lipid accumulation. Microbiology. 153, 2013–2025.

Zhao, G., Chen, J., Yin, A., 2005. Transgenic soybean lines harbouring anti-PEP gene express super-high oil content. Mol. Plant Breed. 3, 792-796.

Zhila, N.O., Kalacheva, G.S., Volova, T.G., 2005. Effect of Nitrogen Limitation on the Growth and Lipid Composition of the Green Alga Botryococcus brauniiKütz IPPAS H-252. Russ. J. . Plant Physiol. 25(3), 311–319.

# 15

# Direct fermentation of sweet sorghum juice by *Clostridium acetobutylicum* and *Clostridium tetanomorphum* to produce bio-butanol and organic acids

B. Ndaba*, I. Chiyanzu, S. Marx

*School of Chemical and Minerals Engineering, North-West University (Potchefstroom Campus), Potchefstroom, South Africa.*

## HIGHLIGHTS

➢Sweet sorghum juice was fermented with single and mixed cultures of *C. acetobutylicum* and *C. tetanomorphum* at varying inoculum ratios.

➢Microbial cell growth as well as ABE and acid concentrations were monitored over 96 h.

➢*C. acetobutylicum* produced high concentrations of bio-butanol, whilst co-fermentation showed a significant increase in the yield of organic acids.

➢Single and co-culture fermentation of sweet sorghum juice using clostridium species was found a promising method for producing bio-butanol and organic acids.

## GRAPHICAL ABSTRACT

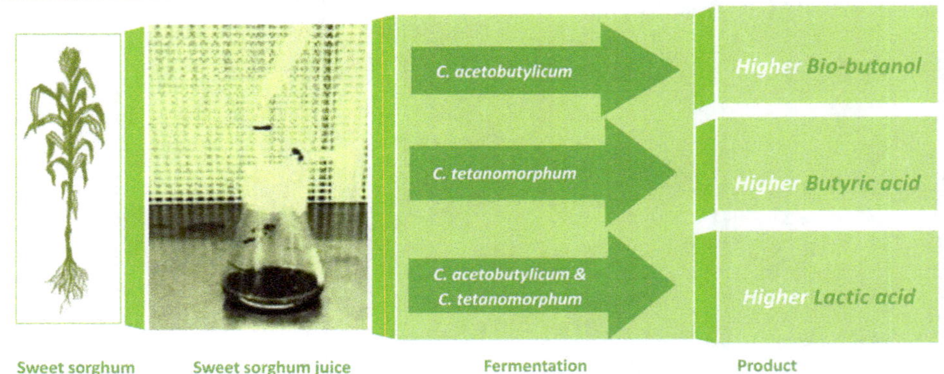


## ABSTRACT

Single- and co-culture clostridial fermentation was conducted to obtain organic alcohols and acids from sweet sorghum juice as a low cost feedstock. Different inoculum concentrations of single cultures (3, 5, 10 v/v %) as well as different ratios of *C. acetobutylicum* to *C. tetanomorphum* (3:10, 10:3, 6.5:6.5, 3:3, and 10:10 v/v %, respectively) were utilized for the fermentation. The maximum butanol concentration of 6.49 g/L was obtained after 96 h fermentation with 10 % v/v *C. acetobutylicum* as a single culture. The fermentation with 10% v/v *C. tetanomorphum* resulted in more than 5 g/l butyric acid production. Major organic acid concentration (lactic acid) of 2.7 g/L was produced when an inoculum ratio of 6.5: 6.5 %v/v *C. acetobutylicum* to *C. tetanomorphum* was used.

**Keywords:**
Bio-butanol
*Clostridium* species
Fermentation
Organic acids
Sweet sorghum juice

## 1. Introduction

Bio-butanol has been considered as a better transportation fuel than bio-ethanol mainly due to its higher number of carbon atoms and consequently higher energy content, its miscibility in diesel, and higher blending capacity. There is a plethora of methods employed by the bio-chemical industries to produce bio-butanol. Over the past decades, direct fermentation of sugars derived from enzymatic conversion of starchy crops or by acid/ enzymatic hydrolysis of lignocellulosic feedstock have dominated (Ranjan et al., 2013; Yang et al., 2014). In the butanol production processes, the production of other byproducts, e.g. acetone, ethanol, and acetic, butyric, and lactic acids, are undesirable; however, these byproducts are considered as main and

* Corresponding author
E-mail address: ndabab@gmail.com

valuable products in a number of processes (Zamani, 2015). Sweet sorghum is one of the common feedstock for bio-butanol production. This is ascribed to the fact that it is grown in diverse temperature climates in both dry and wet areas (Goshadrou and Karimi, 2010). However, the application of this plant as biofuel feedstock has also been the subject of the famous food *vs.* fuel debate.

To avoid such concerns, research has deviated from the starchy biomass to non-edible parts of the plant i.e. the lignocellulosic biomass which could contribute to improving the socio-economic conditions of sweet sorghum as a dedicated energy crop.

Fresh sweet sorghum comprises of sugars, such as; sucrose, glucose and fructose that can be extracted from the stalks. According to Datta Mazumdar et al. (2012), Sobrinho et al. (2011) and Kundiyana et al. (2010), high quality sweet sorghum juice contains approximately 14-22 °Bx, but the values could vary depending on the soil origin of the plant. These sugars can be directly fermented using *Clostridium* spp. to produce acetone, butanol and ethanol (ABE). ABE fermentation involves two stages, namely; i) exponential growth stage (acidogenesis) and, ii) stationary stage (solvontogenesis). During the acidogenesis, mainly acetic and butyric acids are produced while during the latter ABE solvents are generated (Borner et al., 2014).

Improving the cost effectiveness aspects of bio-butanol production has been widely investigated. Several studies have strived to use different *Clostridium* spp. for fermentation of sugars into bio-butanol (Ezeji et al., 2004; Kominkiat and Cheirsilp, 2013; Li et al., 2014). In addition, co-culture fermentation has also been explored aiming to enhance ABE production. Examples of such include co-cultivation of *C. beijerinckii* and *C. tyrobutyricum* (Li et al., 2013), *C. beijerinckii* and *C. cellulovorans* (Wen et al., 2014), *C. thermocellum* and *C. saccharoperbutylacetonicum* N1-4 (Nakayama et al., 2011) and *Bacillus subtilis* in a co-culture with *C. butylicum* (Tran et al., 2010).

Nevertheless, the aforementioned studies have only focused on bio-butanol as a product with organic acids as intermediates and few studies have paid attention to the synergetic effect of inoculum concentrations and the ratios at which microorganisms were added, on acid formation during ABE fermentation. To the authors' best of knowledge, the production of organic acids from sweet sorghum juice through co-culture fermentation has not been reported before. Hence, this study was set to investigate bio-butanol and organic acid production from sweet sorghum juice using *C. acetobutylicum* and *C. tetanonomorphum*. Different parameters were taken into account during the course of this study including the type of microorganisms (*C. acetobutylicum* and *C. tetanomorphum*), inoculum ratios and the synergistic effect of the two micro-organisms in co-fermenting the sugars contained in sweet sorghum juice. All final products were analysed and identified by the high performance liquid chromatography (HPLC) technique.

## 2. Materials and methods

### 2.1. Biomass

Sweet sorghum stalks were harvested in May 2013 at the test farm of the Agricultural Research Council - Grain Crops Institute of South Africa (ARC-GCI), Potchefstroom (26°43'43.16"S - 27°04'47.71"E). The juice was extracted from the stalks without grains using a mechanical press roller. Approximately 4 L of the juice was extracted from 26.80 kg of fresh stalks and was stored at 4°C until used.

### 2.2. Long-term storage of the sweet sorghum juice

The prolonged storage protocol of sweet sorghum juice was adopted from Datta Mazumdar et al. (2012). Briefly, the juice was initially filtered using vacuum filtration to remove all the unwanted solid particles. Prior to heating, Brix and pH of the sample were determined using a refractometer and pH meter, respectively. Concentration of the juice was performed at 70 °C for 45 min in a hot plate with continuous stirring. The heating was controlled to allow gradual temperature increases to avoid charring of the sugars. The juice was then cooled down to 40 °C resulting in clarified sweet sorghum juice of 73° Bx. Thereafter, the cooked juice was stored in PET bottles at 4°C till further use. **Figure 1** provides an overview of the sweet sorghum juice processing from harvest to juice concentration.

### 2.3. Strains and medium

*C. acetobutylicum* ATCC 824 and *C. tetanomorphum* ATCC 49273 were purchased from the American Type Culture Collection (ATCC). The stock cultures were maintained in the form of cell suspensions in 25% (v/v) sterile glycerol at -80°C. The organisms were grown on a Reinforced Clostridial Medium (RCM) for 24-48 h at 37°C before being sub-cultured into Clostridial Growth Medium (CGM) which was used to inoculate sweet sorghum juice for fermentation (**Table 1**). Both media were sterilized at 121°C for 15 min. Short term stock cultures were prepared on Clostridium Growth Agar (CGA) and single colonies were revived every two weeks.

**Table 1.**
The compositions of the growth media used.

| Reinforced Clostridial Medium (RCM) | | Clostridial Growth Medium (CGM) | |
|---|---|---|---|
| *Nutrients* | *Composition (g/L)* | *Nutrients* | *Composition (g/L)* |
| Tryptose | 10 | NaCl | 1 |
| Beef extract | 10 | $(NH_4)SO_4$ | 2 |
| Yeast extract | 3 | Yeast extract | 5 |
| Sodium Chloride | 5 | $KH_2PO_4$ | 0.75 |
| Sodium acetate | 3 | $K_2HPO_4$ | 0.75 |
| Cystein hydrochloride | 0.5 | Asparagine | 2 |
| Soluble starch | 1 | $MgSO_4.7H_2O$ | 0.70 |
| | | $MnSO_4.H_2O$ | 0.01 |
| | | $FeSO_4.7H_2O$ | 0.01 |

### 2.4. Fermentation of sweet sorghum juice

Prior to fermentation, the concentrated sweet sorghum juice was diluted with distilled water to revert to the original Brix index of 14°Bx (16.3 g/L) (Sobrinho et al., 2011). High purity nitrogen (99.9 %) was sparged through the modified flasks before the inoculum addition to remove oxygen from the flasks and to maintain anaerobic conditions. The flask was modified using a tight rubber stopper for sealing the opening, a micro filter outlet, and sampling syringe to avoid closing and opening of the flask. A single colony from the plates was inoculated into 10 mL CGM and was incubated for 12 h. After the incubation the $OD_{600}$ reached 0.798 and 0.810 for *C. tetanomorphum* and *C. acetobutylicum*, respectively. Fermentation was done by transferring certain volumes of the starter-culture (3, 5 or 10 % v/v) into the medium (sweet sorghum juice + nutrients) contained in the modified 250mL Erlenmeyer flasks with a 100 mL working volume.

During fermentation, the temperature was maintained at 37°C, the pH was adjusted between 6-6.5 by addition of NaOH or HCl, and the mixture was agitated at 150 rpm. Fermentation was conducted for 92 h and samples were taken at set time intervals during the experiment. In addition to the single species fermentation, co-culture fermentation of sweet sorghum juice was also conducted. Different inoculum concentration ratios (3:10, 10:3, 6.5:6.5, 3:3, and 10:10% v/v of *C. acetobutylicum* to *C. tetanomorphum*) were used to ferment sweet sorghum juice to investigate the influence of co-culture inocula variables on the product yields. All experiments were conducted in triplicates.

### 2.5. Analytical techniques used

Cell growth was analysed by measuring $OD_{600}$ using a spectrophotometer (UV 7300, Jenway). The samples were then centrifuged for 5 min and the supernatants were used to determine the concentrations of glucose, fructose and solvent after filtration with a 0.22-0.45 μm syringe filter. The reducing sugars, acids, and solvent concentrations were measured using HPLC with an HPX-87H aminex column at 55°C RID and 30°C column temperature. The mobile phase used was 0.005M $H_2SO_4$ at a flow rate of 0.6 $mL.min^{-1}$ with an injection volume of 5 μL.

## 3. Results and discussion

Sweet sorghum juice was subjected to fermentation with different cultures using *C. tetanomorphum* and *C. acetobutylicum*. The effect of different parameters on the fermentation of the juice containing 16.31 g/l sugars (6.25 g/l glucose, 0.31 g/l fructose and 9.75 g/l sucrose), was evaluated.

### *3.1. Fermentation with C. tetanomorphum*

Fermentable sugars, i.e., glucose, fructose and sucrose, would normally undergo conversion to a range of intermediates after being consumed by the bacteria. It was observed that after fermentation with *C. tetanomorphum*, the pH of the culture medium decreased from 6.5 to 4.6 as a result of organic acids formation.

Single culture fermentation using different inoculum loadings (3, 5, 10 %v/v) of *C. tetanomorphum* was investigated for the production of bio-butanol. **Figure 1** shows the effect of inoculum concentration on bio-butanol production within 96 h fermentation. Even with 3 % inoculum, the optical density of the culture increased to more than 2 in the first 24 h of fermentation. Butanol production was accompanied by ethanol production, in the absence of acetone formation which is consistent with the findings of Gottwald et al. (1984). Prior to the fermentation of sweet sorghum juice, broth medium containing nutrients was inoculated with the organism and was incubated for 12 h. The initial optical density after 12 h of incubation and at 0 h after the initiation of the juice fermentation was at 0.798.

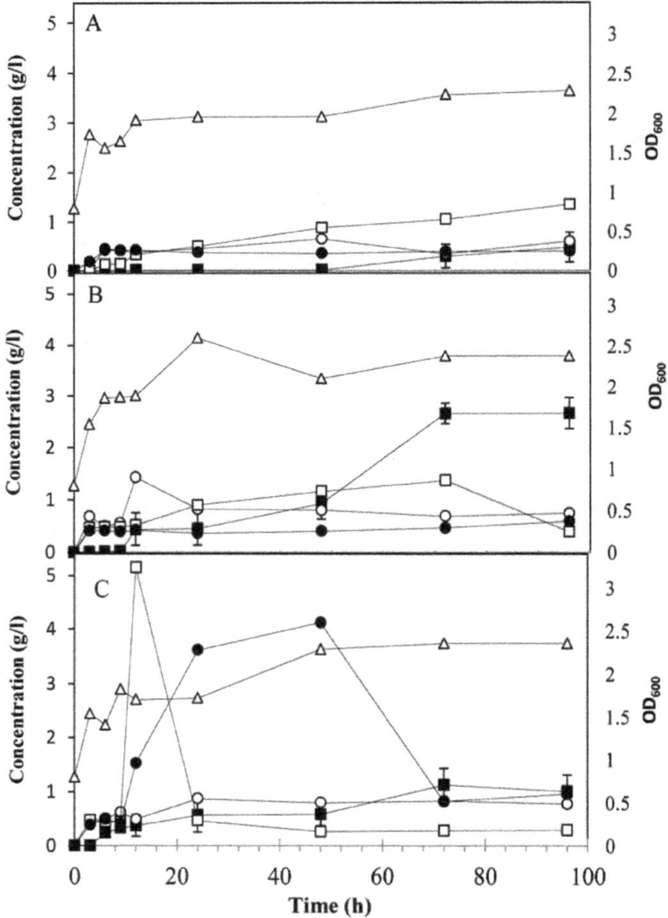


**Fig.1.** Concentration of butanol (■), ethanol (●), acetic acid (○), and butyric acid (□) as well as optical density (Δ) using 3% (A), 5% (B), and 10% *C. tetanomorphum* inoculum concentrations.

Through the fermentation with *C. tetanomorphum*, organic acids i.e. acetic and butyric acids were initially generated followed by the production of alcohols, i.e., ethanol and butanol. The maximum acetic acid concentrations of 0.64, 1.43, and 0.87 g/l, were obtained in the fermentation with 3, 5, and 10% *C. tetanomorphum* after 48, 12, and 24 h, respectively. In addition, butyric acid concentration increased to 1.35, 1.37, and 5.15 g/L after 96, 72, and 12 h fermentation with 3, 5, and 10% inoculum loadings, respectively.

With 3% inoculation, no detectable level of butanol was produced within the first 48 h of fermentation. The butanol concentration of less than 0.3 g/l and ethanol concentration of 0.42 g/l were obtained after 72 h fermentation, which were produced mostly during the stationery phase. Organic acids showed a rapid increase after 12 h. Inoculating with 5 %v/v inoculum resulted in obtaining higher butanol and ethanol concentrations. During 96 h fermentation, more than 2.5 g/L butanol and 0.5 g/L ethanol were produced. With increasing the initial inoculation volume to 10% of the medium, the final butanol and ethanol concentration increased to 1.14 and 0.96 g/L, respectively. Using different inoculation regimes, no acetone was produced during the fermentation with *C. tetanomorphum*. Butanol was the main alcohol produced in the fermentation with 5 % inoculation, whereas ethanol production was prevailed when fermentation was conducted with 10 % inoculum.

Optical density of the medium increased within the initial 12 h of the fermentation. Moreover, as a result of developing high concentrations of acids in the fermentation broth, the cell growth gradually decreased (Wu et al., 2013).

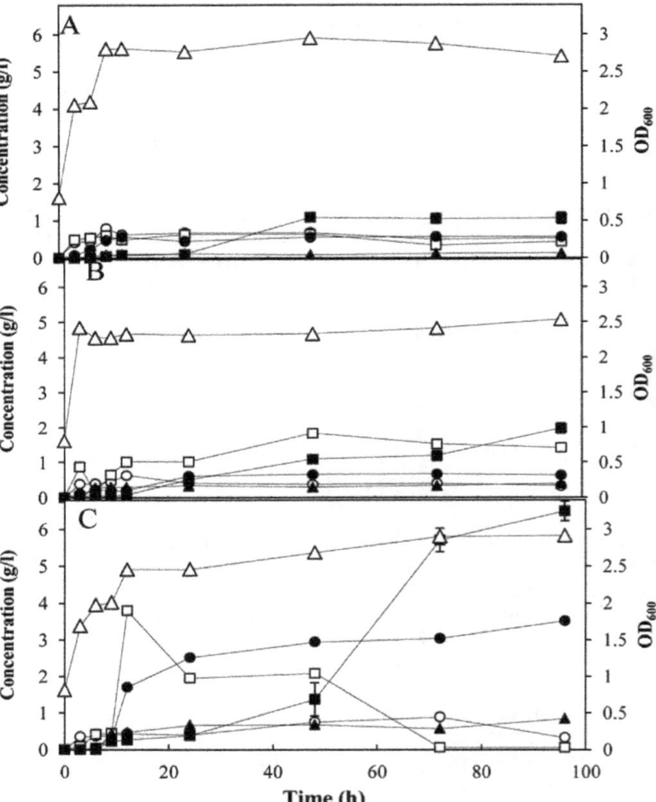


**Fig.2.** Concentration of acetone (▲), butanol (■), ethanol (●), acetic acid (○), and butyric acid (□) as well as optical density (Δ) using 3% (A), 5% (B), and 10% *C. acetobutylicum* inoculum concentrations.

### *3.2. Fermentation with C. acetobutylicum*

The sweet sorghum juice was subjected to the fermentation with *C. acetobutylicum* with different inoculation ratios of 3, 5, and 10 % v/v for 96 h. The profiles of acids and alcohols as well as their respective concentrations and also the optical density of the medium are shown in **Figure 2** Through the first 9 h of fermentation, along with the growth phase of the clostridia, the sugars were utilized and fermented to acetic and butyric acids. After the organic acids concentration increased during the growth phase, alcohols formation was initiated by the cells during their stationary phase. The maximum acetic acid concentration observed for 3, 5 and 10%v/v inoculation were 0.79, 0.63 and 0.89 g/L, respectively. Butyric acid with maximum concentrations of 0.63, 1.83, and 3.30 g/L were produced by using 3, 5 and 10%v/v inoculum, respectively. The pH of the medium decreased from 6.5 to 4.1 through the fermentation by *C. acetobutylicum*. These results were in line with those of Jiang et al. (2014), who reported rapid pH reduction during clostridial fermentation indicating the formation of products.

A gradual increase in the concentrations of both alcohols was generally observed. By increasing the volume of the initial inoculum, the production of acids and alcohols was improved. Using 3 % inoculum of *C. acetobutylicum*, maximum ethanol concentration of 0.56 g/L was obtained within 24 h. In addition, the maximum butanol and acetone concentrations of 1.09 and 0.12 g/L were obtained after 48 h. By increasing the inoculum loading to 5 and 10%, ethanol concentration increased to more than 0.6 and 3.5 g/L, respectively. Moreover, by increasing the inoculum volume to 5%, the final butanol and acetone concentrations were increased to more than 1.9 and 0.3 g/l, respectively. Further increase in the inoculation volume resulted in higher butanol and acetone concentrations of 6.5 and 0.8 g/l, respectively.

Sugar consumption through the fermentation was highly dependent on the culture and its initial concentration. The profiles of sugar concentration during the fermentation with different cultures are presented in **Figure 3**. The sugars contained in the sweet sorghum juice (i.e. 6.25 g/l glucose, 0.31 g/l fructose, and 9.75 g/l sucrose) were consumed by the cultures resulting in 0.3 to 2.5 g/l residual sugar. Through the fermentation with 10% v/v *C. acetobutylicum* more than 98 % of the sugars contained in the juice was consumed.

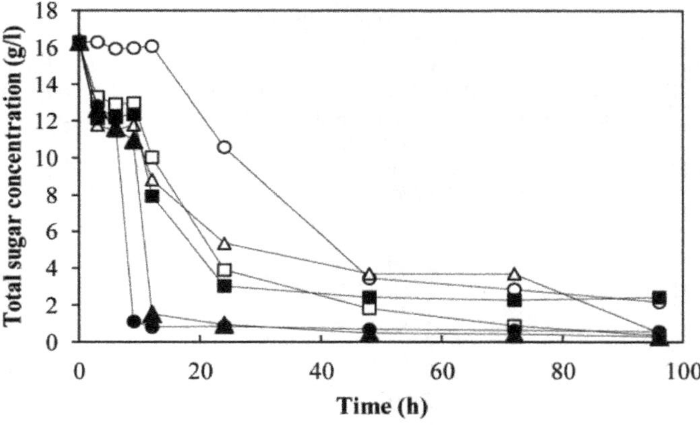


**Fig.3.** Total concentration of sugars (g/L) during the fermentation using different inoculum volumes 3 % (○), 5% (□), and 10% (Δ) *C. tetanomorphum* and 3% (■), 5 % (●), and 10% (▲) *C. acetobutylicum*.

In comparison with the *C. tetanomorphum* fermentation, the fermentation with *C. acetobutylicum* resulted in higher butanol production. Increasing the inoculum volume in the clostridial fermentation improved the butanol production. The highest butanol concentration obtained was with 10% v/v *C. acetobutylicum* inoculation. On the other hand, through fermentation with 10% v/v *C. tetanomorphum,* the highest concentration of butyric acid was generated. After single culture fermentation, co-culture experiments were conducted using a combination of *C. acetobutylicum* and *C. tetanomorphum*.

### *3.3. Co-culture fermentation for acid production*

*C. acetobutylicum* and *C. tetanomorphum* were inoculated simultaneously into the juice medium under anaerobic conditions. The two strictly anaerobic *C. acetobutylicum* ATCC 824 and *C. tetanomorphum* ATCC 49273 were used in these experiments to investigate their synergistic effect on the product yields. The concentrations of organic acids, i.e., acetic acid, lactic acid, and succinic acid, formed after 96 h fermentation using the co-culture are shown in **Figure 4**. As shown, different amounts of acids were formed depending on the initial inoculation regime of the medium.

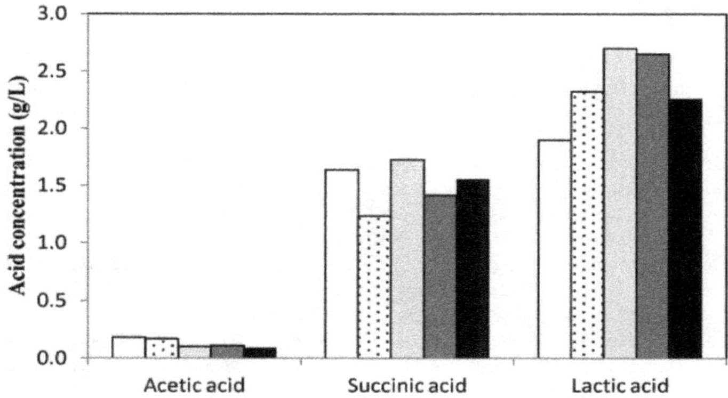


**Fig.4.** Acid concentrations (g/L) obtained for inoculation with 3:3 (white), 3:10 (dotted withe, 6.5:6.5 (light gray), 10;3 (dark gray), 10:10 %v/v (black) of C.acetobulycum to C. tetanomorphum in 96 hrs.

Fermentation with the co-culture resulted in significantly higher acid formation from the sweet sorghum juice compared to to the pure cultures of *C. acetobutylicum* and *tetanomorphum*. Lactic acid was the most prominent acid produced by all co-cultures. The highest lactic acid concentration of 2.7 g/l was produced after 96 h fermentation with a co-culture of 6.5:6.5 %v/v *C. acetobutylicum* to *C. tetanomorphum*. However, the acetic acid concentration using this treatment was low at 0.1 g/l. The concentrations of the remaining sugars after fermentation with different co-cultures are listed in **Table 2**. Less than 0.2 g/l sugars remained unconverted after 96 h fermentation with the 6.5:6.5 and 10:10 % v/v co-cultures.

**Table 2.**
Remaining Reducing sugar concentration after co-culture fermentation in 96 h.

| *C. acetobutylicum* to *C. tetanomorphum* (% v/v) | Residual sugar (g/L) |
|---|---|
| **3:3** | 1.44 |
| **3:10** | 1.41 |
| **10:3** | 1.21 |
| **6.5: 6.5** | 0.18 |
| **10:10** | 0.18 |

Relatively high amounts of organic acids consisting of acetic acid, succinic acid, and lactic acid were obtained through fermentation of sweet sorghum juice with the co-culture of *C. acetobutylicum* and *C. tetanomorphum*. Therefore, co-culture of these strains is recommended for the production of organic acids.

## 4. Conclusions

The fermentation with single- and co-culture of *C. acetobutylicum and C. tetanomorphum* were evaluated for organic alcohols and acids production from sweet sorghum juice. The results obtained showed that an inoculum loading of 10% v/v *C. acetobutylicum* led to the maximum concentration of 6.49 g/L for butanol. An inoculum loading of 10% v/v *C. tetanomorphum* was found to produce a high acid (butyric acid) concentration of 5.15 g/L. In addition, fermentation using a co-culture approach resulted in the production of lactic acid with a concentration of 2.7 g/L, higher than that obtained by single culture fermentations.

## 5. Acknowledgments

This research work was conducted through the support provided by the South African Research Chairs Initiative of the Department of Science and Technology and National Research Foundation of South Africa. Any opinions, findings and conclusions or recommendations expressed in this article are those of the author(s) and the NRF does not accept any liability in this regard.

**References**


Abd-Alla, M.H., El-Enany, A.E., 2012. Production of acetone-butanol-ethanol from spoilage date palm (*Phoenix dactylifera L.*) fruits by mixed culture of *Clostridium acetobutylicum* and *Bacillus subtilis*. Biomass Bioenergy. 42, 172-178.

Börner, R.A., Zaushitsyna, O., Berillo, D., Scaccia, N., Mattiasson, B., Kirsebom, H., 2014. Immobilization of *Clostridium acetobutylicum* DSM 792 as macroporous aggregates through cryogelation for butanol production. Process Biochem. 49, 10-18.

Datta Mazumdar, S., Poshadri, A., Srinivasa Rao, P., Ravinder Reddy, C.H., Reddy, B.V.S., 2012. Innovative use of Sweet sorghum juice in the beverage industry. Int. Food Res. J. 19, 1361-1366.

Du, T.F., He, A.Y., Wua, H., Chen, J.N., Kong, X.P., Liu, J.L., Jiang , M., Ouyang, P.K., 2013. Butanol production from acid hydrolyzed corn fiber with *Clostridium beijerinckii* mutant. Bioresour. Technol. 135, 254-261.

Ezeji, T., Qureshi, N., Blaschek, H.P., 2004. Production of acetone–butanol–ethanol (ABE) in a continuous flow bioreactor using degermed corn and *Clostridium beijerinckii*. Process Biochem. 42, 34-39.

Goshadrou, A., Karimi, K., 2010. Bioethanol Production from Sweet Sorghum Bagasse. Chem. Eng. Congr.13, 5-8.

Gottwald, M., Hippe, H., Gottschalk, G., 1984. Formation of n-Butanol from d-Glucose by strains of the "*Clostridium tetanomorphum*" group. Appl. Environ. Microbiol. 48, 573-576.

Jiang, W., Wen, Z., Wu, M., Li, H., Yang, J., Lin, J., Lin, Y., Yang, L., Cen, P., 2014. The Effect of pH Control on Acetone–Butanol–Ethanol Fermentation by *Clostridium acetobutylicum* ATCC 824 with Xylose and D-Glucose and D-Xylose Mixture. Chin. J. Chem. Eng. 22, 937-942.

Komonkiat, I., Cheirsilp, B., 2013. Felled oil palm trunk as a renewable source for bio-butanol production by *Clostridium spp.* Bioresour Technol. 146, 200-207.

Kovács, K., Willson, B.J., Schwarz, K., Heap, J.T., Jackson, A., Bolam, D.N., Winzer, K., Minton, N.P., 2013. Secretion and assembly of functional mini-cellulosomes from synthetic chromosomal operons in *Clostridium acetobutylicum* ATCC 824. Biotechnol. Biofuels. 6, 117.

Li, L., Ai, H., Zhang, S., Li, S., Liang, Z., Wua, Z.Q., Yang, S.T., Wang, J.F., 2013. Enhanced butanol production by coculture of *Clostridium beijerinckii* and *Clostridium tyrobutyricum*. Bioresour. Technol. 143, 397-404.

Li,J., Chen, X., Qi,B., Luo,J., Zhang,Y., Sub, Y., Wan, Y., 2014. Efficient production of acetone–butanol–ethanol (ABE) from cassava by a fermentation–pervaporation coupled process. Bioresour. Technol. 169, 251-257.

Lin Li, L., Ai,H., Zhang S., Li, S., Liang, Z., Wua, Z., Yang, S., Wang, J., 2013. Enhanced butanol production by coculture of *Clostridium beijerinckii* and *Clostridium tyrobutyricum*. Bioresour. Technol. 143, 397-404.

Moon, C., Lee, C.H., Sang, B.I., Uma, Y., 2011. Optimization of medium compositions favoring butanol and 1,3-propanediol production from glycerol by *Clostridium pasteurianum*. Bioresour. Technol. 102, 10561-10568.

Nakayama, S., Kiyoshi, K., Kadokura, T., Nakazato, A., 2011. Butanol Production from Crystalline cellulose by cocultured *Clostridium thermocellum* and *Clostridium saccharoperbutylacetonicum* N1-4. Appl. Environ. Microbiol. 77, 6470-6475.

Qureshi, N., Saha, B.C., Dien, B., Hector, R.E., Cotta, M.A., 2010. Production of butanol (a biofuel) from agricultural residues: Part I-use of barley straw hydrolysate. Biomass Bioenergy. 34, 559-565.

Ranjan, A., Mayank, R., Moholkar, V.S., 2013. Process optimization for butanol production from developed rice straw hydrolysate using *Clostridium acetobutylicum* MTCC481 strain. Biomass Conv. Bioref. 3, 143-155.

Tran, H.T.M., Cheirsilp, B., Hodgson, B., Umsakul, K., 2010. Potential use of *Bacillus subtilis* in a co-culture with *Clostridium butylicum* for acetone–butanol–ethanol production from cassava starch. Biochem. Eng. J. 48, 260-262.

Wen, Z., Wu, m., Yang, Y.L.L., Lin, j., Cen, P., 2014. Artificial symbiosis for acetone-butanol-ethanol (ABE) fermentation from alkali extracted deshelled corn cobs by co-culture of *Clostridium beijerinckii* and *Clostridium cellulovorans*. Microbial Cell Fact. 13, 92.

Yang, M.K., Keinänen, S., Vepsäläinen,M., Romar, J., Tynjälä, H., Lassi, P.U., Pappinen, A., 2014. The use of (green field) biomass pretreatment liquor for fermentative butanol production and the catalytic oxidation of bio-butanol. Chem. Eng. Res. Design. 92, 1531-1538.

Zamani, A., 2015. Lignocellulose-based Bioproducts, in: Karimi, K., Springer International Publishing, Switzerland.

# Meeting the U.S. renewable fuel standard: a comparison of biofuel pathways

Marc Y. Menetrez

*United State Environmental Protection Agency, Office of Research and Development, National Risk Management Research Laboratory, Air Pollution Prevention and Control Division, Research Triangle Park, NC 27711, USA.*

## HIGHLIGHTS

- *The year 2014 will see a significant increase in the U.S. production of cellulosic ethanol to a total of 104 Mgal/yr (394 x$10^3$ $m^3$/yr) when three new plants are brought under full operation.*
- *The accomplished increase in U.S. cellulosic ethanol production although significant, will remain far below the mandated 2014 Renewable Fuel Standard of 1.75 Bgal/yr (6.6 x$10^6$ $m^3$/yr).*
- *Potential pathways for producing biofuels from algae were evaluated and compared for their feedstock and footprint demands, as well as productivity potential.*
- *The U.S. commercial production of renewable fuels is increasingly behind levels mandated by the Renewable Fuel Standard. Algae based fuels could be developed to help fill this growing gap.*

## GRAPHICAL ABSTRACT

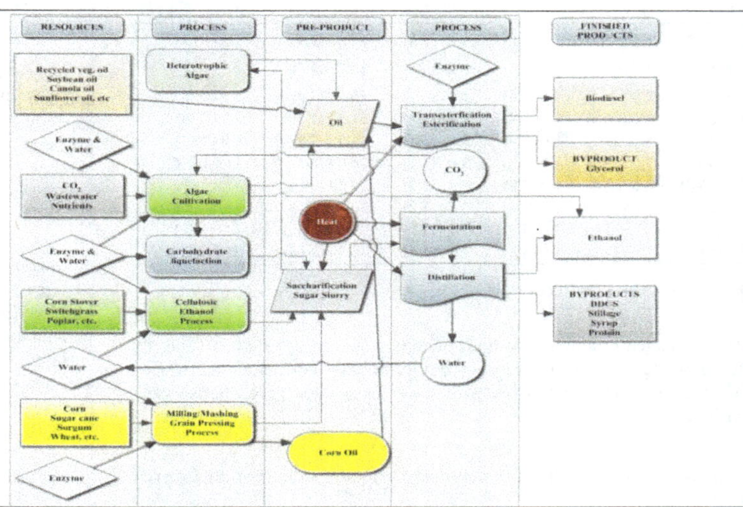


## ABSTRACT

The production of renewable energy is undergoing rapid development. Ethanol primarily derived from corn and biodiesel made from recycled cooking oil and agricultural grains are established sources of renewable transportation fuel. Cellulosic ethanol production is increasing substantially, but at a rate below expectations. If future renewable fuel projections are to be accomplished, additional sources will be needed. Ideally, these sources should be independent of competing feedstock use such as food grains, and require a minimal footprint. Although the uses of algae seem promising, a number of demonstrations have not been economically successful in today's market. This paper identifies efforts being conducted on ethanol and biodiesel production and how algae might contribute to the production of biofuel in the United States. Additionally, the feedstock and land requirements of existing biofuel pathways are compared and discussed.

**Keywords:**
Renewable energy
Ethanol
Biodiesel
Algae
Biofuel pathway

* Corresponding author
E-mail address: menetrez.marc@epa.gov

## 1. Introduction

The 2007 Energy Independence and Security Act (EISA) (H.R. 6) and the 2005 federal renewable fuel standard (RFS) stipulates that, by 2022, the United States must produce 15 billion gallons per year (Bgal/yr) or 56.8 million cubic meters (x10$^6$ m$^3$/yr) of corn-based ethanol (CBE), 16 Bgal/yr (60.6 Mm$^3$/yr) of cellulosic biofuels, 1 Bgal/yr (3.8 x10$^6$ m$^3$/yr) of biodiesel, and 4 Bgal/yr (15.1 x10$^6$ m$^3$/yr) of advanced biofuels (other than corn-based ethanol) (Environmental News Service, 2011; Public Law 110–140, 2007; RFS Renewable Fuels Association, 2012). Achieving a total production goal of 36 Bgal/yr (136 x10$^6$ m$^3$/yr) of renewable fuel by 2022 is a substantial challenge, largely due to the established RFS requirement for cellulosic ethanol (see Table 1). However, U.S. biofuel production is substantial and growing.

**Table 1.**
Renewable Fuel Projections (Public Law 110–140, 2007; Thompson et al., 2010; Environmental News Service, 2011; RFS Renewable Fuels Association, 2012).

| Year | Corn-based Ethanol (CBE) Bgal (x10$^6$ m$^3$/yr) | Cellulosic Ethanol Bgal (x10$^6$ m$^3$/yr) | Biodiesel [a] Bgal (x10$^6$ m$^3$/yr) | Advanced Biofuel Bgal (x10$^6$m$^3$/yr) | Total RFS Bgal (x10$^6$m$^3$/yr) | Total biofuel (minus CBE) Bgal (x10$^6$m$^3$/yr) |
|---|---|---|---|---|---|---|
| 2013 | 13.8 (52.2) | 1.0 (3.8) | 1.0 (3.8) | 0.75 (2.8) | 16.6 (62.6) | 2.8 (10.4) |
| 2014 | 14.4 (54.5) | 1.8 (6.6) | 1.0 (3.8) | 1.0 (3.8) | 18.2 (68.7) | 3.8 (14.2) |
| 2015 | 15.0 (56.8) | 3.0 (11.4) | 1.0 (3.8) | 1.5 (5.7) | 20.5 (77.6) | 5.5 (20.8) |
| 2016 | 15.0 (56.8) | 4.2 (15.9) | 1.0 (3.8) | 2.0 (7.6) | 22.3 (84.2) | 7.3 (27.4) |
| 2017 | 15.0 (56.8) | 5.5 (20.8) | 1.0 (3.8) | 2.5 (9.5) | 24.0 (90.8) | 9.0 (34.1) |
| 2018 | 15.0 (56.8) | 7.0 (26.5) | 1.0 (3.8) | 3.0 (11.4) | 26.0 (98.4) | 11.0 (41.6) |
| 2019 | 15.0 (56.8) | 8.5 (32.2) | 1.0 (3.8) | 3.5 (13.2) | 28.0 (106.0) | 13.0 (49.2) |
| 2020 | 15.0 (56.8) | 10.5 (39.7) | 1.0 (3.8) | 3.5 (13.2) | 30.0 (113.6) | 15.0 (56.8) |
| 2021 | 15.0 (56.8) | 13.5 (51.1) | 1.0 (3.8) | 3.5 (13.2) | 33.0 (124.9) | 18.0 (66.1) |
| 2022 | 15.0 (56.8) | 16.0 (60.6) | 1.0 (3.8) | 4.0 (15.1) | 36.0 (136.3) | 21.0 (79.5) |

[a] EISA is not specific on biodiesel after 2012.

The RFS projected production of multiple renewable fuels listed in Table 1 establishes guidance in consideration of environmental consequences while advocating fuel production practices with lower associated green-house gas (GHG) emission.

## 2. Meeting the RFS challenge for cellulosic ethanol - current to future production

As of 2014, the production of CBE (from more than 200 ethanol plants) was 15 Bgal/yr (56.8_x10$^6$ m$^3$/yr) (equal to the 2022 RFS) (Parker, 2012a; Ethanol Producers Digest, 2013). The majority of this ethanol (or approximately 13.3 Bgal [50.3 x10$^6$ m$^3$/yr]) was used for the 10% additive to gasoline (Parker, 2012a). Biodiesel production, largely from yellow grease (recycled vegetable oil), soy bean oil, and canola oil, was 1.0 Bgal/yr (3.8 x10$^6$ m$^3$/yr) (equal to the 2022 RFS), from more than 200 biodiesel plants (U.S. Biodiesel Digest, 2011; U.S. Energy Information Administration, 2012a; Biodiesel Industry Directory, 2013). Cellulosic ethanol plant production capacity is currently 20 million gallons per year (75.7x10$^3$ m$^3$/yr), as listed in Table 2. Thus, production is already at or above the 2022 RFS levels for biodiesel and CBE biofuels (U.S. Energy Information Administration, 2012a; Ethanol Producers Digest, 2013).

As of early 2014, there were twelve commercial cellulosic ethanol plants in operation, and they have a combined capacity of producing a total of 20 Mgal/yr (75.7x10$^3$ m$^3$/yr), (listed in Table 2), more than triple the 2013 capacity of 6.25 Mgal/yr (23.7x10$^3$ m$^3$/yr) (Ethanol Producers Digest, 2013). As listed in Table 2, the Indian River Bioenergy Plant has the capability of producing 8 Mgal/yr (30.3x10$^3$ m$^3$/yr) (as well as 5 megawatts of electric power), and the Fiberight Blairstown Plant can produce 6 Mgal/yr (22.7x10$^3$ m$^3$/yr) however, the other ten facilities are demonstration pilot plants, with a production limit of less than 1.4 Mgal/yr (5.3x10$^3$ m$^3$/yr) (Shaffer, 2012; Ethanol Producers Digest, 2013). These demonstration projects have proven that their technology is successful. The RFS challenge (listed in Table 1) to produce 1.75Bgal/yr (6.6 x10$^6$ m$^3$/yr) of cellulosic ethanol in 2014 is unlikely, and the possibility of meeting future RFS demands seems increasingly distant.

Five major cellulosic ethanol plants, listed in Table 2, are currently under construction that when in operation will produce in excess of 104 Mgal/yr (393.7x10$^3$ m$^3$/yr). The Poet and Royal plant in Emmetsburg, Iowa is referred to as 'Project Liberty' and is located near other Poet CBE plants (Shaffer, 2012). The plant was constructed at a cost of $250 million, will produce 20 to 28 Mgal/yr (75.7 to 106 x10$^3$ m$^3$/yr) from corn-stover, and is expected to be operational in 2014 (Shaffer, 2012). Poet is an established ethanol manufacturer that owns and operates 27 CBE plants in the U.S. (Shaffer, 2012). The successful operation of the Project Liberty plant and the established availability of corn stover could lead to further expansion and additional plants. (Poet is also currently extracting 250,000 t/yr [227 x10$^6$ kg/yr] or 68 Mgal/yr [257 x10$^3$ m$^3$/yr], of corn oil from 25 equipped CBE plants [Biodiesel Magazine, 2013; Ethanol Producers Digest, 2013].)

As mentioned, in addition to the Project Liberty plant that will be completed this year, there are four other major projects underway, i.e. Abengoa Bioenergy's cellulosic ethanol plant using corn stover in Kansas; Blue Fire Renewables, LLC.'s plant, in Mississippi (that will use wood waste); Dupont Danisco's plant, in Iowa that will use corn stover; and Enerkem Alberta's plant, in Alberta Canada that will use municipal solid waste (Shaffer, 2012; Ethanol Producers Digest, 2013). Table 2 lists the commercial cellulosic ethanol plants and their stages of construction/operation. There are three stages: (1) "Existing Plants", consisting of nine existing plants in intermittent operation with a combined capacity of 20 Mgal/yr (75.7x10$^3$ m$^3$/yr); (2) "Under Construction" consisting of nine plants with a combined capacity of 104 Mgal/yr (394 x10$^3$ m$^3$/yr); and; (3) "Proposed Plants" consisting of 16 proposed plants with a combined capacity of 368 Mgal/yr (1.4 x10$^6$ m$^3$/yr) (Shaffer, 2012; Ethanol Producers Digest, 2013). It should be understood that the existing plants are demonstration plants, many of which may not maintain production on a continuous basis since their purpose may not be to produce ethanol economically. These small scale plants are used to study the details of the process, demonstrate the feasibility of the technology and conduct process optimization studies. An acceptable level of proof of the viability and sustainability of a major cellulosic ethanol plant (producing a minimum of 10 Mgal/yr [37.9x10$^3$ m$^3$]) entails the successful operation of a year or more, with an acceptable level of profit.

Based on data compiled by the U.S. Energy Information Administration and reported in Bloomberg News, the cellulosic ethanol industries will increase production significantly in 2014 due to refinery startups (Shaffer, 2012; U.S. Energy Information Administration, 2012a; Ethanol Producers Digest, 2013). The 'Ethanol Producers Digest' has estimated that the 2013 level of 20 Mgal/yr (75.7x10$^3$ m$^3$/yr) will increase to a total of 104 Mgal/yr (394 x10$^3$ m$^3$/yr) when plants that already are under construction are brought into operation (Shaffer, 2012; Ethanol Producers Digest, 2013). Even if these optimistic predictions for cellulosic ethanol production are achieved, production still will remain far below the 2014 RFS of 1.75 Bgal/yr (6.6 x10$^6$ m$^3$/yr) (Herndon, 2012).

The total capacity of all three stages of cellulosic ethanol plants listed in Table 2 (Existing 20 MMgal/yr (75.7x10$^3$ m$^3$/yr), Under Construction 104 Mgal/yr (394 x10$^3$ m$^3$/yr), and Proposed Plants 368 Mgal/yr [1.4 x10$^6$ m$^3$/yr]) would yield an estimated maximum production rate total of 492 Mgal/yr (1.9 x10$^6$ m$^3$/yr) at some future date. As a theoretical exercise, the production rate of 492 Mgal/yr (1.9 x10$^6$ m$^3$/yr) of cellulosic ethanol if attained by 2017, would greatly improve U.S. capacity, however, it would make up only 11% of the 5.5 Bgal/yr (20.8 x10$^6$ m$^3$/yr), which is specified in the 2017 RFS listed in Table 1. Additionally, until an established cellulosic ethanol plant has been fully operational, production costs, operational and maintenance requirements, and environmental impacts will not be known (Herndon, 2012; Shaffer, 2012; Ethanol Producers Digest, 2013).

The cellulosic ethanol production requirement of the RFS presents a great commercial challenge. This is largely due to the fact that the development of a new industry based on cutting-edge research is a complex matter, and the associated timelines are difficult to predict. Additionally, the cost of an average size cellulosic ethanol plant of 20 to 25 Mgal/yr (75.7 x10$^3$ to 94.6x10$^3$ m$^3$/yr), can range from $200 to $250 million, a daunting investment (Menetrez, 2012). Still, the RFS production standard represents an important national goal that encourages the development of additional sources of renewable ethanol and biodiesel. While the cellulosic ethanol industry

**Table 2.**
Renewable Fuel Projections (Public Law 110–140, 2007; Thompson et al., 2010; Environmental News Service, 2011; RFS Renewable Fuels Association, 2012).

| Plants | Location | Feedstock | | Capacity Mgal/yr (x10$^6$m$^3$/yr) |
|---|---|---|---|---|
| **Existing Plants** | | | | |
| American Process Inc/Alpena Biorefinery | Alpena MI | Wood Sugars | | 0.8 (3.0) |
| BP Biofuels Demonstration Plant, | Jennings LA | Energy Grasses | | 1.4 (5.3) |
| DupontDanisco Cellulosic Ethanol LLC | Vonore TN | Corn Stover, Switchgrass | | 0.3 (1.0) |
| Enerkem Inc. | Westbury QC | Treated Wood | | 1.0 (3.8) |
| Fiberight Demonstration Plant | Lawrenceville VA | MSW | | 0.5 (1.9) |
| Fiberight Blairstown LLC | Blairstown IA | MSW | | 6.0 (22.7) |
| ICM Inc. Pilot Integrated Cellulosic Bio | St. Joseph MO | Corn fiber, switchgrass | | 0.3 (1.2) |
| Indian River Bioenergy Center | Vero Beach FL | Veg., Agric Waste, MSW | | 8.0 (30.3) |
| Iogen Inc. | Ottawa ON | Straw | | 0.5 (2.0) |
| Mascoma Corporation | Rome NY | Woody Biomass | | 0.2 (0.8) |
| Western Biomass Energy, LLC | Upton WY | Cellulosic | | 0.5 (1.9) |
| ZeaChem Boardman Biorefinery LLC | Boardman OR | Poplar | | 0.3 (1.0) |
| | | | **Total** | 19.8 (74.8) |
| **Plants Under Construction** | | | | |
| Abengoa Bioenergy Biomass | Hugoton KS | Corn Stover, Switchgrass | | 25.0 (94.6) |
| American Process Inc. Demonstration Plant | Thomaston GA | Sugarcane bagasse, wood | | 0.3 (1.1) |
| Blue Fire Renewable LLC | Fulton MS | Wood Waste | | 19.0 (71.9) |
| DupontDanisco Cellulosic Ethanol LLC | Nevada IA | Corn Stover | | 30.0 (113.6) |
| Enerkem Alberta Biofuels LP | Edmonton AB | Sorted MSW | | 10.0 (37.9) |
| Freedom Pines Biorefinery | Soperton GA | Woody Biomass | | 2.0 (7.6) |
| Poet-DSM Advanced Biofuels LLC | Emmetsburg IA | Corn Stover | | 20.0 (75.7) |
| Quad County Cellulosic Ethanol Plant | Galva IA | Corn Fiber | | 2.0 (7.6) |
| Woodland Biofuels Inc | SarniaON Canada | Wood Waste | | 0.5 (2.0) |
| | | | **Total** | 108.8 (411.9) |
| **Proposed Plants** | | | | |
| Advanced Biofuels Corp | Moses Lake WA | Cellulose | | 6 (22.7) |
| Agresti LLC | Pikeville KY | MSW | | 20 (75.7) |
| Atlantic Ethanol Inc. | Providence RI | Wood Waste | | 10 (37.9) |
| Canergy LLC | Brawley CA | Energy cane | | 25 (94.6) |
| Chemtex International Inc., Project Alpha | Clinton NC | Energy Grasses | | 20 (75.7) |
| Enerkern Mississippi Biofuels LLC | Pontotoc MS | RDF, Wood residue | | 10 (37.9) |
| Enerkern Green Field | Varennes QC | RDF, C&D debris | | 10 (37.9) |
| Fulcrum BioEnergyInc.Sierra Biofuels | McCarran NV | RDF | | 10 (37.9) |
| Mascoma Corporation | Drayton Valley AB | Hardwood | | 20 (75.7) |
| Mascoma Corp/Frontier Renewable Res. | Kinross MI | Hardwood | | 20 (75.7) |
| Mendota Bioenergy LLC | Tranquility CA | Energy beets | | 1 (3.8) |
| Nipawin Biomass Ethanol Co-operative | Nipawin SK | Waste Wood, Straw | | 26 (98.4) |
| Sunset Ethanol Inc | Fernley NV | Switchgrass, Sorgum | | 5 (18.9) |
| The Green Fuel Association Bieber II | Bieber CA | Switchgrass | | 40 (151.4) |
| The Green Fuel Association Corning II | Corning CA | Switchgrass | | 40 (151.4) |
| The Green Fuel Association Dorris II | Dorris CA | Switchgrass | | 40 (151.4) |
| Woodland Biofuels Inc. | Newton Falls NY | Wood Waste | | 20 (75.7) |
| World Ethanol Institute, LLC | Lenox GA | Paulownia | | 20 (75.7) |
| ZeaChem Boardman Biorefinery LLC | Boardman OR | Poplar, Straw | | 25 (94.6) |
| | | | **Total** | 368 (1393) |

establishes itself, the RFS goal illustrated in Table 1, continues to climb with each year. The development of additional renewable fuel sources seems necessary to achieve the 2022 RFS goal of producing a total of 36 Bgal/yr (136 x10$^6$ m$^3$/yr) of renewable fuel.

## 3. Biofuel production pathways

The following discussion of biofuel pathways illustrated in Figure 1, compares the parallel nature of biofuel production. The extraction of oils and sugars from a large variety of feedstocks can be utilized for final product development. The production of biofuel can be accomplished using different organisms and processes while producing a range of end products. Out of the many variables, process scenarios were chosen to serve as examples of possible pathways for the production of biofuel (Fig. 1). The pathways chosen represent ideal, but realistic alternatives. The generation of ethanol was diagramed by autotrophic algae pathways, taking into consideration both direct generation and indirect generation through the conversion of carbohydrates (Andersen and Andersen, 2006; Shen et al., 2009). The generation of biodiesel was diagramed by way of autotrophic and heterotrophic algae as well as agricultural grain oils and recycled vegetable oils (Chisti, 2007; Gouveia et al., 2009; Meng et al., 2009).

The generation of biofuels is often referred to as first or second (also known as advanced) generation. First generation biofuels are derived from the sugars or vegetable oils from grown crops, such as corn, sugar cane or rapeseed oil, palm oil. These feedstocks can easily be converted into ethanol or biodiesel. However, second generation biofuels are made from cellulosic feedstock or woody crops, agricultural or municipal waste which are comparatively more involved to convert into sugars or oils. Once extracted, these sugars and oils can also be converted into ethanol and biodiesel, as depicted in Figure 1.

Biofuels, such as biodiesel, ethanol, and various petroleum products, can be produced by a large variety of biologically-dependent processes (Menetrez, 2012). Both natural and genetically modified organisms (GMOs) (algae, bacteria, fungi, and yeast) have been used to generate biofuels directly or indirectly by producing biofuel intermediate products such as oil,

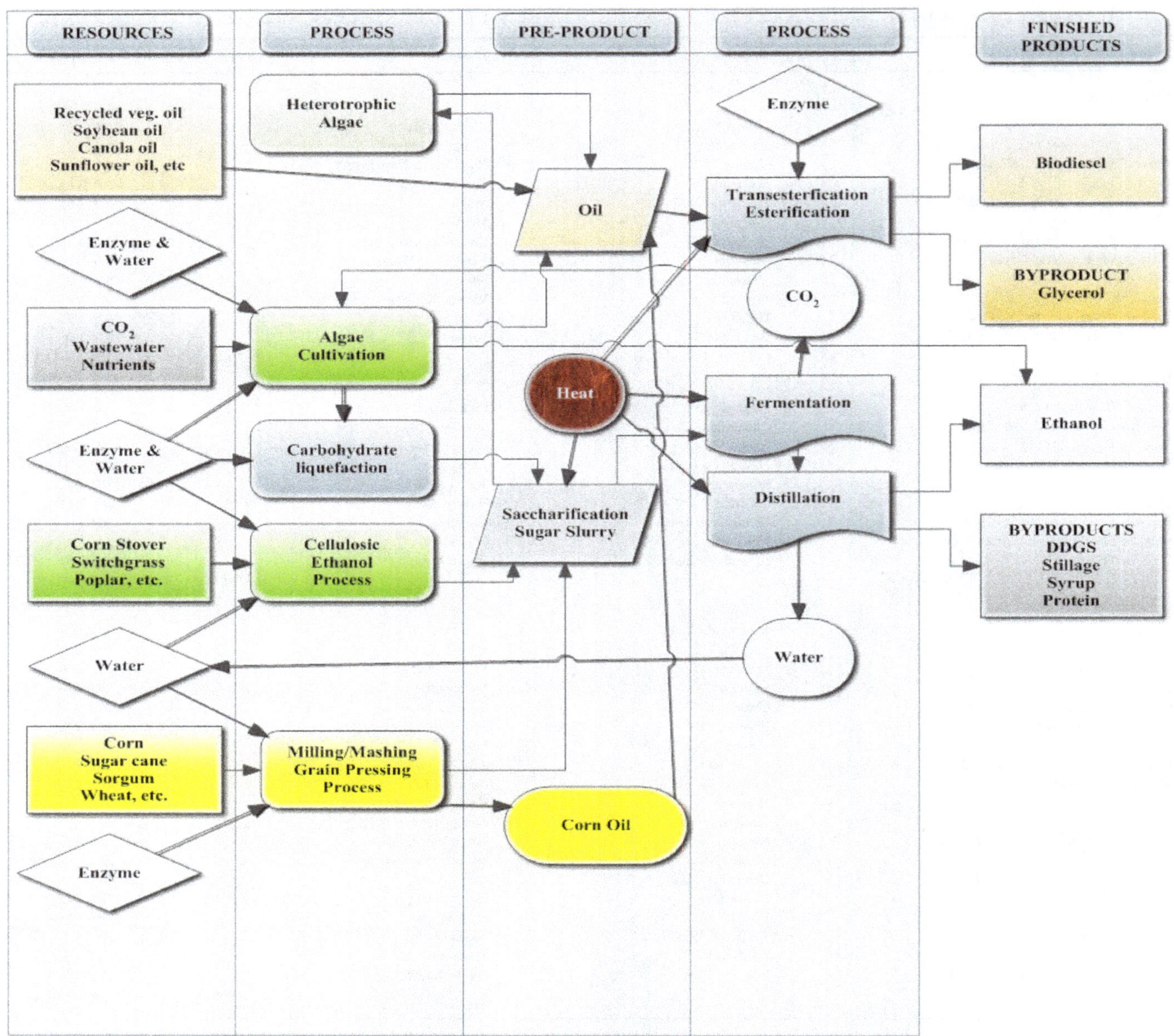


**Fig.1.** Biofuel Pathways (conventional bioethanol, cellulosic bioethanol, autotrophic and heterotrophic algae)

carbohydrates, and sugars or enzymes (Briggs, 2004; Hu et al., 2006; Becker, 2007; Li et al., 2007; Adams, et al., 2008; Evangelista et al., 2008; Graham et al., 2008; Hu et al., 2008; Lundquist et al., 2010; González- Fernández et al., 2011; Venkata Mohan et al., 2011; Menetrez, 2012).

The composition of many dry mass plant feedstocks (corn kernels, corn stover, switchgrass, poplar hybrid) commonly used for biofuel generation can vary (Table 3). Carbohydrates are either soluble sugars (made of glucose, sucrose, or fructose), or storage carbohydrates (composed of starch or fructans) (Sukenik et al., 1991; Burkholder, 1998; American Heritage, 2005; Dien et al., 2006; Dien and Bothast, 2009; Gao et al., 2010; U.S. Department of Agriculture, 2011; U.S. Department of Agriculture, 2012).

**Table 3.**
Constituents of plant feedstock biomass by % of dry matter (Dien et al., 2006; Dien and Bothast, 2009).

| Constituent | Corn kernels | Corn stover | Switchgrass | Poplar hybrid |
|---|---|---|---|---|
| Either Ext. (nonpolar) | 4.6 | 4.6 | 1.0 | 4.2 |
| Protein | 9.1 | 4.0 | 3.2 | 1.2 |
| Starch | 72.0 | 0.0 | 3.9 | 0.0 |
| Cellulose | 2.0 | 36.0 | 28.3 | 42.4 |
| Hemicellulose | 3.6 | 23.4 | 24.5 | 19.0 |
| Klasonlignon | trace | 18.6 | 15.4 | 25.7 |
| Ash | 1.5 | 12.5 | 5.4 | 1.8 |

## 4. Ethanol pathway

As apposes to ethanol derived from a petrochemical process, biologically derived ethanol sometimes referred to as bioethanol (the word ethanol is used in this paper), is a renewable fuel. Ethanol is an alcohol obtained from the fermentation of sugars and starches or by chemical synthesis. All four of the feedstocks shown as column headings in Table 4 can be processed to yield simple sugars and ultimately ethanol. Ethanol is used as a solvent, disinfectant, and as an additive to or replacement for petroleum-based fuels (American Heritage, 2005). Ethanol made from any sources (such as ethylene, algae, or cellulose) is identical to ethanol from other sources, such as corn starch, sugar, or sugarcane. An advantage that ethanol has over petroleum sources is that it can be produced from diverse renewable raw materials that are abundant. As mentioned above, U.S. ethanol production is currently at 15 Bgal/yr (56.8 x$10^6$ $m^3$/yr) from corn, and the U.S. Department of Energy (DOE) and the U.S. Department of Agriculture (USDA) are supporting the development of ethanol production from cellulosic feedstocks (plant materials, such as wood and switchgrass) (Demain et al., 2005; Perlack et al., 2005; U.S. Dept. Of Agric., 2000, 2005, 2006; U.S. Dept. Of Energy, 2004, 2007a, 2007b; Donner et al., 2008; Farrell and Morris, 2008; National Research Council, 2008; Tannura et al., 2008).

Ethanol production in the U.S. relies almost completely on sugar-platform feedstocks in the form of corn, and it consumes approximately one-third of the U.S. corn harvest, of which one-third is converted into distillers grains with solubles (DDGS) (Monceaux, 2011; Schroeder, 2003). The DDGS protein content (20 to 30%) has increased by a factor of three from the original corn grain and is a valuable byproduct ($208-327/ton [$0.29-0.36/kg] when dehydrated to 100% DW) used for animal feed (Schroeder, 2003; Monceaux, 2011; Broderick, 2013).

As of June 2012 the price of one bushel of corn (kernels) was $6.04/bu, ($171.40/$m^3$) and the price of ethanol was $2.14/gal ($0.57/L) (Parker, 2012a, b). At the yield of 2.32 gal/bu (17.4 L/$m^3$) (low) to 2.93 gal/bu (21.9 L/$m^3$) (high) the conversion of one bushel (0.035$m^3$) of corn will yield from $4.95 to $6.26 worth of ethanol at a cost per gallon (3.79 L) of $2.06 to $2.60 (Haefele and Ross, 2009; Parker, 2012b). The feedstock to product price straddles the break-even point, ranging from a loss of $0.08/gal ($0.02/L) to a profit of $0.46/gal ($0.12/L). The actual cost of production will vary for each plant, dependent on staff, operating and maintenance cost and financial commitments which are specific for the plant, but would be expected to raise cost and lower profit. Additionally, the renewable identification number (RIN) (see RIN section below) value to the manufacturer will vary and is also variable, but could add substantially to profits. Actual profitability would be affected by all factors which are beyond the capability of this paper. A number of examples of corn and ethanol prices are included in Table 4, to exemplify the variability and marginal profitability of the corn based ethanol industry. The trend of decreasing corn prices from Feb. to Aug. 2013 has created a favorable economic climate for ethanol producers despite the fluctuation in ethanol prices.

The sale of DDGS, and corn oil (used for biodiesel) can account for 25% additional revenues (3 to 4 cents/gal (0.8 to 1 cents/L) ethanol assuming a 0.3 lb/bu (4.8gm/$m^3$) corn press yield) (Emberland, 2013). Incentives such as tax subsidies, and the revenues from the RIN mechanism created by the 2007 Energy Independence and Security Act can also improve profitability (RIN values are discussed in more detail below, which as of May 2013 had a value of $0.79, approximately three to four times the profit margin) (Parker, 2013c, d). The profitability of producing corn-based ethanol is subject to the dictates of market volatility. The prices of corn, ethanol, and petroleum based products change daily, as do the opening and closing of ethanol manufacturing plants (Gaub, 2013; Parker 2013a). These factors create a difficult and unstable position for ethanol manufacturers that are saddled with enormous investments, long-term commitments, and ever increasing competition for feedstocks.

The trend from 2012 to 2014 as shown in Table 4, has seen corn prices decrease from $8.07 to $3.30 per bu ($0.23 to $0.09/L) and ethanol prices raise from $2.14 to $3.57 per gal ($0.57 to $0.94/L), and produced an increase in profits from near zero to between $2.45 to $2.15 per gallon ($0.65 to $0.57/L). During this same period, petroleum prices have decreased to approximately $91 per barrel and correspondingly, gasoline has decreased to less than the price of ethanol (Marketwatch, 2014). Over the two year period decreasing gasoline prices have absorbed what was initially the blending of ethanol from a less-expensive constituent to a cost addition. However temporary, this exemplifies the nature of fluctuating market pressures.

Climate change has brought about higher temperatures, droughts and floods, all of which contribute to lower yields of corn and higher prices. The marginal success of ethanol production from corn is likely to continue to vary with time, which emphasizes the need for using co-production techniques, such as using cellulosic corn stover and grass to produce ethanol. It also makes it clear why it is important to locate feed-lots that use DDGS near corn-based ethanol plants.

In addition to using corn to produce ethanol, a large variety of other types of biomass can be used, including cellulosic biomass, which has already been mentioned, and feedstocks that can be used to produce ethanol. Feedstocks that are high in carbohydrates (starches and sugars) are barley, cassava, sugar cane, sugar beet, sorghum, bagasse, grain, potatoes, sweet potatoes, sunflowers, fruit, molasses, and wheat. These feedstocks can be processed (using naturally occurring amylase enzymes to hydrolyze carbohydrates) to yield simple sugars (similar to other sugar platform processes). Subsequently, *Saccharomyces cerevisiae* (Brewer's yeast) can be used to convert these sugars to ethanol (Monceaux, 2009). In addition, many types of algae produce carbohydrates that could contribute significantly to the production of ethanol (Chisti, 2007; Mabee et al., 2011).

Carbohydrate containing feedstocks are milled and liquified into a mash or slurry, or they are pressed to yield a liquid (syrup) that has a high sugar content, and this syrup is cooked until it is gelatinous, after which it undergoes enzymatic hydrolyesis using the enzyme glucoamylase (Monceaux, 2009). The yeast *S. cerevisiae* is added to this sugar rich

**Table 4.**
Corn feedstock to ethanol profit/loss ($/bu or $/$m^3$) (Haefele and Ross, 2009).

| Date | Corn Price $/bu ($/$m^3$) | Ethanol Price $/gal ($/L) | Feedstock to Ethanol High to Low Yield $/bu ($/$m^3$) | Range of Profit/Loss High to Low Yield $/bu ($/$m^3$) |
|---|---|---|---|---|
| 6/2012 | 6.04 (171.40) | 2.14 (0.57) | 2.06 to 2.60 (58.46 to 73.78) | 0.08 to -0.46 (2.27 to -13.05) [a] |
| 8/2012 | 8.07 (229.00) | 2.63 (0.70) | 2.75 to 3.48 (78.04 to 98.75) | -0.12 to -0.85 (-3.41 to -24.12) [b] |
| 1/2013 | 6.86 (194.67) | 2.21 (0.58) | 2.34 to 2.96 (66.40 to 84.00) | -0.13 to -0.75 (-3.69 to -21.28) [c] |
| 2/2013 | 6.90 (195.80) | 2.36 (0.62) | 2.35 to 2.97 (66.69 to 84.28) | 0.01 to -0.61 (0.28 to -17.31) [d] |
| 4/2013 | 6.59 (187.00) | 2.42 (0.64) | 2.25 to 2.84 (63.85 to 80.59) | 0.17 to -0.42 (4.82 to -11.92) [e] |
| 5/2013 | 6.37 (180.76) | 2.52 (0.67) | 2.17 to 2.75 (61.58 to 78.04) | 0.35 to -0.23 (9.93 to -34.90) [f] |
| 7/2013 | 4.89 (138.76) | 2.23 (0.60) | 1.67 to 2.11 (47.39 to 59.88) | 0.56 to 0.12 (15.89 to 3.41) [g] |
| 8/2013 | 4.69 (133.09) | 2.18 (0.58) | 1.60 to 2.02 (45.40 to 57.32) | 0.58 to 0.16 (16.46 to 4.54) [h] |
| 9/2014 | 3.30 (93.64) | 3.57 (0.94) | 1.12 to 1.42 (31.78 to 40.30) | 2.45 to 2.15 (69.52 to 61.01) [i] |

[a] Parker,2012b [b] Parker,2012c [c] Parker, 2013a [d] Parker, 2013b [e] Parker, 2013c [f] Parker, 2013d
[g] Parker, 2013e [h] Parker, 2013f [i] AAA Fuel Gauge Report, 2014

mixture, and the fermentation process produces ethanol and carbon dioxide ($CO_2$). The fermented mash contains 10 to 20 vol% ethanol and is heated to 82 to 84 °C in a distillation process (Actual vaporization points of the constituents of the process are; methanol [wood alcohol], 64 °C; ethanol 78 °C; water 100 °C) (Monceaux, 2009). Distillation of the mash and the following distillations strip the ethanol into a condensate that contains 95 vol % ethanol. Further heating of the remaining solids evaporates water from the mixture and produces either stillage, DDGS, or a syrup-like solution, depending on the original feedstock (Monceaux, 2009). A comparison of alternative feedstocks by unit mass (such as ton per unit ton) that is required to produce the same unit of mass of ethanol (i.e., a ton equivalent is approximately 304 gal [1,150 L] of ethanol) is provided in Table 5, along with the amounts of input water and output byproducts.

An ethanol production strategy should take into account feedstock availability, feedstock cost, processing cost, ethanol yield, process efficiency, and the sales prices of ethanol and byproducts of the process. Implicit to each stage is the cost of transportation, i.e., transporting corn to the plant and transporting ethanol and DDGS out of the plant. Transportation costs are significant, and they must be taken into account during the planning stage when making decisions concerning the location of the plant. One example of minimizing transportation costs is the location of animal feedlots near corn based ethanol plants to accommodate the large amounts of DDGS. The cost of producing feedstock and its availability are of great importance to the long-term sustainability of the plant.

Securing a stable, long-term supply of cellulosic feedstock or algae biomass feedstock is also of eminent importance and is uniquely difficult. The growth requirements for the production of feedstock can be effected by many factors, such as climate, seasonal variations and environmental limitations, which can limit or preclude growth in some areas or make it economically infeasible to transport the product. The importance of acquiring a dependable supply of feedstock is emphasized by the cost required to construct a plant, which was $250 million in the case of the Project Liberty plant (Shaffer, 2012).

## 5. Cellulose to ethanol pathway

Milling and pretreatment (milling, including steam explosion, pH adjustment and enzymatic hydrolysis saccharification) of cellulosic, hemicellulosic and lignocellulosic feedstocks, such as those in Table 4, can generate sugars. Feedstocks that have demonstrated the ability to produce various sugars from cellulose are corn stover, switchgrass, miscanthus, straw, hemp, cotton, and kena. In addition, ethanol can be produced from a variety of cellulosic waste materials, such as paper and cardboard municipal waste and agricultural and wood products (Dien and Bothast, 2011; Mabee et al., 2011; Monceaux, 2009). These feedstocks can be processed using naturally occurring enzymes derived either from biological interaction directly with the substrate feedstock or from a biological process that has harvested the enzyme and made it available for the cellulosic process. This processing yields simple sugars, i.e., the six carbon sugars D-glucose, fructose, and sucrose, and five carbon sugar D-xylose (Monceaux, 2009; Dien and Bothast, 2011). Then the slurry that results from the processing contains the sugars that are required to undergo fermentation with *S. cerevisiae* in a manner that is similar to that of other biologically induced processes (Olsson and Hahn-Hägerdal, 1996; Palmqvist and Hahn-Hägerdal, 2000; De Maagd et al., 2001; Kuiper, 2001; Letourneau, 2003; High, 2004; Iogen Technology, 2005; U.S. EPA, 2008).

Several processes commonly use both natural organisms and varieties of GMOs such as the fungus *Trichoderma reesei* for producing commercial cellulases (cellulose specific enzymes), which are used to convert cellulosic biomass to sugar (Palmqvist and Hahn-Hägerdal, 2000; U.S. EPA, 2008; Dien and Bothast, 2009). Other biologically dependent cellulosic processes use various types of fungi and bacteria (including GMO varieties) to produce cellulase, xylanase, and hemicellulase enzymes for the feedstocks listed in Table 4 to produce fermentable sugars, which may then be used to produce cellulosic ethanol (Palmqvist and Hahn-Hägerdal, 2000; U.S. Environmental Protection Agency, 2008; U.S. Department of Agriculture, 2013). After the process has yielded five and/or six carbon sugars, processing continues through yeast fermentation and distillation, similar to any other carbohydrate rich feedstock, to generate ethanol.

Generating ethanol from cellulosic feedstocks involves processes that are similar to, but uniquely different from generating ethanol from other feedstocks. For example, converting corn to sugar is inherently different from converting corn stover to sugar, however, the conversion of sugar to ethanol is similar to that in other processes. A comparison of feedstocks and their resource and footprint requirements for ethanol production are discussed below and listed in Tables 5 and 6.

**Table 5.**

Ethanol production process – mass of input feedstock required per unit mass of ethanol product produced and equivalent mass of byproduct output (i.e. ton of feedstock/ton of ethanol) (Monceaux, 2009).

| Feedstock Alternatives | Input Materials Feedstock and $H_2O$ | Input Water input/unit ethanol | Output DDGS DDGS and stillage | Output Stillage output/unit ethanol |
|---|---|---|---|---|
| Corn kernels (maize) | 3.08 | 0.98 | 0.99 | - |
| Suger cane juice | 14.41 | 0.0 | - | 11.98 ( biomass) |
| Wheat | 3.34 | 1.81 | 1.32 | - |
| Barley | 4.03 | 2.89 | 2.0 | - |
| Rye | 3.72 | 2.57 | 1.7 | - |
| Grain sorghum | 3.05 | 1.64 | 1.04 | - |
| Cassava chips | 2.77 | 5.94 | - | 6.07 ( land application) |
| Potatoes | 10.80 | 8.71 (recycle) | 1.25 (cake) | - |
| Sugar beet syrup | 3.42 | 0.0 | 0.49 (syrup) | - |

Cellulosic processes must deal with the mass and volume of the feedstock which are larger than most commonly used feedstocks such as corn. As listed in Table 6, one ton of ethanol (303.8 gal or 1,150 L) would require three tons of corn, or up to 10 tons (9,072 kg) of corn stover (U.S. Department of Energy, 2007b; Dien and Bothast, 2009; Monceaux, 2011). Process estimations predict that a 20 Mgal/yr (75,708 $m^3$) plant using energy grasses, such as switchgrass and miscanthus, will require approximately 300,000 to 600,000 tons per year (272.2 x$10^6$ to 544.3 x$10^6$ kg) of feedstock delivered by 100 trucks per day. The location of the plant will impact the mean truck-route for the incoming feedstock, but transportation costs remain largely unknown, and they include the cost of transporting the outgoing waste solids, which are likely to contain large quantities of lignin. Lignin is commonly used for heating, or converted to liquid fuel by thermochemical processing (Dien and Bothast, 2009; Pedroso et al., 2011).

Techniques for producing ethanol from cellulosic materials are currently being applied to industrial scale plants. The techniques that use corn stover, such as the Project Liberty plant, depend indirectly on the continued success of the corn ethanol industry. Although this industry is likely to continue, it should be recognized that this dependence brings with it inherent instabilities due to varying economic condition. There is little doubt that the success of the cellulosic ethanol industry is vital to the future of renewable energy (Menetrez, 2010). However, until industrial plants, such as the Project Liberty plant, have established a history of proven success, the viability of this industry will remain in question. Therefore, it also is necessary to explore the potential of other forms of ethanol production.

## 6. Algae ethanol pathway

Development of the algae biofuel industry has the potential to generate a variety of fuels. Commercial facilities that produce algae exist worldwide, and they are used predominantly for manufacturing food, cosmetics, and health-related products, not biofuel. Commercial enterprises have invested heavily and established industrial-scale, micro-algae farms that have the potential to bring algae to a stage similar to that of cellulosic ethanol (Menetrez, 2012; Milledge, 2001). These commercial processes use a variety of technologies, produce many different products, and usually are located in the lower latitudes of the U.S., where temperature, climate, and solar irradiance are favorable for the growth of algae.

**Table 6.**
Ethanol production – mass of feedstock input required per equal ton mass of ethanol produced (303.8 gal [1150 L] @ $2.63/gal [$0.69/L] = $800.00), and acres/ton ($m^2/kg$) $C_2H_5OH$.

| Feedstock Alternatives and Ethanol Yield | Intake feedstocks (mass/mass) | Footprint required [acres/ton ($m^2/kg$)] |
|---|---|---|
| Corn kernels (154 bu/acre or 1.34 $m^3/km^2$ maize, 39.4 bu/ton or 1.5 L/kg) | 3.08 [b] | 0.75 (3.35) [g] |
| Wheat (70 bu/acre, or 609 $L/km^2$, 62 lb/bu or 0.8 kg/L) | 3.34 [b] | 1.80 (8.03) [h] |
| Barley (72 bu/acre, or 626 $L/km^2$, 48 lb/bu or 0.6 kg/L) | 4.03 [b] | 2.33 (10.39) [i] |
| Rye (12 bu/acre, or 104 $L/km^2$, 56 lb/bu or 0.7 kg/L) | 3.72 [b] | 4.74 (21.14) [j] |
| Potatoes (401 bu/acre, or 3,488$L/km^2$, 52 lb/bu or 0.7 kg/L) | 10.80 [b] | 1.04 (4.64) [j] |
| Sugar beet (SB) root (30 ton SB/acre, or 6.7 $kg/m^2$) | 12.323 [c] | 0.379 (1.69) [c] |
| Sugar cane (SC) stalk (36 ton/acre SC or 8.1$kg/m^2$) | 13.54 [c] | 0.372 (1.66) [c] |
| Grain sorghum (1.83 ton/acre or 0.4 $kg/m^2$) | 3.05 [b] | 1.67 (7.45) [j] |
| Sweet sorghum stalk (39 ton/acre SSS or 8.7 $kg/m^2$)* | 21.30 [c] | 0.548 (2.44) [c] |
| Switchgrass (4.6 ton/acre, or 1.0 $kg/m^2$ non-irrigated) | 5-10 [d] | 0.3-1 (1.34-4.46) [d,k,l] |
| Corn Stover (80 gal/ton, 0.16 L/kg byproduct) | 5-10 [d] | 0** |
| Microalgae (30% yield DW by PBR, 23.85 ton/acre, or 5.3 $kg/m^2$carbohydrate, byproduct) [a] | 2.5 [e,f] | 0*** |
| Microalgae (7,000gal/acre or 6,546 $L/km^2$ by PBR) | *in situ* | 0.044 (0.195) [m] |

*Based on two crops per year ** Byproduct of corn kernels *** Byproduct of microalgae oil production
[a] Chisti, 2007 [b] Monceaux, 2009 [c] Amorim, 2009 [d] Dien and Bothast, 2009 [e] Kim et al., 2011 [f] Lee et al., 2011
[g] Haefele and Ross, 2009 [h] Kansas Wheat Harvest Reports, 2012 [i] USDA, 2012 [j] USDA, 2011 [k] Pedroso et. al., 2011 [l] Mononoa et al., 2012 [m] Algenol, 2013

Algae can be used to produce ethanol and biodiesel in ponds that are open to the atmosphere [i.e. shallow ponds or tanks that can be circular or parallel raceway ponds (PRPs)] or in closed photobioreactors (PBRs). Most of the cultivation of algae is done in PRPs because of their low construction and operating costs (Briggs, 2004; Li et al., 2007; Venkata Mohan et al., 2007a,b Evangelista et al., 2008; Graham et al., 2008; González- Fernández et al., 2011). PBRs are contained, closely controlled systems in which an ideal environment is maintained to ensure high and stable production levels of algae (Chisti, 2007; Hu et al., 2006; 2008). Algae require water, efficient exposure to light, carbon dioxide, optimal temperature, culture density, appropriate pH levels, and a reasonable mixing regime. The most significant advantage that closed PBRs have over open PRPs is the ability to eliminate the introduction of unwanted microorganisms. Controlling contamination is necessary to achieve a stable, optimum culture and maximum yield. PBRs are often utilized for growing pure inoculant populations of microalgae in the early stages of the process for supplying large open, raceway, paddle wheel mixed pond PRPs where high growth rate cultivation can occur (Rosenberg et al., 2008; Lundquist, 2010).

Forms of cyanobacteria, commonly known as blue-green algae (including GMO varieties such as those harbouring genes from *Zymomonas mobilis*), can manufacture ethanol within the microalgae cell and (lacking true cell membranes) diffuse it into the culture media and headspace. Using a unique autotrophic PRP, $CO_2$ and inorganic nutrients such as those found in wastewater, ethanol can be synthesized, concentrated and recovered directly from the PRP (Badger, 2002; Demain et al., 2005; U.S. DOE, 2008; Algenol, 2011; Snow and Smith, 2012). Algenol Biofuels in cooperation with Dow Chemical Company, the Linde Group, the National Renewable Energy Laboratory (NREL), Georgia Institute of Technology, and Membrane Technology and Research, Inc., is in the process of developing an operational pilot-scale plant that uses this technology (Deng et al., 1999; Algenol, 2011).

Algae can produce substantial concentrations of lipids which can be used to produce biodiesel, jet-fuel and other petroleum products; as well as carbohydrates which can be processed into ethanol; and proteins (such as *Spirulina maxima* which is composed of 60-71% protein content) with high nutritional quality that commonly are used for human and animal consumption (Dien et al., 2006). Microalgal preparations are commonly marketed as health food, cosmetics, and animal feed (Dien et al., 2006; Becker, 2007; Adams et al., 2009; Dien et al., 2009). The potential value of algae's co-products is evident when comparing the protein content to that of other biofuel feedstocks such as corn kernels, which contains 9.1% protein (used as distillers grains with solubles or DDGS), corn stover which contains 4% protein, switchgrass which contains 3.2% protein, and a hybrid Poplar at 1.2% protein (Dien et al., 2006; Dien et al., 2009).

## 7. Algae biodiesel pathway

Biodiesel is a renewable fuel that is currently produced from various feedstocks, such as, soybean oil (90% of current production), recycled vegetable oil, sunflower oil, canola oil, cottonseed oil, animal fats, and lipids produced by algae (Chisti, 2007; Gouveia et al., 2009; Meng et al., 2009). Lipids are long carbon-chain molecules that serve as a structural component of the membrane of the algal cell. The lipid production of algal species varies from 20% to 80% DW. In addition, temperature, solar irradiance, the species-specific speed of growth and process time, and the manipulation of nutrients supplied per stage of growth can all affect lipid yield (Becker, 2007; Adams, et al., 2008; Gouveia, et al., 2009; Wageningen University, 2011). Becker (2007) listed the lipid, carbohydrate and protein constituents of 17 common algal species, and they are presented in Table 3. Algae that belong to several different families possess the ability to produce and accumulate a large fraction of their dry mass as lipids. As listed in Table 3, autotrophic lipid production from the algae *Botryococcus braunii* can produce 86% DW of lipid (oil) that can be separated and converted to biodiesel (Chisti, 2007; Gouveia et al., 2009; Meng et al., 2009; Chisti, 2013).

Biodiesel is defined as a mono-alkyl ester of a long-chain fatty acid that conforms to the American Society for Testing and Materials (ASTM) D6751 specifications for use in diesel engines (ASTM D6751 – 12, 2013). After extraction and separation, algae oil is processed into biodiesel by either of two process methods. In the classical method, the oil is mixed with an alcohol (usually methanol) and a basic catalyst (usually potasium hydroxide (KOH), or sodium ethanolate ($CH_3COONa$) and heated to approximately 70 °C (at 20 psi) for several hours in a process called transesterification. The products are biodiesel (67%) and glycerol (33%) (Chisti, 2007; Coppola et al., 2009; Gouveia et al., 2009; Meng et al., 2009).

In the second method, the oil and alcohol are mixed with the enzyme lipase, which can be produced from a number of organisms, such as the fungus *Metarhizium anisopliae, Aspergillus oryzae* (and *A. niger*) or a varieties of Gram-negative bacteria, i.e. *Chromobacterium viscosum,* causing transesterification at room temperature (Adachi et al, 2011; Fiametti et al., 2011; Foley et al., 2011; Talukder, 2011). Energy consumption for producing biodiesel is reduced in the enzyme method, and the process is made more energy efficient and less sensitive to process problems (Lam and Lee, 2011). In addition, this process has other advantages such as: creating a high-grade glycerin byproduct with improved value, reduced water consumption (less water washing), and reduced methanol consumption. There are no caustic catalysts used, no soap formed and no ion exchange, or adsorbents are used (Piedmont Biofuels, 2013). The enzyme cost of $0.15/gal ($0.04/L) of processed biodiesel is currently outweighed by the positive attributes of the process (Piedmont Biofuels, 2013; AAA Daily Fuel Gauge Report, 2014).

As of 2012, biodiesel manufacturers in North Carolina were paying the sources of yellow grease, such as restaurants, as much as \$1.50/gal (\$0.40/L) (U.S. Energy Information Administration, 2012a; Piedmont Biofuels, 2013). Yellow grease is the preferred, economical choice, although limited in quantity, and corn oil, soybean oil and canola oil are additional feedstocks being used (U.S. Energy Information Administration. 2012a,b,c; Piedmont Biofuels, 2013). If algae derived oil is to be economically successful, it must be produced and supplied at a price approaching \$1.50/gal (\$0.40/L). With the inclusion of transportation costs (which can vary greatly) and overall processing costs a gallon of B100 biodiesel can be produced for approximately \$3.00 to \$3.10 (\$0.79 to \$0.82/L) (Piedmont Biofuels, 2013). Comparing this price to petroleum-based diesel (of approximately \$4.00/gal or \$1.06/L) indicates an approximate profit of \$1.00/gal (\$0.26/L).

The profitable nature of using yellow grease for biodiesel has allowed small plants to survive, but their production of biodiesel is limited by the supply of recycled cooking oil. There also are major biodiesel plants in operation, such as the Louis Dreyfus Agricultural Industries plant in Claypool, IN, which produces 80 Mgal/yr (0.3 x$10^6$ $m^3$) from soybean oil (Biodiesel Industry Directory, 2013). A biodiesel plant under construction by Archer Daniels Midland in Lloydminster, AB, Canada, will produce 265 Mgal/yr (1.0 x$10^6$ $m^3$) from canola oil (Biodiesel Industry Directory, 2013). When the Lloydminster plant becomes operational, it will yield more than double the output of any other plant. This scale of operation will ensure that overall profitability is sustainable even though the profit per gallon will be less due to the high cost of the feedstock oil. It is not clear whether the plant will satisfy the process-dependent RFA requirement for advanced biofuel of 50% GHG emission reduction. The RFA qualification could affect RIN applicability and further marginalize profits.

(303.8 gal or 1.15 x$10^3$ L) produced, also produced 0.96 tons (871 kg) of $CO_2$ as a byproduct of fermentation, almost a one to one ratio by weight (Monceaux, 2009). Presently in the U.S., the $CO_2$ produced from the annual production of 15 Bgal (50 Mton or 45.4 x$10^6$ kg) of corn based ethanol is vented to the atmosphere or 48 Mton (43.5 x$10^9$ kg) of $CO_2$, a known green-house gas (GHG). As depicted in Figure 1, $CO_2$ should be viewed as a resource, not a waste product. Producing biofuel from growing algae with the use of $CO_2$ is an example of a complete resource utilization, renewable fuel generation and GHG emission reduction. If algae cultivation is located to utilize industrial sources of $CO_2$, heat, and nutrients could improve growth and decrease costs. Locating algae cultivation sites near industrial sources of wastewater, $CO_2$, and waste heat could add value to the process itself. Algae oil may never be competitively priced at approximately \$1.50/gal (\$0.40/L) (Menetrez, 2012). However, if linked to a $CO_2$ emission mitigation price to offset the algae cultivation cost a multi-product mindset as depicted in Figure 1, can be achieved.

$CO_2$ injection from industrial sources such as coal-burning power plants has been demonstrated to increase the yield of algae while decreasing atmospheric emissions (Bhatnagar and Bhatnagar, 2001; Benemann et al., 2003; Chisti, 2006; Woertz et al., 2009; Bhatnagar et al., 2010; Weissman and Benemann, 2011).

Wastewater (agricultural or human), waste heat and $CO_2$ (from biological fermentation or power plant flue-gas) should be viewed as resources for growing algae (Venkata Mohan et al., 2007b; Venkata Mohan et al., 2008). Contributing to the capture and conversion of $CO_2$ into a marketable commodity may be the most important asset of algae cultivation, regardless of the economic value assigned that commodity. All products of lipids and carbohydrates for biofuels and protein for animal or human consumption should be utilized.

**Table 7.**
Biodiesel production – mass of feedstock input required per equal mass of biodiesel produced (assuming 33% loss of oil to glycerol, a density of 7.6 lb/gal [0.9 kg/L] oil, 350 gal [1,300 L] of oil equivalent) (Chisti, 2007; Jatropha for Biodiesel Figures, 2012; NC State University, 2012).

| Feedstock Alternatives and Biodiesel Yield | Intake feedstocks (mass/mass) | Footprint required [acres/ton ($m^2$/kg)] |
|---|---|---|
| Corn kernels (maize, 3.8 ton/acre, or 0.85 kg/$m^2$, 18.4 gal/acre oil, or 17.2 L/$km^2$) [a] | 72.3 | 19.00 (84.7) |
| Soybeans (1.32 ton/acre, or 0.30 kg/$m^2$, 66 gal/acre oil, or 61.72L/$km^2$) [c] | 9.69 | 7.34 (32.7) |
| Canola (1.71 ton/acre, or 0.38kg/$m^2$, 150 gal/acre oil or 140 L/$km^2$) [c] | 4.71 | 2.75 (12.3) |
| Jatropha (2.8 ton/acre, 0.63kg/$m^2$, 1.012 ton/acre oil, or 230 kg/$m^2$) [b] | 4.9 | 1.73 (7.7) |
| Microalgae (70% yield DW by PBR, 14,636 gal/acre, or 13,588L/$km^2$)[a] | 2.14 | 0.024 (0.1) |
| Microalgae (30% yield DW by PBR, 6,276 gal/acre, or 5,869 L/$km^2$) [a] | 5.0 | 0.056 (0.3) |
| Microalgae (25% yield DW by PBR, 2,100 gal/acre or 1,964 L/$km^2$) [d] | 14.9 | 0.166 (0.74) |
| Microalgae (45% yield DW by PRP, 12-15 ton/acre oil, 2,724-3,405 kg/$m^2$) [e] | 10.5 | 0.1 (0.4) |
| Recycled vegetable.oil (recycled yellow grease) | 1.5 | 0 |

[a] Chisti, 2007 [b] Jatropha For Biodiesel Figures, 2012 [c] NC State University, 2012 [d] Lundquist et al., 2010 [e] Chisti, 2013

In addition, similar to corn-based products, questions remain regarding the use of feedstocks that could be used for human consumption, such as soybeans and canola. The area of land required (or footprint) to generate a given quantity of biofuel feedstock is provided in Tables 6 and 7.

Questions remain regarding the balance of resource inputs and outputs which can only be answered after the commercial success of biofuel production facilities has been demonstrated. In order to achieve commercial success it will be necessary for algae farms to take advantage of every available resource to enhance growth potential, while simultaneously recycling the byproducts of the processes. This process interdependency is addressed separately for its symbiotic nature.

## 8. Industrial process symbiosis

Symbiotic processes currently can be found in the biofuel industry. The production of ethanol from corn also produces corn stover, which is being used to produce cellulosic ethanol. Ethanol manufacturers, such as Poet, also are generating corn oil and using the inedible fraction to produce biodiesel. In 2012, Poet extracted 250,000 tons of corn oil from 25 of its plants, which is equivalent to 68 Mgal (0.26 x$10^6$ $m^3$) (Biodiesel Magazine, 2014). This example demonstrates how the harvest of corn can be used to generate bioethanol from carbohydrates (starches/sugars), cellulosic and bioethanol from corn stover, biodiesel from corn oil, as well as DDGS for animal feed.

An example of a potential benefit from the ethanol manufacturing process is the recycling of $CO_2$ from fermentation (see Figure 1). Every ton of ethanol

## 9. Resource balance

Resources that are universally required to develop feedstocks into renewable fuels are nutrients, water, sunlight, land, mechanical processing (harvesting, washing, milling, etc.), heat (cooking, fermentation, distilation, etc.), and mechanical transport (feedstock to plant, process and product to market) energy. Tables 6 and 7 provide examples of sources of feedstocks that can be used to produce renewable fuel. It is however difficult to generalize resource inputs quantitatively due to process differences, recycle and byproduct utilization, such as corn stover for ethanol generation (which comes from the same harvest as grain-corn feedstock used to generate ethanol) or algae products/byproducts of oil for biodiesel and ethanol. The resources that were calculated are tons of feedstock and acres required per ton of ethanol (Table 6) or biodiesel (Table 7).

The land required to grow any feedstocks can be a significant resource investment. A comparison of the land area required for a feedstock is a useful indicator of the energy productivity of that pathway. For example, Table 6 lists the amount of land required to grow enough sugar cane (0.372 acres or 1.5 $km^2$) to produce a ton of ethanol (303.8 gal or 1.15 x$10^3$ L), which is half of that required for corn (0.75 acres or 3.0 $km^2$) (Monceaux, 2009). However, the growth requirements and land attributes differ (such as land value and arability), as does the availability of feedstock. Irrespective of the inherent differences, land area, as a resource input, provides a valuable comparison of renewable fuels. Tables 6 and 7 were developed to approximate the land resource requirement that was sufficient to produce enough feedstock to produce one ton of fuel (ethanol in Table 6: biodiesel in Table 7). Each intake of feedstock material or area of land listed in Tables 6 and 7 can produce a ton of ethanol or biodiesel respectively (Kansas Wheat Report 2012; USA Biodiesel Prices.com, 2012; U.S. Energy Information Administration, 2012b). Commodity prices of corn, ethanol, biodiesel, petroleum, and petroleum-based diesel are constantly changing based on global economic conditions and forces.

As mentioned above, every ton of corn based ethanol (303.8 gal or 1.15 $x10^3$ L) produced, also produced 0.96 tons (871 kg) of $CO_2$ as a byproduct of fermentation (Monceaux, 2009). The U.S. production of 15 Bgal (50 Mton or 45.4 $x10^6$ kg) of corn based ethanol produced an average 48 Mton (43.5 $x10^6$ kg) of $CO_2$. The biomass of microalgae contains approximately 50% carbon, which is obtained from the atmosphere or other sources of $CO_2$. The injection of the additional carbon in the form of $CO_2$ has demonstrated an increase in algae yield while sequestering the carbon (Benemann, 2003; Weissman and Benemann, 2003; Chisti, 2006; Menetrez, 2012). Theoretically, applying the yearly rate of $CO_2$ production to growing algae could produce 26 Mtons (23.6 $x10^6$ kg) of algae biomass (assuming 1.83 tons $CO_2$ per ton algae biomass), which could be converted into biofuel, such as; 10 Mtons (9.1 x $10^6$ kg) of algae crude oil, or 6.5 Mgal (24,603 $m^3$) of biodiesel (assuming a 40% DW yield, and 95% separation recovery) (Christi, 2007). A substantial goal of sustainable biofuel generation (considering the present biodiesel production level is 1Bgal/yr or 3.8 $x10^6$ $m^3$), and a substantial displacement of a GHG which is currently being vented to the atmosphere. There are multiple advantages to both the $CO_2$ supplier (for resource recycling) and the algae/biofuel industry (for growth enhancement by $CO_2$ sequestration). This algae pathway could be promoted through RIN additions for both sides to offset additional costs.

In order to achieve a cost-efficient biofuel process, using waste products, such as wastewater or $CO_2$ emissions, and using land that has marginal agricultural potential should be priorities. It is understood that the availability of land, $CO_2$ and nutrient rich wastewater can be a difficult set of assets to find. However, it is important that every attempt at economic advantages be made, especially in the initial phase of industrial start-up. In addition, the use of byproducts and recycled materials should be an immediate or planned priority. The footprint of byproducts such as corn stover or algae carbohydrate (after lipid extraction) was listed as zero due to it not being a primary or virgin product. In each of these examples the footprint was listed for the primary product corn kernels and algae oil for biodiesel production (see Table 7). Eventually, the economic feasibility will be decided by the site plan (combining wastewater, $CO_2$, and land resources) the process plan, and operational success (including transporting resource inputs and outputs, and containing potential contaminants).

Table 6 lists two sources of ethanol produced from microalgae. The first example utilizes PBR technology as a byproduct after the lipid fraction has been removed, while the second uses a PRP system. Although PRP algae cultivation is the economical choice for biodiesel production purposes, PBR cultivation is required for the direct production of algae to ethanol as currently employed by Algenol. PBR technology is often used in hybrid PBR/PRP systems for biodiesel production. PBR systems are also used in pilot scale $CO_2$ capture systems. Often considered to be "niche systems" the importance of carbon capture could increase its future importance beyond biofuel production.

A form of marine algae (*Laminaria japonica*) contains hydrolysate solids comprised of up to 31.0% mannitol and 7.0% glucose. GMO *Escherichia coli* (KO11) can be used to convert the mannitol and glucose to ethanol, producing about 0.4 g of ethanol per gram of carbohydrate (Kim et al., 2011). Another form of marine algae, *Chlorella vulgaris* was used by Lee et al. (2011) to produce ethanol with a pretreatment of GMO *E. coli* (strains W3110, and SJL2526), achieving 0.4 g ethanol/g biomass (Lee et al., 2011). These are two examples of unique processes which utilize marine algae with high sugar contents in combination with GMO strains of *E. coli* that are able to produce ethanol at a high efficiency. Although feedstock availability may be a limiting factor, these processes illustrate an innovative approach to using many forms of algae.

Numerous species of microalgae can be used to produce lipids, proteins and carbohydrates. One such algae, *Prymnesium parvum*, produces approximately equal portions of lipids, proteins and carbohydrates, i.e., about 33% of each. For the purpose of generating the numbers listed in Tables 6 and 7, the yield from algae cultivation would agree with that measured from the growth of *P. parvum*. In this example, the primary product was lipids, while carbohydrates were listed as a secondary byproduct. Actual yields will depend on the specific microalgae and the conditions of the process. Recycled vegetable oil is listed as having a footprint of zero (similar to the byproducts listed in Table 6) due to it not being a primary and virgin product. The production of one ton of biodiesel (263.0 gal [995 L] at a density of 7.6 lb/gal [0.9 kg/L]) would require 2,666 lb (1,209 kg) of oil (350 gal [1,300 L] of oil, assuming a loss of approximately 33% to glycerol).

The versatility of algae for producing lipids, carbohydrates, and proteins will be needed to create multiple products in multiple markets to satisfy economic considerations successfully. Currently, biotechnology firms and the algae industry are focused on producing relatively low volumes of high-value products, such as pharmaceuticals or nutritional supplements. These same industries must refocus on high volumes of biofuel production at low competitive prices, as well as utilizing byproducts, such as protein for DDGS and carbohydrates for ethanol (Gladue and Maxey, 1994; Unkefer et al., 2004; Foley et al., 2011; Mononoa et al., 2012).

The cultivation of algae requires nutrients that can be accessed by using various wastewater streams, including water from the same recycled streams. It also requires heat for temperature control, which can be obtained from sunlight (when clouds are absent, and sunlight is abundant), industrial waste heat, or geothermal sources. In addition, algae can be utilized simultaneously for both the lipids (to produce biodiesel) and carbohydrates (to produce ethanol).

## 10. Renewable identification number (RIN)

The renewable identification number (RIN) mechanism created by the 2007 Energy Independence and Security Act documents the production of biofuel at refineries by registering every gallon (3.79 L) of biofuel that is produced. The assigned RIN value also reflects the energy content of the biofuel (Table 8). Corn based ethanol is issued a RIN of 1 which as of May 2013 had a value of $0.79, and by July 17 2017 had risen to $1.43, much greater than the profit margin (Haefele et al., 2009; Parker, 2012a,b,c; Emberland, 2013; Parker, 2013a,b,c,d,e).

One gallon of biodiesel derived from oil from grown and pressed feedstock (such as soybean-based biodiesel) or recycled cooking oil (yellow grease) is issued a RIN of 1.5, and as of July 17 2013 had a RIN value of $1.30 (Parker, 2013b). Advanced biofuels, such as cellulosic ethanol (produced from biomass such as corn stover or switchgrass), are issued a RIN of 2.5 (Weihrauch, 2007; Weisner, 2009). The EPA sets a yearly quota which dictates the amount of biofuel blended into petroleum based transportation fuel. Refiners, importers and blenders must comply to their EPA assigned RFS quota. The biofuel manufacturer registers every gallon produced with the EPA at the time of sale, transfer, or export/import for the purpose of ensuring that the biofuel is actually blended into motor fuel. The ethanol or biodiesel company issues the RIN for a month's production of biofuel, and this RIN is reported to EPA using a unique, 38 character number (Weihrauch, 2007). The RIN values are transferred with the biofuel to the purchaser (refiners, exporter, importers, and blenders of the fuel). Eventually, the bioethanol is blended into gasoline (E10, E15, E25 E85, E100), and the biodiesel is blended into petrodiesel (B2, B5, B10, B20, B100) (Weihrauch, 2007; Weisner, 2009).

**Table 8.**
Energy content of transportation fuels (Weihrauch, 2007; Consumer Energy Report, 2013; Zfacts, 2013).

| Fuel | Btu/gal (Btu/L) | Etoh /Fuel Ratio | RIN |
|---|---|---|---|
| Ethanol (E100) | 76,000 (2,008) | 1 | 1 |
| Biodiesel (B100) | 118,296 (3,125) | 0.655 Etoh/Gasoline | 1.5 |
| Gasoline | 116,090 (3,067) | 0.642 Etoh/Biodiesel | 0.0 |
| Cellulosic Ethanol | 76,000 (2,008) | 1 | 2.5 |

During economic periods in which blending biofuel is unprofitable, the RINs can be separated from the fuel and used as a commodity on the open market (Weihrauch, 2007; Weisner, 2009). During those years in which more ethanol is produced than is set by the RFS, RINs can accumulate for up to one year, as noted by the expiration dates. Blenders with excess RINs can sell them to other blenders to be used as the market dictates (Weihrauch, 2007; Weisner, 2009; Wall Street Journal, 2013).

From manufacturing to blending and consumption, a RIN documents the compliance of the biofuel blenders with the RFS mandates of 2005 (Thompson et al., 2010). The buying, selling, and trading (swapping) of RINs in the marketplace is a complex undertaking, because it is influenced by market trends, agricultural fluctuations and regional needs (Weihrauch, 2007; Weisner, 2009). An example of the volatility of this system is seen in the three month variation in the corn grain RIN (D6), changing from $1.44 on July 16, 2013 to $0.675 on August 9, 2013, a 76.5 cent decrease (Platts McGraw Hill Financial, 2013). As described above, the economic position held by the biofuel industry is one of marginal profitability. The RIN system could lend a degree of flexibility to survive unfavorable economic periods, but market influences have demonstrated the vulnerability of the present system (Emery, 2012). A more complete discussion of the RIN system is beyond the scope of this paper.

## 11. Land requirement

The efficiency of algae to produce ethanol and biodiesel listed in Tables 6 and 7 were used to calculate the area of land required per billion gallons of biofuel (see Table 9). Depending on the efficiency of the process and overall yield, each billion gallons of ethanol would require a commitment of 175 miles$^2$ (453 km$^2$) to produce. Similarly, each billion gallons of biodiesel would require from 107 to 741 miles$^2$ (277 to 1,919 km$^2$) to produce (a wide range of variation). This range represents the collective area which would be dedicated to algae cultivation for the primary purpose of biofuel production. Although the dedication of this much land area to algae cultivation is significant, it is less than any other feedstock by at least an order of magnitude.

**Table 9.**
Microalgae Ethanol/Biodiesel production footprint, acres/Bgal (km$^2$/m$^3$) from Tables 6 and 7.

| | Acres/ton (m$^2$/kg) | Acres/Bgal (km$^2$/m$^3$) |
|---|---|---|
| | **Ethanol** | **Ethanol** |
| Microalgae (30% yield DW by PBR, byproduct)[a] | 0.0* | 0.0* |
| Microalgae (7,000gal/acre, or 6,547 L/km$^2$ by PBR)[b] | 0.044(48.4) | 142,800 (152.8) |
| | **Biodiesel** | **Biodiesel** |
| Microalgae (70% yield DW by PBR, 14,636 gal/acre, or 13,588 L/km$^2$)[a] | 0.024 (26.4) | 68,570 (73.4) |
| Microalgae (30% yield DW by PBR, 6,276 gal/acre, or 5,869 L/km$^2$)[a] | 0.056 (61.6) | 160,000 (171.2) |
| Microalgae (25% yield DW by PBR, 2,100 gal/acre, or 1,964 L/km$^2$)[c] | 0.166(182.6) | 474,286 (507.5) |
| Microalgae (45% yield DW by PRP, 12-15 ton/acre oil, or 2,724-3,405 km/m$^2$)[c] | 0.1 (11.0) | 285,714 (305.7) |

[*] Recycled Byproduct [a] Chisti, 2007 [b] Algenol Biofuel, 2011 [c] Chisti, 2013

Actual sites may vary in size, efficiency, capacity and purpose. An example of a dual purpose algae production facility might have a primary purpose of using algae to reduce $CO_2$ emissions produced by coal-fired power plants. In addition to emissions from the combustion of coal, natural gas, and petroleum products, carbon capture to removal $CO_2$ with algae cultivation can be applied to many additional commercial sources such cement manufacturing and biological fermentation. In these examples reducing $CO_2$ emissions is the primary purpose, while the production of biofuels is integral but secondary.

## 12. Conclusion

Currently, biofuel production in the U.S. is limited to approximately 15 Bgal/yr (56.8 x10$^6$ m$^3$/yr) of corn, based ethanol and 1 Bgal/yr (3.8 x10$^6$ m$^3$/yr) of biodiesel (derived mainly from recycled vegetable oil). Feedstock limitations, i.e., sources of oil, are expected to limit the expansion of biodiesel production. Efforts are being made to quickly develop a cellulosic ethanol industry in the U.S. However, the development of this new (and largely unproven) industry with high investment requirements ($200 to $250 million for a plant that produces 20-25 Mgal/yr [75.7 x10$^3$ -94.6 x10$^3$ m$^3$]) and large resource needs is not likely to occur rapidly. As of early 2014, the industrial cellulosic ethanol plants in operation have a combined capacity of 19.75 Mgal/yr (74.8 x10$^3$ m$^3$/yr) (listed in Table 2), more than triple the 2013 capacity of 6.25 Mgal/yr (23.7 x10$^3$ m$^3$/yr), but far below the 2014 RFS of 1.75 Bgal/yr (6.6 x10$^6$ m$^3$/yr). Additionally, if 2014 is as successful as anticipated, the Abengoa Bioenergy plant (25 Mgal/yr [94.6 x10$^3$ m$^3$]), Poet-DSM plant (20 Mgal/yr [75,708 m$^3$]) and Dupont Danesco plant (30 Mgal/yr [113.6 x10$^3$ m$^3$]) will start production. With the addition of these three cellulosic ethanol plants the total capacity will exceed 100 Mgal/yr (378.5 x10$^3$ m$^3$), approximately five times the 2014 capacity, another major increase which falls far short of the RFS.

Many efforts are being made in the U.S. to develop cellulosic ethanol as an integral contributor to biofuels. Although numerous cellulosic ethanol plants are anticipated to initiate operation, the actual gap between available commercial supply and the RFS demand is widening. The 2022 RFS of 36 Bgal (136 x10$^6$ m$^3$) of total biofuel is a difficult goal to achieve, specifically due to the timeline necessary for the development of the cellulosic ethanol industry. It is for this reason that additional advantages should be created which promote other sources of transportation fuels, among them, algae biofuel which are highly flexible in process requirements and versatile in product output. Examples of advantages to promote algae biofuel production are: (1) Incentives to carbon capture processes by algae cultivation through the use of increased RIN values; and (2) Similarly, increased RIN values to algae derived biofuels produced through wastewater reclamation.

Potential pathways for producing biofuels from algae were evaluated for their feedstock and footprint demands. The present focus of the existing algae industries is on producing low volumes of high-value products for pharmaceuticals or nutritional supplements. There also are many developmental efforts backed by significant financial resources that are focused on the production of large volumes of biofuel products. However, little information regarding the present status of these process operations has been made available.

The RFS is a guidance tool for the commercial production of a new generation of renewable biofuels (biodiesel, bioethanol, and petroleum). These sustainable energy sources contribute to our energy needs much like other established technologies such as solar, wind, and geothermal sources. Currently, the development of cellulosic and algae feedstocks for biofuel production has great potential to help diminish our dependence on petroleum-based fuels. The versatility of algae to produce multiple products in multiple markets is unique. The established focus of the algae industry to produce low volumes of high-value products (for pharmaceuticals or nutritional supplements) is evolving. There are many efforts backed by significant financial resources that are focused on the production of large volumes of biofuel products. The biofuel industry is young and growing quickly, it is hoped that this paper can provide a basis for the development of guidance that will assist this industry in growing in an environmentally friendly manner.

## References

AAA Daily Fuel Gauge Report, 2014. Available at http://fuelgaugereport.aaa.com. (accessed on 28 November 2014).

Adachi, D., Hama, S., Numata, T., Nakashima, K., Ogino, C., Fukuda, H., Kondo, A., 2011. Development of an *Aspergillus oryzae* whole-cell biocatalyst coexpressing triglyceride and partial glyceride lipases for biodiesel production. Bioresour. Technol. 102, 6723-6729.

Adams, J.M., Gallagher, J.A., Donnison, I.S., 2009. Fermentation study on Saccharina latissima for bioethanol production considering variable pre-treatments. J. Appl. Phycol. 21, 569-574.

Algenol Biofuels, 2011. Available at http://www.algenolbiofuels.com. (accessed on 28 November 2014).

Amorim, H.V., Basso, L.C., and Lopes, M.L. 2009. Sugar cane juice and molasses, beet molasses and sweet sorghum: Composition and usage, in: Ingledew, W.M., The Alcohol Textbook: A reference for the beverage, fuel, and industrial alcohol industries. Lallemand Ethanol Technology. 1, 39-46.

American Heritage, 2005. Science Dictionary, copyright by Houghton Mifflin Company, Published by Houghton Mifflin Company. pp. 217.

Andersen, T., Andersen, F.O., 2006. Effects of $CO_2$ concentration on growth of filamentous algae and Littorella uniflora in a Danish softwater lake. Aquat. Bot. 84, 267-271.

ASTM D6751 – 12, 2013. Standard Specification for Biodiesel Fuel Blend Stock (B100) for Middle Distillate Fuels. Available at http://www.astm.org/Standards/D6751.htm. (accessed on 28 November 2014).

Badger, P.C., 2002. Ethanol from cellulose: A general review, in: Trends in New Crops and New Uses, ed. J. Janick and A. Whipkey, ASHS Press, Alexandria, VA, 17–21.

Badger, P.C., 2002. Ethanol from cellulose: A general review. Trends in new crops and new uses. 17-21.

Becker, E.W., 2007. Micro-algae as a source of protein. Biotechnol. Adv. 25, 207–210.

Benemann, J., Pedroni, P.M., Davison, J., Beckert, H., Bergman, P., 2003. Technology Roadmap for Biofixation of $CO_2$ and Greenhouse Gas Abatement with Microalgae, Second National Conference on Carbon Sequestration. Available at http://www.netl.doe.gov/publications/proceedings/03/carbon-seq/PDFs/017.pdf. (accessed on 28 November 2014).

Bhatnagar, A., Bhatnagar, M., 2001. Strategies to employ algae and cyanobacteria for wastewater remediation, in: Maheshwari, D.K., Dubey, R.C. (Eds.), Innovative Approaches in Microbiology. Bishen Singh Mahendra Pal Singh, Dehra Dun, India, 379-403.

Bhatnagar, A., Bhatnagar, M., Chinnasamy, S., Das, K.C., 2010. Chlorella minutissima - a promising fuel alga for cultivation in municipal wastewaters. Appl. Biochem. Biotechnol. 161, 523-536.

Biodiesel Industry Directory, 2013. Available at http://store.bbiinternational.com/2012-Biodiesel-Industry-Directory-P9.aspx. (accessed on 28 November 2014).

Biodiesel Magazine, 2013. Twenty-five poet plants extract enough corn oil for 68 Mgal/yr of biodiesel, Poet, Available at http://www.biodieselmagazine.com/articles/8912/25-poet-plants-extract-enough-corn-oil-for-68-mmgy-of-biodiesel. (accessed on 28 November 2014).

Biomass Magazine, 2014. The Year is Here, visited 2014. Available at http://www.biomassmagazine.com/articles/10086/the-year-is-here. (accessed on 28 November 2014).

Briggs, M., 2004. Widescale Biodiesel Production from Algae. University of New Hampshire.

Broderick, S., 2013. DDGS export demand driving prices for now. Ethanol Producers Magazine.

Burkholder, J.M., 1998. Implications of harmful microalgae and heterotrophic dinoflagellates in management of sustainable marine fisheries. Ecol. Appl. 8(spl), S37-S62.

Chisti, Y., 2006. Microalgae as sustainable cell factories. Environ. Eng. Manage. J. 5, 261-274.

Chisti, Y., 2007. Biodiesel from microalgae. Biotechnol. Adv. 25, 294-306.

Chisti, Y., 2013. Raceways-based production of algal crude oil. Green. 3(3-4), 195-216.

Consumer Energy Report, 2013. Available at http://www.energytrendsinsider.com/2006/03/28/biodiesel-king-of-alternative-fuels. (accessed on 28 November 2014).

Coppola, F., Simoncini, E., Pulselli, R.M., Brebbia, C.A., Tiezzi, E., 2009. Bioethanol potentials from marine residual biomass: an Emergy evaluation. Ecosyst. Sustain. Dev. VII. 379-387.

Crickmore, N., 1998. Revision of the nomenclature for the Bacillus thuringiensis pesticidal crystal proteins. Microbiol. Mol. Biol. Rev. 62, 807-813.

De Maagd, R.A., Bravo, A. and Crickmore, N., 2001. How *Bacillus Thuringiensis* Has Evolved Specific Toxins to Colonize the Insect World. Trends Genet. 17, 193-199.

Demain, L.D., Newcomb, M., and Wu, J.H., 2005. Cellulase, Clostridia and Ethanol. Microbiol.Mol. Biol. Rev. 69(1), 124-154.

Deng, M.D., Coleman, J.R., 1999. Ethanol synthesis by genetic engineering in cyanobacteria, Appl. Environ. Microbiol., 65(2), 523-528.

Dien, B.S., Bothast, R.J., 2009. A primer for lignocellulose biochemical conversion to fuel ethanol, in: Ingledew, W.M. (Eds.), The Alcohol Textbook. Nottingham University Press, Nottingham, U.K. Available at http://www1.eere.energy.gov/biomass/feedstock_databases.html. (accessed on 28 November 2014).

Dien, B.S., Jungb, H-J.G., Vogelc, K.P., Caslerd, M.D., Lamb, J.F.S., Itena, L., Mitchellc, R.B., and Sarathc, G., 2006. Chemical composition and response to dilute-acid pretreatment and enzymatic saccharification of alfalfa, reed canarygrass, and switchgrass. Biomass Bioenergy. 30, 880-891.

Donner, S.D., and Kucharik, C.J., 2008. Corn-based ethanol production compromises goal of reducing nitrogen export by the Mississippi River. Proc. Natl. Acad. Sci. U.S.A. 105(11), 4513-4518.

Emberland, P., 2013. Corn oil adds significant profitability to ethanol. Ethanol Producers Magazine, Benchmarking Business Analyst, Christianson and Associates.

Emery, T., 2012. Fraud Case Shows Holes in Exchange of Fuel Credits. The New York Times. Available at http://www.nytimes.com/2012/07/05/us/biofuel-fraud-case-shows-weak-spots-in-energy-credit-program.html. (accessed on 28 November 2014).

Environmental News Service, 2011. Cellulosic Ethanol Production Far Behind Renewable Fuel Standard, Washington, DC, (ENS) - The United States is not likely to reach cellulosic ethanol production mandates. Available at www.ens-newswire.com/ens/oct2011/2011-10-11-093.html. (accessed on 28 November 2014).

Ethanol Producers Digest, 2013. Ethanol Producers Magazine. BBI International, Grand Forks, North Dakota. Available at www.ethanolroducer.com. (accessed on 28 November 2014).

Evangelista, V., Barsanti, L., Frassanito, A.M., Passarelli, V., Gualieri, P., 2008. Algal Toxins: Nature, Occurrence, Effect and Detection Proceedings of the NATO Advanced Study Institute on Sensor Systems for Biological Threats: The Algal Toxins Case. Pisa, Italy, ISBN: 978-1-4020-8479-9.

Farrell, J., and Morris, D., 2008. Rural Power, Community-Scaled Renewable Energy and Rural Economic Development, New Rules Project. Minneapolis, MN. Available at http://www.newrules.org/de/ruralpower.pdf. (accessed on 28 November 2014).

Fiametti, K.G., Sychoski, M.M., De Cesaro, A., Furigo, A., Bretanha, L.C., Pereira, C.M.P., Treichel, H., de Oliveira, D., Oliveira, J.V., 2011. Ultrasound irradiation promoted efficient solvent-free lipase-catalyzed production of mono- and diacylglycerols from olive oil. Ultrason. Sonochem. 18(5), 981-987.

Foley, P.M., Beach, E.S., Zimmerman, J.B., 2011. Algae as a source of renewable chemicals: opportunities and challenges. Green Chem. 13(6), 1399-1405.

Gao, C., Zhai, Y., Ding, Y., Wu, Q., 2010. Application of sweet sorghum for biodiesel production by heterotrophic microalga *Chlorella protothecoides*. *Appl. Energy*. 87, 756-761.

Gaub, A., 2013. Ethanol company slashes staff, production comes to halt, Casa Grande Dispatch. Available at http://www.trivalleycentral.com/news_premium/ethanol-co-slashes-staff-production-comes-to-halt/article_ee5f041e-6edd-11e2-b0ed-001a4bcf887a.html. (accessed on 28 November 2014).

Gladue, R.M., Maxey, J.E., 1994. Microalgal feeds for aquaculture. J. Appl. Phycol. 6(2), 131-141.

González-Fernández, C., Molinuevo-Salces, B., García-González, M.C., 2011. Nitrogen transformations under different conditions in open ponds by means of microalgae–bacteria consortium treating pig slurry. Bioresour. Technol. 102(2), 960-966.

Gouveia, L., Marques, A.E., da Silva, T.L., Reis, A., 2009. Neochloris oleabundans UTEX #1185: a suitable renewable lipid source for biofuel production. J. Ind. Microbiol. Biotechnol. 36(6), 821-826.

Graham, L.E., Graham, J.E., Wilcox, L.W., 2008. Algae, Benjamin-Cummings Publishing, Menlo Park, CA, 2nd ed., ISBN: 0321559657.

Haefele, D.M., and Ross, A.J., 2009. Corn: Genetics, composition and quality, in: Ingledew W.M., The Alcohol Textbook. Lallemand Ethanol Technology.

Herndon, A., 2012. Cellulosic Biofuel to Surge in 2013 as First Plants Open, Bloomberg. Available at http://www.bloomberg.com/news/print/2012-12-11/cellulosic-biofuel-to-surge-in-2013-as-first-plants-open.html. (accessed on 28 November 2014).

High, S.M., Cohen, M.B., Shu, Q.Y. and Altosaar, I., 2004. Achieving successful deployment of Bt rice. Trends Plant Sci. 9(6), 286-292.

Hu, Q., Sommerfeld, M., Jarvis, E., Ghirardi, M.. Posewitz, M., Seibert, M., Darzins, A., 2008. Microalgal triacylglycerols as feedstocks for biofuel production: perspectives and advances. Plant J. 54(4), 621-639.

Hu, Q., Zhang, C., Sommerfeld, M., 2006. Biodiesel from algae: lessons learned over the past 60 years and future perspectives. Annual Meeting of the Phycological Society of America, Juneau, AK, July 7-12, pp. 40-41.

Iogen Technology, 2005. Makes it Possible (Process Overview), Iogen. Available at http://www.iogen.ca/cellulose_ethanol/what_is_ethanol/process.htm. (accessed on 28 November 2014).

Jatrophia for Biofuel, 2012. Figures, Look at the financial costs of commercial Jatrophia growing for Biodiesel. Available on http://www.reuk.co.uk/Jatropha-for-Biodiesel-Figures.htm. (accessed on 28 November 2014).

Kansas Wheat Harvest Reports, 2012. Kansas Wheat.

Kim, N.J., Li, H., Jung, K., Chang, H.N., Lee, P.C., 2011. Ethanol production from marine algal hydrolysates using Escherichia coli KO11. Bioresour. Technol. 102(16), 7466-7469.

Kuiper, H.A., Kleter, G.A., Noteborn, H.P., Kok, E.J., 2001. Assessment of the food safety issues related to genetically modified foods. Plant J. 27(6), 503-528.

Lam, M.K., Lee, K.T., 2011. Mixed methanol–ethanol technology to produce greener biodiesel from waste cooking oil: a breakthrough for $SO_4^{2-}/SnO_2–SiO_2$ catalyst. Fuel Process. Technol. 92(8). 1639-1645.

Lee, S., Oh, Y., Kim, D., Kwon, D., Lee, C., Lee, J., 2011. Converting Carbohydrates Extracted from Marine Algae into Ethanol Using Various Ethanolic Escherichia coli Strains. Appl. Biochem. Biotechnol. 164(6), 878-888.

Letourneau, D.K., Robinson, G.S. and Hagen, J.A., 2003. *Bt* crops: predicting effects of escaped transgenes on the fitness of wild plants and their herbivores. Environ. Biosaf. Res. 2(04), 219-246.

Li, X, Xu, H., Wu, Q., 2007. Large-scale biodiesel production from microalga Chlorella protothecoides through heterotrophic cultivation in bioreactors. Biotechnol. Bioeng. 98(4), 764-771.

Lundquist, T.J., Woertz, L.C., Quinn, N.W.T., and Benemann, J.R., 2010. A Realistic Technology and Engineering Assessment of Algae Biofuel Production. Energy Biosci. Inst. 1.

Mabee, W.E., McFarlane, P.N., and Saddler, J.N., 2011. Biomass availability for lignocellulosic ethanol production. Biomass Bioenergy. 35(11), 4519-4529.

Marketwatch, 2014. Available at http://www.marketwatch.com/story/oil-rebounds-on-positive-china-manufacturing-data-2014-09-23-1103542. (accessed on 28 November 2014).

McAloon, A., Taylor, F., Yee, W., Ibsen, K., Wooley, R., 2000. Determining the cost of producing ethanol from corn starch and lignocellulosic feedstocks. National Renew. Energy Lab. Rep.

Menetrez, M.Y., 2010. The Potential Environmental Impact of Waste from Ethanol Production. J. Air Waste Manage. Assoc. 60(2), 245-250.

Menetrez, M.Y., 2012. An Overview of Algae Biofuel Production and Potential Environmental Impact. Environ. Sci. Technol.46 (13), 7073-7085.

Meng, X., Yang, J., Xu, X., Zhang, L., Nie, Q., Xian, M., 2009. Biodiesel production from oleaginous microorganisms. Renewable Energy. 34(1), 1-5.

Milledge, J., 2011. Commercial application of microalgae other than as biofuels: a brief review. Rev. Environ. Sci. Biotechnol. 10(1), 31-41.

Monceaux, D.A., 2009. Alternative Feedstocks for fuel ethanol production, in: Ingledew W.M., The Alcohol Textbook. Lallemand Ethanol Technology.

Mononoa, E.M., Nyrenb, P.E., Berti, M.T., Pryora, S.W., 2012. Variability in biomass yield, chemical composition, and ethanol potential of individual and mixed herbaceous biomass species grown in North Dakota. Ind. Crops Prod. 41, 331-339.

National Research Council, 2008. Water Implications of Biofuels Production in the United States. National Academy Press, Washington, DC.

NC State University A&T State University, Cooperative Extension, 2012. Oilseed Crops for Biodiesel Production, Available at http://www.extension.org/pages/28006/oilseed-crops-for-biodiesel-production. (accessed on 28 November 2014).

Olsson, L., and Hahn-Hägerdal, B., 1996. Fermentation of lignocellulosic hydrolysates for ethanol fermentation. Enzyme Microb. Technol. 18(5), 312-331.

Palmqvist, E., Hahn-Hägerdal, B., 2000. Fermentation of lignocellulosic hydrolysates. I : Inhibition and detoxification. Bioresour. Technol. 74(1), 17-24.

Parker, M., 2012a. Ethanol Rebounds from One-Week Low as Heat Threatens Corn Crop. Bloomberg News. Available at http://www.businessweek.com/news/2012-05-23/ethanol-rebounds-from-one-week-low-as-heat-threatens-corn-crop(accessed on 28 November 2014).

Parker, M., 2012b. Ethanol Output in U.S. Rose to Three-Month High, Report Shows. Bloomberg News. Available at http://www.businessweek.com/news/2012-05-23/ethanol-output-in-u-dot-s-dot-rose-to-three-month-high-report-shows . (accessed on 28 November 2014).

Parker, M., 2012c. Ethanol Caps Biggest Monthly Gain Since 2008 on Higher Corn. Bloomberg News. Available at http://www.businessweek.com/news/2012-07-31/ethanol-caps-biggest-monthly-gain-since-2008-on-higher-corn. (accessed on 28 November 2014).

Parker, M., 2012d. Ethanol Falls Against Gasoline on Concern Margins to Spur Output. Bloomberg News. Available at http://www.businessweek.com/news/2013-05-14/ethanol-falls-against-gasoline-on-concern-margins-to-spur-output. (accessed on 28 November 2014).

Parker, M., 2013a. Ethanol Stronger Against Gasoline as Shutdowns Curtail. Bloomberg News. Available at http://www.businessweek.com/news/2013-02-04/ethanol-stronger-against-gasoline-as-shutdowns-curtail. (accessed on 28 November 2014).

Parker, M., 2013b. Cellulosic biofuel RINs generated in June. Bloomberg News. Available at http://www.ethanolproducer.com/articles/10095/cellulosic-biofuel-rins-generated-in-june. (accessed on 28 November 2014).

Parker, M., 2013c. Ethanol Weakens Against Gasoline on Higher Inventory and Imports. Bloomberg News. Available at http://www.businessweek.com/news/2013-01-22/ethanol-weakens-against-gasoline-on-higher-inventory-and-imports. (accessed on 28 November 2014).

Parker, M., 2013d. Ethanol Weakens as Dry Midwestern Weather Boosts Corn Planting. Bloomberg News. Available at http://www.bloomberg.com/news/2013-05-06/ethanol-weakens-as-dry-midwestern-weather-boosts-corn-planting.html. (accessed on 28 November 2014).

Parker, M., 2013e. Ethanol's Discount to Gasoline Tightens on Lower Stockpiles. Available at http://www.bloomberg.com/news/2013-08-05/ethanol-s-discount-to-gasoline-tightens-on-lower-stockpiles.html. (accessed on 28 November 2014).

Pedroso, G.M., De Ben, C., Hutmacher, R.B., Orloff, S., Putnam, D., Six, J., van Kessel, C., Wright, S.D., Linquist, B.A., 2011. Switchgrass is a promising, high-yielding crop for California biofuel. California Agric. 65(3):168-173.

Perlack, R.D., Erbach, D.C.,Graham, R.L., Stokes, B.J., Turhollow, A.F., Wright, L.L., 2005. Biomass as Feedstock for a Bioenergy and Bioproducts Industry: the Technical Feasibility of a Billion-Ton Annual Supply. Environmental Science Division, Oak Ridge National Laboratory.

Piedmont Biofuels, 2013. Enzymatic Biodiesel. Available at http://www.biofuels.coop/wp-content/uploads/2011/01/2012-FAeSTER-Letter1.pdf. (accessed on 28 November 2014).

Platts McGraw Hill Financial, 2013. US RINs prices fall amid selloff, ending six-session ascent. Available at http://www.platts.com/latest-news/agriculture/houston/us-rins-prices-fall-amid-selloff-ending-six-session-21447191. (accessed on 28 November 2014).

Public Law 110-140-Dec. 19, 2007, 121 STAT. 1492, Energy Independence and Security Act of 2007. Available at http://www.gpo.gov/fdsys/pkg/PLAW-110publ140/content-detail.html. (accessed on 28 November 2014).

RFS Renewable Fuels Association, 2012. Renewable Fuels Standard. Available at http://www.ethanolrfa.org/pages/renewable-fuels-standard (accessed on 28 November 2014).

Rosenberg J.N., Oyler GA, Wilkinson L., Betenbaugh M.J., 2008. A green light for engineered algae: Redirecting metabolism to fuel a biotechnology revolution. Curr. Opin. Biotechnol. 19(5), 430-436.

Schroeder, J.W., 2003. Distillers Grains as a Protein and Energy Supplement for Dairy Cattle. NDSU Extension Service, AS-1241, North Dakota State University, Fargo, North Dakota.

Shaffer, D., 2012. Poet breaks ground on cellulosic ethanol plant in Iowa, Star Tribune Updated. Available at http://www.startribune.com/business/142515155.html. (accessed on 28 November 2014).

Shen, Y., Pei, Z., Yuan, W., Mao, E., 2009. Effect of nitrogen and extraction method on algae lipid yield. Int. J. Agric. Biol. Eng. *2*(1), 51-57.

Snow, A.A., Smith, V.S., 2012. Genetically Engineered Algae for Biofuels: A Key Role for Ecologists. BioScience. 62(8), 765-768.

Sukenik, A., Levy, R.S., Levy, Y., Falkowski, P.G., Dubinsky, Z., 1991. Optimizing algal biomass production in an outdoor pond: a simulation model. J. Appl. Phycol. 3(3), 191-201.

Talukder, M.R., Das, P., Shu Fang, T., Wu, J.C., 2011. Enhanced enzymatic transesterification of palm oil to biodiesel. Biochem. Eng. J. 55(2), 119-122.

Tannura, M., Irwin, S., and Good, D., 2008. Are corn trend yields increasing at a faster rate? Marketing and Outlook Briefs, Department of Agriculture and Consumer Economics, University of Illinois at Urbana-Champaign. MOBR 08-02.

The Wall Street Journal, 2013. January 28, 2013, 7:06 P.M. ET Zero Dark Ethanol, Available at http://online.wsj.com/article/SB10001424127887324329204578270043967301294.html?KEYWORDS=biofuel&mod=dist_smartbrief#printMode. (accessed on 28 November 2014).

Thompson, W., Meyer, S., Westhoff, P., 2010. The New Markets for Renewable Identification Numbers. Appl. Econ. Perspect. Policy. *32*(4), 588-603.

Unkefer P.J., Knight T.J., Martinez R.A., 2004. Use of prolines for improving growth and other properties of plants and algae. U.S. Patent No. 6,831,040.

USA Biodiesel Prices.com , 2012. Biodiesel Stations and Prices for the entire US. Available on http://www.usabiodieselprices.com/. (accessed on 28 November 2014).

U.S. Biodiesels Digest, 2011. US biodiesel production sets all-time record. Available at http://www.biofuelsdigest.com/bdigest/2011/11/29/2011-us-biodiesel-production-sets-all-time-record/. (accessed on 28 November 2014).

U.S. Department of Agriculture, 2011. Principal Crops Area Planted and Harvested – States and United States: 2009-2011. Available at http://usda01.library.cornell.edu/usda/current/CropProdSu/CropProdSu-01-12-2012.pdf. (accessed on 28 November 2014).

U.S. Department of Agriculture, 2012. Press Release, National Agricultural Statistics Service, United States Department of Agriculture. Washington, DC, Available at http://www.nass.usda.gov/Statistics_by_State/Washington/Publications/Current_News_Release/wwhtaug.pdf. (accessed on 28 November 2014).

U.S. Department of Agriculture, 2006. Rural Development. Section 9006, Renewable Energy Systems and Energy Efficiency Improvements Program. (Available at http://www.farmenergy.org/ )

U.S. Department of Energy. Biomass Program Multi-Year Technical Plan. (Available at http://www1.eere.energy.gov/biomass/pdfs/mytp.pdf). (accessed on 28 November 2014).

U.S. Department of Agriculture, 2013. Statement on the Detection of Genetically Engineered Wheat in Oregon. Available at http://www.usda.gov/wps/portal/usda/usdahome?contentid=2013/06/0127.xml. (accessed on 28 November 2014).

U.S. Department of Energy, 2004. Assumptions for the Annual Energy Outlook, with Projections to 2025, Energy Information Administration, DOE/EIA-0554.

U.S. Department of Energy, 2007a. National Renewable Energy Laboratory (NREL), Research Advances Cellulosic Ethanol, NREL/BR-510-40742.

U.S. Department of Energy, 2007b. DOE Selects Six Cellulosic Ethanol Plants for Up to $385 Million in Federal Funding, DOE Office of Public Affairs, Washington, D.C.

U.S. Department of Energy (DOE), 2008. Office of Science, Genomics: GTL Systems Biology for Energy and Environment, Cellulosic Ethanol: Fuel Ethanol Production. Available at http://genomicsgtl.energy.gov. (accessed on 28 November 2014).

U.S. Energy Information Administration, 2012a. Biodiesel Performance, Costs, and Use. Available at http://www.eia.gov/oiaf/analysispaper/biodiesel/. (accessed on 28 November 2014).

U.S. Energy Information Administration, 2012b. Biodiesel demand estimates now provided in petroleum supply and demand balances. Available at http://www.eia.gov/biofuels/biodiesel/production/table1.pdf. (accessed on 28 November 2014).

U.S. Energy Information Administration, 2012c. Petroleum & Other Liquids, Gasoline and Diesel Fuel Update, Sept. 17, 2012. Available at http://www.eia.gov/petroleum/gasdiesel/. (accessed on 28 November 2014).

U.S. Environmental Protection Agency, 2008. Fact Sheet: Final Changes for Certain Ethanol Production Facilities Under Three Clean Air Act Permitting Programs, EPA-HQ-OAR-2006-0089. Available at http://www.epa.gov/nsr/fs20070412.html. (accessed on 28 November 2014).

Venkata Mohan, S., Bhaskar, Y.B., Krishna, T.M., Chandrasekhara Rao, N., Lalit Babu, V., Sarma, P.N., 2007a. Biohydrogen production from chemical wastewater as substrate by selectively enriched anaerobic mixed consortia: Influence of fermentation pH and substrate composition. Int. J. Hydrogen Energy. 32(13), 2286-2295.

Venkata Mohan, S., Mohanakrishna, G., Raghuvulu, S.V., Sarma, P.N., 2007b. Enhancing biohydrogen production from chemical wastewater treatment in anaerobic sequencing batch biofilm reactor (AnSBBR) by bioaugmenting with selectively enriched kanamycin resistant anaerobic mixed consortia. Int. J. Hydrogen Energy. 32(15), 3284-3292.

Wageningen University, 2011. Research on microalgae within Wageningen UR. Available at http://www.algae.wur.nl/UK/technologies/production/heterotrophic_organisms/. (accessed on 28 November 2014).

Wallace, R., Ibsen, K., McAloon, A., Yee, W., 2005. Feasibility Study for Co-Locating and Integrating Ethanol Production Plants from Corn Starch and Lignocellulosic Feedstocks (Revised) (No. NREL/TP-510-37092). National Renew. Energy Lab., Golden, CO (US).

Weihrauch, J., 2007. Registration, Recordkeeping, and Reporting Requirements [for RINs], U.S. EPA Office of Transportation and Air Quality. Presentation at a Renewable Fuels Standard Workshop, May 10, 2007.

Weissman, J.C., Benemann, J.R., 2003. Comparison of Marine Microalgae Culture Systems, Second National Conference on Carbon Sequestration. Available at http://www.netl.doe.gov/publications/proceedings/03/carbon-seq/PDFs/217.pdf. (accessed on 28 November 2014).

Wisner, R., 2009. Renewable Identification Numbers (RINs) and Government Biofuels Blending Mandates. AgMRC Renewable Energy Newsletter.

Woertz, I., Feffer, A., Lundquist, T., Nelson, Y., 2009. Algae grown on dairy and municipal wastewater for simultaneous nutrient removal and lipid production for biofuel feedstock. J. Environ. Eng. 135(11), 1115-1122.

Zfacts, 2013. Available at http://zfacts.com/p/436 html. (accessed on 28 November 2014).

# 17

# Advances in biofuel production from oil palm and palm oil processing wastes

Jundika C. Kurnia [1,*], Sachin V. Jangam [2], Saad Akhtar [3], Agus P. Sasmito [3], Arun S. Mujumdar [2,3,*]

[1] *Mechanical Engineering Department, Universiti Teknologi PETRONAS, 32610 Bandar Seri Iskandar, Perak Darul Ridzuan, Malaysia.*

[2] *Department of Chemical and Biomolecular Engineering, National University of Singapore, 4 Engineering Drive 4, 117575 Singapore.*

[3]*Department of Mining and Materials Engineering, McGill University, 3450 University Street, Frank Dawson Adams Bldg, Montreal Quebec H3A 2A7, Canada.*

## HIGHLIGHTS

- ➢Technologies used for processing oil palm and palm oil wastes are reviewed.
- ➢Major challenge in biofuel production from oil palm wastes is remote locations of palm plantations complicating transportation and distribution.
- ➢Among phases in producing biofuel from oil palm wastes, oil palm plantation has the most severe environmental impacts.
- ➢Development of cost-effective, environmentally friendly, and profitable biofuel production technologies from oil palm wastes is required.

## GRAPHICAL ABSTRACT

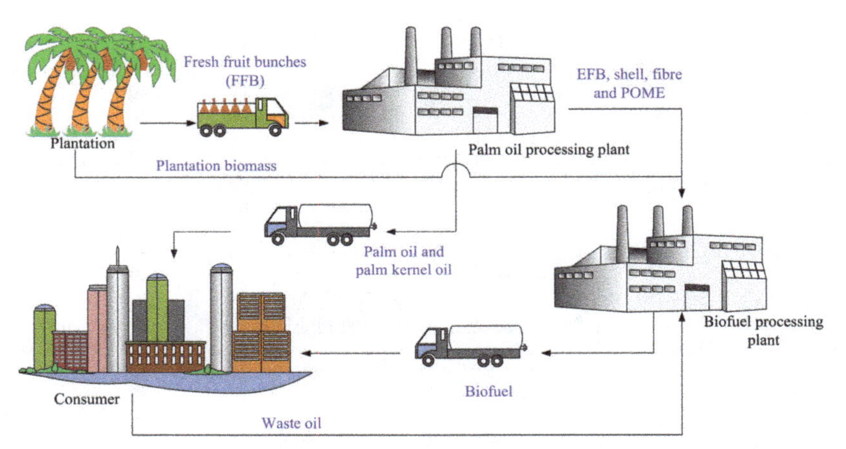


## ABSTRACT

Over the last decades, the palm oil industry has been growing rapidly due to increasing demands for food, cosmetic, and hygienic products. Aside from producing palm oil, the industry generates a huge quantity of residues (dry and wet) which can be processed to produce biofuel. Driven by the necessity to find an alternative and renewable energy/fuel resources, numerous technologies have been developed and more are being developed to process oil-palm and palm-oil wastes into biofuel. To further develop these technologies, it is essential to understand the current stage of the industry and technology developments. The objective of this paper is to provide an overview of the palm oil industry, review technologies available to process oil palm and palm oil residues into biofuel, and to summarise the challenges that should be overcome for further development. The paper also discusses the research and development needs, technoeconomics, and life cycle analysis of biofuel production from oil-palm and palm-oil wastes.

**Keywords:**
Oil palm wastes
Palm oil
Biofuel
Life cycle analysis
Technoeconomics

* Corresponding authors E-mail address: jundika.kurnia@petronas.com.my (J.C. Kurnia) E-mail address: arunmujumdar123@gmail.com (A.S. Mujumdar)

**Contents**



## 1. Introduction

The degradation of global environment and the prediction of the depletion of the fossil fuel resources have all encouraged the global community to search for alternative sustainable and environmentally-friendly energy resources. One of the most promising candidates is biomass energy. Despite its wide availability and relatively low cost in some locations, biomass energy has inherent drawbacks which hinder its wide utilization: low energy conversion, difficulty to transport and to store, and harmful effects of direct combustion of biomass. Often referred as traditional energy, biomass energy is commonly utilized in rural areas where other energy resources are not accessible due their availability or cost. Biomass is generally used for cooking and heating. To minimize the complexity of biomass transportation and storage as well as to avoid harmful effects of direct combustion of biomass, its conversion into biofuels is suggested (Baratieri et al., 2008). Biomass can be used to produce biofuels *via* different thermochemical and biochemical process such as biomethanation, fermentation, pyrolysis, and gasification (Verma et al., 2012; Akia et al., 2014).

Biomass sources can easily be found in our daily life including plant/crop roots, seeds, by-products/wastes, forest residues, municipal wastes, as well as cattle and human wastes (Verna et al., 2012). One tremendous source for biomass is palm oil industry. Palm oil itself is considered as a promising candidate to produce biofuel. Aside from producing palm oil, the industry also generates a huge quantity of residues (dry and wet) which can be processed to produce biofuels as well. In fact, the produced oil only contributes to 10 % of total biomass generated from plantations (Chew and Bhatia, 2008; Sulaiman and Taha, 2015). The other 90 % is disposed of as waste materials (e.g., empty fruit bunches, oil palm trunks, oil palm fronds, palm shells, palm pressed fibres, palm oil mill effluent, and old trees). In a specific location, the potential of biomass generated from oil palm industry is amounted up to seven times that of natural timber industry (Basiron and Chan, 2004). In addition to the biomass generated during palm oil production, the increasing rate of cooking oil consumption worldwide has also generated a huge amount of waste cooking oil which could trigger complex problems if not handled carefully. Currently, the waste cooking oil is discarded to the waste water stream, complicating waste water treatment, contaminating environmental water, and undermining its potential as biofuel feedstock. As such, disposing waste cooking oil to water drainage has been banned in the majority of the developed country (Kulkarni and Dalai, 2006).

Over last few years, there has been a growing interest to produce biofuel from vegetable oil especially palm oil. This is mainly driven by the desire to reduce greenhouse gas emission. The problems associated with the production of biofuel from palm oil are (i) biofuel from palm oil is not sufficient to compensate for global fuel consumption, (ii) it triggers food and fuel competition which may lead to high food price, and (iii) environmental degradation due to conversion of forests to oil palm plantations to excel the oil production (Sheil et al., 2009; Mukherjee and Sovacool, 2014). As such, an initiative has been put forth to produce the biofuel from oil palm and palm oil wastes. In line with that, numerous studies have been conducted and various processes have been proposed to produce biofuel from the oil palm and palm oil wastes (Amin et al., 2007; Geng, 2013; Awalludin et al., 2015). The methods presented vary according to the waste used as feedstock for biofuel production. The characteristics of the resultant biofuel also vary depending on the feedstock and method used. Hence it is important to summarize and discuss the main findings of these studies.

Therefore, the present paper is intended to comprehensively review the production of biofuel from oil palm and palm oil wastes and to investigate the various aspects that could potentially influence future advancements in the field. To achieve that, an overview on the palm oil industry, production processes, and current waste management scenarios is presented and discussed. Moreover, the technologies used for biofuel production from oil palm and palm oil wastes including their techno-economical aspects are also presented. Finally, the research and development needs for further advancements of the field are highlighted.

## 2. Palm oil industry

Grown in tropical regions, oil palm tree has been cultivated to produce palm oil which is widely consumed for food and other products. Here, the essential information on the oil palm and palm oil, production of palm oil, growth of oil palm industry, as well as the management and utilization of the wastes generated by the palm oil industry are presented.

### *2.1. Oil palm and palm oil*

Palm oil is an edible vegetable oil extracted from the mesocarp of the fruit of oil-palm tree (*Elaeis guineensis*). The origin of this type of palm tree can be tracked to a region along the coastal strip of Africa between Liberia and Angola (Sheil et al., 2009). The tree can be raised in places with abundant rainfalls and heat such as tropical countries in Southeast Asia and South America. As such, large oil palm plantations can be easily found in these regions. Belonging to the subfamily Arecoideae, the morphology of oil palm is similar to the other palm species with a height up to 30 m (Edem, 2002). Generally, an oil palm tree starts to bear fruit after 3-4 years (Awalludin et al., 2015). The farmers need to wait for 5-6 months for the fruit to mature before they can harvest them. The fruit is plump-size, reddish in colour and is collated in a bunch weighting 10 to

40 kg on average (Shuit et al., 2009). The fruit comprises exocarp, mesocarp, endocarp (shell), and endosperm (kernel). The mesocarp and endosperm contains 45-55% edible oil (Edem, 2002; Sumathi et al., 2008).

Oil palm is considered as the most efficient oilseed crop in the world due to its high productivity per hectare. Among the major oilseeds and oil plant (e.g., soybean, sunflower, rapeseed, groundnut, cotton), oil palm has higher oil production efficiency (oil produced/land area) of 4000 kg/ha (Yusoff and Hansen, 2007; Zulkifli et al., 2010; Salim et al., 2012). In addition, oil-palm has a long lifespan of over 200 years with relatively long economic life span of 25-30 years (Amin et al., 2007; Abdullah and Wahid, 2010; Rupani et al., 2010; Abdullah and Sulaiman, 2013), providing a reliable supply for oil production. Along with the high production efficiency, this has driven the rapid expansion of oil palm plantation around the globe.

*2.2. Rapid growth of palm oil industry*

Due to its affordable price, efficient production, and high oxidative stability, palm oil has been widely used in food, cosmetic, and hygienic products. From 2005, palm oil has replaced soybean oil as the most consumed edible oil globally. In 2012, consumption of palm oil reached 52.1 million tonnes worldwide (Sime Darby Plantation, 2014). Major palm oil consuming countries include China, India, Indonesia, and The European Union. In fact, driven by high market demands especially in the developing countries, the palm oil industry has grown rapidly over the last decades. During the 1950s to the early 60s, the average production of palm oil was roughly 1.26 million tonnes (Abdullah and Wahid, 2010). This increased to 5 million tonnes in 1980 and doubled to 11 million tonnes in 1990 (Abdullah and Sulaiman, 2013). Within 1995-2010, palm oil production expanded to 46.7 million tonnes (Mahat, 2012).

Although originated from Africa, oil-palm is widely cultivated in almost 43 countries in the tropical regions of Southeast Asia, Africa, and South America (Koh and Wilcove, 2008). Indonesia and Malaysia dominate the global production of palm-oil, contributing to around 85% of the palm oil production world-wide (Sime Darby Plantation, 2014; Siregar et al., 2014). Other major palm oil producing countries are Thailand, Columbia, Nigeria, Ecuador, and Papua New Guinea. Malaysia was the leading palm oil producer for a long period until 2006 when Indonesia overtook Malaysia to become the world largest palm oil producer. This is mainly attributed to the fast expansion of oil palm plantation areas in Indonesia and the stagnation of the oil palm plantation areas in Malaysia (Mahat, 2012).

This growing palm oil industry has changed the economy scenario especially in Malaysia and Indonesia as palm oil is one of the main export commodities for both countries. In fact, the palm oil industry has been a source of income and employment for the indigenous communities residing near the plantations and has led to substantial improvements in their life quality (Basiron, 2007; Mukherjee and Sovacool, 2014). The industry has also provided access to healthcare and education for the indigenous communities (Sheil et al., 2009). A study revealed that millions of people currently working in the oil palm industry, used to live in poverty (Wakker, 2006; Zen et al., 2006). In addition to that, the industry continues to generate huge revenues for the producing countries. Therefore, it is not surprising that the oil palm industry is expected to grow further in the coming years.

Despite its economic benefits and role as tool in poverty alleviation programs, the palm oil industry has received intense critics and negative reviews due to its land utilization expansion. The fast expansion of oil palm plantations has raised issues about the industry sustainability and its impact on the environment: destruction of old-growth rainforest and its biodiversity, air, soil and water pollutions as well as land disputes and social challenges. One way to address these issues is to increase the efficiency of the mills and plantations so that no or minimum further plantation expansion is required. Another way is to maximize the utilization of biomass produced in the plantations and mills to meet energy demands. This will reduce the cost of waste treatment and increase the profitability through the energy generated.

*2.3. Palm oil production*

Two distinct types of oil can be produced from oil palm fruit, i.e., crude palm oil (CPO) which is produced from the mesocarp and palm kernel oil which is produced from the kernel or endosperm (kernel) (Abdullah and Wahid, 2010; Mba et al., 2015). After harvested, the oil palm fruit should be transported quickly to the palm oil mill to be processed into palm oil. **Figure 1** shows the palm oil production process. Once the fresh fruit bunches (FFB) reach the processing plant, they will be sterilized by using steam. The FFB will then be stripped to separate the fruit from the stalk. The fruit will be directed to digesters and then pressers to extract the crude oil while the empty bunches will be collected to be used as fertilizer or dried before being fed into boilers. The oil extracted through the pressing process will be purified by using centrifugal and vacuum dryers before it is stored in storage tanks. CPO will be further processed in a refinery plant to produce cooking oil and other products. The other oil (i.e., palm kernel oil) is extracted from the nuts obtained from the pressing process. After fibre/nut separation, the nuts are sent to nut crackers and then to crushers to extract the kernel oil. Meanwhile, the shells and fibres are sent to boilers as fuel.

As can be observed in **Figure 1**, a palm oil mill plant is generally energy self-sufficient processing plant. The palm oil mill is commonly equipped with low pressure boilers. The wastes generated during the oil production process, mainly fibres and shells, are burnt as fuel in boilers to generate steam or hot gas for drying, sterilization, and power generation. Nevertheless, for the start-up process, a back-up diesel generator is generally installed to provide the initial power (Mahlia et al., 2001; Yusoff, 2006). It should be noted that not all wastes are burnt in boilers. Although the efficiency of boilers installed in mills is relatively low, some mills still have excess generated power which is distributed to the residential areas nearby. These areas are generally located in remote area where no electricity grid is available.

*2.4. Oil palm and palm oil wastes: current disposal and utilization scenario*

Palm oil industry generates a huge quantity of residues which can be processed to produce biofuel. As stated previously, in oil palm plantations, the extracted oil constitutes only 10% of the total biomass generated while the other 90% is considered as wastes. With rapid growing of palm oil industry, more residues will be generated, adding complexity to the current waste management procedures. On average, 50 to 70 tonnes of biomass residues are produced from each hectare of oil palm plantation (Shuit et al., 2009). The by-products or wastes generated from palm oil production includes oil palm trunk (OPT), oil palm frond (OPF), empty fruit bunch (EFB), mesocarp fruit fibre (MF), palm kernel shells (PKS), and palm oil mill effluent (POME). Except POME, these wastes have high fibre content.

In palm oil plantations, OPF is steadily available in the plantation throughout the year as harvesting is generally followed by pruning. In contrast, OPT is available only during the replanting season. As stated previously, oil palm trees have a relatively long lifespan and when they reach the end of their economic lifespan they should be replaced by new plants. The current practice is to leave the dead tress between the rows of palm trees to naturally decompose for soil conservation, erosion control, and in the long term nutrients recycling purpose. However, this practice poses the risk of attracting harmful insect to live and breed. In addition, leaving the trunk in the plantation will obstruct re-plantation activity. The other method is to utilize them as soil fertilizer by burning. This will minimize the risk of attracting insects; however, it results in air pollution. Open burning is commonly practiced in plantations in Indonesia causing hazardous air pollution not only in Indonesia but also in the neighbouring countries. Thick hazardous smoke generated from such open burning activities paralyses socio-economic activities in the nearby areas and therefore, many countries have raised their concern on this annual issue. In return, the Indonesian authorities stated that they would investigate and prosecute the plantation owners who practice open burning (The Jakarta Post, 2015). Nevertheless, it is believed that legal prosecutions alone would be insufficient and that to eliminate this problem, a more efficient and environmentally-friendly utilization of the generated OPT and OPF during the replanting session is urgently needed.

Other than its application as fertilizer, OPF can also be chopped into small pieces, mixed with other ingredients, and utilized as livestock feed (Abu Hassan et al., 1996). Several studies were conducted to examine this possibility and proposed an integrated crop-livestock system where the livestock farm should be located inside the oil palm plantation (Abu Hassan et al., 1996).

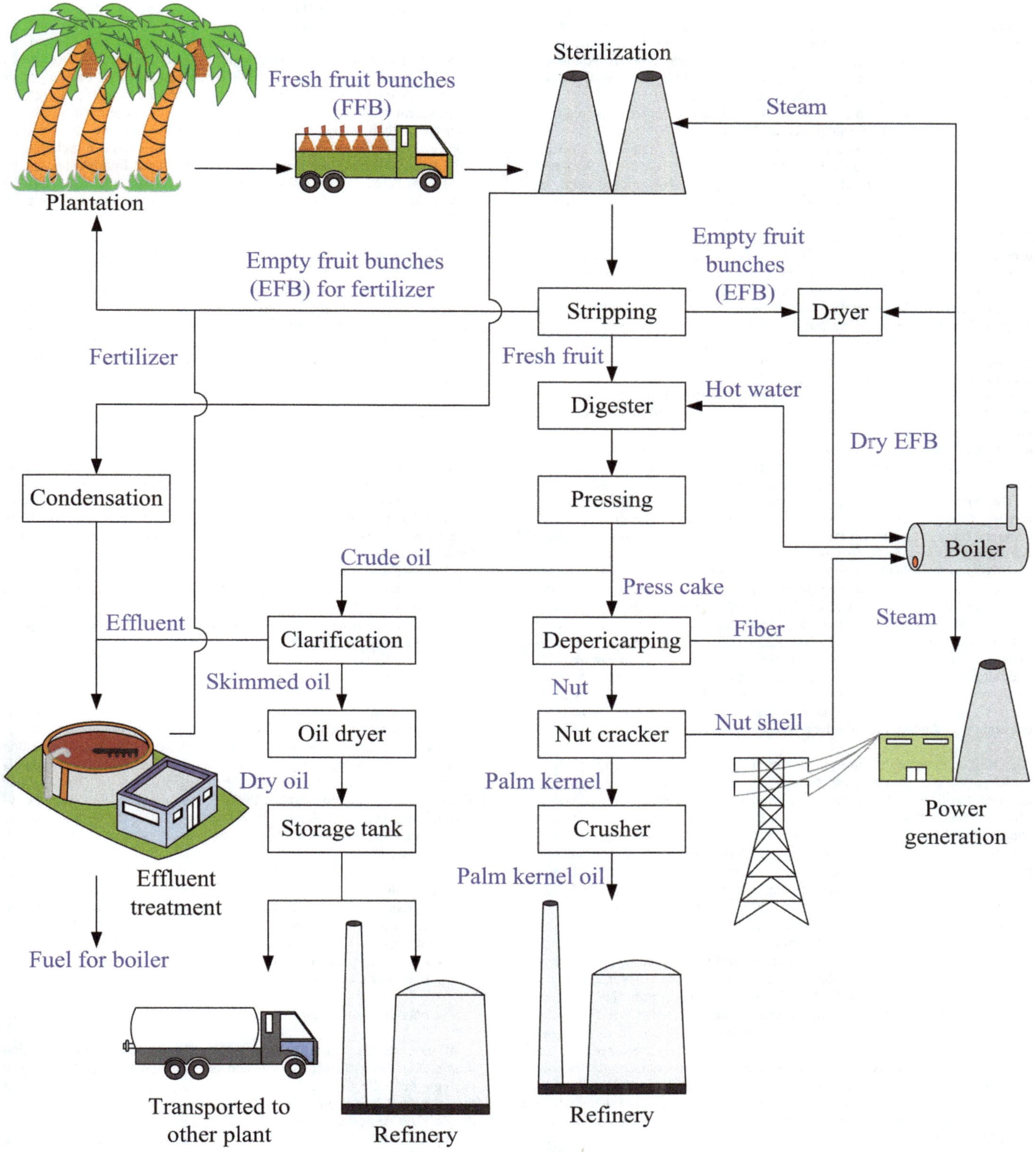


**Fig.1.** Palm oil production process (adapted from Abdullah and Sulaiman, 2013).

EFB, MF, and PKS are generally used as fuel in mill boilers. The ash generated in boilers is transported back to plantations as fertilizer (Shuit et al., 2009). It is worth mentioning that EFB cannot be burned directly due to its high moisture content resulting in low heating value and air pollutions (Abdullah and Sulaiman, 2013), and therefore, it should be dried using hot air until its moisture content is significantly decreased. Hence, MF and PKS are more desirable as boiler fuel while EFB is usually dumped in plantations (Chew and Bhatia, 2008; Awalludin et al., 2015).

Aside from being used as soil fertilizer and boiler fuel, there is a growing interest to use EFB to produce bioplastics (Abdullah et al., 2011; Siyamak, 2012; Tan et al., 2014). The characteristics of bioplastic are similar to those of fossil fuel-derived plastics, making them suitable to produce biodegradable foil, moulds, tins, cups, bottles, and other packaging materials (Shuit et al., 2009). Palm fibres, i.e., MF produced during the palm oil processing can be used as fillers in the production of thermo-plastics and thermoset composites which have wide applications in furniture and automobile components (Shuit et al., 2009). The oil palm biomass can also be utilized to produce absorbents for toxic gas and heavy metal. For instance, the waste generated through burning PKS and MF in boilers can be converted into absorbents for pollutant removal. This waste has been found to contain high concentrations of silica, calcium, potassium, and alumina which are essential in absorbents production (Zainudin et al., 2005; Mohamed et al., 2006).

Currently, relatively low efficient boilers are installed in mills to produce steam for sterilization, drying, and power generation. Installation of more efficient co-generation plants is strongly advisable to generate more energy and reduce emissions. However, the challenge is that under the current conditions, mill cannot sell their excess electricity to the grid and hence, the installation of new plants is not economically justified.

The only liquid waste produced from the palm mill is POME. It mainly consists of water with small amounts of solid and oil. The processes that generate huge amounts of POME in palm oil processing plants include sterilization, crude oil clarification, and cracked mixture separation (Rupani et al., 2010). In fact, this huge amount of POME is the result of the tremendous amount of water used to clean up the palm fruit and to extract the oil from the mesocarps. To extract 1 ton of crude palm oil, approximately 5-7.5 tons of water is used, out of which more than half (i.e., >2.5-3.75 tons) ends up as POME (Ma, 1999; Ahmad et al., 2003). Even though it is considered as non-toxic material, POME cannot be discharged to the environment directly without treatment as it is acidic and contains residual oil which cannot be easily separated using the gravitational method (Madaki and Seng, 2013). If the raw or untreated POME is discharged to rivers, it will extensively consume and deplete the dissolved oxygen content essential for the aquatic life.

POME contains high concentrations of organic compounds such as protein, carbohydrate, nitrogenous compounds, lipids, and minerals, making it suitable as plant fertilizer provided that it is properly treated (Habib et al., 1997; Muhrizal et al., 2006). The current disposal scenario for POME is to store it in anaerobic and aerobic digestion ponds before being discharged into rivers. Anaerobic ponds if covered are more desirable as they use less energy, produce minimum sludge, does not results in unpleasant odour, and offer efficient breakdown of organic substances to produce methane-rich biogas which can be used as fuel (Rincon et al., 2006; Rupani et al., 2010). Before it can be deposited into digestion ponds, however, POME has to be passed through several physical pre-treatment processes including screening, sedimentation, and oil removal. Due to its generally low cost, the pond system has been widely adopted by palm oil mills. This method however, requires a large area of land and relatively long hydraulic retention time (HRT) which often creates a problem of discharging incompletely-treated POME into water bodies. In addition to that, in most open pond systems, due to the difficulty in collecting the generated biogas, the gas is directly released to the environment, wasting its potential as an alternative environmentally-friendly fuel and contributing to the greenhouse gas emissions. To overcome this, there are initiatives proposed to install closed anaerobic pond systems where the high quality methane-rich biogas can be collected (Abdullah and Sulaiman, 2013).

In addition to the biomass waste produced during the palm oil production, the palm oil consumption also generates a huge deal of waste in the form of used cooking oil. In recent years, the demand for palm oil has grown significantly especially by the developing countries due to their rapid population and per capita income growth. In fact, palm oil is mainly used in food industries and households for cooking (frying). During frying, the fatty acids contained in palm oil undergo multiple reactions such as oxidation, polymerization, and hydrolysis, and therefore, should be disposed of to avoid human health and nutrition problems (Naghshineh and Mirhosseini, 2010; Stier, 2013). The disposal of used cooking oil is tricky as its direct discharge into the water drainage system pose serious environmental threats. A more economical and environmentally-friendly disposal method is by its collection and conversion into biofuel (i.e., biodiesel) to be used as an alternative to fossil-derived diesel fuel. It should also be highlighted that the biodiesel produced from waste cooking oil is considered to be carbon neutral as the carbon emissions released from biodiesel combustion are compensated by those absorbed by palm oil trees during the photosynthesis process (Sheil et al., 2009).

### *2.5. Challenges in utilization of oil palm and palm oil wastes*

In the current waste management scenario, the biomass residues generated by oil palm plantations and mills are underutilized. Therefore, there is a need to explore and evaluate various strategies to maximize the utilization of these biomass wastes. However, there are several roadblocks that hinder its further advancement and need to be overcome. Some of these are summarized as follows:

- *Location of plantation in remote area*

As mentioned earlier, plantations are commonly located in remote areas where no electricity grid is available and hence, the excess electricity generated by the power plants in the palm oil mills cannot be sold. This makes the installation of power plants with higher capacities and efficiencies not economically feasible. Consequently, the utilization of biomass generated in plantations and mills for power generation is also hindered.

Moreover, aalthough OPF could be potentially used as animal feed but it needs to be transported from plantations to livestock farms. This will lead to additional carbon emissions through the transportation. Therefore, as mentioned earlier, integrate livestock farming inside oil palm plantations should be considered.

- *Large open digestion ponds*

Another challenge currently faced is the large amount of biogas released into the atmosphere from the POME treatment ponds. This is ascribed to the fact that the main purpose of the currently in-use digestion ponds is not to produce biogas but to decompose the organic compounds of the POME so that it can be safely discharged into rivers. This practice also undermines the potential of POME to produce an environmentally-friendly fuel, i.e., biomethane. The obstacle to collect the biogas from the current open ponding systems is the large area of ponds making it difficult to collect the biogas. In addition, the conditions inside the ponds cannot be thoroughly controlled; hence the production of biogas fluctuates. Another problem is the utilization of the produced biogas because the boilers installed in palm oil mills are commonly designed to be fuelled with mesocarp fibres and palm kernel shell. To change the boilers will impose additional cost on the mills. A possible utilization strategy is to sell the produced gas to other parties. However, the compression and transportation of the gas will be challenging.

- *Collection process and quality of used cooking oils*

In case of the waste cooking oil utilization, although producing biodiesel is the most economical and environmentally-friendly disposal strategy, the collection of the oil from various locations (restaurant, food factories, and households) is challenging. Currently, in most countries, there are no dedicated pipelines to collect the waste cooking oil and it is mostly collected manually from every households. The other challenge faced in the utilization of waste cooking oil as biodiesel feedstock is the varying characteristics of the oil since it is exposed to various cooking conditions leading to different oil compositions and structures. Hence, additional pre-treatment may be required before it can be converted into biodiesel.

## 3. Production of biofuel from oil palm and palm oil wastes

The different types of palm oil biomass along with the waste palm oil itself are effective resources to produce biofuel. Palm oil makes up 33% of the global vegetable oil production catering for the domestic and export needs of many countries such as Malaysia, Indonesia, and Thailand (Pool, 2014). The extensive use of palm oil in cooking, lubrication, cosmetics, etc. generates a huge quantity of waste palm oil as well. Furthermore, the biomass generated during the palm oil production is also a potential source for sustainable energy production. It has been reported that for every kilogram of palm oil produced, four kilograms of wastes in the form of fibrous strands of empty fruit bunch are also generated (Law et al., 2007). Several attempts have been made over the last couple of decades to convert these wastes into useful products such as hydrogen, transportation fuels, liquid and gaseous hydrocarbons, briquettes, etc. (Marquevich et al., 1999; Demirbaş, 2005; Huber and Corma, 2007; Nasrin et al., 2008; Pütün et al., 2008; Balat et al., 2009; Misson et al., 2009; Sulaiman and Abdullah, 2011). Chew and Bhatia (2008) reviewed the literature extensively with regards to different catalytic technologies involved in utilizing the palm oil and palm oil biomass as well.

This section attempts to review different energy conversion technologies for the conversion of liquid palm oil wastes as well as solid waste fibers.

### *3.1. Production of biofuel from palm oil*

**Figure 2** shows an outline of different processing technologies used for biofuel production from palm oil. The biofuel production from palm oil can be divided into two main categories, i.e., catalytic cracking and transesterification. Historically, transesterification has been used for centuries to produce glycerin from vegetable oil which is used in the manufacturing of soap. Initial attempts to use transesterification for biodiesel production date back to the early twentieth century (Mamilla et al., 2012). However, due to the increasing environmental concerns and exhaustion of fossil fuels, the focus has been shifted significantly to vegetable oil derived biofuels.

Alternatively, catalytic cracking is also used to convert high molecular weight vegetable oils into lighter and more useful hydrocarbons. Amongst the two technologies, the catalytic cracking process is more developed since it has been extensively utilized to get the desired petroleum products such as diesel, gasoline, olefins, etc. from crude oil. The following discussion presents a technological evaluation of the transesterification and the catalytic cracking.

#### *3.1.1. Transesterification*

Palm oil is well-known vegetable oil feedstock to produce biodiesel through the transesterification process. Transesterification is a process by which triglycerides (vegetable oil) react with an alcohol (methanol or ethanol) to form fatty acid methyl/ethyl esters and glycerol (Korus, 1993). The esters derived from vegetable oils are very similar to petro-diesel in terms of cetane number, viscosity, and energy content (Darnoko and Cheryan, 2000), thus aptly named as 'biodiesel'. Amongst different types of vegetable oils, palm oil holds significant potentials in meeting energy demands owing to its high yield (Pool, 2014). Due to this, many countries located in the Association of South East Asian Nations (ASEAN) region like Malaysia, Indonesia, Thailand, etc. have focused on utilizing palm oil to produce biodiesel.

There are different operational parameters which could impact the overall efficiency and yield of the transesterification process. These include, 1) temperature of the mixture, 2) moisture quantity in the mixture, 3) mass transport (intensity of mixing), 4) molar ratio of alcohol to vegetable oil, and 5) type of catalyst (Korus, 1993; Mamilla et al., 2012).

A detailed study of chemical kinetics is important in optimizing the yield of the reaction and reaction time. Unlike diesel produced from crude oil, very limited kinetic data are available on biodiesel produced from vegetable oils. Darnoko and Cheryan (2000) were amongst the pioneers who developed the chemical kinetics for the 3-step transesterification of palm oil. The study was followed by a series of experimental investigations aimed at determining the impact of the catalyst type, temperature, and alcohol to oil ratio on the overall yield of the process.

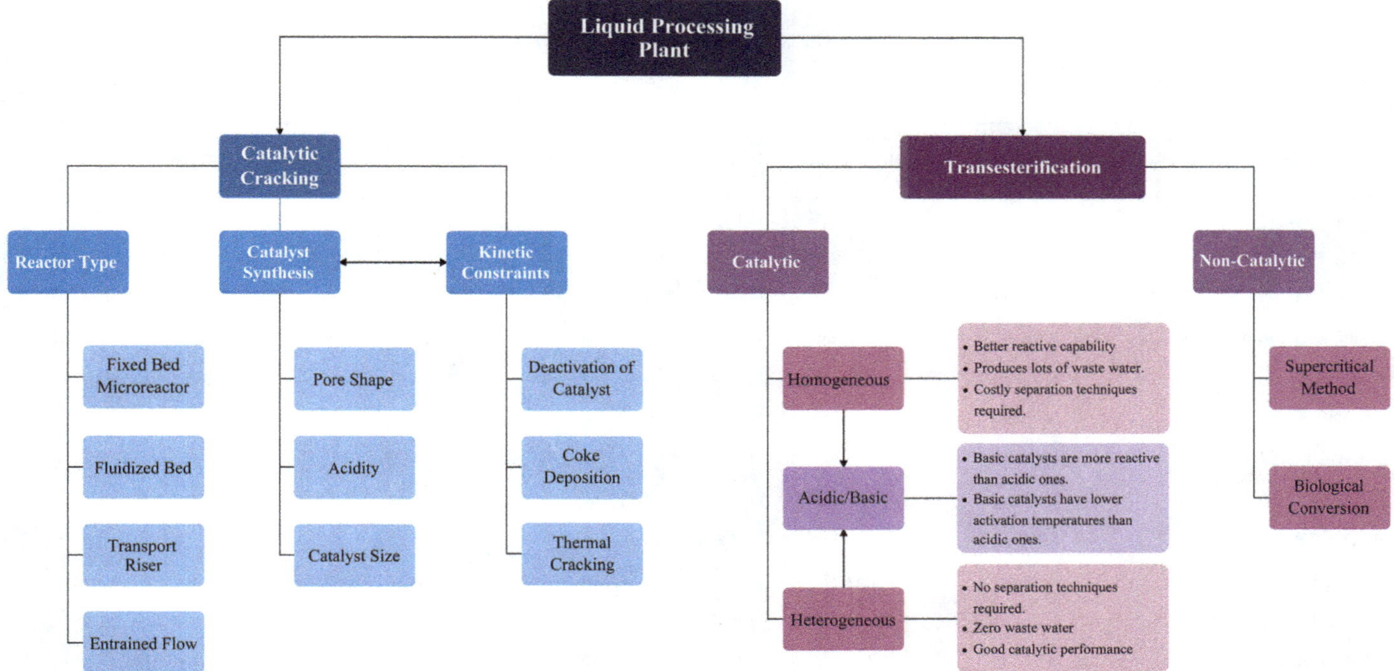


**Fig.2.** Overview of different processing technologies used for biofuel production from palm oil.

- *Optimization studies*

**Table 1** presents a comparative overview of different experimental studies dedicated to biodiesel production from palm oil. Most of these studies used methanol as acid in the transesterfication reaction since the physical and chemical properties of methyl esters are very close to those of petro-diesel. All of the above-mentioned studies have attempted to optimize the impacts of different factors such as catalyst loading, alcohol to oil ratio, reactor temperature, reaction time, and type of catalyst used on biodiesel production. Several studies have also considered the impact of different process variables involving preparation of the catalyst but it is beyond the scope of this review and hence not discussed further (Kansedo et al., 2009; Chen et al., 2015)

Since the yield of palm oil transesterification process is dependent upon a wide variety of parameters pertaining to reactor system configuration, catalytic synthesis. and operating conditions, optimization of all these parameters is not straightforward. Hence, researchers have used different statistical techniques such as Taguchi method (Chongkhong et al., 2007) and Response Surface Methodology (RSM) (Mootabadi and Abdullah, 2015) to obtain the optimum values of the kinetic parameters leading to the maximum yield of fatty acid methanol ester (FAME) was. In a recent study, Mootabadi and Abdullah (2015) used the RSM to optimize an ultrasound-assisted transesterification process.

- *Effect of catalyst type on transesterification*

From the studies tabulated in **Table 1**, it can be inferred that process yield depends on catalyst type, reaction conditions, and catalyst treatment parameters. Different classes of catalysts have been utilized to determine the optimum reaction conditions for biodiesel production from palm oil. Traditionally, homogenous base catalysts are used due to their high catalytic activity and wide availability. However, their use limits the overall yield and reusability of the catalyst since they require additional processing for separation. The downstream process used to wash away the base catalyst residues (typically NaOH or KOH) also results in lots of toxic waste (Ma and Hanna, 1999; Gao et al., 2008; Noiroj et al., 2009). An additional disadvantage is that the homogeneous base catalysts are not suitable for raw materials containing high water and free fatty acids contents (Gao et al., 2008). These factors resulted in a lookout for a new and improved catalytic technology which can remove the additional purification and separation process without compromising the overall yield. It is worth mentioning that although there are some studies in which using acid based catalyst was investigated (Al-Widyan and Al-Shyoukh, 2002; Chongkhong et al., 2007), but due to the low catalytic activity of these catalysts as well as their requirement for high reactor temperatures, their use has been discouraged. So, in most of the biodiesel production studies from palm and other vegetable oils, basic catalysts are still used because of their enhanced catalytic reactivity.

Given the above-mentioned disadvantages of the homogeneous catalyst, the use of heterogeneous (mostly solid) catalysts for the transesterification process of palm oil was adopted (Bo et al., 2007; Kawashima et al., 2008; Gao et al., 2008; Kansedo et al., 2009; Chen et al., 2015). Although the use of solid catalysts simplified the process, however, they have several technological shortcomings. To address these challenges over the last decade, the research in the area of catalytic development has been steered towards investigating the impacts of different combinations of metallic oxides and zeolites along with analysing catalyst synthesis techniques and conditions on the process yield and durability of the catalyst. Kawashima et al. (2008), for instance, analysed the catalytic performance of a wide variety of metallic oxides including Calcium, Magnesium, Barium, and Lanthanum. They concluded that oxides of Calcium enhanced catalytic performance compared with the other metallic oxides investigated. They attributed their findings to the surface structure of the catalyst, i.e., favourable porosity and basicity compared with the other metallic oxides. Other commonly used metal oxides reported in the literature are NaO and $TiO_2$ (Kawashima et al., 2008).

- *Use of renewable resources to synthesize catalysts*

Due to the increasing attention towards the use of renewable resources to meet our energy demands, the research in this area has also been recently shifted towards developing catalysts from renewable resources and enhancing the reusability of the catalysts while maintaining the yield.

**Table 1.**
Optimum parameters for different transesterification studies.

| Catalyst Used | Alcohol Used | Optimum Catalyst Loading (%; wt./wt. oil) | Optimum Alcohol to Oil Molar Ratio | Reactor Temperature (°C) | Optimum Reaction Time (h) | Yield (%) | Reference |
|---|---|---|---|---|---|---|---|
| Rice husk ash | Methanol | 7 | 9 to 1 | 65 | 4 | 91.5 | Chen et al. (2015) |
| CaO | Methanol | 5 | 6 to 1 | 65 | 3 | 93.2 | Chen et al. (2015) |
| $KF/Al_2O_3$ | Methanol | 4 | 12 to 1 | 65 | 3 | 90 | Bo et al. (2007) |
| 13 different metal oxides of Ca, Ba, Mg, La | Methanol | - | 6 to 1 | 60 | 10 | 79-92 | Kawashima et al. (2008) |
| KF/Hydrotalcite | Methanol | 3 | 12 to 1 | 65 | 3 | 90 | Gao et al. (2008) |
| CaO/ZnO | Methanol | 10 | 30 to 1 | 60 | 1 | 94 | Ngamcharussrivichai et al. (2008) |
| $KOH/Al_2O_3$ | Methanol | 25 | 15 to 1 | 60 | 2 | 91.07 | Noiroj et al. (2009) |
| KOH/NaY | Methanol | 10 | 15 to 1 | 60 | 3 | 91.07 | Noiroj et al. (2009) |
| $H_2SO_4$ and HCl | Ethanol | - | 100 % excess ethanol | 90 | 3 | - | Al-Widyan and Al-Shyoukh (2002) |
| NaOH | Methanol | 0.38 | 5 to 1 | 60 | 3 to 4 | 92 | Mamilla et al. (2012) |
| $SO4^{2-}/ZrO_2$ | Methanol | 1 | 6 to 1 | 200 | 1 | 90.3 | Jitputti et al. (2006) |
| Montmorillonite KSF | Methanol | 3 | 8 to 1 | 190 | 3 | 79.6 | Kansedo et al. (2009) |
| $H_2SO_4$ | Methanol | 1.834 | 4.3 to 1 | 70 | 1 | 93.9 | Chongkhong et al. (2007) |

**Table 2.**
Reactor type and yield comparison for catalytic cracking of palm oil.

| Reactor Type | Catalyst Used | Operating Temperature (°C) | Conversion of Palm Oil (%) | Yield (product) (%) | Reference |
|---|---|---|---|---|---|
| Batch reactor | $Na_2CO_3$ | 450 | - | 65.86 (Organic Liquid Products) | Da Mota et al. (2014) |
| Fixed bed micro-reactor | Various zeolite catalysts | 350-450 | 99 | 28 (Gasoline) | Twaiq et al. (1999) |
| Transport riser reactor | Zeolite REY | 450 | 74.9 | 59.1 (Gasoline) | Tamunaidu and Bhatia (2007) |
| Fixed bed micro-reactor | HZSM-5 (microporous) MCM-41 (mesoporous) | 450 | 99 | 48% (Gasoline) | Sang (2003) |
| - | $V_2O_5$, $MoO_3$, ZnO, $CO_3O_4$, $ZnCl_2$ | 320 | 77.6 | 33.62% (Gasoline) | Yigezu and Muthukumar (2014) |
| Fixed bed micro-reactor | Nanocrystalline zeolite beta and zeolite Y | 450 | 84 | 53% (OLP), 35% (Gasoline) | Taufiqurrahmi et al. (2010) |

Chen et al. (2003) demonstrated the use of renewable resources such as rice husk ash in synthesizing the CaO catalyst for biodiesel production from palm oil. Moreover, Shan et al. (2015) investigated the impact of sodium poly styrenesulfonate induced mineralization and calcination of CaO on the enhancement of the catalytic activity and reusability of the catalyst. In their study, not only the trasesterification yield of palm oil was improved, the reusability was also enhanced as compared with the traditional CaO catalyst. In another study by Wong et al. (2015), a biodiesel yield of 95% was achieved by using a combination of Calcium and Cerium oxides. Moreover, the synthesized catalyst could be reused 6 times without significant losses in the yield.

- *Non-catalytic approaches to transesterification*

Besides the catalytic approach, other methods have also been studied to carry out transesterification reaction of palm oil effectively. One of such novel approaches is using supercritical methanol for the alcoholysis of the palm oil (Joelianingsih et al., 2008; Song et al., 2008). This approach can significantly reduce the reaction time and also eliminate any complex pre- and post-treatment steps for the catalyst and the reaction mixture, respectively. Supercritical transesterification uses methanol in a supercritical state (high temperature and high pressure) to react with the triglycerides. This results in very fast reaction kinetics and a high final yield of biodiesel. Furthermore, as mentioned earlier, it simplifies the overall process by eliminating pretreatment, soap removal, and catalyst removal processes altogether (van Kasteren and Nisworo, 2007). One of the main disadvantages of this process is the harsh reaction conditions required (i.e., high temperature and pressure) which complicate the reactor design.

### *3.1.2. Catalytic cracking*

One of the main disadvantages of using biodiesel as fuel is that it generally cannot be used in its pure form in engines, gas turbines, etc. and it needs to be blended with petro-diesel. Furthermore, the technology is still not economically competitive with the petro-diesel (Pool, 2014). Unlike transesterification, catalytic cracking is a mature technology since it has been being used to convert crude oil into useful olefins and paraffins for almost a century now. Another obvious advantage is that cracking of oil yields gasoline, diesel, and kerosene directly. Catalytic cracking of palm oil to obtain bio-gasoline has been a subject of many studies in the literature (Chew and Bhatia, 2008). The process involves breaking the heavier chains of fatty acids contained in palm oil into lighter and more useful products such as olefins, paraffins, ketones, and aldehydes. According to the literature, the important factors affecting the catalytic cracking process are, 1) type of reactor, 2) catalyst synthesis, and 3) reaction conditions (temperature, residence time, etc.) (Twaiq et al., 1999; Sang, 2003; Taufiqurrahmi et al., 2011). It should be mentioned that all the three factors are dependent upon each other making the optimization of the process complex.

- *Choice of reactor for catalytic cracking*

There are several key process variables and operational constraints which influence the choice and design of the reactor. Factors such as process chemistry, kinetics of the cracking process, deactivation of catalyst due to coke deposition, thermal cracking, cracking temperature, and adjustment of residence *vs.* contact time of the catalyst, etc. must be considered carefully when constructing the reactor (Avidan and Shinnar, 1990; Ong and Bhatia, 2010).

At laboratory scale, there are different reaction setups that have been analysed in the literature for catalytic cracking of palm oil. The most commonly used reaction systems are fixed bed, fluidized bed, transport-riser, and entrained flow reactors. Almost all of the afore-mentioned reactor types are designed for heterogeneous operation (i.e., solid-gas interface) (Miller and Jackson, 2004), aiming at enhancing the contact area between the solid catalyst and the liquid fuel. Most of the experimental setups carrying out fluid catalytic cracking (FCC) employ a fluidized bed system due to various reasons. Firstly, it allows for continuous operation of the reactor and ensures uniformity of the product. Secondly, if employed on a large scale, it lowers the production cost as compared with the other technologies (Ong and Bhatia, 2010).

The problems of coke deposition, residence, and contact time optimization are the major issues driving the design of chemical reactors. Coke deposition in particular is very detrimental since it significantly limits the catalytic activity and therefore, excessive regeneration of the catalyst will be required making the continuous production difficult (Bhatia et al., 2007; Chew and Bhatia, 2008). Hence, the kinetics of coke formation should be known and reactors should be designed in such a way so as to limit its production. Research in this area indicates that in order to achieve a trade-off between the gasoline yield and coke production, the reactor should be designed to have short contact times between the catalyst and the atomized fuel while operating at high temperatures (Tamunaidu and Bhatia, 2007; Kansedoet al., 2009; Taufiqurrahmi et al., 2010). For such an application, a transport riser reactor serves the purpose well since it allows for continuous operation while ensuring short contact times. This leads to lower coke deposition and in turn highest gasoline production amongst all other reactor types as demonstrated in **Table 2**.

- *Effect of catalyst*

The efficiency and economic feasibility of the FCC process is a strong function of type and synthesis of catalyst. Key properties that influence the catalytic activity in the cracking reaction are acidity, size, pore shape, and selectivity. **Table 2** compares the conversion and yield (mostly bio-gasoline) from the cracking of palm oil. A review of these studies indicates that zeolites are the most widely used catalysts for fluid cracking. In fact, a wide variety of studies in the literature assessed the performance of different zeolite catalysts such as HZSM-5, zeolite-β and

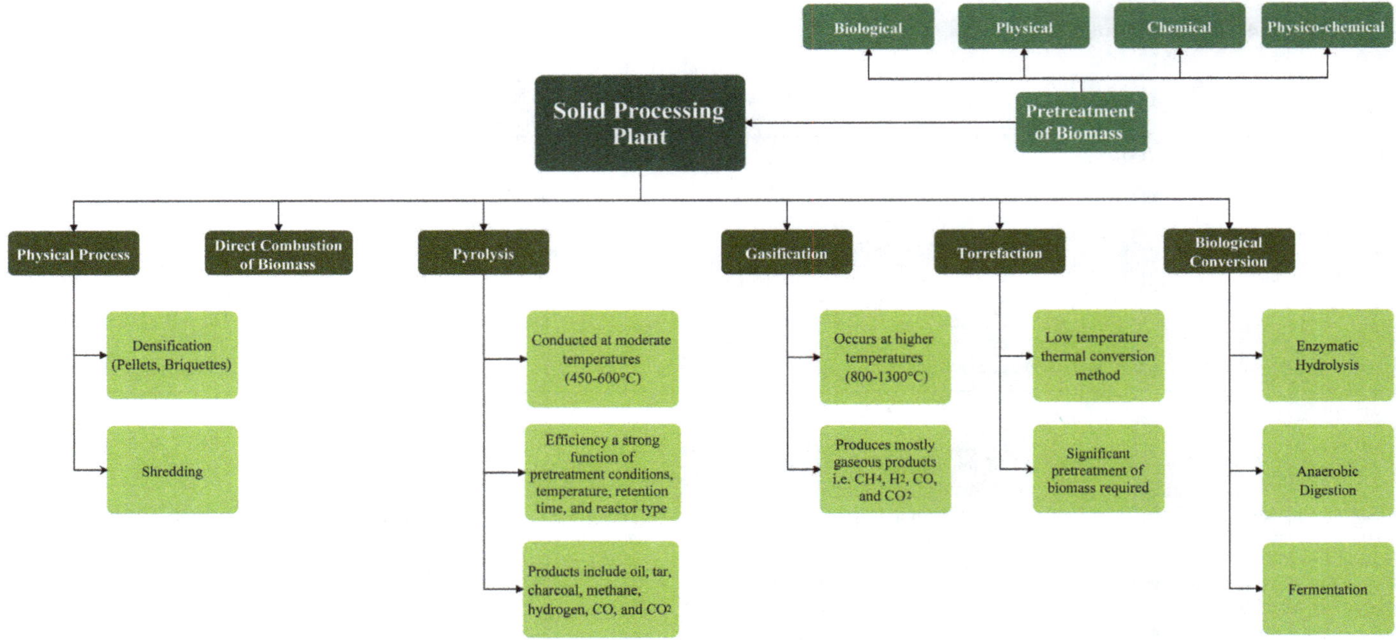


**Fig.3.** Overview of solid processing technologies for palm oil biomass.

USY, on overall palm oil conversion and gasoline yield (Adjaye et al., 1996; Leng et al., 1999; Twaiq et al., 1999). Amongst these catalysts, HZSM-5 reportedly led to the best results in terms of palm oil conversion, yield of bio-gasoline, and lower coke formation. The other zeolite catalyst enhanced the coke formation kinetics due to their bigger channel intersections. HZSM-5 on the other hand due to its higher acidity and shape selectivity yielded better results.

Apart from zeolites, another important class of catalysts used in the catalytic cracking of palm oil are microporous and mesoporous type such as CZM and MCM-41, etc. Sang (2003) analysed the impact of microporous, mesoporous, and composite (micromesoporous) catalysts on the overall conversion of palm oil and bio-gasoline yield. It was found that micromesoporous catalyst yielded best results in terms of the desired gasoline and palm oil conversion since it enhanced both acidity and pore size of the catalyst (Sang, 2003)

- *The state of the art and future avenues*

According to different comparative studies, palm oil has the highest oil yield and is the most economic source amongst all other vegetable oils. As elaborated, there are two major technologies by which palm oil is converted into biofuels namely transesterification and catalytic cracking.

The state of the art along with future directions regarding transesterification of palm oil can be summarized as follows:

- Catalytic transesterification is more technologically developed and hence widely used.
- Use of heterogeneous catalysts is encouraged for catalytic transesterification since they make the catalyst separation from the biodiesel easier preventing a lot of water from being wasted in the process.
- The current research in this area is directed towards enhancing the reactivity of the heterogeneous catalysts using novel pretreatment techniques. The techniques include mineralization with different metal.
- oxides and zeolites, temperature treatment, and combining oxides of different metals in optimum proportion.
- Use of renewable resources to synthesize catalysts and to improve their reusability is also a research area of interest.
- In the transesterification reaction, the use of supercritical methanol makes the overall process simple and improves the overall yield. However, the high pressure and high temperature conditions required for supercritical methanol transesterification require sophisticated reactor design and high energy input.

As for catalytic cracking process, the state of the art and the potential improvements in the future can be summarized as follows:

- Choice of reactor and catalyst synthesis are two major points of interests in catalytic cracking process.
- Transport-riser reactor is recommended for catalytic cracking process since it allows for continuous operation and reduces the coke deposition in the reactor.
- Future research in this area should aim for enhancing the catalyst regeneration through improved reactor design.
- Development of microporous, mesoporous, and micromesoporous catalysts to improve acidity and to optimize pore size is an active area of research.

### *3.2. Biofuels production using oil palm biomass*

As mentioned earlier, lots of biomass is produced as wastes from the production of palm oil. Originating from different parts of the palm tree such as empty fruit bunches, fronds, trunks, palm pressed fibres and palm shells, the lignocellulosic materials of the palm oil biomass have a lot of energy content which if utilized properly can meet a part of the present energy needs. In order to achieve this objective, several challenges associated with different biomass sources such as lower energy content and higher energy consumption for collection, difficulty in transport and uneven composition, etc., have to be addressed.

Biofuel production from oil palm biomass involves a wide range of methods. Most of these technologies involve turning the biomass into liquid-gaseous mixture form and then upgrading (decreasing water and oxygenated compounds contents) the liquid mixture to render it suitable for burning purposes. **Figure 3** gives an overview of different solid processing methods that have been employed for utilizing the energy of palm oil biomass. Methods such as gasification, torrefaction, pyrolysis, and direct combustion come under the category of thermochemical energy conversion methods. Fermentation, enzymatic hydrolysis, and anaerobic digestion are classified under biological conversion methods whereas densification and shedding come under physical processes. The thermochemical processes require more energy input than the other categories. However, in terms of the process yield and large scale production, thermochemical technologies are more suitable. Most of the current efforts are focused on thermochemical energy conversion domain since the resultant liquid products possess higher energy densities (Bridgwater and Bridge, 1991). Accordingly, these methods are discussed here in further detail.

### *3.2.1. Pyrolysis of palm oil biomass*

The process of pyrolysis involves converting the organic matters into bio oils by burning it with very little or no oxygen. The resulting products comprise of a wide variety of solid, gaseous, and liquid materials such as char, coke, bio-oil, $CO_2$, CO, $CH_4$, and $H_2$, etc. Amongst these products, bio-oil, $CH_4$, and $H_2$ have high heating values and thus, can be used as replacement of fossil fuels (Bridgwater and Bridge, 1991). Literature shows that the maximum yield of bio-oil can be obtained by operating the reactors at high heating rates and short gas residence times (Chen et al., 2003; Chew and Bhatia, 2008). Furthermore, the maximum yield is also dependent upon the cellulosic content of the palm oil biomass and the type of reactor (Demirbaş, 2001). The pyrolysis process itself does not require any catalysts. However, catalysts play a very important role in deoxygenation and upgrading of the liquid fuel obtained by the pyrolysis. This ultimately impacts the fuel quality of the resulting biofuels such as bio-oil, $H_2$, and $CH_4$, etc.

**Table 3** presents a comparison of the reactor technology, yield of bio-oil, and pretreatment conditions employed by different palm oil biomass studies. Most of the studies shown in the table only reported fast pyrolysis reaction of biomass without catalytic upgrading of the resultant bio-oil. The highest yield was reported for the biomass of the EFB conducted by Asadieraghi and Daud (2015). This could be attributed to the high volatile content of the EFB biomass.

hydrogen. Furthermore, using Fischer tropsch synthesis, the synthesis gas obtained from direct gasification can further be processed to yield transportation fuels like bio-gasoline, diesel, naphta, etc. (Chew and Bhatia, 2008). However, the further processing of the synthesis gas requires effective catalytic systems and reactor designs. Accordingly, a wide number of studies have examined the conversion efficiency of various combination of catalysts, such as zeolites, metallic oxides, microporous and mesoporous surfaces, etc., as well as different reactor designs (Kelly-Yong et al., 2007; Lahijani and Zainal, 2011; Mohammed et al., 2011). Lahijani and Zainal (2011) in their study achieved a carbon conversion efficiency of 93% and 85% for EFB and sawdust biomass, respectively, in a bubbling fluidized bed reactor setup. For EFB, particle agglomeration was a major problem at higher reactor temperatures (>1000 °C). However, for saw dust biomass, agglomeration never occurred even at high temperatures. Chew and Bhatia (2008) investigated different types of catalysts used in the process of higher alcohol synthesis (HAS) from biomass. The HAS process was operated at 250°C - 425°C and 30 - 330 bar, depending upon the catalyst type. Different metal oxides and alkaline catalysts were tested and their operating conditions along with final product types were reported. Thy found out that metal catalyst were prone to deactivation in case of sulphur impurities in the water gas. Amongst the different metallic catalysts tried, such as Rhodium, $MoS_2$, $Cr_2O_3$, ZnO etc., no clear 'leader' in terms of catalytic performance was found. Most of the catalysts suffered the problem of different types of undesirable alcohols as opposed to only higher alcohol which is desired product of the process. That is why novel catalytic synthesis techniques are an active area of research in this area.

Hydrogen production from the gasification of palm oil biomass is one of the most sought after technologies in this domain. Supercritical water gasification technology is amongst the latest gasification techniques to produce hydrogen. The cost of hydrogen production from supercritical water gasification is reported to be the least amongst the different pyrolysis and gasification technologies since it utilizes high moisture content biomass without the expensive pre-processing (Matsumura, 2002; Shuit et al., 2009). The process, however, produces fermentation sludge which is difficult to deal with. Nevertheless, there are still lots of room for improving the overall efficiency of this process.

**Table 3.**
Overview of pyrolysis technologies for palm oil biomass.

| Reactor Type | Catalyst Used | Pretreatment of Biomass | Biomass Used | Yield of Bio-oil | Reference |
|---|---|---|---|---|---|
| Fixed and stirred bed | - | - | Oil palm shell | - | Salema and Ani (2011) |
| Fixed bed reactor | - | Acid washed red mud with biomass | Empty fruit bunch | 52% | Lim et al. (2014) |
| Fixed bed reactor | Catalyzed by minerals in biomass itself | - | Empty Fruit Bunch (EFB), Palm Kernel Shell (PKS), Palm Meso carp Fiber (PMF) | 58.2 % for EFB, 49.8% for PKS and 53.1% for PMF | Asadieraghi and Daud (2015) |
| Fixed bed reactor | - | - | Trunk, Frond, palm leaf and palm leaf rib | 40.87% (Trunk), 43.50% (Frond), 16.58% (Palm leaf) and Palm leaf rib (29.02%) | Abnisa et al. (2011) |
| Fluidized bed reactor | - | Grinding, sieving and oven-drying of biomass | Oil palm shell | 58% (at 500 °C) | Islam et al. (1999) |
| Turbular reactor | - | Grinded and screened | Palm shell waste | 46.40% | Abnisa et al. (2011) |

### *3.2.2. Gasification*

Gasification as opposed to pyrolysis occurs in the presence of oxygen and under high temperatures in the range of 800-1300 °C. Unlike pyrolysis, the oxidation of biomass results in gaseous and solid products such as charcoal, water gas, and $CO_2$ (Geng, 2013). Gasification process has several advantages such as high thermal efficiency, availability of well developed equipment, and reduced emissions. However, the process has some drawbacks which need to be addressed in order to improve the biofuel yield. Minimizing energy content and improving reactor design are key areas which require further attention of the researchers in this field.

As far as the palm oil industry is concerned, biomass from all sources such as EFB, PKS, PMF, fronds, leaves, etc. can be burned in a gasifier to produce

### *3.2.3. Other Technologies*

Besides pyrolysis and gasification, there are other thermochemical, physical and biological technologies which can be employed to utilize the biomass. Torrefaction in conjunction with gasification or co-firing can improve the process yield from palm oil biomass significantly. Torrefaction is a slow roasting process carried out at 200-300 °C which destroys the fibrous structure of the biomass while enhancing its calorific value (van der Stelt et al., 2011; Sabil et al , 2013). The torrefaction process efficiency is a strong function of chemical composition of the biomass and decomposition temperature. Torrefaction, although enhances the overall yield of the energy conversion process, comes with economic disadvantages as it increases the overall cost of the biomass pretreatment process.

Bioconversion methods involve the break-down of cellulose and hemicellulose structure of the biomass into fermentable sugars using enzymes. This 'enzymatic hydrolysis' process is followed by fermentation which produces bioethanol, biogas, and biobutanol. Numerous studies have investigated different pretreatment methods, such as acid/alkali, steam, etc., for enhancing the digestibility of the EFB biomass (Han et al., 2011; Jung et al., 2011; Shamsudin et al., 2012). The results have shown considerable improvements in overall yield of useful products such as bioethanol. Although the bioconversion process is less energy intensive as opposed to the thermochemical energy conversion processes, however, the process yield is significantly less compared with the pyrolysis and gasification methods. That is why it has still a long way to go in terms of large scale production of biofuels.

Another method to utilize the energy of palm oil biomass is by compacting the high energy content areas of biomass through physical processes. It has been reported that the densification of the biomass enhances the material handling and combustion property. The EFB biomass in terms of dust and powder can be transformed into briquettes under high pressure and temperature. Several studies have reported the mechanical and combustion properties of the briquettes and pellets made from such biomass (Husain et al., 2002; Nasrin et al., 2008). This method however, has less energy conversion efficiency as opposed to thermochemical and bioconversion methods.

### *3.2.4 Future prospects of oil palm biomass conversion technologies*

Moisture content plays an important role in determining the energy conversion process to be used. Higher moisture content favours the use of biochemical methods such as fermentation and enzymatic hydrolysis. However, if the moisture content is lower, thermochemical methods such as pyrolysis will be more suitable for the biomass (Asadieraghi and Daud, 2015).

Thermochemical conversion of oil palm biomass is still a developing area which has a lot of room for improvements in terms of pretreatment of oil palm biomass, reactor technology for biooil production, catalytic upgrading of the resultant biooil, and hydrogen production. Although some novel approaches such as microwave-induced pyrolysis and plasma-induced pyrolysis/gasification have been introduced (Wan et al., 2009; Salema and Ani, 2011; Salema and Ani, 2012), however, their yields are yet to become competitive with those of the traditional gasification and pyrolysis processes executed through conventional methods. Also the moisture and oxygen contents present in the bio-oil lower its heating value. This problem can be addressed by refining the catalytic cracking process through reactor design and improved catalyst synthesis.

Plasma-induced gasification is a novel technique to utilize the solid waste from palm oil in an environmentally-friendly manner. Standard gasification operates at a lower temperature and produces tars and other contaminants which need to be removed. Plasma-induced gasification on the other hand converts most of the carbon into fuel since it uses an external heat source resulting in little combustion (Mountouris et al., 2006). It should be mentioned that although this technique is not currently being employed for gasification of palm oil biomass, the authors believe that this technology is worth investigating for energy conversion of oil palm biomass. Furthermore, the use of renewable resources to synthesize catalysts for the pyrolysis and gasification reactions which can enhance the regeneration of the catalysts is also a potential area of future investigations in this area.

## 4. Life cycle assessment and technoeconomics aspects

In recent years, the sustainability issue almost in all sectors has received a lot of attention worldwide. The terms "sustainability" and "sustainable development" have been defined in different ways by different researchers. Although sustainability can cover several aspects, the important goals are to minimize the use of natural resources, production of toxic materials, emissions of hazardous pollutants, and to improve energy efficiency, economic growth, and social standards. Environmental sustainability assessment, which involves evaluating major environmental impacts of a production process throughout the life cycle of the product is carried out using different tools, one commonly used tool is life cycle assessment or analysis (LCA). It is an environmental assessment tool used to evaluate and quantify the impacts of a product over its life cycle (which includes extraction of raw material, processing, product supply, recycling, etc). The acceptance and reliability of the LCA depends on several factors among which the important one is the selection of the system boundary which defines which production processes are included in or excluded from the analysis.

As pointed out earlier, biofuels do have a main impact on food security, water quality, biodiversity, and the environment. The extent of these impacts depends on what raw materials are selected for biofuel production, the plantation and harvesting of the raw material, the synthesis route, and the methods used to supply the produced biofuels. More information on the sustainability and its importance in the biofuel sector could be found elsewhere (Lee and Ofori-Boateng, 2013). There are a number of studies carried out to understand the sustainability of palm-based biofuels produced using a variety of raw materials and synthesis routes. Some researchers have also compared the sustainability of palm-based biofuels with those obtained using other raw materials. This section discusses selected sustainability studies on oil palm biofuels mainly using LCA tools.

Mukherjee and Sovacool (2014) provided a concise review of palm-oil based biofuels in Indonesia, Malaysia, and Thailand as well as also some information on the sustainability implications of palm-oil based biofuels in the Southeast Asia region. The review provides a detailed analysis of the environmental, ecological and socio-economic considerations. They finally recommended three policies which include implementation of standards for oil palm plantation to address the environmental sustainability, recognition and revision of the traditional land use rights and support, and finally, encouragement for the development of new biofuel technologies that uses different feedstock with improved energy efficiency of processes to avoid sole dependence on palm oil-based biofuels. Chiew and Shimada (2013) carried out an interesting study on the environmental impacts of utilizing oil palm EFBs for various applications such as fuel, fiber, and fertilizer. They reported that the technology with the least emissions was composting while the emissions associated with fuel production was comparatively higher. However, they reported that the most favourable technology based on the product was combined heat and power system (Chiew and Shimada, 2013). Johari et al. (2015) in their review pointed out the challenges and prospects of palm oil-based biodiesel in Malaysia. The production sustainability was highlighted as one of the most important factors in the use of palm oil based biofuel. They did conclude that further research is needed to improve the sustainability of biodiesel and to improve the socioeconomic aspects of Malaysian biodiesel.

Yusoff and Hansen (2007) investigated the feasibility of performing LCA on crude palm oil production. Their LCA analysis included three steps of plantation, transportation, and milling of biomass as the most significant steps according to the authors. Based on their analysis and the eco-indicators calculated, they pointed out that the most important aspect concerning the environmental impact was the way the land was prepared for plantation, i.e., burning used as the easiest way which. The transportation and milling also had considerable impacts on the environment, less severe than the plantation though. Yusoff and Hansen (2007) also provided suggestions to improve the sustainability of the palm oil industry such as compulsory use of LCA tool for environment assessment, incentives for introduction of cleaner technologird, and execuation of the LCA on plantation land use in Malaysia. In a different study, Peng (2015) carried out a comparison of the exhaust emissions using three types of biodiesels with the pure petro-diesel fuel. This experimental study was carried out using a water cooled diesel engine. The results showed the fuel consumption was higher for all biodiesels compared with the petro-diesel. However, the CO, hydrocarbons, and smoke emissions were much lower for all the biodiesels compared with petro-diesel.

Although the palm based biodiesel is mainly produced in the Southeast Asia region, there are a number of research articles providing a perspective on the other oil-palm growing countries such as Brazil. Queiroz et al. (2012) carried out the LCA of palm oil biodiesel in the Amazon. The analysis was carried out for the three phases of plantation, oil production, and biodiesel production using transesterification reaction.

Based on their energy performance study, it was suggested that all the three phases could be potentially improved. The most energy intensive phase was found to be the plantation. This observation was similar to what reported by the other researchers. For instance, de Souza et al. (2010) also carried out greenhouse gas (GHG) emission study of palm oil biofuel and also concluded the agriculture (plantation) to be the phase with the highest GHG emissions.

On the other hand, it has been reported that the palm-oil based biofuels produced using the traditional processes are not acceptable because of the certain unfavourable properties such as high viscosity. There have been several attempts to improve such properties; one of them was the use of microemulsion fuels. It has been reported that the microemulsion based biofuels had favourable combustion performance compared to petro-fuels resulting in lower exhaust emissions (Arpornpong et al., 2015). Arpornpong et al. (2015) carried out a comparative LCA study of microencapsulation-based biofuel produced from palm oil-diesel blends with ethanol with neat biodiesel and biodiesel-diesel blend. The LCA analysis was divided into five stages which included cultivation, palm oil production, microemulsion stage, transportation, and exhaust emissions of the fuel application stage. It was found that the microemulsion fuel production had the lowest impact on the environment except in terms of land use and fossil depletion which were mainly the results of the use of surfactant for microemulsion. Another alternative method used was biodiesel production in supercritical alcohols as it has been found to generate only a traceable amount of waste and pure glycerol as a by-product; the details can be found elsewhere (de Boer and Bahri, 2011). Sawangkeaw et al. (2012) studied another novel process with supercritical alcohols using Hysys simulations and carried out the LCA analysis. It was shown that the novel process which was carried out at higher temperature (400 °C) than the previously-proposed biodiesel production in supercritical alcohols (carried out at 300 °C) generated lower environmental impacts.

In general the LCA analysis of palm-oil based biofuels obtained using various feedstock and processing routes suggest that the highest environmental impacts are attributed to the plantation stage.

## 5. Research and development needs

As discussed in the introduction section, the use of renewable fuels received a lot of attention in recent years for several reasons. The previous sections also provided information on the importance of oil palm and palm oil processing wastes in biofuel production while explaining several recent technologies for production of biofuel from oil palm-based raw materials. However, there are certain challenges which also provide opportunities for further research and development in this area.

A rapid depletion of crude oil reserves and fluctuating oil prices were always important reasons for adoption of other fuel options such as biofuels. However, considering the recent trends in oil prices (a continuous decrease in the oil prices from USD 105 per barrel in August 2014 to USD 25 per barrel in February 2016), it is difficult to justify the use of biofuel solely based on the cost considerations. Johari et al. (2015) showed that for palm biodiesel to breakeven, the crude oil price must be around USD 100 per barrel. Therefore, it will be challenging to come up with cost effective production routes for biofuels to compete with this reverse trend in the crude oil prices. The choice of feed is also important for the success of biofuel production as the price of biofuel is significantly affected by the cost of raw materials.

On the other hand, as discussed in the previous section, the use of oil palm and palm wastes for production of biofuels has a long term effect on the environment. It has been reported by several researchers that the use of palm feedstock for biofuel can result in several impacts on biodiversity as well. Therefore, A number of measures have to be taken especially about the plantation phase in the palm-based biodiesel production cycle. In line with that, detailed LCA analyses can provide useful information for the governments to take the necessary actions.

The sustainability analysis of the various production routes for palm-based biodiesel also shows that there is a need for more research and innovative ideas in order to reduce the environmental impacts and reduce energy consumption. The improvement of processes is also needed in order to produce biofuels with certain properties which are more compatible with the existing diesel engines, especially the properties such as viscosity, cetane number, and calorific value.

## 6. Concluding remarks

Production of biofuel from biomass wastes has received considerable attention worldwide amid the efforts to find alternative sustainable and environmently-friendly energy resources. Among the sources of biomass, palm oil industry is one promising source as it generates a huge quantity of biomass residues which are currently underutilized. Despite the efforts devoted to maximizing the utilization of biomass potentials in oil palm plantations and mills, the progress is slow. This slow development is mainly attributed to the remote location of palm oil plantations and mills making it difficult to transport and distribute the products (electricity, biogas, and biofuels) or the feedstock to produce biofuels from the plantation to the end-user. On the other hand, the main issue in utilizing the palm oil wastes to produce biofuels, is the unavailability of dedicated pipelines to collect the waste cooking oil.

On the production technology, various processes have been developed and evaluated in producing biofuels from palm oil and oil palm wastes. In general, these technologies can be classified into liquid processing technologies and solid processing technologies. The main issue with the current biofuel production scenario from palm oil wastes is the high moisture content limiting the energy conversion efficiency of thermochemical methods. Should the moisture content be lowered (by an energy efficient drying method), thermochemical methods such as pyrolysis/gasification can be applied effectively. One issue in the utilization of pyrolysis/gasification method is that these methods produce tars and other contaminants which need to be removed. Therefore, some novel thermochemical conversion techniques have been proposed and evaluated such as microwave-induced pyrolisis and plasma-induced pyrolysis. The later, for example, converts most of the carbon into fuel since it uses an external heat source which results in little combustion. These technologies, however, require further investigations prior to their utilization in real life.

On top of the potentials and technologies of biofuel production from oil palm and palm oil wastes, LCA studies have been conducted to evaluate the sustainability aspects of the different scenarios, i.e., biofuels produced using a variety of raw materials and synthesis routes. These analyses revealed that in producing biofuel from oil palm and palm oil wastes, the plantation phase has the highest environmental impacts.

Finally, with the current drop of oil price, it is difficult to justify the use of biofuel solely based on the cost considerations. More specifically, it will be difficult to design cost effective production routes for biofuels to contend the current low prices of fossil fuels. Hence, more studies are required to (i) develop an efficient transport and distribution system to connect plantation, biofuel plants, and end users, (ii) design an efficient conversion method to produce biofuel which has no or minimum impacts on the environment, and (iii) implement the improvement gained from the LCA studies with the main goal to develop cost-effective, environmentally-friendly and profitable biofuel production from oil palm wastes.

## References


[1] Abdullah, M.A., Nazir, M.S., Wahjoedi, B.A., 2011. Development of value-added biomaterials from oil palm agro wastes. IPCBEE. 7, 32-35.

[2] Abdullah, N., Sulaiman, F., 2013. The oil palm wastes in Malaysia, in: Matovic, M.D. (Ed.), Biomass Now - Sustainable Growth and Use. InTech, Croatia, pp. 75-100.

[3] Abdullah, R., Wahid, M.B., 2010. World Palm Oil Supply, Demand, Price and Prospects: Focus on Malaysian and Indonesian Palm Oil Industry. Malaysian Palm Oil Board Press, Malaysia.

[4] Abnisa, F., Daud, W.W., Husin, W.N.W., Sahu, J.N., 2011. Utilization possibilities of palm shell as a source of biomass energy in Malaysia by producing bio-oil in pyrolysis process. Biomass Bioenergy. 35, 1863-1872.

[5] Abu Hassan, O.A., Ishida, M., Shukri, I.M., Tajuddin, Z.A., 1996. Oil-palm fronds as a roughage feed source for ruminants in Malaysia. ASPAC, Food and Fertilizer Technology Center.

[6] Adjaye, J.D., Katikaneni, S.P.R., Bakhshi, N.N., 1996. Catalytic conversion of a biofuel to hydrocarbons: effect of mixtures of

HZSM-5 and silica-alumina catalysts on product distribution. Fuel Process. Technol. 48(2), 115-143.

[7] Ahmad, A., Ismail S., Bhatia, S., 2003. Water recycling from palm oil mill effluent (POME) using membrane technology. Desalination. 157, 87-95.

[8] Akia, M., Yazdani, F., Motaee, E., Han, D., Arandiyan, H., 2014. A review on conversion of biomass to biofuel by nanocatalysts. Biofuel Res. J. 1, 16-25.

[9] Al-Widyan, M.I., Al-Shyoukh, A.O., 2002. Experimental evaluation of the transesterification of waste palm oil into biodiesel. Bioresour Technol. 85(3), 253-256.

[10] Amin, N.A.S., Misson, M., Wan Omar, W.N.N. Misi, S.E.E., Haron, R., Kamaruddin, M.F.A., 2007. Biofuel from oil palm and palm oil residues, in: Manan, Z.A., Nasef, M.M., Setapar, S.H. (Eds.), Advances in Chemical Engineering. Penerbit UTM, Johor, pp. 41-66.

[11] Arpornpong, N., Sabatini, D.A., Khaodhiar, S., Charoensaeng, A., 2015. Life cycle assessment of palm oil microemulsion-based biofuel. Int. J. Life Cycle Ass. 20, 913-926.

[12] Asadieraghi M., Daud W.M.A.W., 2015. In-depth investigation on thermochemical characteristics of palm oil biomasses as potential biofuel sources. J. Anal. Appl. Pyrol. 115, 379-391.

[13] Avidan, A.A., Shinnar, R., 1990. Development of catalytic cracking technology. A lesson in chemical reactor design. Ind. Eng. Chem. Res. 29(6), 931-942.

[14] Awalludin, M.F., Sulaiman, O., Hashim, R., Wan Nadhari, W.N.A., 2015. An overview of the oil palm industry in Malaysia and its waste utilization through thermochemical conversion, specifically via liquefaction. Renew. Sust. Energ. Rev. 50, 1469-1484.

[15] Balat, M., Balat, M., Kirtay, E., Balat, H., 2009. Main routes for the thermo-conversion of biomass into fuels and chemicals. Part 1: Pyrolysis systems. Energy Convers. Manage. 50(12), 3147-3157.

[16] Baratieri, M., Baggio, P., Fiori, L., Grigiante, M., 2008. Biomass as an energy source: thermodynamic constraint on the performance of the conversion process. Bioresour. Technol. 99(15), 7063-7073.

[17] Basiron, Y., 2007. Oil palm production through sustainable plantations. Eur. J. Lipid Sci. Tech. 109, 289-295.

[18] Basiron, Y., Chan, K.W., 2004. The oil palm and its sustainability. J. Oil Palm Res. 16, 87-93.

[19] Bhatia, S., Leng, C.T., Tamunaidu, P., 2007. Modeling and Simulation of Transport Riser Reactor for Catalytic Cracking of Palm Oil for the Production of Biofuels. Energy Fuels. 21(6), 3076-3083.

[20] Bo, X., Guomin, X., Lingfeng, C., Ruiping, W., Lijing G., 2007. Transesterification of palm oil with methanol to biodiesel over a $KF/Al_2O_3$ heterogeneous base catalyst. Energy Fuels. 21(6), 3109-3112.

[21] Bridgwater, A.V., Bridge, S.A., 1991. A Review of Biomass Pyrolysis and Pyrolysis Technologies, in: Bridgwater, A.V., Grassi, G., (Eds), Biomass Pyrolysis Liquids Upgrading and Utilization, The Netherlands, Springer, pp. 11-92.

[22] Chen, G., Andries, J., Spliethoff, H., 2003. Catalytic pyrolysis of biomass for hydrogen rich fuel gas production. Energy Convers. Manage. 44(14), 2289-2296.

[23] Chen, G.Y., Shan, R., Shi, J.F., Yan, B.B., 2015. Transesterification of palm oil to biodiesel using rice husk ash-based catalysts. Fuel Process. Technol. 133, 8-13.

[24] Chew, T.L., Bhatia, S., 2008. Catalytic processes towards the production of biofuels in a palm oil and oil palm biomass-based biorefinery. Bioresour. Technol. 99(17), 7911-7922.

[25] Chiew, Y.L., Shimada, S., 2013. Current state and environmental impact assessment for utilizing oil palm empty fruit bunches for fuel, fiber and fertilizer – A case study of Malaysia. Biomass Bioenergy. 51, 109-124.

[26] Chongkhong, S., Tongurai C., Chetpattananondh, P., Bunyakan, C., 2007. Biodiesel production by esterification of palm fatty acid distillate. Biomass Bioenergy. 31(8), 563-568.

[27] Da Mota, S.A.P., Mancio, A.A., Lhamas, D.E.L., de Abreu, D.H., da Silva, M.S., dos Santos, W.G., de Castro, D.A.R., de Oliveira, R.M., Araújo, M.E., Borges, L.E., Machado, N.T., 2014. Production of green diesel by thermal catalytic cracking of crude palm oil (*Elaeis guineensis* Jacq) in a pilot plant. J. Anal. Appl. Pyrolysis. 110, 1-11.

[28] Darnoko, D., Cheryan, M., 2000. Kinetics of palm oil transesterification in a batch reactor. J Am. Oil Chem. Soc. 77(12), 1263-1267.

[29] de Boer, K., Bahri, P.A., 2011. Supercritical methanol for fatty acid and methyl ester production: a review. Biomass Bioenergy. 35, 983-991.

[30] de Souza, S.P., Pacca, S., de Avila, M.T., Borges, J.L.B., 2010. Greenhouse gas emissions and energy balance of palm oil biofuel. Renew. Energ. 35, 2552-2561.

[31] Demirbaş, A., 2005. Bioethanol from cellulosic materials: a renewable motor fuel from biomass. Energ. Source. 27(4), 327-337.

[32] Demirbaş, A., 2001. Biomass resource facilities and biomass conversion processing for fuels and chemicals. Energy Convers. Manage. 42(11), 1357-1378.

[33] Edem, D., 2002. Palm oil: biochemical, physiological, nutritional, hematologica, and toxicological aspects: A review. Plant Food Hum. Nutr. 57, 319-341.

[34] Gao, L., Bo, X., Guomin, X., Lv, J., 2008. Transesterification of palm oil with methanol to biodiesel over a KF/Hydrotalcite solid catalyst. Energy Fuels. 22(5), 3531-3535.

[35] Geng, A., 2013. Conversion of oil palm empty fruit bunch to biofuels, in Fang, Z. (Ed.), Liquid, Gaseous and Solid Biofuels - Conversion Techniques, InTech, Croatia, pp.479-490.

[36] Habib, M.A.B., Yusoff, F.M., Pahng, S.M., Ang, K.J., Mohamed, S., 1997. Nutritional values of chironomid larvae grown in palm oil mill effluent and algal culture. Aquaculture. 158, 95-105.

[37] Han, M., Kim, Y., Kim, S.W., Choi, G.W., 2011. High efficiency bioethanol production from OPEFB using pilot pretreatment reactor. J. Chem. Technol. Biotechnol. 86(12), 1527-1534.

[38] Huber, G.W., Corma, A., 2007. Synergies between bio- and oil refineries for the production of fuels from biomass. Angew. Chem. Int. Ed. 46(38), 7184-7201.

[39] Husain, Z., Zainac, Z., Abdullah, Z., 2002. Briquetting of palm fibre and shell from the processing of palm nuts to palm oil. Biomass Bioenergy. 22(6), 505-509.

[40] Islam, M.N., Zailani, R., Ani, F.N., 1999. Pyrolytic oil from fluidised bed pyrolysis of oil palm shell and itscharacterisation. Renew. Energ. 17(1), 73-84.

[41] Jitputti, J., Kitiyanan, B., Rangsunvigit, P., Bunyakiat, K., Attanatho, L. and Jenvanitpanjakul, P., 2006. Transesterification of crude palm kernel oil and crude coconut oil by different solid catalysts. Chem. Eng. J. 116, 61-66.

[42] Joelianingsih, Maeda, H., Hagiwara, S., Nabetani, H., Sagara, Y., Soerawidjaja, T.H., Tambunan, A.H., Abdullah, K., 2008. Biodiesel fuels from palm oil via the non-catalytic transesterification in a bubble column reactor at atmospheric pressure: a kinetic study. Renew. Energ. 33(7), 1629-1636.

[43] Johari, A., Nyakuma, B.B., Nor S.H.M., Mat, R., Hashim, H., Ahmad, A., Zakaria, Z.Y., Abdullah, T.A.T., 2015. The challenges and prospects of palm oil based biodiesel in Malaysia. Energy. 81, 255-261.

[44] Jung, Y.H., Kim, I.J., Han, J,I., Choi, I.G., Kim, K.H., 2011. Aqueous ammonia pretreatment of oil palm empty fruit bunches for ethanol production. Bioresour. Technol. 102(20), 9806-9809.

[45] Kansedo, J., Lee, K.T., Bhatia, S., 2009. Biodiesel production from palm oil via heterogeneous transesterification. Biomass Bioenergy. 33(2), 271-276.

[46] Kawashima, A., Matsubara, K., Honda, K., 2008. Development of heterogeneous base catalysts for biodiesel production. Bioresour. Technol. 99(9), 3439-3443.

[47] Kelly-Yong, T.L., Lee, K.T., Mohamed, A.R., Bhatia, S., 2007. Potential of hydrogen from oil palm biomass as a source of renewable energy worldwide. Energ. Policy. 35(11), 5692-5701.

[48] Koh, L.P., Wilcove, D.S., 2008. Is oil palm agriculture really destroying tropical biodiversity?. Conserv. Lett. 1, 60-64.

[49] Korus R.A., Hoffman, D.S., Bam, N., Peterson, C.L., Drown, D.C., 1993. Transesterification process to manufacture ethyl ester of rape oil. The Proceedings of the First Biomass Conference of the Americas: Energy, Environment, Agriculture, and Industry, Burlington, VT, 2, 815-822.

[50] Kulkarni, M.G., Dalai, A.K., 2006. Waste cooking oil – An economical source for biodiesel: a review. Ind. Eng. Chem. Res. 45, 2901-2913.

[51] Lahijani, P., Zainal, Z.A., 2011. Gasification of palm empty fruit bunch in a bubbling fluidized bed: a performance and agglomeration study. Bioresour. Technol. 102(2), 2068-2076.

[52] Law, K.N., Daud, W.R.W., Ghazali, A., 2007. Morphological and chemical nature of fiber strands of oil palm empty-fruit-bunch (OPEFB). Bioresources. 2(3), 351-362.

[53] Lee, K.T., Ofori-Boateng, C., 2013. Sustainability of Biofuel Production from Oil Palm Biomass, Singapore, Springer-Verlag Singapore.

[54] Leng, T.Y., Mohamed, A.R., Bhatia, S., 1999. Catalytic conversion of palm oil to fuels and chemicals. Can. J. Chem. Eng. 77(1), 156-162.

[55] Lim, X., Sanna, A., Andrésen, J.M., 2014. Influence of red mud impregnation on the pyrolysis of oil palm biomass-EFB. Fuel. 119, 259-265.

[56] Ma, A.N., 1996. Treatment of palm oil mill effluent, in: Singh, G., Lim, K.H., Leng, T., David, L.K. (Eds.), Oil palm and the environment: a Malaysian perspective, Malaysian Oil Palm Growers' Council, Malaysia, p. 113-123.

[57] Ma, F., Hanna, M.A., 1999. Biodiesel production: A review. Bioresour. Technol. 70(1), 1-15.

[58] Madaki, Y.S., Seng, L., 2013. Pollution control: How feasible is zero discharge concepts in Malaysia palm oil mills. Am. J. Eng. Res. 2, 239-252.

[59] Mahat, S.B.A., 2012. The palm oil industry from the perspective of sustainable development: A case study of Malaysia palm oil industry, Master Thesis, Ritsumeikan Asia Pacific University, Japan.

[60] Mahlia, T.M.I., Abdulmuin, M.Z., Alamsyah, T.M.I., Mukhlishien, D., 2001. An alternative energy source from palm waste industry for Malaysia and Indonesia. Energy Convers. Manage. 42, 2109-2118.

[61] Mamilla, V.R., Mallikarjun, M.V., Lakshmi N.R.G., 2012. Biodiesel production from palm oil by transesterification method. Int. J. Curr. Res. 4(8), 83-88.

[62] Marquevich, M., Czernik, S., Chornet, E., Montane, D., 1999. Hydrogen from biomass: steam reforming of model compounds of fast-pyrolysis oil. Energy Fuels. 13(6), 1160-1166.

[63] Matsumura, Y., 2002. Evaluation of supercritical water gasification and biomethanation for wet biomass utilization in Japan. Energy Convers. Manage. 43(9-12), 1301-1310.

[64] Mba, O.I., Dumont, M.J., Ngadi, M., 2015. Palm oil: Processing, characterization and utilization in the food industry – a review. Food Biosci. 10, 26-41.

[65] Miller, D.J., Jackson, J.E., 2004. Catalysis for Biorenewables Conversion. In: National Science Foundation Workshop Report.

[66] Misson, M., Haron, R., Kamaroddin, M.F.A., Amin, N.A.S., 2009. Pretreatment of empty palm fruit bunch for production of chemicals via catalytic pyrolysis. Bioresour. Technol. 100(11), 2867-2873.

[67] Mohamed, A.R., Zainuddin, N.F., Lee, K.T., Kamaruddin, A.H., 2006. Reactivity of adsorbent prepared from oil palm ash for flue gas desulfurization: effect of $SO_2$ concentration and reaction temperature. Stud. Surf. Sci. Catal. 159, 449-452.

[68] Mohammed, M.A.A., Salmiaton, A., Wan Azlina, W.A.K.G., Amran, M.S.M., Fakhru'l-Razi, A., 2011. Air gasification of empty fruit bunch for hydrogen-rich gas production in a fluidized-bed reactor. Energy Convers. Manage. 52(2), 1555-1561.

[69] Mootabadi, H., Abdullah, A.Z., 2015. Response Surface Methodology for Simulation of Ultrasonic-assisted Biodiesel Production Catalyzed by SrO/Al2O3 Catalyst. Energy Sources Part A. 37(16), 1747-1755.

[70] Mountouris, A., Voutsas, E., Tassios, D., 2006. Solid waste plasma gasification: Equilibrium model development and exergy analysis. Energy Convers. Manage. 47(13–14), 1723-1737.

[71] Muhrizal, S., Shamsuddin, J., Fauziah, I., Husni, M.A.H., 2006. Changes in iron-poor acid sulfate soil upon submergence. Geoderma. 131, 110-122.

[72] Mukherjee, I., Sovacool, B.K., 2014. Palm oil-based biofuels and sustainability in Southeast Asia: A review of Indonesia, Malaysia and Thailand. Renew. Sust. Energ. Rev. 37, 1-12.

[73] Naghshineh, M., Mirhosseini, H., 2010. Effect of frying condition on physicochemical properties of palm olein-olive oil blends. Int. J. Food. Agric. Environ. 8, 175-178.

[74] Nasrin, A.B., Ma, A.N., Choo, Y.M., Mohamad S., Rohaya, M.H., Azali, A., 2008. Oil palm biomass as potential substitution raw materials for commercial biomass briquettes production. Am. J. Appl. Sci. 5(3), 179-183.

[75] Ngamcharussrivichai, C., Totarat, P. and Bunyakiat, K., 2008. Ca and Zn mixed oxide as a heterogeneous base catalyst for transesterification of palm kernel oil. Appl. Catal. A. 341(1), 77-85.

[76] Noiroj, K., Intarapong, P., Luengnaruemitchai, A., Jai-in, S., 2009. A comparative study of $KOH/Al_2O_3$ and KOH/NaY catalysts for biodiesel production via transesterification from palm oil. Renew. Energ. 34(4), 1145-1150.

[77] Ong, Y.K., Bhatia, S., 2010. The current status and perspectives of biofuel production via catalytic cracking of edible and non-edible oils. Energy. 35(1), 111-119.

[78] Peng, D.X., 2015. Exhaust emission characteristics of various types of biofuels. Adv. Mech. Eng. 7(7), 1-7.

[79] Pool, R., 2014. The Nexus of Biofuels, Climate Change, and Human Health: Workshop Summary, Washington, DC, The National Academies Press.

[80] Pütün, E., Ateş, F., Pütün, A.E., 2008. Catalytic pyrolysis of biomass in inert and steam atmospheres. Fuel. 87(6), 815-824.

[81] Queiroz, A.G., Franca, L., Ponte, M.X., 2012. The life cycle assessment of biodiesel from palm oil ("dende") in the Amazon. Biomass Bioenergy. 36, 50-59.

[82] Rincon, B., Raposo, F., Dominguez, J.R., Millan, F., Jimenez, A.M., Martin, A., Borja, R., 2006. Kinetic models of an anaerobic bioreactor for restoring wastewater generated by industrial chickpea protein production. Int. Biodeterior. Biodegrad. 57, 114-120.

[83] Rupani, P.F., Singh, R.P., Ibrahim, M.H., Esa, N., 2010. Review of current palm oil mill effluent (POME) treatment methods: Vermicomposting as a sustainable practice World Appl. Sci. J. 11 (1), 70-81.

[84] Sabil, K. M., Aziz, M. A., Bhajan, L., Uemura, Y., 2013. Effects of torrefaction on the physiochemical properties of oil palm empty fruit bunches, mesocarp fiber and kernel shell. Biomass Bioenergy. 56, 351-360.

[85] Salema, A.A., Ani, F.N., 2011. Microwave induced pyrolysis of oil palm biomass. Bioresour. Technol. 102(3), 3388-3395.

[86] Salema, A.A., Ani, F.N., 2012. Microwave-assisted pyrolysis of oil palm shell biomass using an overhead stirrer. J. Anal. Appl. Pyrol. 96, 162-172.

[87] Salim, N., Hashim, R., Sulaiman, O., Ibrahim, M., Sato, M., Hirizoglu, S., 2012. Optimum manufacturing parameters for compressed lumber from oil palm (*Elaeis guineensis*) trunks: respond surface approach. Compos. Part B-Eng. 43, 988-996.

[88] Sang, O.Y., 2003. Biofuel Production from Catalytic Cracking of Palm Oil. Energ. Source. 25(9), 859-869.

[89] Sawangkeaw, R., Teeravitud, S., Piumsomboon, P., Ngamprasertsith, S., 2012. Biofuel production from crude palm oil with supercritical alcohols: Comparative LCA studies. Bioresour. Technol. 120, 6-12.

[90] Shamsudin, S., Shah, U.K.M., Zainudin, H., Abd-Aziz, S., Kamal, S.M.M., Shirai, Y., Hassan, M.A., 2012. Effect of steam pretreatment on oil palm empty fruit bunch for the production of sugars. Biomass Bioenergy. 36, 280-288.

[91] Shan, R., Chen, G., Yan, B., Shi, J., Liu, C., 2015. Porous CaO-based catalyst derived from PSS-induced mineralization for biodiesel production enhancement. Energy Convers. Manage. 106, 405-413.

[92] Sheil, D., Casson, A., Meijaard, E., van Noordwijk, M., Gaskell, J., Sunderland-Groves, J., Wertz, K., Kanninen, M., 2009. The impacts and opportunities of oil palm in Southeast Asia, Center for International Forestry Research (CIFOR), Indonesia.

[93] Shuit, S.H., Tan, K.T., Lee, K.T., Kamaruddin, A.H., 2009. Oil palm biomass as a sustainable energy source: A Malaysian case study. Energy. 34, 1225-1235.

[94] Sime Darby Plantation, 2014. Palm Oil Fact and Figures.

[95] Siregar, M.A., Sembiring, S.A., Ramli, 2014. The price of palm-cooking oil in Indonesia: Antecendents and consequences on the international price and the export volume of CPO. J. Econ. Sustain. Dev. 5, 227-234.

[96] Siyamak, S., Ibrahim, N.A., Abdolmohammadi, S., Wan Yunus, W.M.Z., Ab Rahman, M.Z., 2012. Enhancement of mechanical and thermal properties of oil palm empty fruit bunch fiber poly(butylene adipate-co-terephtalate) biocomposites by matrix esterification using succinic anhydride. Molecules. 17, 1969-1991.
[97] Song, E.S., Lim, J.W., Lee, H.S., Lee, Y.W., 2008. Transesterification of RBD palm oil using supercritical methanol. J. Supercrit. Fluids. 44(3), 356-363.
[98] Stier, R.F., 2013. Ensuring the health and safety of fried foods. Eur. J. Lipid Sci. Technol. 115(8), 956-964.
[99] Sulaiman A.A., Taha, F.F.F., 2015. Drying of oil palm fronds using concentrated solar thermal power. Appl. Mech. Mater. 599, 449-454.
[100] Sulaiman, F., Abdullah, N., 2011. Optimum conditions for maximising pyrolysis liquids of oil palm empty fruit bunches. Energy. 36(5), 2352-2359.
[101] Sumathi, S., Chai, S.P., Mohamed, A.R., 2008. Utilization of oil palm as a source of renewable energy in Malaysia. Renew. Sust. Energ. Rev. 12, 2404-2421.
[102] Tamunaidu, P., Bhatia, S., 2007. Catalytic cracking of palm oil for the production of biofuels: Optimization studies. Bioresour. Technol. 98(18), 3593-3601.
[103] Tan, K.L., Lim, S.K., Chan J.H., Low, C.Y., 2014. Overview of poly(lactic acid) production with oil palm biomass as potential feedstock. Int. J. Eng. Appl. Sci. 5, 1-10.
[104] Taufiqurrahmi, N., Mohamed, A.R., Bhatia, S., 2010. Deactivation and coke combustion studies of nanocrystalline zeolite beta in catalytic cracking of used palm oil. Chem. Eng. J. 163(3), 413-421.
[105] Taufiqurrahmi, N., Mohamed, A.R., Bhatia, S., 2011. Nanocrystalline zeolite beta and zeolite Y as catalysts in used palm oil cracking for the production of biofuel. J. Nanopart. Res. 13(8), 3177-3189.
[106] The Jakarta Post, Editorial: Justice for forest burners, 16 September 2015.
[107] Twaiq, F.A., Zabidi, N.A.M., Bhatia, S., 1999. Catalytic Conversion of Palm Oil to Hydrocarbons: Performance of Various Zeolite Catalysts. Ind. Eng. Chem. Res. 38(9), 3230-3237.
[108] van der Stelt, M.J.C., Gerhauser, H., Kiel, J.H.A., Ptasinski, K.J., 2011. Biomass upgrading by torrefaction for the production of biofuels: A review. Biomass Bioenergy. 35(9), 3748-3762.
[109] van Kasteren, J.M.N., Nisworo, A.P., 2007. A process model to estimate the cost of industrial scale biodiesel production from waste cooking oil by supercritical transesterification. Resour. Conserv. Recycl. 50(4), 442-458.
[110] Verma, M., Godbout, S., Brar, S.K. Solomatnikova, O., Lemay, S.P., Larouche, J.P., 2012. Biofuels production from biomass by thermochemical conversion technologies. Int. J. Chem. Eng. doi:10.1155/ 2012/542426.
[111] Wakker, E., 2006. The Kalimantan border oil palm mega-project, Friend of the Earth.
[112] Wan, Y., Chen, P., Zhang, B., Yang, C., Liu, Y., Lin, X., Ruan, R., 2009. Microwave-assisted pyrolysis of biomass: Catalysts to improve product selectivity. J. Anal. Appl. Pyrol. 86(1), 161-167.
[113] Wong, Y.C., Tan, Y.P., Taufiq-Yap, Y.H., Ramli, I., Tee, H.S., 2015. Biodiesel production via transesterification of palm oil by using CaO–CeO2 mixed oxide catalysts. Fuel. 162, 288-293.
[114] Yigezu, Z.D. Muthukumar, K., 2014. Catalytic cracking of vegetable oil with metal oxides for biofuel production. Energy Convers. Manage. 84, 326-333.
[115] Yusoff, S., 2006. Renewable energy from palm oil – innovation on effective utilization of waste. J. Clean. Prod. 14, 87-93.
[116] Yusoff, S., Hansen S.B., 2007. Feasibility study of performing a life cycle assessment on crude palm oil production in Malaysia. Int. J. Life Cycle Assess. 12(1), 50-58.
[117] Zainudin, N.F., Lee, K.T., Kamaruddin, A.H., Bhatia, S., Mohamed A.R., 2005. Study of adsorbent prepared from oil palm ash (OPA) for flue gas desulfurization. Sep. Purif. Technol. 45, 50-60.
[118] Zen, Z., Barlow, C., Gondowarsito, R., 2006. Oil palm in Indonesia socio-economic improvement: a review of options. Working Paper (Australian National University, Canberra, Australia).
[119] Zulkifli, H., Halimah, M., Chan, K.W., Choo, Y.M., Mohd Basri, W., 2010. Life cycle assessment for oil palm fresh fruit bunch production from continued land use for oil palm planted on mineral soil (Part 2). J. Oil Palm Res. 22, 887-894.

# 18

# Maximising high solid loading enzymatic saccharification yield from acid-catalysed hydrothermally-pretreated brewers spent grain

Stuart Wilkinson[1], Katherine A. Smart[2], Sue James[2], David J. Cook[1,*]

[1] *International centre for Brewing Science (ICBS), Division of Food Sciences, The University of Nottingham, Sutton Bonington Campus, Loughborough, Leicestershire LE12 5RD, U.K.*

[2] *SABMiller Plc, SABMiller House, Church Street West, Woking, Surrey, GU21 6HS, U.K.*

## HIGHLIGHTS

- ➢Cellulolytic enzyme saccharification of pre-treated brewers spent grains was investigated at high solids loading.
- ➢Aerated high-torque mixing offered enhanced glucose yields to 53 g/L.
- ➢Fed-batch protocols enhanced glucose yields to 59 g/L.
- ➢Supplementary carbohydrate degrading enzymes boosted achieved glucose yields to 64 g/L.
- ➢A novel consolidated saccharification protocol further increased glucose yields to 78 g/L at 25% w/v solids loading.

## GRAPHICAL ABSTRACT

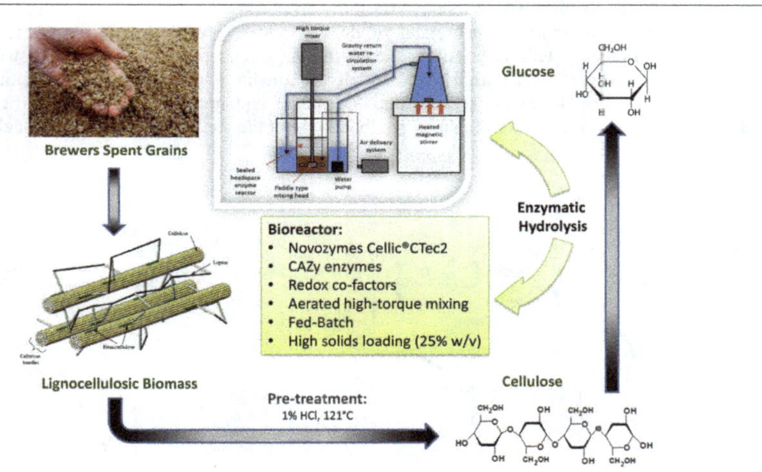


## ABSTRACT

Enzyme saccharification of pretreated brewers spent grains (BSG) was investigated, aiming at maximising glucose production. Factors investigated were; variation of the solids loadings at different cellulolytic enzyme doses, reaction time, higher energy mixing methods, supplementation of the cellulolytic enzymes with additional enzymes (and co-factors) and use of fed-batch methods. Improved slurry agitation through aerated high-torque mixing offered small but significant enhancements in glucose yields (to 53 ± 2.9 g/L and 45% of theoretical yield) compared to only 41 ± 4.0 g/L and 39% of theoretical yield for standard shaking methods (at 15% w/v solids loading). Supplementation of the cellulolytic enzymes with additional enzymes (acetyl xylan esterases, ferulic acid esterases and α-L- arabinofuranosidases) also boosted achieved glucose yields to 58 – 69 ± 0.8 - 6.2 g/L which equated to 52 - 58% of theoretical yield. Fed-batch methods also enhanced glucose yields (to 58 ± 2.2 g/L and 35% of theoretical yield at 25% w/v solids loading) compared to non-fed-batch methods. From these investigations a novel enzymatic saccharification method was developed (using enhanced mixing, a fed-batch approach and additional carbohydrate degrading enzymes) which further increased glucose yields to 78 ± 4.1 g/L and 43% of theoretical yield when operating at high solids loading (25% w/v).

**Keywords:**
Brewers Spent Grains
Bioethanol
Enzymatic saccharification
High solids loading

* Corresponding author
E-mail address: david.cook@nottinham.ac.uk

## 1. Introduction

Production of bioethanol from brewers spent grains (BSG) is a current area of research interest as higher value uses are sought for this co-product derived from the beer brewing process. The recalcitrant nature of BSG (like other lignocellulosic biomass types) renders some forms of pre-treatment essential before saccharification enzymes can be used to liberate fermentable sugars. Numerous effective chemical and thermal pre-treatments for BSG have been developed that can enhance the subsequent enzymatic saccharification yields (Wilkinson et al., 2014a; Wilkinson, 2014b). However, large doses of expensive commercial enzymes are still usually required to achieve high conversion efficiencies of cellulose to glucose (Leathers, 2003; Chundawat et al., 2008; Qing et al., 2010). Therefore, optimisation of the enzymatic saccharification stage is a key objective in the cost-effective production of lignocellulosic biofuels.

Maximising the operational solids loading used (a high solid to liquid ratio) during the enzymatic saccharification step, whilst minimising the required enzyme dose are likely to be key factors which need to be addressed in order to produce high glucose concentrations cost effectively and minimise water usage (Hodge et al., 2008; Kristensen et al., 2009). Commercial scale bioethanol production would likely require a minimum glucose concentration of ca. 100 g/L (ca. 10% w/v) in order to realistically produce the benchmark minimum ethanol concentration of ca. 50 g/L (ca. 5% v/v) which may then facilitate economically viable distillation (Lau and Dale, 2009). Operation at high solids loading (>10% w/v) during the enzymatic saccharification step would likely be essential in order to achieve this through limiting the dilutional effect of surplus liquid water (Kristensen et al., 2009). However, operation at high solids loading would likely drop the conversion efficiency of cellulose to glucose resulting in % theoretical yields well below acceptable limits of process efficiency. This is possibly due to rheology related mass transfer limitations because of the high viscosity of the media impeding enzymatic access to all of the available substrate (cellulose) and impeding dispersal of the hydrolysis products which may feedback inhibit the cellulases (Gan et al., 2003; Chundawat et al., 2008). There is also a minimum requirement of some available water for the successful catalytic function of many lignocellulolytic enzymes due to their hydrolytic mechanism of glycosidic bond fracturing through addition of a water molecule (Horn et al., 2012). This would suggest that an upper limit exists with regards to the maximum functional solids loading that can be used during the enzyme saccharification step as going beyond this would detrimentally limit the available free water present.

Various enzymes from the CAZy database (http://www.cazy.org/) of carbohydrate degradation specific enzymes are indicated as being potentially involved in the successful degradation of lignocellulosic substrates such as BSG, many of which come from fungal enzymatic systems such as brown and white rot fungi (Evans et al., 1994; Delmas et al., 2012). These enzymes include members of the glycoside hydrolase (GH) family, carbohydrate esterase (CE) family, and polysaccharide lyase (PL) family (**Fig. 1**). Other non-CAZy classified candidates could include ferulic acid esterases (feruloyl esterases) which have been shown to release ferulic acid (and other hydroxycinnamic acids) from substrates such as BSG (Bartolome and Gómez-Cordovés, 1999; Faulds et al., 2002). Also various metal ions (in particular copper) have been suggested to bind to the active site of GH61 class enzymes and play a crucial role in their activity (Quinlan et al., 2011). In addition various redox active cofactors such as gallate and ascorbate have been suggested to function as possible enhancers of GH61 enzyme activity as they may act as electron donors or chemical reductants (Quinlan et al., 2011; Horn et al., 2012).

Most commercial enzyme preparations designed for advanced generation biofuel processes are tailored to be as broad spectrum as possible so as to be effective on the wide variety of lignocellulosic substrates of interest such as woody biomass, energy crops/grasses, and industrial waste streams etc. (Harris et al., 2010). However, the highly substrate specific nature of the format of the lignocellulose (such as the specific side chain decorations of the xylan backbone which composes hemicellulose) may render generic enzyme preparations sub-optimal in

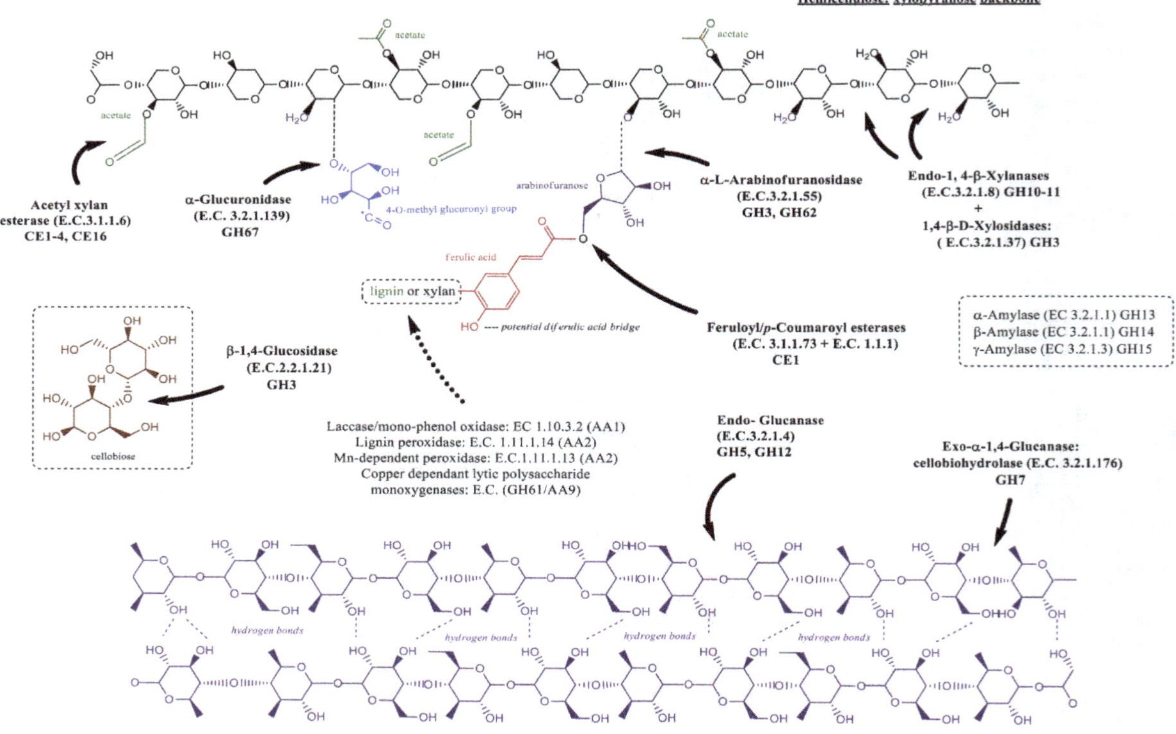


**Fig.1.** Schematic representation of the attack sites of various enzymes involved in deconstruction and saccharification of lignocellulosic material. (adapted from Faulds and Williamson (1995); Beg et al. (2001); Faulds et al. (2004); Dodd and Cann (2009); Rai (2009); Gupta et al. (2010); Kuhad et al. (2011), and http://www.cazy.org/).

terms of the specific content of their constituent enzymes or the dose of any one particularly crucial enzyme for any particular substrates (Gilbert et al., 2008; Dodd and Cann, 2009). Therefore, addition of specific ancillary enzymes (used in conjunction with a broad spectrum commercial enzyme mixture) could prove effective in maximising glucose yields from a specific lignocellulosic substrate such as BSG through the cleaving of a particular linkage or bond that is critical to the recalcitrance of a particular substrate (Chundawat et al., 2008).

Here we present an investigation looking at a range of parameters associated with the efficacy of enzymatic saccharification of pretreated BSG (with regards to their significance in terms of achieved glucose yields) which has not been previously published. Factors investigated were; simultaneous effect of variation of the solids loadings at several different cellulolytic enzyme doses, the effect of extension of hydrolysis incubation time, a range of different mixing methods, the effect of supplementation of the cellulolytic enzymes with various additional enzymes and enzymatic cofactors, and also the effectiveness of fed-batch methods. Finally the development of a novel high solids loading enzymatic saccharification method is demonstrated. This was a consolidated approach of a combination of a selection of the previously optimised parameters in an attempt to further maximise achieved glucose yields. This could then provide the basis for a commercial industrial scale saccharification step as part of the process of biofuel production from BSG.

## 2. Materials and Methods

### 2.1. Reagents

All reagents were of AnalaR grade and obtained from Sigma-Aldrich (UK) and Fisher Scientific (UK). All water used was deionised reverse osmosis and of ≥18 mega-ohm purity (Purite Select Ondeo IS water system, Purite, UK). All enzymes used for the saccharification experiments are described in section 2.5.

### 2.2. BSG

Brewers spent grains (BSG) were sourced from the SABMiller 10 hL Research Brewery (Sutton Bonington, UK) from a high gravity brewing process using 100% malted barley (**Table 1**). The BSG was dried in an oven at 105°C overnight and ground to a particle size of less than 212 µm to ensure homogeneity prior to any sampling (KG49 grinder, Delonghi, UK).

**Table 1.**
Typical compositional analysis of brewers spent grains (Wilkinson et al., 2014a).

| BSG component | % (w/w) |
|---|---|
| starch* | 4.8 ± 0.46 |
| protein | 26.6 ± 0.38 |
| ash | 2.7 ± 0.070 |
| lipid | 5.2 ± 2.1 |
| lignin | 9.9 ± 1.4 |
| cellulose (glucose) | 19.2 ± 1.4 |
| hemicellulose | 18.4 ± 3.7 |
| of which xylose | 11.3 ± 1.2 |
| other | 7.7 |

Data are the mean ±SD of 3 replicate measurements.
*high residual starch present due to the use of a 2-roller mill.

### 2.3. Pre-treatment of BSG

All enzymatic saccharification experiments were conducted using pretreated BSG (dilute acid catalysed hydrothermal pre-treatment) according to the optimised method described by Wilkinson et al. (2014a). Pretreatment was conducted at 121°C (for 30 min) and at 25% (w/v) solids loading (with 1% HCl) using a 40-L bench-top autoclave (Priorclave, Tactrol 2; RSC/E, UK). After pre-treatment, all pretreated slurries were neutralized to pH 7.0 (±0.5) via 40% NaOH (w/w). The biomass was then centrifuged at 5000 rpm for 10 min (Heraeus Megafuge 16; Thermo Scientific, UK) and the supernatant was removed. The remaining solid residue was then exhaustively washed with water, conducted by re-suspension and centrifugation at 5000 rpm for 10 min, discarding the supernatant each time. The remaining residues were then dried in an oven overnight at 60°C prior to any enzyme saccharification. Whilst the pre-treatment step did liberate a significant quantity of sugars directly into the supernatant or hydrolysate (particularly xylose from hemicellulose depolymerisation), this was set aside as the focus of this research was more specifically on optimising the subsequent enzymatic saccharification of the remaining insoluble residue after pre-treatment.

### 2.4. Total glucose composition of pretreated BSG

The total glucose concentration in the remaining insoluble residue (after pre-treatment) was quantified using the method described by Wilkinson et al. (2014b). A total acid hydrolysis method (using 12M $H_2SO_4$ at 37°C for 1 h then diluted to 1M at 100°C for 2 h) was used to liberate glucose which was then quantified using HPAE-PAD (described in section 2.6). This was required for determination of accurate achieved % theoretical glucose yields liberated after enzymatic hydrolysis. All analyses were conducted in triplicate.

### 2.5. Enzymatic saccharification of pretreated BSG

All enzyme hydrolysis reactions (the saccharification of the insoluble residues after pre-treatment of the BSG as described in section 2.3) were conducted using the commercial enzyme preparation Novozymes Cellic® CTec2 (kindly supplied by Novozymes A/S, Demark) with a 50°C incubation temperature used for all saccharification reactions. The efficacy of all enzymatic saccharifications was evaluated exclusively via the achieved glucose yields (both g/L and achieved % theoretical yields). Whilst many of the experiments evaluated the effect of modification of the hemicellulose component (on the subsequent degree of cellulose saccharification achieved), the liberation of pentose sugars was not a primary objective. As such, only liberated glucose levels were quantified. All analyses were conducted in triplicate.

The Cellic® CTec2 used for saccharification had a total cellulase activity of 200 FPU/mL in undiluted format and was determined according to Ghose, 1989. An enzyme dosing range of 10 - 160 FPU/g (biomass) was used. The correct quantity of pretreated BSG (0.5 - 15 g) and 50 mM sodium citrate buffer (pH 5.0) were combined to produce a slurry and achieve the various desired solids loadings (5% - 50% w/v). After the correct incubation period (10-72 h), all enzyme saccharification preparations were centrifuged at 5000 rpm for 10 min and the supernatant was removed and sampled for HPLC based quantification of glucose concentrations. Prior to all enzymatic saccharification experiments, the glucose concentration present in the Cellic® CTec2 enzyme preparation was determined by HPLC and subtracted from final enzymatic hydrolysis glucose yields to allow calculation of accurate achieved % of theoretical yields (which were expressed as a percentage of the total glucose content of the pre-treatment generated residue prior to saccharification as described in section 2.4).

#### 2.5.1. Evaluation of the effect of higher energy mixing methods during enzymatic saccharification

A standard shaking incubator method using agitation at 150 rpm (MaxQ 4358 shaking incubator, Thermo Scientific, UK) was used as a benchmark with various alternative mixing methods then compared against it. All analyses were conducted in triplicate. Efficacy of each different mixing method was determined by the liberated glucose yields achieved. Roller bed mixing methods were conducted at 50 rpm using an SRT6D roller bed (Stuart Scientific, UK) whilst magnetic stirring methods were conducted at 150 rpm using a multi-plate magnetic stirrer (Variomag poly 15, Thermo Scientific, UK). Both the roller bed and multi-plate magnetic stirrers were housed within an MIR-253 incubator (Panasonic, Japan) to achieve the 50°C temperature optima. A high torque mixing (HTM) method was conducted using a custom made, 2-piece, sealed headspace reaction vessel housed within a re-circulating water bath which acted as an incubator (**Fig. 2**). A 6-bladed paddle type mixer (at

150 rpm) was used for the actual agitation. The effect of supplemental aeration on enzymatic saccharification was investigated using the custom HTM system with an additional 2 mm air delivery line which supplied air from an external air pump (TetraTec APS100 Air Pump: 100 l/h, 2.5w). All analyses were conducted in triplicate.

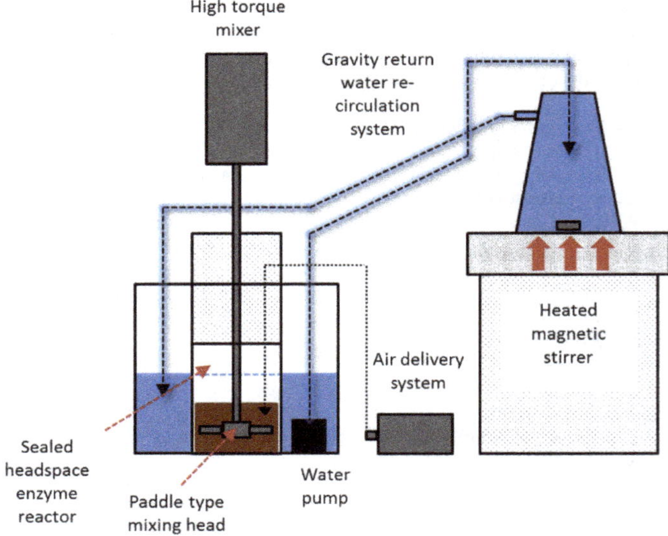


**Fig.2.** Schematic diagram of high-torque mixing (HTM) system and optional additional aeration.

*2.5.2. Effect of supplementation of the Cellic® CTec2 with additional enzymes and cofactors on achieved glucose yields*

The effectiveness of additional hydrolytic enzymes or various cofactors (in conjunction with Novozymes Cellic® CTec2) was investigated using the aerated HTM method as described in section 2.5.1. All analyses were conducted in triplicate. For preliminary proof of principle experiments, all enzymes and cofactors were dosed at levels which were considered to be in excess (100 µL aliquot of each enzyme preparation as supplied) order to ascertain any additional improvement in achieved glucose yields compared with the use of Novozymes Cellic® CTec2 alone. Additional xylanases were tested using Novozymes Cellic® HTec2 (100 µL/10 g biomass). Ferulic acid esterase was tested using Prozomix Feruloyl esterase (CAZy CE1 carbohydrate esterase family, from *Clostridium thermocellum*: 100 µL/10 g biomass; activity: 112.5 Units/mL). One unit was defined as the amount of enzyme required to release 1 µM of ferulic acid per minute from 32 µM methyl ferulate in 50 mM sodium phosphate buffer, pH 6.0, at 37°C, and at 335 nm. Acetyl xylan esterase was tested using Prozomix Acetyl xylan esterase (CAZy CE3, carbohydrate esterase family 3, from *Clostridium thermocellum ATCC 2740,* 100 µL/10 g biomass; activity: 112.5 Units/mL). One unit was defined as the amount of enzyme required to release 1 µM of p-nitrophenol per minute from p-nitrophenyl acetate (1 mM in the assay) in 50 mM phosphate buffer, pH 7, at 50°C, containing 1 mg/mL of BSA. Arabinofuranosidase was tested using Prozomix α-L-Arabinofuranosidase (CAZy GH51, glycoside hydrolase family 51, from *Streptomyces coelicolor A3,* 100 µL/10 g biomass; activity: 125 Units/mL). One unit was defined as the amount of enzyme required to release 1 µM of p-nitrophenol per minute from p-nitrophenyl α-L-arabinofuranoside (1 mM in the assay) in 50 mM phosphate buffer, pH 7, at 60°C, containing 1 mg/mL of BSA. The addition of starch degradation enzymes was tested using α-amylase and glucoamylase (Megazyme, Ireland) with each enzyme tested individually (100 µL/10 g biomass) and also both enzymes simultaneously (100 µL of each/10 g biomass). The specific activity of α-amylase was 3000 Units/ml. One Unit was defined as the amount of enzyme required to release one µM of p-nitrophenol from blocked p-nitrophenyl-maltoheptaoside per minute (in the presence of excess α-glucosidase) at pH 6.0 and 40°C. The specific activity of glucoamylase was 160 Units/mg (60°C, pH 4.5 on soluble starch). One unit of glucoamylase activity was defined as the amount of enzyme required to release one µg of β-D-glucose reducing-sugar equivalents per minute from soluble starch (10 mg/mL) in sodium acetate buffer (100 mM) at pH 4.5.

The effect of addition of proteolytic enzymes was investigated using the commercial preparations Alphalase™ NP (Danisco) and Pronase E (Sigma-Aldrich, UK). Alphalase™ NP is a commercial enzyme mixture (derived from *Bacillus amyloliquefaciens*) used within the beer brewing industry in order to assist in the hydrolysis of the protein component found within malted barley (the pre-cursor of BSG) in order to boost free amino nitrogen (FAN) levels to assist in high gravity fermentations thus rendering it potentially effective on the native protein fraction found in BSG. Alphalase™ NP was dosed at 100 µL/10 g biomass. Alphalase™ NP has a suggested dosing of 0.1-0.3 kg/MT grist when used during the mashing stage of commercial beer brewing, therefore, the dose used here was also considered to be in excess. The Alphalase™ NP was prepared in 50 mM sodium citrate buffer at pH 6.5. Pronase E is a commercial mixture of ca. 10 proteases from *Streptomyces griseus* K-1 and includes five serine-type proteases, two zinc endopeptidases, two zinc leucine aminopeptidases, and one zinc carboxypeptidase (with an overall very broad spectrum of activity). Pronase E was prepared as a 1% w/w solution; 1 g/100 mL 50 mM sodium citrate buffer (pH 7.5). The specific activity of Pronase E was 4 Units/mg. One unit of Pronase E was defined as the amount of enzyme required to hydrolyse casein to produce colour equivalent to 1.0 µM (181 µg) of tyrosine per min at pH 7.5 at 37°C (colour by Folin-Ciocalteu reagent). For the evaluation of the effect of the supplementary protease enzyme preparations, both BSG and pretreated BSG were subjected to a 5 h pre-incubation with each of the protease solutions (150 rpm agitation; MaxQ 4358 shaking incubator, Thermo Scientific, UK) using 15% w/v solids loading (at 60°C for the Alphalase™ NP and 37°C for the Pronase E) and prior to the saccharification stage using Novozymes Cellic® CTec2 (as described in section 2.5). The effect of a protein denaturation step after the protease pre-incubation was also investigated. The denaturation was conducted by incubation at 90°C for 10 min in a water bath. This was designed to denature and inactivate the proteases prior to dosing with Novozymes Cellic® CTec2 (to minimise any possible degradation of the lignocellulolytic enzymes found within the CTec2 by the proteases). After denaturation, samples were cooled to 50°C prior to dosing with Cellic® CTec2 and then further incubated at 50°C (150 rpm agitation; MaxQ 4358 shaking incubator, Thermo Scientific, UK) for the required time period. In addition, the supplementation with the potential GH61 enzyme family cofactors, i.e., copper ($CU(NO_3)_2$) and ascorbic acid were tested (in conjunction with Cellic® CTec2) using a concentration range of 2.5 – 10 mM incorporated into the 50 mM sodium citrate buffer used for saccharification.

*2.5.3. Effect of variation of solids loading and enzyme dose during hydrolysis on achieved glucose yields*

The individual and combined impacts of solids-loadings (5-25% w/v) and enzyme dose (Novozymes Cellic® CTec2; 10-160 FPU/g biomass) during enzymatic saccharification of pretreated BSG were investigated using a D-optimal designed experiment (**Table 1S**). The D-optimal experimental design was created using DesignExpert (Stat-Ease, USA) using a collection of reaction combinations (solids loading and enzyme doses) from which the D-optimal algorithm chose the treatment combinations to include in the design. This reduced the total number of experiments needed to be conducted from that of a full factorial experimental design yet still provided statistically valid data. Response data (glucose yields) were then modelled against these factors using Design Expert with analysis of variance (ANOVA) automatically incorporated into the modelling. For this preliminary investigation, a simplified shaking incubator method was utilised due to its high-throughput nature in comparison with the optimised aerated HTM system (described in section 2.5.1) which consisted of a 'single-shot' bioreactor.

*2.5.4. Effect of variation of solids loading and hydrolysis time during hydrolysis on achieved sugar yields*

The individual and combined impacts of solids-loadings (5-25% w/v) and reaction time (10-72 h) during enzymatic saccharification of pretreated BSG (as described in section 2.3) were also investigated using a D-optimal designed experiment (**Table 2S**). Novozymes Cellic® CTec2 was used as the enzyme preparation (20 FPU/g biomass). Response data (glucose yields) were then modelled against these factors using Design Expert software v 7.0 (Stat-Ease, Mn, USA).

*2.5.5. Fed-batch enzymatic saccharification experiments*

Fed-batch enzymatic saccharification experiments were conducted to evaluate their effectiveness against non-fed-batch saccharification methods to determine whether the staggered addition of biomass achieved higher glucose yields. Preliminary experiments utilised the standard shaking incubator method (48 h incubation at 50°C with agitation at 150 rpm, MaxQ 4358 shaking incubator, Thermo Scientific, UK) due to its high throughput nature. The fed-batch system used the correct amount of biomass (0.5 -5 g pretreated BSG) to achieve the desired solids loading rates (5-50% w/v) but with the biomass split into two aliquots with the second aliquot added after 24 h incubation. All analyses were conducted in triplicate.

*2.5.6. Consolidated (combined) enzymatic saccharification method*

A consolidated enzymatic saccharification method was formulated from a combination of all of the previous individual experimental optimisations in an attempt to further maximise the achieved glucose yields when operating at high solids loading. This method was based on all the parameters that were suggested to play a significant role in enhancing the glucose yields achieved. The consolidated method involved using the aerated HTM system operating at 50°C for 72 h incubation (considered an excess reaction time and thus non rate limiting) with a fed-batch approach (tested over a final solids loading range of 5-25% w/v) using a low dose of Novozymes Cellic® CTec2 (10 FPU/g biomass; to simulate a commercially viable large scale industrial process) and with the additional enzymes; ferulic acid esterase, acetyl xylan esterase, arabinofuranosidase, Cellic® HTec2 (all dosed at 100 µL and prepared as described previously in section 2.5.4), and also with the additional enzymatic cofactors ($Cu(NO_3)_2$, and ascorbic acid (both dosed at 10 mM concentrations). All analyses were conducted in triplicate.

*2.6. HPLC*

Glucose concentrations were quantified using the method described by Wilkinson et al. (2015). This was then quantified using an ICS 3000 system from Dionex (fitted with a CarboPac PA20 column, 150 mm × 3.0 mm; Dionex, USA) and a high performance anion exchange pulsed amperometric electrochemical detector (HPAE-PAD; Dionex, USA). The system used isocratic elution with 10 mM NaOH at 0.5 mL/min flow rate with a subsequent column regeneration step (using 200 mM NaOH at 0.5 mL/min). Inhibitory compounds (pre-treatment generated sugar and lignin degradation products) were quantified *via* HPLC according to the method described by Wilkinson et al. (2014b). The system (2695 HPLC system and 996 Photodiode Array Detector, Waters, USA) used UV detection at 280 nm with UV spectra for secondary confirmation of identity. A Techsphere ODS C18 column (5 µm, 4.6 mm × 250 mm; HPLC Technologies, UK) was used at ambient temperature. The mobile phase was a mixture of 1% acetic acid (solvent A) and methanol (solvent B) with a flow rate of 1.0 mL/min. Gradient elution was used ramping from 20% to 50% methanol over 30 min with a final 100% methanol column cleaning phase.

*2.7. Quantification of free amino nitrogen (FAN) levels*

FAN was determined using the ninhydrin colorimetric assay using glycine as the standard (European Brewery Convention, 1998).

*2.8. Measurement of the protein content of the insoluble residue following pre-treatment of the BSGA*

Thermo Flash Nitrogen Analyzer (ThermoFisher Scientific, Waltham, Massachusetts, USA) was used to determine protein content using the method described by Wilkinson et al. (2014a). The sample was heated to 900°C in the presence of an oxidation catalyst and helium carrier gas. Oxygen was then added and the temperature was increased to 1800°C. The resulting gases were then passed through a reduction reactor that converted nitrogen oxides to elemental nitrogen for detection. Protein was then calculated using the × 6.25 conversion factor. All analyses were conducted in triplicate.

*2.9. Statistical analysis*

Experimental design, response surface modeling, and ANOVA were all performed using Design Expert v 7.0 (Stat-Ease Inc., Minneapolis, USA). Additional ANOVA (followed by the Tukey HSD test) were performed using the software package SPSS 16.0 for Windows (SPSS, Germany). All experiments were conducted in triplicate (biological replication).

## 3. Results and discussion

For all subsequent experiments, the efficacy of enzymatic saccharification is presented as both % theoretical yields and g/L yields. Whilst high % theoretical yields would indicate a high degree of process efficiency (high rate of conversion of substrate to product; glucose), they do not provide a measure of the 'usability' of a feedstock. For example, a feedstock with a high % theoretical glucose yield but low g/L yield (low absolute glucose yield) would not be viable to use for biofuel production. This is due to the ca. 100 g/L minimum glucose concentration required to achieve the minimum ethanol concentration that would render distillation economically viable (Lau and Dale, 2009). Therefore, both units are included for consideration.

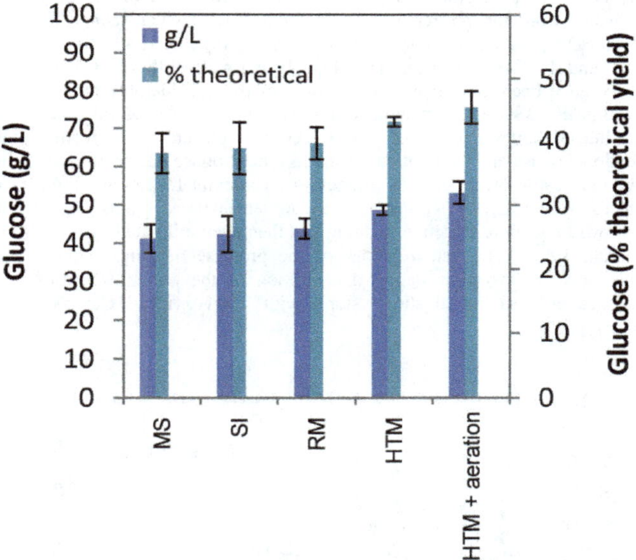


**Fig.3.** Impact of mixing methods during enzymatic saccharification of pretreated BSG on the resultant glucose concentrations (g/L and % theoretical yields). Enzymatic saccharification conducted using Novozymes Cellic® CTec2 (excess enzyme dose: 160 FPU/g biomass with 72 h incubation at 50°C, 15% w/v solids loading) on pretreated BSG (1% HCl 121°C 30 min pre-treatment, 25% w/v solids loading). Key: MS: magnetic stirring, SI: shaking incubator, RM: roller bed mixing, HTM: high torque mixing, HTM + aeration: high torque mixing with additional aeration supplied from an external air pump. Data are the mean ± SD of 3 replicate measurements.

*3.1. Effect of variation of the mixing method on enzymatic saccharification yields from pretreated BSG*

The use of different mixing (agitation) systems for the enzymatic saccharification revealed that magnetic stirring (MS), shaking incubation (SI), and roller mixing (RM) produced similar results with 40 - 44 (± 2.6 – 4.6) g/L final glucose concentrations and achieved 38 - 40% theoretical glucose yields after 72 h incubation when using an excess dose (160 FPU/g biomass) of Novozymes Cellic® CTec2 (**Fig. 3**). However, the use of the HTM was shown to give a small but significant boost in glucose yields (increased to 49 ± 1.3 g/L and 44% of theoretical yield; P<0.05: one way ANOVA). This suggested the other methods were sub-optimal in terms of their agitation and the enhanced mixing of the enzyme and substrate that occurred with the HTM system was beneficial. Improved mixing technology may play some part in ensuring better enzyme substrate interactions (especially at very high solids loading; >20% w/v), however, some research has suggested there may only be minimal improvements in achieved glucose yields through additional mixing (Kristensen et al., 2009).

In addition, achieving effective mixing at very high solids loadings may also prove prohibitively energy intensive due to the amount of torque likely required to agitate highly viscous slurries of the biomass. However, supplementation of the HTM system with additional aeration further increased the glucose yields (to 53 ±2.9 g/L and 45% of theoretical yield) suggesting the additional oxygen present in the air supplied externally may have enhanced the activity of some saccharification component, possibly the GH61 oxidative enzymes found within the Novozymes Cellic® CTec2. Previous published work has also suggested that the provision of supplementary oxygen or aeration may be a low cost option which could boost the catalytic activity of GH61 enzymes (Beeson et al., 2011; Horn et al., 2012).

### *3.2. Effect of supplementation of the Cellic® CTec2 with additional enzymes and enzymatic cofactors on saccharification yields from pretreated BSG*

Enzymatic saccharification experiments utilising the aerated HTM method (again using Novozymes Cellic® CTec2) indicated that supplementation with the commercial protease mixtures (Alphalase NP and Pronase E with a 5 h pre-incubation prior to the saccharification step with Novozymes Cellic® CTec2) did not result in any subsequent enhancement of the liberation of glucose from the pretreated BSG compared with when using Novozymes Cellic® CTec2 alone (yields were actually slightly lower; 49 - 51 (±1.6 – 2.7) g/L and 41 - 43% of theoretical; **Fig. 4**). It was hypothesized that proteases may have been able to hydrolyse some of the considerable protein content within the BSG which may have indirectly improved physical access of any cellulase enzymes to the crystalline cellulose present. The several kinds of endopeptidase and exopeptidase found in the Pronase has been used to digest various cattle feeds such as grasses in an attempt to produce low nitrogen feeds (Abe et al., 1979). Pronase may target dextrans (glucan) and has been shown to be more effective in doing this than even mild acid hydrolysis (Kato et al., 1991). The use of either of the protease mixtures without a heat mediated inactivation stage ( denaturation of the proteases ) prior to the subsequent saccharification step with Novozymes Cellic® CTec2 significantly decreased the glucose yields achieved (38 - 39 ± 1.9 – 2.5 g/L and 32 - 33% of theoretical yield) compared with when using Cellic® CTec2 alone. This suggested that some proteolytic attack of the carbohydrate degrading enzymes within the Cellic® CTec2 occurs as a direct result of the presence of the proteases. Overall, the lack of any enhancement in glucose yields indicates that the considerable protein fraction which is still present within the BSG after hydrothermal pre-treatment (ca. 22% crude protein) does not impede the activity of lignocellulolytic enzymes contained within Novozymes Cellic® CTec2. This would confirm as expected that proteinaceous tissues within the BSG are generally morphologically discreet from the lignocellulosic matrix. However, whilst glucose yields may not have been boosted by the addition of the proteases, their inclusion could still be considered as it may also be desirable to liberate free amino acids (or low molecular weight peptides) into the hydrolysate which is generated from the saccharification step (the actual feedstock subsequently fermented to bioethanol), thus increasing the free amino nitrogen (FAN) content. This could potentially improve the subsequent fermentation performance which could be particularly useful for very high gravity fermentation systems (using very high initial sugar concentrations) as it may assist in reducing the osmotic stress on the yeast (or bacterial species) used and allow for production of high ethanol titres.

Therefore, the lack of any observed improvement of the enzymatic saccharification yields with the use of additional proteases may not definitively exclude their incorporation into enzymatic cocktails used (for the saccharification step which would be required for biofuel production) due to the considerable, currently under-utilised protein fraction found within pretreated BSG. Both protease mixtures were indicated to be effective at significantly increasing FAN liberation from pretreated BSG as levels were indicated to rise from approximately 23 mg/L to >100 mg/L when using the 5 h pre-incubation. Due to this success of FAN liberation from pretreated BSG, an identical method was conducted but this time testing the protease mixtures on the starting BSG biomass (before any pre-treatment). This resulted in an even greater increase in FAN from approximately 51 mg/L to >180 mg/L. This data would suggest that hydrothermal pre-treatment does either liberate or modify a significant proportion of the protein present in the BSG, thus explaining the lower FAN concentrations liberated by the proteases on pretreated BSG. Overall, this could allow the generation of a FAN enriched solution before the saccharification step using Cellic® CTec2 and is the subject of current on-going research. This could either be used for incorporation into any subsequent feedstock after saccharification or form an additional high value product stream which could be fractionated.

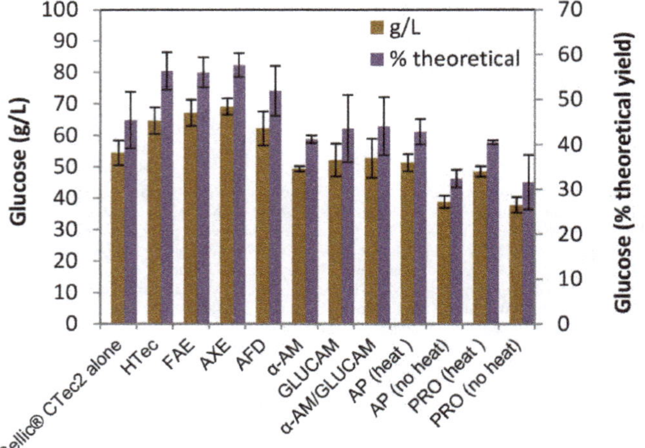


| **Key** | |
|---|---|
| HTec | Xylanases |
| FAE | Ferulic acid esterases |
| AXE | Acetyl xylan esterases |
| AFD | α-L-Arabinofuranosidase |
| α-AM | Amylase |
| GLUCAM | Glucoamylase |
| α-AM/GLUCAM | Amylase + Glucoamylase |
| AP (heat ) | Alphalase protease with heat inactivation |
| AP (no heat) | Alphalase protease without heat inactivation |
| PRO (heat ) | Pronase protease with heat inactivation |
| PRO (no heat) | Pronase protease without heat inactivation |

**Fig.4.** Impact of using a range of supplementary enzymes in conjunction with Cellic® CTec2 during enzymatic saccharification of pretreated BSG on achieved glucose concentrations (g/L and achieved % theoretical yields). Enzymatic saccharification conducted using Novozymes Cellic® CTec2 (excess enzyme dose: 160 FPU/g biomass with 72 h incubation at 50°C, 15% w/v solids loading) using aerated high-torque mixing (HTM) on pretreated BSG (1% HCl 121°C 30 min pre-treatment, 25% w/v solids loading) with all supplementary enzymes dosed at 100 μL (all considered in excess dosages). Data are the mean ± SD of 3 replicate measurements.

In a similar fashion, supplementation with either (or both) of the starch degrading enzymes, glucoamylase (amyloglucosidase) and α-amylase, did not significantly increase enzymatic saccharification yields (glucose yields) beyond that achieved by Novozymes Cellic® CTec2 alone. Some starch does usually remain present in BSG (although typically <5% w/w) unless a very intense and effective mashing method has been used during the initial beer brewing process such as use of mash filter brewing technology and the associated brewhouse apparatus such as a hammer mill (e.g., extra finely milled grist used initially and also possibly with additional saccharification enzymes added during mashing; both of which aid extraction efficiency of starch removal from the malted barley). However, the hydrothermal temperatures (121°C) that were used during the pre-treatment step initially would have very likely already gelatinised and solubilised any residual starch that may have been present (in the BSG) into the pre-treatment generated hydrolysate. Therefore, starch degradative enzymes did not directly boost glucose yields through starch hydrolysis or indirectly through enhanced access of the cellulases (present in the Novozymes Cellic® CTec2) to cellulose through removal of any of the very small quantity of starch grains that may have physically impeded the enzymes. In contrast to this, the supplementation with additional xylanases (Novozymes Cellic® HTec2), ferulic acid esterases, acetyl xylan esterases, and α-L-arabinofuranosidases all resulted in a significant increase (P<0.05: one-way ANOVA) in enzymatic saccharification glucose yields compared with Cellic® CTec2 alone. The ferulic acid esterase, acetyl xylan esterase, and xylanases (Cellic® HTec2) were suggested to achieve the greatest enhancement in glucose yields (65 - 69 ± 2.6 – 4.1 g/L and 56 - 58% of theoretical yields) whilst the α-L-arabinofuranosidase liberated ca. 62 ± 4.1 g/L glucose and 52% theoretical yield. Acetyl xylan esterase enzymes (CE1) target ester linkages (of acetate groups) on the xylose backbone of hemicellulose (Shallom and Shoham, 2003) whereas α-L-arabinofuranosidase enzymes (GH62) hydrolyse the terminal non-reducing α-L-arabinofuranoside residues in α-L-arabinosides that are present in various places as side chain decorations in arabinoxylans and hemicellulose (Schwarz et al., 1995; Poutanen, 1988). The successful de-branching of these residues may be crucial for complete and effective hydrolysis of branched formats of hemicellulose. This suggested the pretreated BSG still contained various lignocellulosic linkages that to a small extent impeded the function of the cellulases found within Novozymes Cellic® CTec2. It further supports the idea that specific lignocellulosic biomass such as BSG requires a tailor made enzyme cocktail in order to maximise the glucose yields achieved. Whilst Novozymes Cellic® CTec2 is a highly effective commercial enzyme preparation on its own (and a complex mixture of different lignocellulolytic enzymes), specific substrates may require larger quantities of specific enzymes, e.g., ferulic acid (feruloyl) esterases, due to the prevalence of particular linkages found within it.

An alternative hypothesis is that specific additional enzymes may be needed due to the specific format of the components present (in BSG) that may be chemically modified as a result of the specific pre-treatment step employed upstream of the enzymatic hydrolysis step. Overall, it suggested a requirement for comprehensive optimisation of the enzymatic saccharification step for each specific lignocellulosic substrate. In the specific case of BSG, it may also suggest that each different batch may also require some degree of optimisation in order to maximise the saccharification yields due to compositional differences that arise as a result of the use of differing barley cultivars used initially in the grist or different mashing conditions used during brewing (wort production).

Additionally, the supplementation of the aerated HTM method (using Novozymes Cellic® CTec2) with the GH61 specific enzymatic cofactors, i.e., copper and ascorbic acid, suggested that a small but significant increase (P<0.05: one-way ANOVA) in glucose yields was achieved when using a 10 mM concentration of both cofactors simultaneously (liberating 58 ± 1.7 g/L glucose and 56% of theoretical yield; **Fig. 5**). The CAZy GH61 family of glycoside hydrolases have been indicated to play an important role in the oxidative cleavage of certain linkages in lignocellulolytic substrates in nature with many GH61 genes being indicated to be present in the genome of various fungal species including the brown rot fungi *Postia placenta* (Quinlan et al., 2011). As a result, additional GH61 enzymes have been included in certain commercial enzyme products such as Novozymes Cellic® CTec2 (Cannella et al., 2012). Various other published work has suggested they may be able to boost the effectiveness of other cellulase class enzymes such as endo and exo-1,4-β-glucanases and in particular cellobiohydrolases (Harris et al., 2010; Beeson et al., 2011; Quinlan et al., 2011; Horn et al., 2012). This suggested that GH61 enzymes found within Novozymes Cellic® CTec2 may be responsive to increased concentrations of redox active cofactors and function sub-optimally without this additional supplementation and confirmed the work of Quinlan et al. (2011) and Beeson et al. (2011).

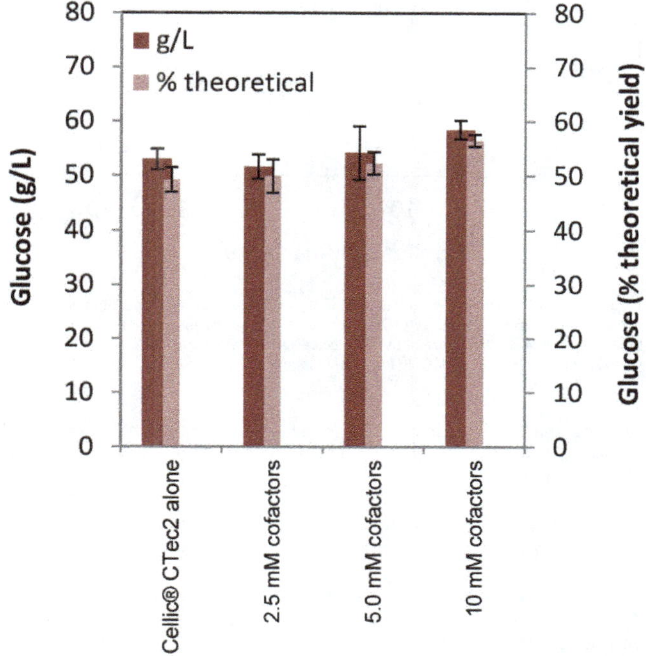


**Fig.5.** Impact of the simultaneous addition of both the enzymatic cofactors $Cu(NO_3)_2$ and ascorbate in conjunction with Cellic® CTec2 during enzymatic saccharification of pretreated BSG on achieved glucose concentrations (g/L and achieved % theoretical yields). Enzymatic saccharification conducted using Novozymes Cellic® CTec2 (excess enzyme dose: 160 FPU/g biomass with 72 h incubation at 50°C, 15% w/v solids loading) using aerated high-torque mixing (HTM) on pretreated BSG (1% HCl 121°C 30 min pre-treatment, 25% w/v solids loading) with a range of $Cu(NO_3)_2$ and ascorbic acid concentrations (2.5-10 mM with equal concentrations of both reagents). Data are the mean ± SD of 3 replicate measurements.

Further work would need to be conducted in order to ascertain whether the presence of copper ions at concentrations of around 10 mM (should those concentrations be retained in the feedstock ultimately generated) would be detrimental to any downstream fermentations conducted as various yeast species have been indicated to be sensitive to high concentrations of copper ions resulting in sub-optimal fermentation performance (Dönmez and Aksu, 1999). However, these suggested GH61 enzymatic cofactors may offer a more cost effective alternative to augmentation of generic lignocellulosic enzyme cocktails with additional expensive exotic enzymes. Whilst some members of the GH61 family exhibit weak *β*-1,4 endoglucanase activity, it is unlikely they act upon single cello-oligomeric chains as it has been indicated that no soluble reaction products such as cellobiose or glucose are detected when they are used alone on lignocellulosic substrates (Harris et al., 2010; Horn et al., 2012). In addition, they do not appear to enhance the activity of cellulases on pure cellulose (Harris et al., 2010). In conclusion, it would then suggest that GH61 class enzymes may act upon a rare bond found only in lignocellulosic materials (which are obviously not found in pure cellulose) that may possibly obstruct the normal function of other primary cellulolytic enzymes in some way. Alternatively, they may function *via* a non-hydrolytical based mechanism possibly similar to the suggested mechanism of carbohydrate binding modules (CBMs) in that they may penetrate micro-cracks in crystalline cellulose and cause localised

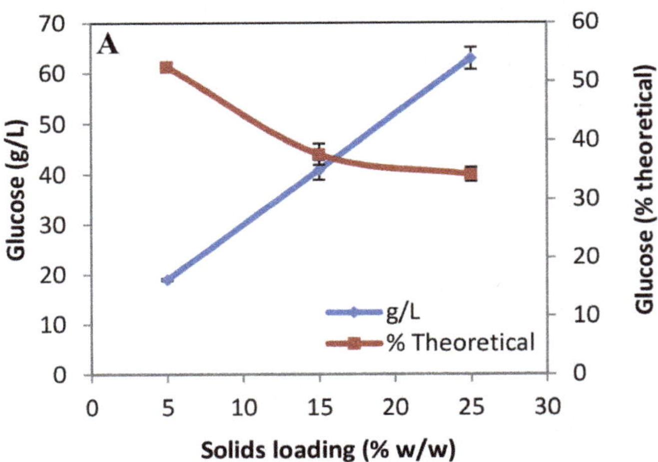


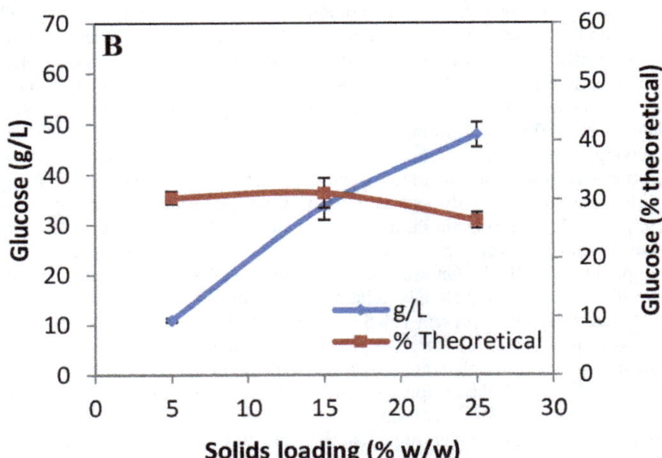


**Fig.6.** Effect on achieved glucose concentrations of variation of the solids loading used during enzymatic saccharification of pretreated BSG. **A:** shows the theoretical maximum yield achieved using an excess enzyme dose (160 FPU/g biomass) whereas **B:** shows the more realistic achieved yield when constraints of enzyme pricing are considered and a lower enzyme dose used (10 FPU/g biomass). Enzymatic saccharification conducted using Novozymes Cellic® CTec2 with 72 h incubation at 50°C, 5-25% w/v solids loading) using standard shaking incubator method (SI) on pretreated BSG (1% HCl 121°C 30 min pre-treatment, 25% w/v solids loading). Data are the mean ± SD of 3 replicate measurements.

swelling that enables free water to penetrate, thus further increasing the porosity allowing better access for additional enzymes to function (Arantes and Saddler, 2010).

### *3.3. Effect of variation of the solids loading and enzyme dose used during enzymatic saccharification on liberated glucose yields*

Two-dimensional variants of the derived models (using a D-optimal design; see section 2.5.3 for experimental details) for glucose release (using only a selection of the data set) suggested that when using an excess of Novozymes Cellic® CTec2 (160 FPU/g biomass) increasing solids loading during enzymatic saccharification from 5% w/v to 25% w/v resulted in a linear increase in g/L glucose concentrations liberated (**Fig. 6A**). However, achieved % theoretical glucose yields dropped as solids loading was increased, particularly sharply with the increase from 5% w/v to approximately 15% w/v. Increasing solids loading beyond 15% w/v resulted in only a minor drop in achieved % theoretical glucose yields. In comparison, when shifting to using a low dose of Novozymes Cellic® CTec2 (10 FPU/g biomass), increasing solids loading from 5% w/v to 25% w/v also resulted in an almost linear increase (albeit lower ramp than when using an excess enzyme dose) in g/L glucose concentrations liberated (**Fig. 6B**). However, in contrast to when using an excess enzyme dose, with the use of a low enzyme dose (10 FPU/g biomass) there was only a very small drop in achieved % theoretical yields (from ca. 30% to 26%) as solids loading was increased from 5% to 25% w/v. This suggested that use of higher solids loading is optimal when using low enzyme doses. A direct comparison of the achieved % theoretical glucose yields at 25% w/v solids loading with both a low and an excess enzyme dose suggests that ×16 enzyme dose increase only resulted in an increase from 26% to 34% of theoretical glucose yield. This suggests that low enzyme doses at high solids loading may be optimal in terms of maximising efficiency of processing large quantities of biomass (i.e., pretreated BSG). If achieving the highest g/L glucose concentration was the primary target for a feedstock and achieved % theoretical yield a secondary objective (which could be sacrificed somewhat), then the use of very high solids loading could prove very effective as high g/L glucose concentrations could be achieved. Whilst this implies that the value of the glucose in the generated feedstock is greater than the value of the spent grains (which may not be the case), it would in addition allow a large quantity of biomass to be processed more rapidly.

Three-dimensional response surface models (3D RSM) of the glucose release data looking at the simultaneous effect of variation of solids loading and enzyme dose (using Novozymes Cellic® CTec2) suggested that increasing enzyme dose from 10 to 160 FPU/g biomass at any given solids loading between 5% and 25% w/v did not significantly increase the g/L glucose concentration liberated (**Fig. 7A**). For achieving optimal g/L glucose concentrations, a low enzyme dose - high solids loading combination was confirmed to be optimal as was previously apparent as with the 2D models (**Fig. 6B**).

However, in contrast to this, increasing solids loading at any given enzyme dose was suggested to be a significant factor affecting glucose yields. 3D RSM of achieved % theoretical glucose yield data indicated that at low enzyme doses (10 FPU/g biomass), the increase in solids loading was not a significant parameter whilst at high enzyme doses (160 FPU/g biomass), increasing solids loading was a significant parameter, resulting in a considerable reduction in achieved % theoretical glucose yields (**Fig. 7B**). Increasing enzyme dose at low loadings loading (5% w/v) was suggested to have a much more significant effect on achieved glucose yield than at high solids loading (25% w/v). For optimal conversion efficiency (optimal % theoretical yields achieved), a low solids loading - high enzyme dose combination was suggested to be most suitable. Overall, the enzyme reaction kinetics seen here did not follow the traditional Michaelis–Menten model. This is likely due to the (relatively) highly insoluble slurries (with significant quantities of hydrophobic components) being used as opposed to highly soluble, dilute enzyme-substrate mixes.

### *3.4. Effect of variation of solids loading and reaction time of enzymatic saccharification on achieved glucose yields*

As described in section 3.3, a higher throughput system using a shaking incubator was utilised for the investigation of whether the optimal hydrolysis (reaction) time varied in response to variation of the solids loading used during enzymatic saccharification.

Evaluation of the 3D RSM data suggested that liberated glucose concentrations (g/L) peaked (achieving 60 g/L) at the maximal solids loading tested (25% w/v) and when the longest reaction time (72 h) was used (**Fig. 7C**). At low solids loading (5% w/v), an extension of the reaction time beyond 6 h did not result in any significant increase in glucose liberation. However, at high solids loading (25% w/v) extension of the reaction time beyond 10 h resulted in a linear increase in glucose liberation. The data suggested that in terms of g/L glucose liberation, the optimal reaction time depends heavily on the actual solids loading used for the enzymatic hydrolysis. It was shown that increasing the solids loading from 5% to 25% w/v when using a shorter reaction time (10 h) did result in a small but significant (model $R^2$: 0.78) increase in liberated glucose concentrations (g/L). However, this enhancement effect was much more pronounced when using longer reaction times (>24 h).

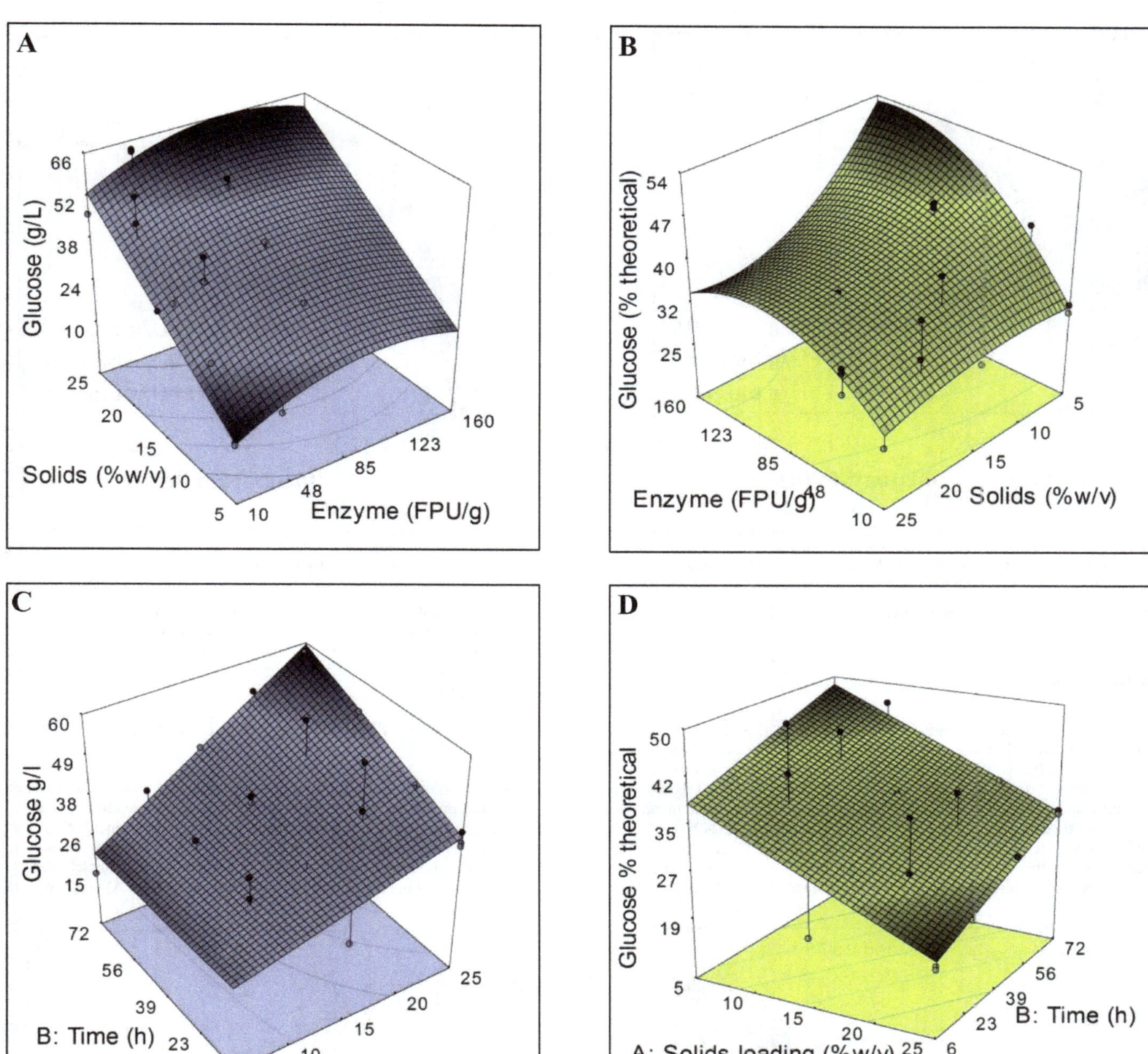


**Fig.7. A-B:** 3D response surface models showing the simultaneous effects of the factors (i) % solids-loading (w/v) and (ii) enzyme dose for saccharification. **A:** Glucose: g/L (model $R^2$: 0.94), **B:** Glucose: % theoretical yield (model $R^2$: 0.86). **C-D:** 3D response surface models showing the simultaneous effects of the factors (i) % solids loading (5-25% w/v) and (ii) reaction time (6-72 h) during enzymatic saccharification of pretreated BSG on the resultant glucose concentrations liberated. **A:** Glucose; g/l ($R^2$: 0.78), **B:** Glucose; achieved % theoretical yield ($R^2$: 0.65). **A-B:** Enzymatic saccharification conducted using Novozymes Cellic® CTec2 (10-160 FPU/g biomass) with 72 h incubation at 50°C, 5-25% w/v solids loading using standard shaking incubator method (SI). **C-D:** Enzymatic saccharification conducted using Novozymes Cellic® CTec2 (20 FPU/g biomass) with 6-72 h incubation at 50°C, 5-25% w/v solids loading using standard shaking incubator method (SI). All experiments conducted using pretreated BSG (1% HCl 121°C 30 min pre-treatment, 25% w/v solids loading).

In direct contrast to g/L glucose liberation, evaluation of the achieved % theoretical glucose yields suggested that a linear drop in conversion efficiency was observed (regardless of reaction time used) as solids loading was increased (**Fig. 7D**). As expected, the highest conversion efficiencies achieved (50% of theoretical yield) were observed when the lowest solids loading was used (5% w/v). At any given solids loading used, the extension of reaction time from 10 h to 72 h was shown to offer only a relatively small increase in achieved % theoretical glucose yield (from 40 - 50% theoretical yield at 5% w/v solids loading). This may suggest that extension of the reaction time from ca. 24 h to 72 h would not be economically viable when conducting an enzyme saccharification step at less than 25% (w/v) solids loading. However, this would depend heavily on the cost of the enzyme bioreactors used. Low cost vessels and the availability of space to house them could then justify this increased

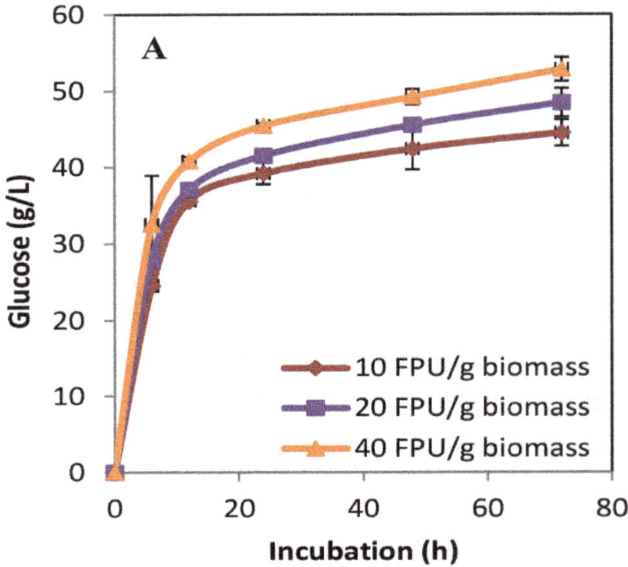


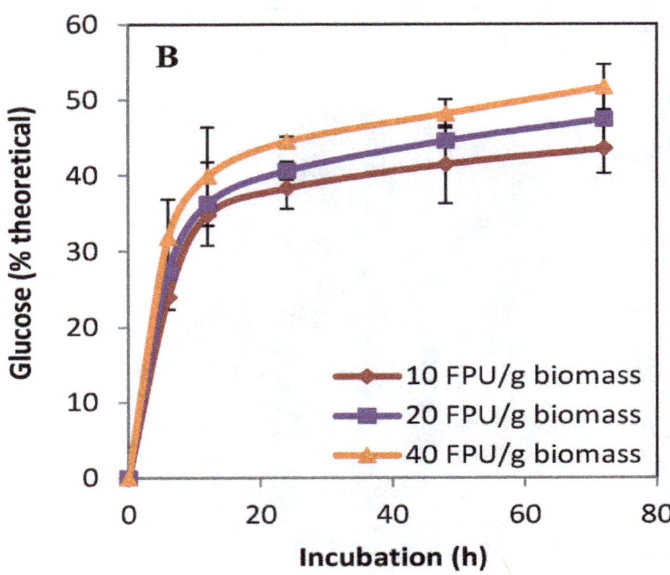


**Fig.8.** Effect of simultaneous variation of the enzyme dose and solids loading used during enzymatic saccharification on achieved glucose concentrations. **A:** Glucose (g/L) and **B:** Glucose (achieved % theoretical yield). Enzymatic saccharification conducted using Novozymes Cellic® CTec2 (10-40 FPU/g biomass) with 6-72 h incubation at 50°C, 15% w/v solids loading using standard shaking incubator method (SI) on pretreated BSG (1% HCl 121°C 30 min pre-treatment, 25% w/v solids loading). Data are the mean ± SD of 3 replicate measurements.

residence time in order to achieve higher glucose yields.

### *3.5. Effect of variation of reaction time of enzymatic saccharification at various low enzyme doses on achieved glucose yields*

Using a 6-72 h time course (again with a standard SI based method for higher throughput than the aerated HTM system), a very similar trend in liberated g/L glucose concentrations was observed across a low range of enzyme doses (10-40 FPU/g biomass; Novozymes Cellic® CTec2) when operating at a static 15% w/v solids loading (**Fig. 8A**). The predominant bulk of the total final saccharification yields were achieved by ca. 24 h (86% of the final glucose yields that were achieved when using a 72 h reaction time). Regardless of the enzyme dose used, any extension of the reaction time beyond ca. 24 h resulted in only a small increase in achieved glucose yields.

A very similar trend was observed with achieved % theoretical yields (**Fig. 8B**). Overall, the data suggested that at 15% (w/v) solids loading and when using low enzyme doses, reaction times of ca. 24 h would be optimal as doubling the reaction time to 48 h only liberated 9% more glucose (g/L).

### *3.6. Investigation of fed-batch methods on enzymatic saccharification glucose yields*

The preliminary screening of fed-batch methods (in which the required amount of pretreated BSG to achieve the desired solids loading was added in two aliquots with the second aliquot added after 24 h incubation) was also initially conducted using a high-throughput SI method as opposed to the optimised aerated HTM system. The use of this fed-batch method resulted in small but significant ($P<0.05$: one-way ANOVA) increases in

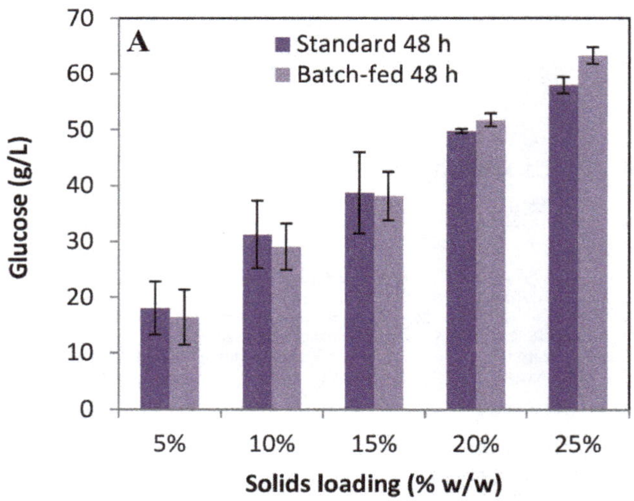


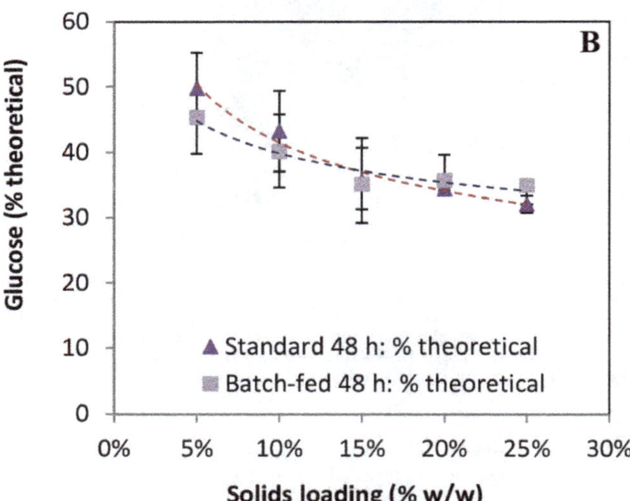


**Fig.9.** Efficacy of fed-batch enzymatic saccharification methods compared with non-fed-batch methods. **A:** Glucose (g/L) and **B:** Glucose (% theoretical yield). Enzymatic saccharification conducted using Novozymes Cellic® CTec2 (50 FPU/g biomass) with 48 h incubation at 50°C, 5-25% w/v final solids loading using standard shaking incubator method (SI) on pretreated BSG (1% HCl 121°C 30 min pre-treatment, 25% w/v solids loading. Fed-batch method had the required biomass quantity split into two equal portions with the second aliquot added after 24 h. Data are the mean ± SD of 3 replicate measurements.

final g/L glucose concentrations (ca. 12% more glucose) when solids loading exceeded 20% (w/v) in comparison with a standard, non-fed-batch method using both an identical enzyme dose and identical incubation period but with all the biomass added entirely at the start of the incubation period (**Fig. 9A**). Whilst the degree of relative enhancement of the fed-batch method at both 20% and 25% w/v solids loadings was similar (12% increases in achieved % theoretical glucose yield compared with the standard non-fed-batch method at similar final solids loading), the effect on final g/L glucose concentrations was more pronounced as solids loading increased obviously due to the presence of ever greater quantities of biomass. The achieved % theoretical glucose yields followed a similar trend with fed-batch methods only functioning slightly more effectively (than non-fed-batch methods) at very high solids loading (≥20% w/v; **Fig. 9B**). Overall, this suggested that when operating at ≥20% w/v solids loading, splitting the biomass into two equal aliquots whilst using an identical enzyme dose boosted both final g/L glucose concentrations and % theoretical yields through effectively halving the operational solids loading for the first 24 h of a 48 h saccharification period. The initial 24 h saccharification period facilitated a considerable degree of liquefaction of the slurry which likely assisted in some reduction of any mass transfer limitations possibly enabling better enzyme mobility and thus better enzyme-substrate linkages.

### *3.7. Comparison of the effect of increasing solids loading during enzymatic saccharification using both an optimised aerated HTM method and a standard SI method*

Both g/L glucose concentrations and achieved % theoretical glucose yields were suggested to be improved through the use of the aerated HTM method compared with the standard SI method with both responses ranging from 10% to 25% higher across all solids loadings tested (**Fig. 10A**). However, the degree of improvement (achieved using the aerated HTM method) for both response parameters increased as solids loading increased with the greatest and only statistically significant improvement observed at the highest solids loading tested (25% improvement in glucose yields at 25% w/v solids loading; $P<0.05$: one-way ANOVA). This suggested the HTM method overcame some of the substrate insolubility issues which limited mass transfer at very high solids loadings. Overall, it suggested higher energy mixing regimes may still play some part in facilitating effective operation of high solids loading enzymatic saccharification of pretreated lignocellulosic biomass such as BSG if very high solids loadings are to be used.

### *3.8. Consolidated (combined) enzymatic saccharification method*

Due to the variability of enzyme activity across numerous batches of experiments, only small relative differences in glucose yields were often observed in response to variation of many of the experimental parameters. As such, a consolidated saccharification method was derived using a combination of the previously suggested marginal enhancements, all together in one single step in an attempt to maximise the achieved glucose yields when operating at high solids loading. The combination of the aerated HTM system but also simultaneously incorporating a fed-batch approach when using a low cellulolytic enzyme dose (Novozymes Cellic® CTec2 dosed at 10 FPU/g biomass) and a synergistic combination of various additional carbohydrate degrading enzymes and enzymatic cofactors previously shown to enhance the saccharification yields (ferulic acid esterase, acetyl xylan esterase, Novozymes Cellic® HTec2, α-L-arabinofuranosidase, $Cu(NO_3)_2$ , and ascorbic acid) resulted in a small but significant ($P<0.05$: one-way ANOVA) improvement of 13% to 19% more glucose (both g/L and achieved % theoretical yields, compared with the use of a non-fed-batch variant of the aerated HTM method without any supplements) at 15% and 25% (w/v) solids loadings tested, respectively (**Fig. 10B**). As previously observed with the aerated HTM system (in comparison with the standard SI method), the degree of improvement for both response parameters (g/L and achieved % theoretical glucose yields) increased as solids loading increased but in this case with the greatest improvement observed at 15% w/v solids loading (19% improvement in glucose yields). Increasing solids loading to 25% w/v then reduced the relative degree of improvement in achieved glucose yields to only 15% more compared with the aerated HTM system alone. This suggested that 15% solids loading was a 'sweet spot' in terms of optimal conversion efficiency (optimal achieved % theoretical yields). This ever smaller relative increase in glucose yields even with the inclusion of all the optimisations suggested the point of diminishing returns had likely been reached. It suggested that simultaneous inclusion of all the additives (supplementary enzymes and enzymatic cofactors) alongside the main cellulolytic enzyme cocktail (Cellic® CTec2) would not likely be economically viable in terms of their incorporation into any large scale enzymatic saccharification process as use of any one supplementary additive alone would likely do the job effectively and facilitate a good degree of enhancement of saccharification. The choice of which particular additive (if any) would likely depend solely on what was



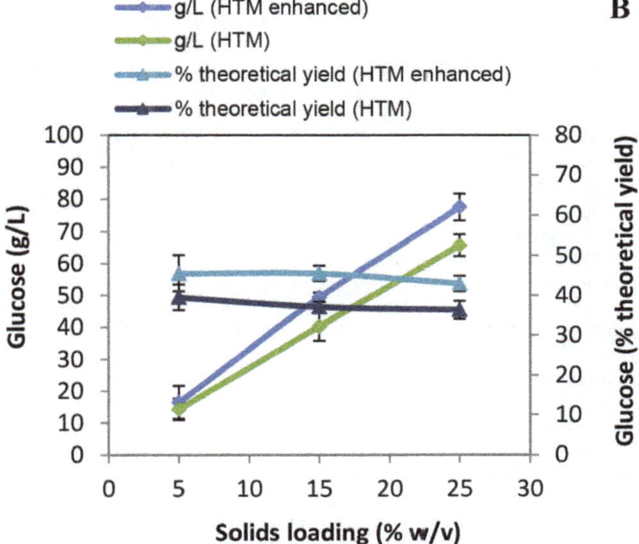


**Fig.10.** Comparison of the effect of variation of solids loading on post enzymatic saccharification glucose concentrations achieved using different methods. **A:** shaking incubator (SI) method compared with the aerated high-torque mixing (HTM) method. **B:** aerated HTM method alone compared with the consolidated method (aerated HTM with additional enzymes and cofactors). Enzymatic saccharification conducted using Novozymes Cellic® CTec2 (10 FPU/g biomass and incubated at 50°C for 72 h) on pretreated BSG (1% HCl 121°C 30 min, 25% w/w solids loading) over a 5%-25% solids loading range. **A:** Standard shaking incubator (SI) system (150 rpm) compared with aerated high-torque mixing (HTM) system (150 rpm). **B:** HTM aerated method (non-fed-batch with no additives) compared with a HTM aerated method fed-batch (biomass split into 2 × aliquots; second added after 24 h) with additional enzymes: FAE; ferulic acid esterase, AXE; acetyl xylan esterase, AFD; α-L-arabinofuranosidase, HTec2; Cellic® HTec2 xylanases (all at 100 μL doses), and $Cu(NO_3)_2$ and ascorbic acid (both at 10 mM). Data are the mean ± SD of 3

the most economically viable and readily available in the quantities required for large scale saccharification. Overall, even with all the suggested optimisation strategies employed simultaneously, only relatively low % theoretical glucose yields were achieved (43% when using 25% w/v solids loading). This could be due to a number of possible factors. Some degree of product inhibition (catabolite repression) is a possibility as cellobiose has been shown to feedback inhibit various endocellulases (Xiao et al., 2004). However, the inclusion of high concentrations of β-glucosidase enzymes within the Novozymes Cellic® CTec2 preparation should in theory combat this issue to a significant degree through cleaving any cellobiose molecules into their constituent monomeric glucose sub-units. As such, this was considered unlikely to be the cause of the low yields. In addition, HPLC analysis of all post enzymatic saccharification supernatants (the hydrolysates produced from all experiments, etc.) did not show any significant cellobiose concentrations to be present that would have possibly been an indicator of feedback inhibition. Alternatively, inhibition of the cellulolytic enzymes from deleterious product formation that may have occurred during the upstream pre-treatment process (such as liberation of lignin based hydroxycinnamic acids or furan based hemicellulose or cellulose degradation products) can be discounted in this case as all experiments were conducted using thoroughly washed biomass prior to the enzymatic hydrolysis reactions. In addition, HPLC analysis was conducted to screen for these inhibitory compounds and concluded negligible quantities were present. Another possible factor is that adsorption of various cellulases to the cellulose substrate has been experimentally shown to decrease with increasing solids loading (Kristensen et al., 2009). This could possibly be due to mass transfer limitations due to the insoluble nature of the slurries used when attempting high solids loading enzyme saccharification (the low free water content and thus low 'mobile phase' in which the enzymes can freely move throughout the media). This is particularly applicable to pretreated lignocellulosic substrates such as BSG as the upstream processing (the thermo-chemical pre-treatment step in particular) generates a wide range of sizes of fragments of the material (some of which are very small; < 106 μm) that saturates the liquid or 'mobile phase'.

Additionally, the build-up of hydrolysis end products such as high concentrations of glucose and cellobiose have been shown to physically impede the adsorption of endocellulases to cellulose (Stutzenberger and Lintz, 1986). The general observed trend of a linear increase in liberated glucose concentrations (g/L) as solids' loading was increased (for much of the experimental data presented here) would suggest that when very high solids loadings were used (>25% w/v), sufficient glucose concentrations may be present to impede any further successful adsorption (or mobility of) the CBMs. This could assist in explaining the sub-optimal achieved % theoretical glucose yields. It is the CBMs that facilitate the adsorption and subsequent activity and mobility of the critical endocellulases that initially cleave internal β-1,4 glycosidic linkages thus generating more free reducing ends that other cellulolytic enzymes can attack (Arantes and Saddler, 2010). The modification of endocellulase CBMs (including their substitution for higher binding affinity variants) in order to enhance both their activity and their resistance to high solid to liquid ratio environments thus improving their function at high solids loading has also been considered (Taylor et al., 2012).

The issue of this pseudo form of non-competitive inhibition by enzyme reaction products such as glucose could possibly be solved through investigation of simultaneous saccharification and fermentation (SSF) approaches. This would enable the selective uptake and metabolism of reaction products such as glucose thus reducing their deleterious feedback inhibitive effects. However, the system would need to be carefully designed as some provision for fermentation product removal would be required, especially in the case of bioethanol production as ethanol is a known inhibitor of cellulolytic enzymes although less so than the degree of feedback inhibition that occurs as a result of cellobiose build-up (Ooshima et al., 1985). The use of a thermophilic fermentative organism in conjunction with the high temperature optima (50°C) of Novozymes Cellic® CTec2 may enable some degree of evaporative recovery of any ethanol produced as an extremophile may tolerate short periods of temperature spikes which could be used to enhance recovery. Alternatively, a sealed system that artificially reduced the atmospheric pressure within the bioreactor to below the vapour pressure of ethanol could then enable lower evaporative temperatures. However, this would not solve the issue of the build-up of any other non-volatile metabolic waste product which may still be equally inhibitory towards various enzymes.

## 4. Conclusions

Different parameters associated with enzymatic saccharification (at high solids loading) of pretreated BSG were investigated in order to attempt to maximise the achieved glucose yields. Supplementation of the cellulolytic enzymes with additional enzymes (ferulic acid esterases, acetyl xylan esterases, and xylanases) and cofactors (ascorbate and copper) could significantly boost achieved glucose yields. Additionally, using fed-batch method and improved slurry agitation (aerated high-torque mixing) could also offer enhancements in glucose yields. From the experimental investigations, a novel, consolidated enzymatic saccharification method was developed which effectively functioned at high solids loading in an attempt to further maximise glucose yields. Glucose yields of 78 ± 4.1 g/L and ca. 43% of theoretical yield were achieved when operating at 25% w/v solids loading when using a commercially applicable low Cellic® CTec2 dose (10 FPU/g biomass). However, additional development is required in order to increase these glucose yields further when increasing the operational solids loading beyond 25% w/v.

## Acknowledgements

We gratefully acknowledge the financial support of SABMiller plc and the University of Nottingham in sponsoring this research. All four authors gratefully acknowledge the support of the BBSRC Sustainable Bioenergy Centre Programme Lignocellulosic Conversion To Ethanol (BB/G01616X/1).

## References

[1] Abe, A., Horii, S., Kameoka, K., 1979. Application of enzymatic analysis with glucoamylase, pronase and cellulase to various feeds for cattle. J. Anim. Sci. 48(6), 1483-1490.

[2] Arantes, V., Saddler, J.N., 2010. Access to cellulose limits the efficiency of enzymatic hydrolysis: the role of amorphogenesis. Biotechnol. Biofuels. 3,4.

[3] Bartolome, B., Gómez-Cordovés, C., 1999. Barley spent grain: release of hydroxycinnamic acids (ferulic and p-coumaric acids) by commercial enzyme preparations. J. Sci. Food Agric. 79(3), 435-439.

[4] Beeson, W.T., Phillips, C.M., Cate, J.H., Marletta, M.A., 2011. Oxidative cleavage of cellulose by fungal copper-dependent polysaccharide monooxygenases. J. Am. Chem. Soc. 134(2), 890-892.

[5] Cannella, D., Hsieh, C.W., Felby, C., Jorgensen, H., 2012. Production and effect of aldonic acids during enzymatic hydrolysis of lignocellulose at high dry matter content. Biotechnol. Biofuels. 5, 26.

[6] Chundawat, S.P., Balan, V., Dale, B.E., 2008. High-throughput microplate technique for enzymatic hydrolysis of lignocellulosic biomass. Biotechnol. Bioeng. 99(6), 1281-1294.

[7] Delmas, S., Pullan, S.T., Gaddipati, S., Kokolski, M., Malla, S., Blythe, M.J., Ibbett, R., Campbell, M., Liddell, S., Aboobaker, A., 2012. Uncovering the genome-wide transcriptional responses of the filamentous fungus *Aspergillus niger* to lignocellulose using RNA sequencing. PLos Genet. 8(8), e1002875.

[8] Dodd, D., Cann, I.K., 2009. Enzymatic deconstruction of xylan for biofuel production. GCB Bioenergy. 1(1), 2-17.

[9] Dönmez, G., Aksu, Z., 1999. The effect of copper(II) ions on the growth and bioaccumulation properties of some yeasts. Process Biochem. 35(1), 135-142.

[10] European Brewery Convention, 1998. Free amino nitrogen of malt by spectrophotometry, Method 4.10. Analytica-EBC, 5th ed., Fachverlag Hans Carl: Nuremburg.

[11] Evans, C.S., Dutton, M.V., Guillén, F., Veness, R.G., 1994. Enzymes and small molecular mass agents involved with lignocellulose degradation. FEMS Microbiol. Rev. 13(2-3), 235-239.

[12] Faulds, C., Sancho, A., Bartolomé, B., 2002. Mono-and dimeric ferulic acid release from brewer's spent grain by fungal feruloyl esterases. Appl. Microbiol. Biotechnol. 60(4), 489-494.

[13] Gan, Q., Allen, S., Taylor, G., 2003. Kinetic dynamics in heterogeneous enzymatic hydrolysis of cellulose: an overview, an experimental study and mathematical modelling. Process Biochem. 38(7), 1003-1018.

[14] Ghose, T., 1987. Measurement of cellulase activities. Pure Appl. Chem. 59(2), 257-268.

[15] Gilbert, H.J., Stålbrand, H., Brumer, H., 2008. How the walls come crumbling down: recent structural biochemistry of plant polysaccharide degradation. Curr. Opin. Plant Biol. 11(3), 338-348.

[16] Harris, P.V., Welner, D., McFarland, K., Re, E., Navarro Poulsen, J.C., Brown, K., Salbo, R., Ding, H., Vlasenko, E., Merino, S., 2010. Stimulation of lignocellulosic biomass hydrolysis by proteins of glycoside hydrolase family 61: structure and function of a large, enigmatic family. Biochemistry. 49(15), 3305-3316.

[17] Hodge, D.B., Karim, M.N., Schell, D.J., McMillan, J.D., 2008. Soluble and insoluble solids contributions to high-solids enzymatic hydrolysis of lignocellulose. Bioresour. Technol. 99(18), 8940-8948.

[18] Horn, S.J., Vaaje-Kolstad, G., Westereng, B., Eijsink, V.G., 2012. Novel enzymes for the degradation of cellulose. Biotechnol. Biofuels. 5, 45.

[19] Kato, A., Shimokawa, K., Kobayashi, K., 1991. Improvement of the functional properties of insoluble gluten by pronase digestion followed by dextran conjugation. J. Agric. Food. Chem. 39(6), 1053-1056.

[20] Kristensen, J.B., Felby, C., Jørgensen, H., 2009. Yield-determining factors in high-solids enzymatic hydrolysis of lignocellulose. Biotechnol. Biofuels. 2, 11.

[21] Lau, M.W., Dale, B.E., 2009. Cellulosic ethanol production from AFEX-treated corn stover using *Saccharomyces cerevisiae* 424A (LNH-ST). Proc. Natl. Acad. Sci. 106(5), 1368-1373.

[22] Leathers, T.D., 2003. Bioconversions of maize residues to value-added coproducts using yeast-like fungi. FEMS Yeast Res. 3(2), 133-140.

[23] Ooshima, H., Ishitani, Y., Harano, Y., 1985. Simultaneous saccharification and fermentation of cellulose: Effect of ethanol on enzymatic saccharification of cellulose. Biotechnol. Bioeng. 27(4), 389-397.

[24] Poutanen, K., 1988. An α-L-arabinofuranosidase of *Trichoderma reesei*. J. Biotechnol. 7(4), 271-281.

[25] Qing, Q., Yang, B., Wyman, C.E., 2010. Xylooligomers are strong inhibitors of cellulose hydrolysis by enzymes. Bioresour. Technol. 101(24), 9624-9630.

[26] Quinlan, R.J., Sweeney, M.D., Lo Leggio, L., Otten, H., Poulsen, J.C.N., Johansen, K.S., Krogh, K.B.R.M., Jørgensen, C.I., Tovborg, M., Anthonsen, A., Tryfona, T., Walter, C.P., Dupree, P., Xu, F., Davies, G.J., Walton, P.H., 2011. Insights into the oxidative degradation of cellulose by a copper metalloenzyme that exploits biomass components. Proc. Natl. Acad. Sci. 108(37), 15079-15084.

[27] Schwarz, W., Bronnenmeier, K., Krause, B., Lottspeich, F., Staudenbauer, W., 1995. Debranching of arabinoxylan: properties of the thermoactive recombinant α-L-arabinofuranosidase from *Clostridium stercorarium* (ArfB). Appl. Microbiol. Biotechnol. 43(5), 856-860.

[28] Shallom, D., Shoham, Y., 2003. Microbial hemicellulases. Curr. Opin. Microbiol. 6(3), 219-228.

[29] Stutzenberger, F., Lintz, G., 1986. Hydrolysis products inhibit adsorption of *Trichoderma reesei* C30 cellulases to protein-extracted lucerne fibres. Enzyme Microb. Technol. 8(6), 341-344.

[30] Taylor, C.B., Talib, M.F., McCabe, C., Bu, L., Adney, W.S., Himmel, M.E., Crowley, M.F., Beckham, G.T., 2012. Computational Investigation of Glycosylation Effects on a Family 1 Carbohydrate-binding Module. J. Biol. Chem. 287(5), 3147-3155.

[31] Wilkinson, S., Smart, K., Cook, D., 2014a. A comparison of dilute acid and alkali catalysed hydrothermal pre-treatments for bioethanol production from Brewers Spent Grains. J. Am. Soc. Brew. Chem. 72(2), 143-153.

[32] Wilkinson, S., Smart, K., Cook, D., 2015. Optimising the (microwave) hydrothermal pre-treatment of Brewers Spent Grains for bioethanol production. J. Fuels. DOI: 10 1155/2015/369283.

[33] Wilkinson, S., Smart, K., Cook, D., 2014b. Optimisation of alkaline reagent based chemical pre-treatment of Brewers Spent Grains for bioethanol production. Ind. Crops Prod. 62, 219-227.

[34] Xiao, Z., Zhang, X., Gregg, D.J., Saddler, J.N., 2004. Effects of sugar inhibition on cellulases and beta-glucosidase during enzymatic hydrolysis of softwood substrates. Appl. Biochem. Biotechnol. 113-116, 1115-1126.

[35] Zhang, J., Siika-aho, M., Tenkanen, M., Viikari, L., 2011. The role of acetyl xylan esterase in the solubilization of xylan and enzymatic hydrolysis of wheat straw and giant reed. Biotechnol. Biofuels. 4(1), 60.

## Supplementary Data

**Table 1S.**
Experimental reaction conditions for evaluation of various solids loadings (% w/v) and various enzyme doses (Novozymes Cellic® CTec2) during enzymatic saccharification of pretreated BSG (1% HCl hydrothermal method; 121°C, 30 min at 25% w/v solids loading) according to a D-optimal design space*.

| Run | Factor 1<br>A: Enzyme dose<br>FPU/g biomass | Factor 2<br>B: Solids loading<br>(% w/v) |
|---|---|---|
| 1 | 40 | 5 |
| 2 | 160 | 5 |
| 3 | 10 | 5 |
| 4 | 160 | 25 |
| 5 | 10 | 25 |
| 6 | 40 | 15 |
| 7 | 10 | 15 |
| 8 | 40 | 25 |
| 9 | 40 | 15 |
| 10 | 40 | 15 |
| 11 | 160 | 5 |
| 12 | 10 | 5 |
| 13 | 160 | 15 |
| 14 | 160 | 25 |
| 15 | 80 | 10 |
| 16 | 80 | 10 |
| 17 | 20 | 10 |
| 18 | 20 | 10 |
| 19 | 80 | 20 |
| 20 | 80 | 20 |
| 21 | 20 | 20 |
| 22 | 20 | 20 |
| 23 | 160 | 15 |
| 24 | 40 | 25 |
| 25 | 10 | 15 |
| 26 | 10 | 25 |
| 27 | 80 | 15 |
| 28 | 20 | 15 |
| 29 | 80 | 15 |
| 30 | 20 | 15 |

*D-optimal experimental design was created by DesignExpert (Stat-Ease, USA) using a collection of reaction combinations (enzyme doses and solids loadings) from which the D-optimal algorithm chose the treatment combinations to include in the design. This reduced the total number of experiments needed to be conducted from that of a full factorial experimental design.

**Table 2S.**
Experimental reaction conditions for evaluation of various solids loadings (% w/v) and various hydrolysis times (10-72 h) during enzymatic saccharification of pretreated BSG (1% HCl hydrothermal method; 121°C, 30 min at 25% w/v solids loading) using Novozymes Cellic® CTec2 (20 FPU/g biomass) according to a D-optimal design space*.

| Run | Factor 1<br>A: Solids loading<br>(% w/v) | Factor 2<br>B: Hydrolysis time<br>(h) |
|---|---|---|
| 1 | 25 | 10 |
| 2 | 5 | 72 |
| 3 | 25 | 72 |
| 4 | 25 | 48 |
| 5 | 5 | 72 |
| 6 | 20 | 48 |
| 7 | 15 | 48 |
| 8 | 15 | 48 |
| 9 | 5 | 48 |
| 10 | 20 | 72 |
| 11 | 15 | 10 |
| 12 | 15 | 24 |
| 13 | 25 | 10 |
| 14 | 10 | 24 |
| 15 | 25 | 72 |
| 16 | 5 | 10 |
| 17 | 5 | 24 |
| 18 | 10 | 24 |
| 19 | 25 | 72 |
| 20 | 5 | 10 |
| 21 | 10 | 48 |
| 22 | 20 | 24 |
| 23 | 15 | 72 |
| 24 | 15 | 24 |
| 25 | 5 | 72 |
| 26 | 25 | 10 |
| 27 | 10 | 72 |
| 28 | 10 | 10 |
| 29 | 20 | 24 |
| 30 | 25 | 24 |

*D-optimal experimental design was created by DesignExpert (Stat-Ease, USA) using a collection of reaction combinations (solids loadings and hydrolysis time) from which the D-optimal algorithm chose the treatment combinations to include in the design. This reduced the total number of experiments needed to be conducted from that of a full factorial experimental design.

# 19

# Evaluation of different lignocellulosic biomass pretreatments by phenotypic microarray-based metabolic analysis of fermenting yeast

Stuart Wilkinson , Darren Greetham , Gregory A. Tucker *

*Brewing Science Section, Division of Food Sciences, The University of Nottingham, Sutton Bonington Campus, Loughborough, Leicestershire LE12 5RD, U.K.*

## HIGHLIGHTS

- ➢Efficacy of pre-treatments on different lignocellulosic materials tested.
- ➢Phenotypic microarray was used to access fermentation.
- ➢Alkaline system liberated more sugar but hydrolysates not as fermentable.
- ➢Acid system had best fermentability.
- ➢Acetic acid and furfural present reduced ethanol production to 70% theoretical yield.

## GRAPHICAL ABSTRACT

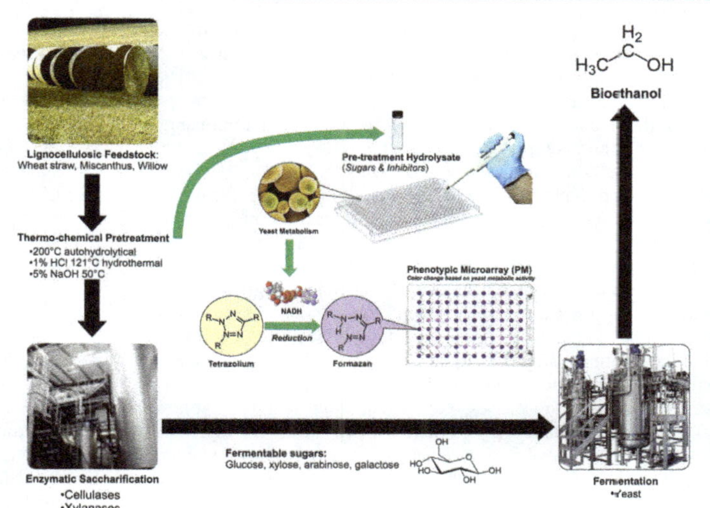


## ABSTRACT

Advanced generation biofuel production from lignocellulosic material (LCM) was investigated. A range of different thermo-chemical pre-treatments were evaluated with different LCM. The pre-treatments included; alkaline (5% NaOH at 50°C), acid (1% $H_2SO_4$ at 121°C) and autohydrolytical methods (200°C aqueous based hydrothermal) and were evaluated using samples of miscanthus, wheat-straw and willow. The liberation of sugars, presence of inhibitory compounds, and the degree of enhancement of enzymatic saccharification was accessed. The suitability of the pre-treatment generated hydrolysates (as bioethanol feedstocks for *Saccharomyces cerevisiae*) was also accessed using a phenotypic microarray that measured yeast metabolic output. The use of the alkaline pre-treatment liberated more glucose and arabinose into both the pre-treatment generated hydrolysate and also the hydrolysate produced after enzymatic hydrolysis (when compared with other pre-treatments). However, hydrolysates derived from use of alkaline pre-treatments were shown to be unsuitable as a fermentation medium due to issues with colloidal stability (high viscosity). Use of acid or autohydrolytical pre-treatments liberated high concentrations of monosaccharides regardless of the LCM used and the hydrolysates had good fermentation performance with measurable yeast metabolic output. Acid pre-treated wheat straw hydrolysates were then used as a model system for larger scale fermentations to confirm both the results of the phenotypic microarray and its validity as an effective high-throughput screening tool.

**Keywords:**
Lignocellulosic ethanol
Pre-treatment
Yeast
Metabolic output
Phenotypic microarray

* Corresponding author
E-mail address: gregory.tucker@nottingham.ac.uk

## 1. Introduction

Current environmental, economic, and social concerns regarding the sustainability of use of fossil fuels have led to considerable research into alternative energy resources such as liquid biofuel production from various biomass types (Balat, 2011; Chundawat et al., 2011a). Whilst first generation biofuel production from energy crops such as sugar cane and corn have had some success, concerns over use of potential food sources for the production of transportation fuels has been highlighted (Rathmann et al., 2010). Second generation biofuel production has been developed to minimise this issue through the use of lignocellulosic biomass (utilising the structural polysaccharide components found within the cell wall material), as this material cannot be directly used for human food production (Carvalheiro et al., 2008; Chundawat et al., 2011a).

However, there are considerable technical difficulties to be overcome in order to utilise lignocellulosic feedstocks due to their recalcitrant nature which resists biotic degradation (such as *via* enzymatic hydrolysis routes). As a consequence of this recalcitrance, the production of second generation biofuels from lignocellulosic biomass (such as wheat straw, willow, and miscanthus) normally requires chemical or thermal pre-treatment prior to enzymatic saccharification in order to boost fermentable sugar yields (Binod et al., 2012; Galbe, 2012). The aim of the pre-treatment is to improve enzymatic access to the cellulosic component within the lignocellulosic matrix through solubilisation or fractionation of various components. Different pre-treatments target different components with autohydrolytical (entirely aqueous based hydrothermal techniques) and acid catalysed hydrothermal pre-treatments primarily targeting hemicellulose removal whilst alkaline (caustic) reagents primarily targets lignin removal (Girio et al., 2010; Banerjee et al., 2011). Pre-treatment has been highlighted as the most energy intensive stage of the second generation biofuel process, and optimising protocols in terms of minimising chemical and energy inputs is crucial for any potential large scale production (Yang and Wyman, 2008).

In addition to energy efficiency, the use of excessive pre-treatment reaction conditions results in the formation of compounds which act as inhibitors to downstream processes including enzymatic saccharification and fermentations (Palmqvist et al., 1998; Chheda et al., 2007; Allen et al., 2010). These compounds are often the result of thermal or chemical degradation of liberated sugars (to furan based inhibitors) or lignin (to phenolic/hydroxycinnamic acid based inhibitors). Hydrolysates are generated directly from the pre-treatment step (the liquid fractions) and they can contain significant concentrations of these compounds. This can present significant problems (such as long yeast lag phases, poor attenuation, and sub-optimal ethanol yields) for any fermentations conducted using these hydrolysates. These hydrolysates can contain a significant quantity of supplementary fermentable sugars (in addition to those liberated after enzymatic saccharification). As such, their use is crucial for maximising the use of lignocellulosic material (LCM) as a biofuel substrate. Consequently, the assessment of the fermentation performance of these hydrolysates is a key factor in identifying issues which may reduce the efficiency of any proposed biofuel production systems using LCM. Additionally, identifying pre-treatment systems that generate excessive quantities of inhibitors is a key factor in the formulation of effective advanced generation biofuel production processes.

Wheat-straw (*Triticum aestivum* L.) is a by-product from wheat production and was chosen due to its status as the largest biomass feedstock in the Europe (Saha and Cotta, 2006). *Miscanthus × giganteus* or miscanthus is an Asian perennial rhizomatous grass and is potentially a dedicated energy crop (Bauen et al., 2010). This was chosen due to its current major use as a fuel for heat generation in power stations (DEFRA, 2007) and also as a representative of herbaceous, perennial biomass (McKendry, 2002). Perennials are often considered superior to annuals in terms of their lower pesticide and fertiliser requirements and their superior usage of nutrients (Jorgensen, 2011). Short rotation coppice (SRC) willow was also chosen as another candidate due to its high growth yields and again low fertiliser requirements as with miscanthus (Ray, 2012) and as a model for the woody biomass type (Sticklen, 2008).

Three different pre-treatments were selected and applied on commonly available LCMs which have all previously been highlighted as potential energy crops in the UK (Glithero, 2013a; Glithero et al., 2013b). The pre-treatments were all chosen as effective for LCM and consisted of an acid hydrothermal system; 1% $H_2SO_4$ at 121°C (Wilkinson et al., 2014a), an alkaline system; 5% NaOH at 50°C (Wilkinson et al., 2014b), and finally an autohydrolytical system; 200°C aqueous based (Wilkinson et al., 2015). Use of sodium hydroxide as an alkali pre-treatment has been well established and successful, as alkali does not cause sugar degradation (Chang and Holtzapple, 2000). Dilute acid pre-treatments reduce hemicellulose to its monomeric sugars making cellulose more accessible (Nguyen et al., 2000) and use of autohydrolytical methods causes hemicellulose to become solubilised making the cellulose more accessible (Chandra et al., 2007).

This paper evaluated the efficacy of different pre-treatment protocols on various LCMs. Efficacy was determined in terms of differences in liberated sugar yield (both directly into the pre-treatment generated hydrolysate and also post cellulolytic enzymatic saccharification), the degree of formation of metabolically inhibitory compounds, and the subsequent fermentation performance of the pre-treatment generated hydrolysates. The fermentation performance was then assessed using a phenotypic microarray (PM) as a novel, rapid screening tool that has previously been used to measure yeast metabolic output (Greetham, 2014; Wimalasena et al., 2014). To the author's knowledge, very few studies have been published using the PM for screening the fermentability of different biofuel feedstocks. The PM results were then confirmed using larger scale fermentations using wheat straw (with acid pre-treatment) as a model system. This validated the use of the PM as a novel, high-throughput screening tool for identifying issues with fermentability of biofuel feedstocks.

## 2. Materials and methods

### *2.1. Yeast strain and growth conditions*

*Saccharomyces cerevisiae* NCYC 2592 (www.ncyc.co.uk) was maintained on agar containing 10 g/L yeast extract, 20 g/L peptone, 20 g/L glucose, and 20 g/L agar (YPD agar) and grown in 10 g/L yeast extract, 20 g/L peptone, and 20 g/L glucose (YPD) in an orbital shaker (180 rpm) at 30°C under aerobic conditions.

### *2.2. Raw materials and inhibitors*

Inhibitory chemicals such as acetic, formic, *p*-coumaric, and ferulic acids, furfural, 5-hydroxymethylfurfural (HMF), syringaldehyde, and vanillin were all supplied by Sigma (Dorset, UK). Other chemicals were standard laboratory reagents. Wheat straw was harvested at the University of Nottingham. Willow and miscanthus were harvested at Rothamsted (BBSRC funded) and details of the harvest have been published previously (Ray, 2012). Willow and miscanthus were harvested as part of field trials conducted under the BBSRC's BSBEC renewable initiative with site permission from the Lawes Trust; grid coordinates 51.8168°N and 0.3798°W.

### *2.3. Pre-treatments of LCM (wheat straw, miscanthus, and willow)*

Acid catalysed hydrothermal pre-treatments were conducted using the protocol described previously (Wilkinson et al., 2014a). Biomass (50 g dry weight) was added to 500 mL 1% $H_2SO_4$ (w/v) in a screw-capped glass bottle (1L) to give the required solids-loading (10% w/v). This was then heated at 121°C for 30 min using a 40L bench top autoclave (Priorclave, Tactrol 2, RSC/E, UK). Alkaline pre-treatments were conducted as described previously (Wilkinson et al., 2014b). 500 mL borosilicate glass bottles with the appropriate amount of biomass (25 g) and caustic reagent (5% NaOH w/w) required to achieve the 10% (w/v) solids loading were incubated at 50°C for 12 h in a GD100 water bath (Grant, UK). Microwave-assisted autohydrolytical pre-treatment was conducted using the described protocol (Wilkinson et al., 2015c). A Monowave 300 microwave synthesis reactor (Anton Paar Gmbh, Gratz, Austria) was used. Glass G30 (30 mL) microwave reaction vessels (Anton Paar Gmbh, Gratz, Austria) with the appropriate amount of biomass (2.0 g) and water (20 mL) required to achieve the 10% (w/v) solids loading were heated at 200°C for 5 min.

After pre-treatment, samples were centrifuged at 3500 × g for 10 min and the supernatant was removed for analysis of sugars and known inhibitory compounds. The residual biomass was then re-suspended in reverse osmosis (RO) water (20 mL). The three different re-suspended pre-treatment samples all had widely different pH values: pH 14 (± 0.3) for the alkaline sample, pH 2 (± 0.3) for the acid sample, and pH 4 (± 0.3) for the autohydrolytical sample. As such, all were adjusted to pH 5.0 (± 0.1) with either glacial acetic acid or 40% NaOH (w/v) and then exhaustively washed with RO water (by repeated re-suspension and centrifugation at 3500 × g for 10 min, discarding the supernatant each time). The remaining residues were then oven-dried overnight at 60°C prior to enzymatic saccharification. All experiments were conducted in triplicate.

### *2.4. Enzymatic saccharification of pre-treated residues*

All enzyme digestions of pre-treatment residues were conducted using 24 h incubation periods at 50°C with agitation at 150 rpm (MaxQ 4358 shaking incubator, Thermo Scientific, UK). The assessment of the efficacy of different pre-treatment processes was conducted using a low solids-loading protocol (0.5% w/v) with lyophilised Celluclast® cellulase from *Trichoderma reesei* (ATTC 26921, Sigma-Aldrich, UK) using an excess of enzyme (50 FPU/g biomass) to determine the maximum sugar concentration obtainable. Pre-treated residue (200 mg) was mixed with 40 mL of a 1 g/L Celluclast® solution in 50 mM sodium citrate buffer (adjusted to pH 4.8 via glacial acetic acid) and incubated at 50°C for 24 h. Following the incubation period, the samples were centrifuged at 3500 × g for 10 min and the supernatant was sampled for quantification of sugars *via* ion chromatography (IC). The FPU was determined according to the method described by Ghose (1987). All experiments were conducted in triplicate.

### *2.5. Quantification of total monosaccharide content of pre-treated residues*

Total glucose, xylose, and arabinose concentrations were quantified from total sugar analysis using IC after complete acid hydrolysis of the pre-treated residues using the protocol described previously (Wilkinson et al., 2014b). Dried biomass (30 mg) was weighed into a heat resistant Pyrex reaction vessel (50 mL). To each tube was added 1 mL of 12 M $H_2SO_4$ and the contents were incubated at 37°C in a water bath for 1 h. Water (11 mL) was added to dilute the acid concentration to 1 M and the contents were further incubated at 100°C for 2 h. The resulting solutions were then syringe-filtered (GF/C 25 mm filter diameter/1.2 µm pore size, Whatman, USA) and the concentration of monosaccharides was quantified by IC as described in the Section 2.8.

### *2.6. High performance chromatography (HPLC) and IC*

#### *2.6.1. Quantification of weak acid based inhibitors present in the pre-treatment generated hydrolysates*

HPLC (utilising an AS-2055 Intelligent Auto-sampler and a PU-1580 Intelligent HPLC Pump; Jasco, Japan) was used for the analysis of acetic and formic acid. An aliquot (20 µL) of the hydrolysate (the liquid fraction generated directly from the pre-treatment step) was injected onto a 250 x 4.6 mm Synergi Hydro-RP column (Phenomenex, Macclesfield UK). The compounds were eluted with 20 mM potassium dihydrogen phosphate buffer (pH 2.5) at a flow rate of 1 mL/min and detected at 220 nm using a Spectro Monitor 3000 UV spectrophotometer (Milton Roy, Stone, UK). The amounts of acetic and formic acid were determined by peak area comparison (Azur software, Jasco, Great Dunmow UK) with authentic standards.

#### *2.6.2. Quantification of furan and phenolic based inhibitors present in the pre-treatment generated hydrolysates*

Furan and phenolic based inhibitors were quantified using HPLC with UV detection at 280 nm using the protocol described previously (Wilkinson et al., 2014b).

#### *2.6.3. Quantification of monosaccharides in the pre-treatment generated hydrolysates and the feedstocks produced after enzymatic saccharification of pre-treated LCM*

Liberated sugars were quantified *via* IC using an ICS 3000 system (Dionex, USA) fitted with a CarboPac PA20 column (150 mm × 3.0 mm; Dionex, USA) with pulsed amperometric electrochemical detection (PAD) using the method described by Wilkinson et al. (2014b).

### *2.7. Phenotypic microarray (PM) analysis*

The Biolog OmniLog (Biolog, Hayward, CA, USA) was used as a rapid screening tool to measure the metabolic output of the yeast when cultured in the pre-treatment generated hydrolysates. This was primarily to evaluate the response of the yeast to the inhibitors present in the hydrolysates and to identify any issues with fermentation performance. The fermentation performance of the feedstocks generated after enzymatic saccharification was not evaluated. Instead, this study concentrated specifically on the pre-treatment generated hydrolysate. As the biomass had been exhaustively washed after pre-treatment (before the enzymatic saccharification step) the feedstocks subsequently produced would have contained negligible concentrations of inhibitors.

Biolog growth medium was prepared using 0.67% (w/v) yeast nitrogen base (YNB) supplemented with 6% (w/v) glucose, and 0.2 µL of tetrazolium redox dye (dye D; specific for fungi) (Biolog, Hayward, CA, USA). Final volume was made up to 30 µL using RO sterile distilled water and aliquoted to individual wells. Microarray analysis experiments on the effects of acetic acid and furfural were set up as above but the amount of RO sterile distilled water added was modified to account for the presence of the inhibitory compounds. Stock solution (1 M) of acetic acid, was prepared using RO water; furfural was prepared as 1 M stock solutions in 100 % ethanol. Hydrolysates were spiked with the appropriate concentrations of glucose to give a 6% final solution and 0.2 µL of dye D was added. Strains were prepared for inoculation into the PM assay plates as follows. Glycerol stocks stored at -80°C were streaked onto YPD plates and incubated at 30°C for approximately 48 h. Two to three colonies from each strain were re-streaked to one section of a fresh YPD plate and incubated overnight at 30°C. Cells were then inoculated into sterile water in 20 × 100 mm test tubes and adjusted to a transmittance of 62% (~$5x10^6$ cells/mL) using a Biolog turbidimeter (Biolog, USA). Cell suspensions for the inoculums were prepared by mixing 125 µl of the above cells with IFY buffer™ (Biolog, USA) and the final volume was adjusted to 3 mL using RO sterile distilled water. Next 90 µl of the above mix was inoculated into each well in a Biolog 96-well plate. Anaerobic conditions were created using Oxygen absorbing packs (Mitsubishi AnaeroPak™System, Pack-Anaero, Mitsubishi Gas Chemicals, Tokyo, Japan) with an anaerobic indicator (Oxoid, Basingstoke, UK) and the plates were placed inside PM gas bags (Biolog, USA). The plates were then placed in the OmniLog reader and incubated for 50 h at 30°C. The OmniLog reader photographed the PM plates at 15 min intervals, and converted the pixel density in each well to a signal value reflecting cell growth and dye conversion. Dye reduction which reflects metabolic activity of cells was defined here as the redox signal intensity. After completion of the run, the signal data was compiled and exported from the Biolog software using Microsoft® Excel. In all cases, a minimum of three replicate PM assay runs were conducted, and the mean signal values are presented. All experiments were conducted in triplicate.

### *2.8. Confirmation of phenotypic microarray results using larger scale fermentations*

Fermentations using pre-treatment generated hydrolysates were conducted in 180 mL fermentation vessels (FVs). Cryopreserved yeast colonies were streaked onto YPD plates and incubated at 30°C for 48 h. These were then transferred to 200 mL of YPD and grown for 48 h in a 500 mL conical flask shaking at 30°C. Cells were harvested and washed three times with sterile RO water and then re-suspended in 5 mL of sterile water. Under control conditions, $1.5 \times 10^7$ cells/mL were inoculated in

92.5 mL of medium containing 4 % glucose, 2 % peptone, 1 % yeast extract with 7.5 mL RO water. 92.5 mL hydrolysate was spiked with 7.5 mL from an 80% glucose stock to give a final glucose concentration of 6% and buffered to a starting pH of 5 using 2M NaOH. Anaerobic conditions were prepared using a sealed butyl plug (Fisher, Loughborough, UK) and aluminium caps (Fisher Scientific). A hypodermic needle attached with a Bunsen valve was purged through rubber septum to facilitate the release of $CO_2$. All experiments were performed in triplicate and weight loss was measured at each time point. Fermentations were conducted at 30ºC, with orbital shaking at 200 rpm. All experiments were conducted in triplicate.

## 3. Results and discussion

Three different pre-treatments were employed on three commonly available LCMs. The LCMs chosen have all previously been highlighted as potential energy crops in the UK (Glithero, 2013a; Glithero et al., 2013b) and this study looked at the efficacy of the pre-treatments in terms of sugars liberated, presence of inhibitory compounds, pH, and yeast fermentation performance.

### *3.1. Liberation of sugars from LCM using a range of pre-treatments.*

The assessment of liberation of monomeric sugars from LCM into the hydrolysate (sugars liberated directly into the liquid fraction generated from the pre-treatment) was conducted following three pre-treatments. Hydrothermal pre-treatment (employing 1% $H_2SO_4$ at 121°C for 30 min) liberated significantly higher concentrations of xylose, arabinose, and glucose when compared with use of alkaline (5% NaOH at 50°C) or autohydrolytical (200°C microwave-assisted) pre-treatment methods regardless of the LCM used (**Fig. 1A**).

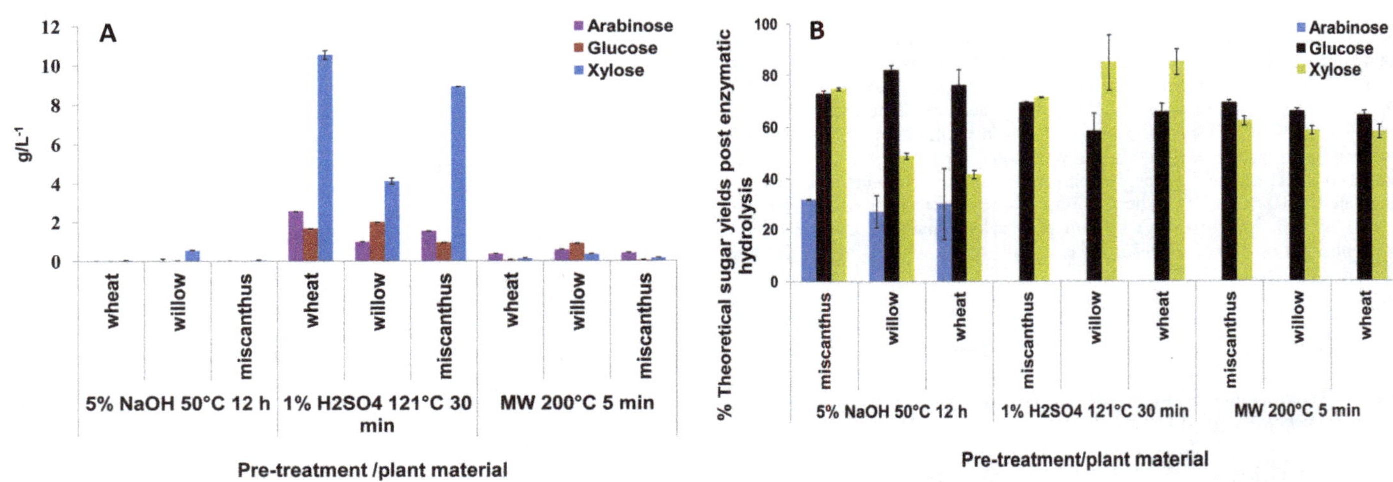


**Fig.1.** Liberation of sugars (directly into the liquid fraction generated from pre-treatment) using different pre-treatments and different lignocellulosic materials: (A) Liberation of monomeric sugars (xylose, arabinose, and glucose) by NaOH, $H_2SO_4$, and hydrothermal pre-treatment methods (200°C) on wheat, willow, or *Miscanthus* and, (B) Enzyme digestion of cellulose fraction after NaOH, $H_2SO_4$, and hydrothermal pre-treatment methods (200°C) on wheat, willow, or *Miscanthus*. Data are represented as % theoretical sugar yields post enzymatic hydrolysis. Data are representative of triplicate values with standard deviation shown.

### *3.2. Sugar yields from pre-treated biomass after enzymatic saccharification*

Assessment of the sugar levels liberated after enzymatic saccharification from the various pre-treated residues indicated that similar theoretical glucose yields (ca. 65%) were achieved from all three biomass types using either acid (1% $H_2SO_4$ at 121°C) or 200°C microwave-assisted autohydrolytical pre-treatment protocols. However, use of the alkaline (5% NaOH at 50°C) pre-treatment liberated the highest glucose yield (ca. 75% theoretical) from all LCM biomass types (**Fig. 1B**). In addition, a significant arabinose concentration was detected only in pre-treatments using the alkaline system (which only equated to ca. 30% theoretical yield). Similar % theoretical xylose concentrations were observed from all biomass types using either the acid or autohydrolytical pre-treatment systems. The alkaline pre-treatment was the only pre-treatment to exhibit specific LCM biomass type related variability in the % theoretical xylose concentrations achieved, with willow and miscanthus liberating 7% and 33% more xylose than wheat-straw, respectively. Although containing high concentrations of fermentable glucose, the fermentation performance of the feedstocks generated after enzymatic saccharification was not evaluated.

### *3.3. Liberation of acetic acid from the pre-treatment process*

Regardless of the LCM or pre-treatment employed, relatively high concentrations of acetic acid (30-75 mM: **Fig. 2A**) were present in all samples which would have been high enough to affect yeast growth rates and reduce glucose consumption (Narendranath et al., 2001). In general, use of autohydrolytical pre-treatment liberated lower concentrations of acetic acid (30-35 mM) when compared to the hydrolysates from the same LCM using either the alkaline or acid pre-treatments (61-69 mM and 45-73 mM respectively) (**Fig. 2A**).

### *3.4. Presence of weak acid, furan, and phenolic based inhibitors in the pre-treatment generated hydrolysates*

The concentrations of weak acids (i.e., *p*-coumaric, ferulic, and formic acid), furans (i.e., HMF and furfural) and additional phenolic compounds (i.e., vanillin and syringaldehyde) present in hydrolysates after pre-treatment of LCM was measured. Very low concentrations of these compounds were detected in hydrolysates generated using alkaline (5% NaOH at 50°C) pre-treatment with the exception of syringaldehyde in hydrolysates from miscanthus and wheat. Presence of 10 mM syringaldehyde has been observed to reduce ethanol productivity by approximately 30% when compared with unstressed controls without the compound present (Taherzadeh and Karimi, 2008). In the present study, 4 mM syringaldehyde was detected in hydrolysates derived from miscanthus and 2 mM in hydrolysates derived from wheat (**Fig. 2B**).

### *3.5. Metabolic profiling of yeast cultured in pre-treatment generated hydrolysates (supplemented with glucose)*

The PM was used for the metabolic profiling of the different pre-treatment generated hydrolysates (**Fig. 3**). The hydrolysates were supplemented with glucose to ensure a suitable carbon source was present (to avoid anystarvation-induced effects) and any effects of the presence of the inhibitors could then be accurately determined. The starting pH of the

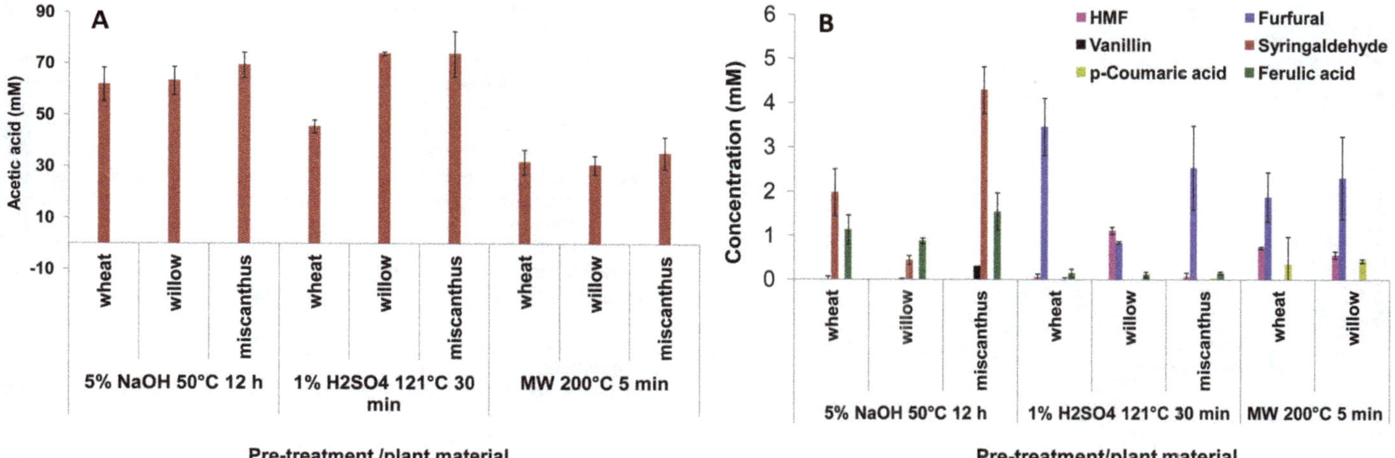


**Fig.2.** (A) Liberation of acetic acid (mM) by NaOH, $H_2SO_4$, and autohydrolytical (200°C) methods (directly into the liquid fraction generated from pre-treatment) on wheat, willow, or *Miscanthus* and (B) Liberation of HMF, furfural, vanillin, syringaldehyde, coumaric acid, and ferulic acid (all mM) by NaOH, $H_2SO_4$, and hydrothermal methods (200°C) of wheat, willow, or *Miscanthus*. Data are representative of triplicate values with standard deviation shown.

hydrolysates was adjusted to pH 5, as typical fermentations start at this pH and then subsequently typically drop to ca. pH 4.1 (Coote and Kirsop, 1976). The pH adjustment was problematic for the hydrolysates derived from the alkaline pre-treatment system as the pH (post pre-treatment) was ca. pH 14. As such, the adjustment down to pH 5 resulted in a significant increase viscosity (significant effects on the colloidal stability) which made it subsequently problematic to work with from a practical point. Alternatively, the pH of acid and autohydrolytical pre-treatment generated hydrolysates was pH 2 and pH 6, respectively, and as such, the pH adjustment of these hydrolysates did not cause a fundamental change in viscosity or physical properties of the media.

Yeast metabolic output when cultured in hydrolysates derived from the alkaline pre-treatment system was severely reduced when compared to controls containing an identical quantity of glucose but no inhibitors (**Fig. 4A**). This was possibly due to the buffering that was required to adjust the pH down to pH 5 rather than the presence of syringaldehyde. There is also the possibility that alkaline pre-treatments generate hydrolysates had reduced nitrogen levels (in particular reduced free amino nitrogen or FAN levels) when compared with other pre-treatments. Overall, yeast metabolic output was shown to be higher when using acid or autohydrolytical pre-treatment derived hydrolysates than those derived from the use of alkaline pre-treatment (**Figs. 4B and 4C**). Additionally

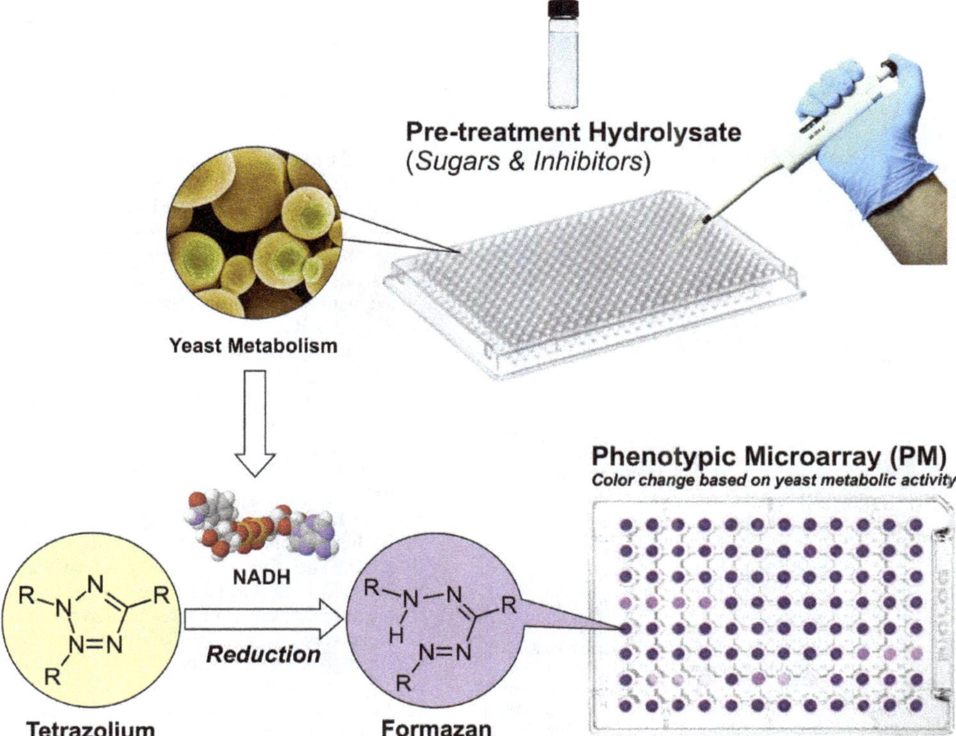


**Fig.3.** Evaluation of different lignocellulosic biomass pretreatments by phenotypic microarray-based metabolic analysis of fermenting yeast

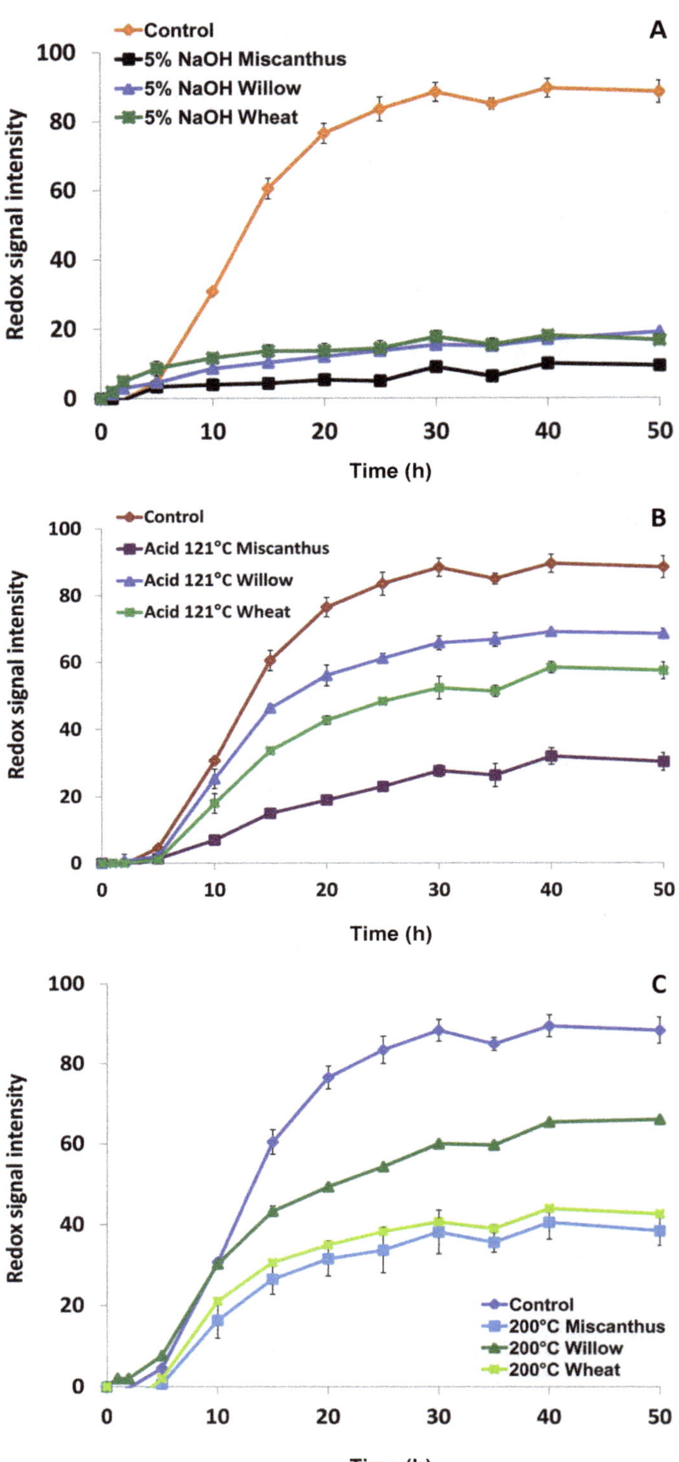


**Fig.4.** Phenotypic microarray analysis (redox signal intensity) for *S. cerevisiae* (NCYC 2592) on hydrolysates derived from NaOH, $H_2SO_4$, and hydrothermal pre-treatment methods (200°C) on wheat, willow, or *Miscanthus*. Plates were incubated at 30°C and read for 50 h, under anaerobic conditions: (A) Metabolic output (redox signal intensity) for hydrolysates from wheat, willow, and *Miscanthus* using NaOH as a pre-treatment method, (B) Metabolic output (redox signal intensity) for hydrolysates from wheat, willow, and *Miscanthus* using acid hydrolysis as a pre-treatment method, and (C) Metabolic output (redox signal intensity) for hydrolysates from wheat, willow, and *Miscanthus* using hydrothermal methods (200°C) as a pre-treatment. Data are representative of triplicate values with standard deviation shown.

willow derived hydrolysates performed better than those from wheat-straw or miscanthus although this maybe an artefact related to these fractions containing 0.8 g/L and 2 g/L glucose (respectively) which may have provided an additional metabolic boost (**Figs. 4B and 4C**). However, the presence of the inhibitory compounds did reduce metabolic output of all hydrolysates when compared with controls (without the inhibitors present). When comparing hydrolysates derived from wheat-straw using the different pre-treatments, it was noted that in terms of metabolic output, yeast performed equally well in hydrolysates derived from either autohydrolytical or acid pre-treatment systems (**Fig. 5A**).

Autohydrolytical (aqueous-based hydrothermal) or steam explosion pre-treatment systems (autohydrolytical with an additional physical, decompressive effect) have been shown to liberate high concentrations of monomeric sugars from LCMs (Tomás-Pejó et al., 2008). However, use of relatively high temperatures has been shown to liberate high concentrations of certain inhibitory compounds (Tomás-Pejó et al., 2008). In this study, we observed lower concentrations of acetic acid in the hydrolysates derived from autohydrolytical pre-treatment when compared with the other pre-treatments employed. However, the hydrolysates produced using autohydrolytical methods did still contain significant concentrations of furfural. Hydrolysates derived from both autohydrolytical and acid pre-treatment methods were characterised by the presence of furfural (0.8-3.45 mM) and HMF (0.05-1.2 mM). Furfural and HMF have been shown to be derived from degradation of sugar molecules when under acid catalysed or high temperature hydrolysis (Taherzadeh and Karimi, 2008). Previous studies have demonstrated that 20 mM furfural inhibits *S. cerevisiae* (Park et al., 2011; Greetham et al., 2014). This would suggest that the concentrations detected in these hydrolysates would not be significantly problematic for *S. cerevisiae*. However, the presence of furan compounds (such as furfural) is an unavoidable consequence of the use of relatively high temperatures (>140°C) to break LCM into fermentable sugars (Tomás-Pejó et al., 2010). Additionally, the higher temperature steam explosion methods (220°C) may generate up to 8 mM furfural (Bailey et al., 2008). This can significantly reduce the ethanol yields achieved or reduce the volumetric productivity (the ethanol output per unit time) thus reducing the process efficiency.

### *3.6. Correlation of fermentation performance of hydrolysates (derived from acid pre-treatment of wheat-straw) with fermentations of control media with an equivalent carbon loading*

The fermentation performance of *S. cerevisiae* NCYC2592 was evaluated using the pre-treatment generated hydrolysate. This hydrolysate containing pentose sugars and inhibitory compounds was derived from the acid pre-treatment of wheat-straw and supplemented with 4% glucose (included as a useable carbon source). This was then compared with the fermentation performance of the same yeast strain when using just YPD media (also containing 4% glucose). The use of the acid pre-treatment paired with wheat-straw was chosen as the model system for further investigation due to a compromise between various factors and practical constraints. Wheat straw was chosen as the LCM biomass as use of this cereal straw has been highlighted as a potential energy crop within the UK with arable farmers actually willing to sell the crop for this purpose (Glithero, 2013a). The alkaline pre-treatment system was discounted as the hydrolysates exhibited poor fermentation performance. The autohydrolytical pre-treatment utilising the microwave reactor was highly effective, but only on a small scale (i.e., limited to a maximum initial working liquid phase volume of ca. 20 mL, and with up to 50% of the initial liquid volume being absorbed by the biomass as the reactions proceed). Therefore, it was impractical for generating suitably large volumes (0.1 L) for larger trial fermentations to be conducted with adequate biological replication as the volume of recoverable hydrolysate was relatively low. Finally, the use of an acid catalysed hydrothermal pre-treatment had the capacity for generating suitably large quantities of hydrolysate. Fermentation progression (of the larger scale system using the hydrolysate from wheat-straw with acid pre-treatment) was monitored by measuring weight loss over time, which has been shown to correlate with sugar utilisation (Powell et al., 2003). It was observed that fermentation profiles from the hydrolysates correlated well with fermentation profiles when using 4% YPD as a control (**Fig. 5B**). However, during the initial stages of the fermentations, there was an approximate 2 h delay (extended yeast lag phase) between the test fermentation vessels (the pre-treatment derived hydrolysates) and the control fermentation vessels. However, even with the extended lag phase, all fermentations were still completed (attenuated) within 16 h. Quantification of ethanol concentrations produced indicated there was a ca. 36% conversion of glucose into ethanol from the hydrolysates which compares with the 51% theoretical maximum for the stoichiometric conversion of glucose into ethanol. Hydrolysates derived from the acid pre-treatment of wheat-straw were shown to contain acetic acid and furfural as the principal inhibitory compounds (**Figs. 2A and 2B**). Through the measurement of the effect of these inhibitors on yeast metabolic output, it was observed that acetic acid and furfural both individually reduced yeast metabolic output when compared with controls (**Figs. 5C and 5D**). Assays were all buffered to pH 5 prior to the start to mimic conditions present at the beginning of fermentation (Verduyn et al., 1990). Therefore, external pH-derived effects could be discounted and the deleterious effects of the inhibitors confirmed.

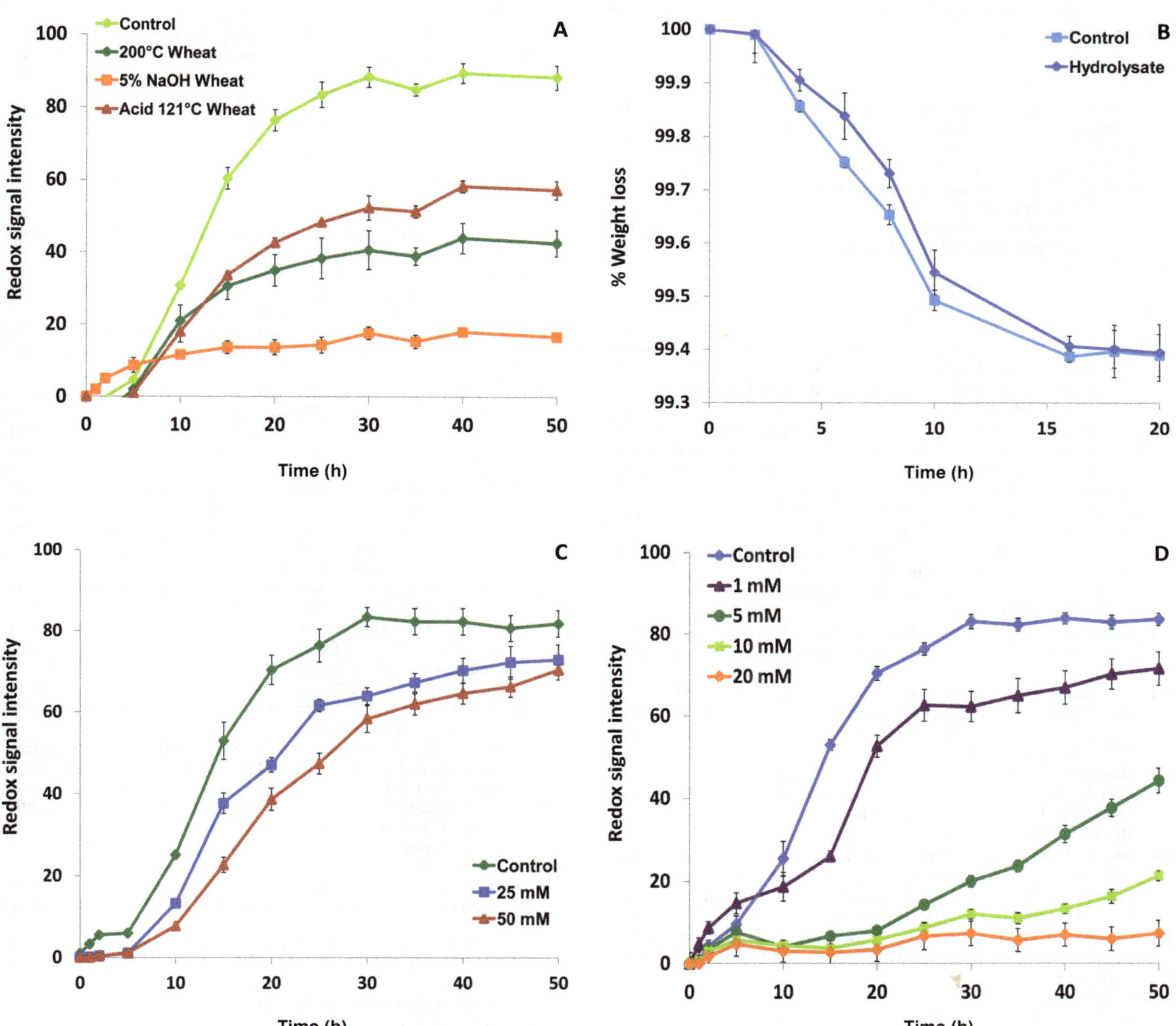


**Fig.5.** (A) Phenotypic microarray for *S. cerevisiae* (NCYC 2592) on hydrolysates derived from wheat using NaOH, $H_2SO_4$, and hydrothermal pre-treatment methods (200°C). (B) Fermentation kinetics analysis of *S. cerevisiae* NCYC 2592 using mini-fermenters on 4 % (w/v) glucose or hydrolysate derived from wheat using acid hydrolysis as a pre-treatment spiked with glucose to give a final glucose concentration of 4% (w/v), (C) Effect of 25-50 mM acetic acid on yeast metabolic output, and (D) Effect of 1-10 mM furfural on yeast metabolic output. Data are representative of triplicate values with standard deviation shown.

### 4. Conclusions

Acid pre-treatment (1% $H_2SO_4$ at 121°C) was concluded to be optimal system (in terms of sugar liberation, inhibitor generation, and fermentability) when compared to the alkaline (5% NaOH at 50°C) or autohydrolytical (200°C) pre-treatment systems. Whilst alkaline pre-treatment was shown to enhance the enzymatic saccharification yields more than the other pre-treatments (and also the generated hydrolysate had the lowest concentration of inhibitors present), the hydrolysate showed poor fermentability. The fermentability of the hydrolysates was determined using a phenotypic microarray (PM) to measure yeast metabolic activity. The PM provided a rapid, high-throughput screening tool to access fermentation performance and could be used to evaluate which pre-treatment systems where optimal for different lignocellulosic biomass.

### Acknowledgements

The research reported here was supported (in full or in part) by the Biotechnology and Biological Sciences Research Council (BBSRC) Sustainable Bioenergy Centre (BSBEC), under the programme for 'Lignocellulosic Conversion to Ethanol' (LACE) [Grant Ref: BB/G01616X/1]. This is a large interdisciplinary programme and the views expressed in this paper are those of the authors alone, and do not necessarily reflect the views of the collaborators or the policies of the funding bodies. This project is in part financed by the European Regional Development Fund project EMX05568.

### References

[1] Allen, S.A., Clark, W., McCaffery, J.M., Cai, Z., Lanctot, A., Slininger, P.J., Liu, Z.L., Gorsich, S.W., 2010. Furfural induces reactive oxygen species accumulation and cellular damage in *Saccharomyces cerevisiae*. Biotechnol. Biofuels. 3(2), 1-10.

[2] Almeida, J.R., Modig, T., Petersson, A., Hähn-Hägerdal, B., Lidén, G., Gorwa-Grauslund, M.F., 2007. Increased tolerance and conversion of inhibitors in lignocellulosic hydrolysates by *Saccharomyces cerevisiae*. J. Chem. Technol. Biotechnol. 82(4), 340-349.

[3] Antoni, D., Zverlov, V.V., Schwarz, W.H., 2007. Biofuels from microbes. Appl. Microbiol. Biotechnol.77(1), 23-35.

[4] Bailey, A.M., Paulsen, I.T., Piddock, L.J., 2008. RamA confers multidrug resistance in Salmonella enterica via increased expression of acrB, which is inhibited by chlorpromazine. Antimicrob. Agents Chemother. 52(10), 3604-3611.

[5] Balat, M., 2011. Production of bioethanol from lignocellulosic materials via the biochemical pathway: a review. Energy Convers. Manage. 52(2), 858-875.

[6] Banerjee, G., Car, S., Scott-Craig, J.S., Hodge, D.B., Walton, J.D., 2011. Alkaline peroxide pretreatment of corn stover: effects of biomass, peroxide, and enzyme loading and composition on yields of glucose and xylose. Biotechnol. Biofuels. 4(1), 16.

[7] Bauen, A.W., Dunnett, A.J., Richter, G.M., Dailey, A.G., Aylott, M., Casella, E., Taylor, G., 2010. Modelling supply and demand of bioenergy from short rotation coppice and Miscanthus in the UK. Bioresour. Technol. 101(21), 8132-8143.

[8] Binod, P., Satyanagalakshmi, K., Sindhu, R., Janu, K.U., Sukumaran, R.K., Pandey, A., 2012. Short duration microwave assisted pretreatment enhances the enzymatic saccharification and fermentable sugar yield from sugarcane bagasse. J. Renew. Energ. 37(1), 109-116.

[9] Carvalheiro, F., Duarte, L.C., Girio, F.M., 2008. Hemicellulose biorefineries: a review on biomass pretreatments. J. Sci. Ind. Res. 67(11), 849-864.

[10] Chandra, R.P., Bura, R., Mabee, W.E., Berlin, A., Pan, X., Saddler, J.N., 2007. Substrate pretreatment: the key to effective enzymatic hydrolysis of lignocellulosics? Adv. Biochem. Eng. Biotechnol. 108, 67-93.

[11] Chang, V.S., Holtzapple, M.T., 2000. Fundamental factors affecting biomass enzymatic reactivity. in: Biotechnology for Fuels and Chemicals. Humana Press. pp. 5-37.

[12] Chheda, J.N., Román-Leshkov, Y., Dumesic, J.A., 2007. Production of 5-hydroxymethylfurfural and furfural by dehydration of biomass-derived mono-and poly-saccharides. Green Chem. 9(4), 342-350.

[13] Chundawat, S.P., Beckham, G.T., Himmel, M.E., Dale, B.E., 2011a. Deconstruction of lignocellulosic biomass to fuels and chemicals. Annu. Rev. Chem. Biomol. Eng. 2, 121-145.

[14] Chundawat, S.P., Bellesia, G., Uppugundla, N., da Costa Sousa, L., Gao, D., Cheh, A.M., Agarwal, U.P., Bianchetti, C.M., Phillips Jr, G.N., Langan, P., 2011b. Restructuring the crystalline cellulose hydrogen bond network enhances its depolymerization rate. J. Chem. Soc. 133(29), 11163-11174.

[15] Coote, N., Kirsop, B., 1976. Factors responsible for the decrease in pH during beer fermentations. J. Inst. Brew. 82(3), 149-153.

[16] DEFRA, 2007. PLanting and growing miscanthus- natural England. Best Practice Guidelines.

[17] Eggeman, T., Elander, R.T., 2005. Process and economic analysis of pretreatment technologies. Bioresour. Technol. 96(18), 2019-2025.

[18] Galbe, M.Z., Zacchi, G., 2012. Pretreatment: The key to efficient utilization of lignocellulosic materials. Biomass Bioenergy. 46, 70-78.

[19] Ghose, T., 1987. Measurement of cellulase activities. Pure Appl. Chem. 59(2), 257-268.

[20] Girio, F.M., Fonseca, C., Carvalheiro, F., Duarte, L.C., Marques, S., Bogel-Lukasik, R., 2010. Hemicelluloses for fuel ethanol: a review. Bioresour. Technol. 101(13), 4775-4800.

[21] Glithero, N., Ramsden, S.J., Wilson P., 2013a. Barriers and incentives to the production of bioethanol from cereal straw: a farm business perspective. Energ. Policy. 59, 161-171.

[22] Glithero, N.J., Wilson, P., Ramsden, S.J., 2013b. Prospects for arable farm uptake of Short Rotation Coppice willow and miscanthus in England. Appl. Energy. 107(100), 209-218.

[23] Greetham, D., 2014. Phenotype microarray technology and its application in industrial biotechnology. Biotechnol. Lett. 36(6), 1153-1160.

[24] Greetham, D., Wimalasena, T., Kerruish, D., Brindley, S., Ibbett, R., Linforth, R., Tucker, G., Phister, T., Smart, K., 2014. Development of a phenotypic assay for characterisation of ethanologenic yeast strain sensitivity to inhibitors released from lignocellulosic feedstocks. J. Ind. Microbiol. Biotechnol. 41(6), 931-945.

[25] Hawkins, G.M., Doran-Peterson, J., 2011. A strain of Saccharomyces cerevisiae evolved for fermentation of lignocellulosic biomass displays improved growth and fermentative ability in high solids concentrations and in the presence of inhibitory compounds. Biotechnol. Biofuels. 4(1), 1-14.

[26] Jorgensen, U., 2011. Benefits versus risks of growing biofuel crops: the case of Miscanthus. Curr. Opin. Environ. Sustain. 3(1), 24-30.

[27] Klinke, H.B., Thomsen, A., Ahring, B.K., 2004. Inhibition of ethanol-producing yeast and bacteria by degradation products produced during pre-treatment of biomass. Appl. Microbiol. Biotechnol. 66(1), 10-26.

[28] McKendry, P., 2002. Energy production from biomass (Part 1): Overview of biomass. Bioresour. Technol. 83(1), 37-46.

[29] Mira, N.P., Teixeira, M.C., Sá-Correia, I., 2010. Adaptive response and tolerance to weak acids in Saccharomyces cerevisiae: a genome-wide view. OMICS. 14(5), 525-540.

[30] Narendranath, N., Thomas, K., Ingledew, W., 2001. Effects of acetic acid and lactic acid on the growth of Saccharomyces cerevisiae in a minimal medium. J. Ind. Microbiol. Biotechnol. 26(3), 171-177.

[31] Nguyen, Q.A., Tucker, M.P., Keller, F.A., Eddy, F.P., 2000. Two-stage dilute-acid pretreatment of softwoods. In: Twenty-first Symposium on biotechnology for fuels and chemicals. Humana Press. pp. 561-576.

[32] Palmqvist, E., Galbe, M., Hahn-Hagerdal, B., 1998. Evaluation of cell recycling in continuous fermentation of enzymatic hydrolysates of spruce with Saccharomyces cerevisiae and on-line monitoring of glucose and ethanol. Appl. Microbiol. Biotechnol. 50(5), 545-551.

[33] Park, S.E., Koo, H.M., Park, Y.K., Park, S.M., Park, J.C., Lee, O.K., Park, Y.C., Seo, J.H., 2011. Expression of aldehyde dehydrogenase 6 reduces inhibitory effect of furan derivatives on cell growth and ethanol production in *Saccharomyces cerevisiae*. Bioresour. Technol. 102(10), 6033-6038.

[34] Powell, C.D., Quain, D.E., Smart, K.A., 2003. The impact of brewing yeast cell age on fermentation performance, attenuation and flocculation. FEMS Yeast Res. 3(2), 149-157.

[35] Rathmann R., Szklo, A., Schaeffer R., 2010. Land use competition for production of food and liquid biofuels: an analysis of the arguments in the current debate. Renew. Energ. 35(1), 14-22.

[36] Ray, M.J., Brereton, N.J.B., Shield, I., Karp, A., Murphy R.J., 2012. Variation in Cell Wall Composition and Accessibility in Relation to Biofuel Potential of Short Rotation Coppice Willows. Bioenerg. Res. 5, 685-698.

[37] Saha, B.C., Cotta, M.A., 2006. Ethanol production from alkaline peroxide pretreated enzymatically saccharified wheat straw. Biotechnol. Progr. 22(2), 449-453.

[38] Sticklen, M.B., 2008. Plant genetic engineering for biofuel production: towards affordable cellulosic ethanol. Nat. Rev. Genet. 9(6), 433-443.

[39] Taherzadeh, M.J., Karimi, K., 2008. Pretreatment of Lignocellulosic Wastes to Improve Ethanol and Biogas Production: a Review. Int. J. Mol. Sci. 9(9), 1621-1651.

[40] Tomás-Pejó, E., Ballesteros, M., Oliva, J., Olsson, L., 2010. Adaptation of the xylose fermenting yeast Saccharomyces cerevisiae F12 for improving ethanol production in different fed-batch SSF processes. J. Ind. Microbiol. Biotechnol. 37(11), 1211-1220.

[41] Tomás-Pejó, E., Oliva, J.M., Ballesteros, M., Olsson, L., 2008. Comparison of SHF and SSF processes from steam-exploded wheat straw for ethanol production by xylose-fermenting and robust glucose-fermenting *Saccharomyces cerevisiae* strains. Biotechnol. Bioeng. 100(6), 1122-1131.

[42] Verduyn, C., Postma, E., Scheffers, W.A., Van Dijken, J., 1990. Energetics of *Saccharomyces cerevisiae* in anaerobic glucose-limited chemostat cultures. J. Gen. Microbiol. 136(3), 405-412.

[43] Wimalasena, T.T., Greetham, D., Marvin, M.E., Liti, G., Chandelia, Y., Hart, A., Louis, E.J., Phister, T.G., Tucker, G.A., Smart, K.A., 2014. Phenotypic characterisation of *Saccharomyces* spp. yeast for tolerance to stresses encountered during fermentation of lignocellulosic residues to produce bioethanol. Microb. Cell Fact. 13(1), 47.

[44] Yang, B., Wyman, C.E., 2008. Characterization of the degree of polymerization of xylooligomers produced by flowthrough hydrolysis of pure xylan and corn stover with water. Bioresour. Technol. 99(13), 5756-5762.

[45] Zhu, L., O'Dwyer, J.P., Chang, V.S., Granda, C.B., Holtzapple, M.T., 2008. Structural features affecting biomass enzymatic digestibility. Bioresour. Technol. 99(9), 3817-3828.

20

# Surfactant-assisted direct biodiesel production from wet *Nannochloropsis occulata* by *in situ* transesterification/reactive extraction

Kamoru A. Salam *, Sharon B. Velasquez-Orta , Adam P. Harvey

*School of Chemical Engineering and Advanced Materials (CEAM), Newcastle University, NE1 7RU, United Kingdom.*

## HIGHLIGHTS

- ➢Surfactant assisted *in situ* transesterification of wet algae was studied.
- ➢A surfactant catalyst ("ZDS") produced high yields in *Nannochloropsis occulata*.
- ➢Inclusion of SDS in $H_2SO_4$ increased FAME production in the wet algae.
- ➢The process was not adversely affected by water in the algae up to 20%.

## GRAPHICAL ABSTRACT

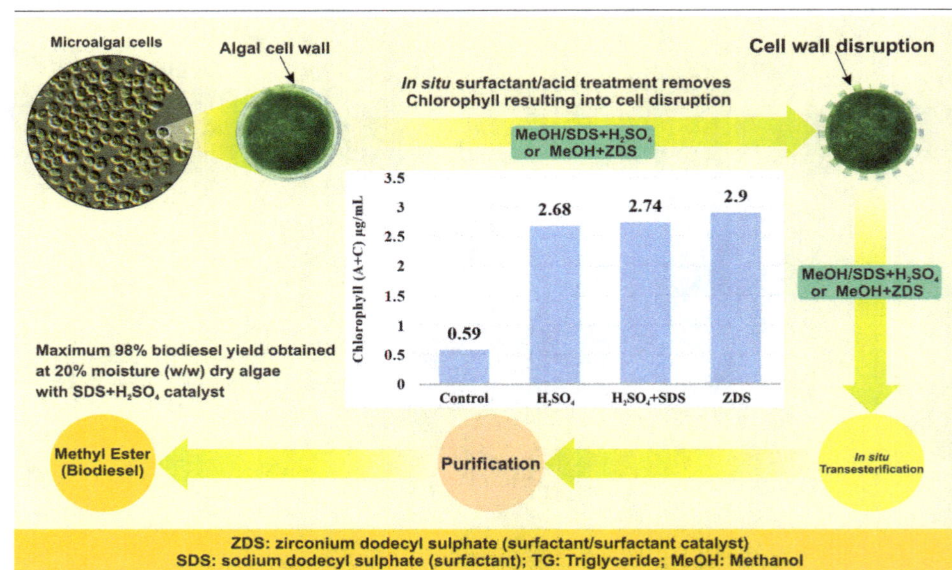


## ABSTRACT

This article reports an *in situ* transesterification/reactive extraction of *Nannochloropsis occulata* for fatty acid methyl ester (FAME) production using $H_2SO_4$, sodium dodecyl sulphate (SDS) plus $H_2SO_4$ and zirconium dodecyl sulphate (ZDS). A maximum 67 % FAME yield was produced by ZDS. Effect of inclusion of sodium dodecyl sulphate (SDS) in $H_2SO_4$ for FAME enhancement and water tolerance was also studied by hydrating the algae with 10 % - 30 % distilled water (w/w) dry algae. Treatment with SDS in $H_2SO_4$ increases the FAME production rate and water tolerance of the process. Inclusion of SDS in $H_2SO_4$ produced a maximum 98.3 % FAME yield at 20 % moisture in the algae. The FAME concentration began to diminish only at 30 % moisture in the algae. Furthermore, the presence of a small amount of water in the biomass or methanol increased the lipid extraction efficiency, improving the FAME yield, rather than inhibiting the reaction.

**Keywords:**
Wet microalgae
Reactive extraction
Biodiesel (Fatty acid methyl ester)
Surfactant
Surfactant catalyst

* Corresponding author
E-mail address: kamorusalam@gmail.com ; kamoru.salam@ncl.ac.uk

## 1. Introduction

The need to produce alternative renewable transport fuels has generated considerable global interest in biodiesel (Meng et al., 2009). Consequently, different biodiesel feedstocks have been explored, including food oil crops (Zeng et al., 2008), non-food oil crops such as *Jatropha curcas* (Kasim and Harvey, 2011), and microalgae (Wahlen et al., 2011; Velasquez-Orta et al., 2012). Food oil crops are not sustainable, as freshwater and considerable hectares of arable land are required for their cultivation (Chisti, 2007). On the other hand, non-food oil crops and waste oil can only supply limited quantities of biofuels, so cannot meet world transport fuels requirements.

There are still a number of challenges for algae to be used as fuel feedstocks including limited supply of concentrated $CO_2$, full utilisation of nitrogen or phosphorous nutrients, adverse effect of small quantity of fresh water even if marine algae is used, and efficient utilisation of algal residues after oil extraction (Chisti, 2013). Additionally, economic construction of large algae photobioreactors, and reducing the drying costs, perhaps by increasing the water tolerance of the reaction step need to be done for micro algal biodiesel to become a commercial reality. Regardless of these challenges, microalgae could still serve as alternative biodiesel feedstock as it has short growing time, high lipid productivity while it is capable of capturing concentrated $CO_2$ and can potentially be used in waste water remediation.

Biodiesel can be made either by reactive extraction ("*in situ* transesterification") (Wahlen et al., 2011; Velasquez-orta et al., 2012) or by two step transesterification of pre-extracted oil (Eze et al., 2014). A major advantage of *in situ* transesterification over the two step transesterification is that it reduces the number of process steps (by eliminating the solvent extraction steps) by contacting the biomass directly with the reactants. This could reduce the cost of biodiesel production. However, the major drawback of *in situ* transesterification is that it requires a high molar ratio of methanol to oil. The need to recycle the unreacted methanol (over 94 % of it) increases the process costs. Additionally, extraction of intracellular lipids from microalgae requires a significant excess of solvent because of the chemical resistance and structural toughness of algal cell walls (Gerken et al., 2012).

The relatively low permeability of polar solvents such as methanol and ethanol, as well as non-polar solvents such as hexane through the walls of dried oil-bearing cells can significantly reduce lipid extraction effectiveness, but it can be improved by addition of a small amount of water (Cohen et al., 2012). The water swells the cellular structure of polysaccharide-containing biomass, which increases the solvent permeability through the cell walls (Cohen et al., 2012). Similarly, the inclusion of water in alcohol such as methanol or ethanol was effective for extraction of polar lipids such as phospholipids or glycolipids (Zhukov and Vereshchagrin, 1981). Polar lipids are the major components of algal cell walls. Their removal from micro algal cell walls compromises their integrity, which can improve fatty methyl ester (FAME) recovery during *in situ* transesterification.

The most common homogeneous catalysts for *in situ* transesterification of microalgae are NaOH and $H_2SO_4$. NaOH is seldom used in microalgae if its lipids contain high free fatty acid (FFA) perhaps due to long term storage (Chen et al., 2012) to prevent soap formation (Canakci and Gerpen, 1999; Ma and Hanna, 1999).When $H_2SO_4$ is used, a high concentration of the catalyst is always required to achieve high yields (Wahlen et al., 2011; Velasquez-Orta et al., 2013). However, the need to neutralise the unreacted acid in the product streams will increase operating costs.

A surfactant catalyst (cerium (III) trisdodecyl trihydrate) has been evaluated for a two-step FAME production from soybean oil and oleic acid (Ghesti et al., 2009). The authors concluded that the surfactant catalyst efficiently promoted the transesterification of triglycerides and the esterification of free fatty acids. Similarly, use of cetyltrimethylammonium bromide (CTAB) (a cationic surfactant) with an alkali catalyst resulted in an increased FAME yield and reduction in catalyst concentration during *in situ* transesterification of *Jatropha curcas* by acting as a phase transfer catalyst (Hailegiorgis et al., 2011).

Park et al. (2014) reported that inclusion of sodium dodecyl benzene sulfonate (SDBS) in $H_2SO_4$-catalysed hot water enhanced extraction of FFA and lipids from *Chlorella vulgaris*. They reported that the inclusion of SDBS in $H_2SO_4$ significantly reduced the amount of $H_2SO_4$ required to convert the pre-extracted algal oil into FAME using a two-step transesterification. Inclusion of sodium dodecyl sulphate (SDS) in water has also been reported to increase oil extraction from canola seeds (Tuntiwiwattanapun et al., 2015). In a different study, SDS has been used for lysing cells to recover intracellular components (Brown and Audet, 2008).

Effect of inclusion of SDS in $H_2SO_4$ for a direct FAME production from wet microalgae has not been investigated. Similarly, an *in situ* transesterification of microalgae by a surfactant catalyst has not been reported in the literature. Therefore, this paper reports on the usage of zirconium dodecyl sulphate ("ZDS") (a surfactant catalyst) to catalyse *in situ* transesterification of *Nannochloropsis occulata.* Cell wall disruption by ZDS was explored for FAME enhancement.

In addition, the inclusion of SDS in $H_2SO_4$ was used in this report for improving water tolerance of the *in situ* transesterification of *N. occulata*. This is entirely a new approach to produce biodiesel from wet microalgae through *in situ* transesterification. It is worth quoting that even small amounts of water have been reported to significantly decrease conversion during a two-step transesterification of vegetable oil (Canakci and Gerpen, 1999). On the other hand, complete drying of algae is energy intensive, which significantly increases the cost of algae pre-treatment. Hence, the findings of the present study are important, as the significant amounts of energy required to dry microalgal biomass or microalgal oil to the levels required in a two-step biodiesel production render the process uneconomic, and is currently one of the major technical challenges to micro algal biodiesel production.

## 2. Materials and methods

### *2.1. Microalgae culture and their major biochemical compositions*

Concentrated wet *N. occulata* was purchased from Varicon Aqua Solutions (London, UK). Guldhe et al. (2014) has shown that there was no significant differences in the lipid extraction yield of *Scenedesmus* sp. dried by three techniques: freeze-drying, oven- drying, and sun-drying. Therefore, a frozen sample was freeze-dried at -40°C for ~24 h in a Thermo Modulyo D Freeze Dryer as this method is faster than the other drying techniques. A moisture analyser was used to further dry the algae at 60°C to preserve its biochemical compositions (Widjaja et al., 2009) until their moisture remained constant. The moisture content of the resulting dry microalgae was taken as 0 % (w/w dry algae). The total lipids content were measured using the method described by Folch et al. (1956). The FFA and cell wall lipids (phospholipids and glycolipids) of the species were measured using the solid phase extraction method of Kaluzny et al. (1985).

### *2.2. Determination of maximum FAME content*

The maximum FAME concentration was quantified using the procedure described by Garces and Mancha (1993). A methylating mixture of methanol, toluene, 2,2-dimethoxypropane, and sulphuric acid at a volumetric ratio of 39:20:5:2 was prepared. The mixture was thoroughly mixed using a vortex mixer. A homogeneous mixture containing 3.3 mL of the methylating mixture and 1.7 mL of heptane was added to 0.2 g microalgae and vortexed well. After this, the mixture was transesterified in an IKA incubator at 60°C; 450 rpm for 12 h. Subsequently, the acid catalyst was neutralised with calcium oxide (CaO) to quench the reaction. The resulting upper FAME layer was carefully pipetted into a pre-weighed centrifuge tube and weighed. After that, it was prepared for FAME analysis and its concentration was measured by gas chromatography. The maximum FAME content in the sample was calculated by multiplying the FAME concentration obtained by the mass of the upper FAME layer.

### *2.3. Catalyst synthesis*

Zirconium (IV) dodecyl sulphate (Zr $[OSO_3C_{12}H_{25}]_4$) was synthesised using the modified method presented by Zolfigol et al. (2007) as follows by inclusion of 4 % KCl (w/w zirconium dodecyl sulphate solution):

(i) 2.86 g (8.9 mmol) of zirconium oxychloride octahydrate (Sigma Aldrich, UK) was dissolved in 100 ml of distilled water at room temperature;

(ii) 12.13 g (42 mmol) of sodium dodecyl sulphate (VWR, UK) was put in a three-neck 500 ml round bottom flask. Then, 300 ml of distilled water was added to this at room temperature;
(iii) A zirconium oxychloride octahydrate solution was added to the sodium dodecyl sulphate solution whilst mixing at 500 rpm and stirred for 30 min;
(iv) 4 % KCl (w/w zirconium dodecyl sulphate solution) was added to increase catalyst recovery;
(v) The precipitate was centrifuged and washed repeatedly with 150 mL distilled water;
(vi) The resulting white solid was further calcined at 80°C for 4 h and was then dried in a desiccator (Duran vacuum desiccator).

### *2.4. Quantification of cell disruption after in situ transesterification*

The amount of chlorophyll extracted from the microalgae has been correlated with cell wall disruption by Gerde et al. (2012). The total chlorophyll A and C obtained after the *in situ* transesterification by the different catalysts was measured using a modified version of the method previously described by Gerde et al. (2012). To study the extent of cell disruption in *N. occulata*, 0.47 mL of methanol was added to a 100 mg of dried microalgae in a 2.5 mL tube followed by the addition of 100 % $H_2SO_4$ (w/w oil). To another tube containing the same amount of microalgae, methanol, $H_2SO_4$, and 9 mg SDS was added to study the effect on cell disruption by including SDS in $H_2SO_4$. A third test tube was used with 100 % ZDS (w/w lipids), 100 mg of microalgae, and 0.47 mL of methanol. The reactions were allowed to progress for 24 h, at 32°C to avoid degradation of the chlorophyll at a stirring rate of 450 rpm using IKA KS 4000 icontrol incubator shaker (IKA, Germany). At the end of the reaction, the samples were centrifuged at 17,000 ×g for 10 min using an Accu Spin Micro 17 centrifuge (Fisher Scientific, UK). Methanol was used as blank. The absorbance of the supernatant obtained was measured at 664, 647, and 630 nm and the chlorophyll concentrations in µg /mL were calculated using the formulae presented by Jeffrey and Humphrey (1975) (**Eqs. 1 and 2**):

$$Chla = 11.93\,A_{664} - 1.93A_{647} \qquad \text{Eq. 1}$$

$$Chlc = -3.73\,A_{664} + 24.36\,A_{630} \qquad \text{Eq. 2}$$

Where $Chla$ is chlorophyll a and $Chlc$ is chlorophyll c.

### *2.5. Experimental designs*

An 8.5 mol. $H_2SO_4$/(mol. lipids) which equals to 100 % (w/w lipids) was used. ZDS was fixed as 100 % ZDS (w/w lipids). These amounts of catalysts used in this study were based on the optimum of 100 % $H_2SO_4$ (w/w oil) reported by Ehimen et al. (2010).

A 9 mg of SDS was added to $H_2SO_4$ to study the effect of combination of a surfactant and homogeneous $H_2SO_4$ catalyst on FAME yield. This amount of SDS was significantly greater than 2 mol. SDS/(mol. oil) reported to be enough to solubilise the phospholipid bilayer (Tan et al., 2002). The molar ratio of methanol to lipid was 600:1, which equals to 0.0047 mL/(mg algae cells). A temperature of 60°C was used for all the experiments as most previous reports on *in situ* transesterification of microalgae were optimised at 60°C (Haas and Wagner, 2011; Li et al., 2011; Velasquez-Orta et al., 2013). An 880 g/(mol.) was the average molecular mass of the oil used to calculate the entire molar ratios. Rehydrated samples of *N. occulata* were prepared by adding 10 %, 20 %, and 30 % of distilled water (w/w dry algae), then allowing the samples to equilibrate for 1 h. The resulting wet biomass was then transesterified using $H_2SO_4$, with or without SDS, to isolate the water tolerance effect.

All *in situ* transesterification were conducted in 15 mL glass tubes containing 100 mg of microalgae. The tubes were loaded in an IKA KS 4000 icontrol incubator shaker (IKA, Germany) and kept at a constant temperature of 60°C. A high stirring rate of 450 rpm was used to prevent mass transfer limitations. The acid catalyst in each sample taken at each specified *in situ* transesterification was neutralised with CaO to quench the reaction. The biomass was separated from the liquid by centrifugation. The biodiesel filtrate (a mixture of methanol, FAME, and by-products) was stored in pre-weighed tubes and weighed. The FAME concentration in the biodiesel filtrate was measured by gas chromatography, as explained in the Section 2.6.

### *2.6. Analytical techniques*

The Standard UNE-EN 14103 (2003) was used to determine the FAME concentration after the *in situ* transesterification. The biodiesel filtrate was mixed with 0.1 mL of an internal standard solution: methyl heptadecanoate (Sigma Aldrich, UK, 10 mg/(mL methanol) in 2 mL vials. Then, 1µL of the homogeneous mixture was injected into the GC and data was collected using the Data Apex Clarity software (UK). The gas chromatograph was operated at the following conditions: carrier gas: helium, 7 psi; air pressure, 32 psi; hydrogen pressure, 22 psi, and capillary column head pressure, 4.5 psi. The carrier gas flow rate was 2 mL/min. The oven temperature was maintained at 230°C for 25 min. Heat rate was 15°C/min; initial temperature was set at 150°C and held for 2 min; final temperature was set at 210°C and held for 20 min; injection temperature was 250°C while detector temperature was 260°C. The column used was CP WAX 52 CB 30 m×0.32 mm (0.25 µm) (Agilent, Netherlands). The mass of FAME obtained in the biodiesel-rich phase from the experiments was calculated by multiplying the mass of the final biodiesel mixture obtained and the FAME concentration measured by the GC. The FAME yield was calculated by dividing the mass of FAME obtained by the maximum FAME available in the algae (**Eq. 3**).

$$\text{FAME Concentration (C)} = \frac{(\sum A) - A_{Ei}}{A_{Ei}} \times \frac{C_{Ei}V_{Ei}}{m} \times 100\ \% \qquad \text{Eq. 3}$$

Where $\sum A$ is the total peak areas from C12 - C20:1, $A_{Ei}$ is the peak area of the methyl heptadecanoate, $V_{Ei}$ stands for the volume in ml of the methyl heptadecanoate used, $C_{Ei}$ is the concentration in mg/(mL of the methyl heptadecanoate solution), and $m$ is the mass of the sample in mg.

The mass of the methyl ester in the sample was calculated by multiplying the FAME concentration (C) by the mass of the biodiesel filtrate from the *in situ* transesterification (**Eq. 4**).

$$\text{Mass of the methyl ester (mg)} = \text{C (\%)} \times \text{w (mg)} \qquad \text{Eq. 4}$$

Where w is the mass of the biodiesel filtrate.

Yield (% w/w) was the determined by comparing the mass of methyl ester obtained with the maximum FAME in the sample as follows (**Eq. 5**):

$$\text{Yield (\% w/w)} = \frac{Mass\ of\ methyl\ ester\ from\ the\ experiments\ (mg)}{Mass\ of\ the\ maximum\ FAME\ in\ the\ sample\ (mg)} \times 100\ \% \qquad \text{Eq.5}$$

## 3. Results and discussion

### *3.1. In situ transesterification using $H_2SO_4$*

The amount of total lipids was determined as 17±0.8 % (w/w dry algae) while the FFAs were determined as 18.3±2.4 % (w/w total lipids).This level of FFA necessitates the use of acid rather than base catalysts. Lotero et al. (2005) reported an upper limit of 0.5 % FFA content to prevent saponification for two-step alkali-catalysed transesterification. **Figure 1** shows that the FAME yield increased with increasing the reaction time as expected. The maximum FAME yield was 53.8±8 % occurring at 24 h.

Increasing the acid concentration to 0.15 µL/(mg algae) resulted in increased FAME yield from 53 to 87 %, in 24 h. El-shimi et al. (2013) observed a 53% increase in FAME yield during $H_2SO_4$-catalysed *in situ* transesterification of *Spirulina platensis* by increasing acid volume from 0.0016 to 0.19 µL/(mg algae). Other researchers also reported increases in the yield of biodiesel with an increase in acid concentration during acid-catalysed *in situ* transesterification of microalgae (Wahlen et al., 2011; Velasquez-Orta et al., 2013). One reason for this is that acids can be involved in other reactions, such as hydrolysis of carbohydrates during acid-catalysed *in situ* transesterification as well. Consequently, higher acid concentrations may be required to achieve high FAME yields.

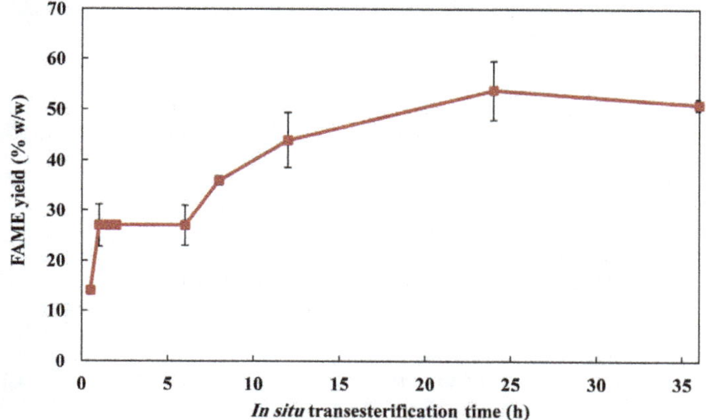


**Fig.1.** Reactively extracted FAME profile of *Nannochloropsis occulata* with $H_2SO_4$ catalyst. Process conditions: 600 mol methanol/(mol lipids) = 0.47 mL methanol/(mg algae), agitation rate = 450 rpm, temperature = 60°C, mass of microalgae = 100 mg, 8.5 mol $H_2SO_4$/ (mol lipids) = 0.087 μL/ (mg biomass).

### 3.2. *In situ transesterification using SDS/$H_2SO_4$*

The total amount of phospholipids and glycolipids in the *N. occulata* was determined as 50±0 % (w/w total lipids). A 3.2 mol SDS/(mol lipids) was added to $H_2SO_4$ to study its effect on FAME enhancement. As mentioned earlier, this amount of SDS in $H_2SO_4$ was significantly greater than 2 mol SDS/(mol phospholipids) required to effectively solubilise the phospholipids bilayers as reported by Tan et al. (2002). The effect of the inclusion of SDS in $H_2SO_4$ on FAME yields for *N. occulata* is shown in **Figure 2**.

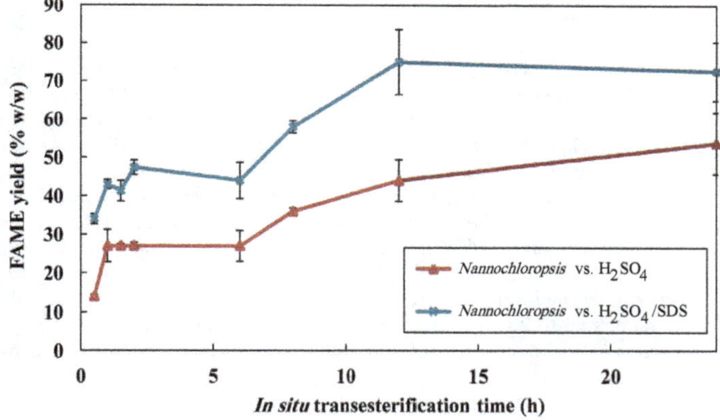


**Fig.2.** Reactively extracted FAME profile of *Nannochloropsis occulata* with $H_2SO_4$ *vs.* SDS plus $H_2SO_4$. Process conditions: 600 mol methanol/(mol lipids) = 0.47 mL methanol/(mg algae), 8.5 mol H2SO4/ (mol lipids) = 0.087 μL/ (mg algae), agitation rate = 450 rpm, temperature = 60°C, mass of microalgae = 100 mg.

It can be seen clearly in the figure that the inclusion of SDS in $H_2SO_4$ caused higher FAME yields compared with the $H_2SO_4$ alone at each data point. At 24 h, a 72.6 ± 7.7 % maximum FAME yield was obtained using $H_2SO_4$/SDS while a 53.8 ± 8 % FAME yield was obtained in this species at the same duration with $H_2SO_4$ alone. This FAME yield represents 35 % increase. This is significantly higher than the 11 % increase obtained by the inclusion of cetyltrimethylammonium bromide (CTAB) (a surfactant) in NaOH for *in situ* ethanolysis of *Jatropha curcas* L (Hailegiorgis et al., 2011), it is difficult to attribute this to the effect of surfactant though, given the different catalysts used.

### 3.3. *In situ transesterification with surfactant catalyst ("ZDS") vs. $H_2SO_4$*

The performance of the synthesized "surfactant catalyst" (zirconium dodecyl sulphate, or "ZDS") for FAME production from *N. occulata* was compared with the FAME yield obtained using $H_2SO_4$ alone as shown in **Figure 3**.

As can be seen in the figure, the FAME yield produced by both catalysts increased with increases in time as expected. FAME production rate by the ZDS was greater than that produced by $H_2SO_4$ between 12-36 h. This result shows that *in situ* transesterification of *N. occulata* could be catalysed by ZDS and that ZDS performed more efficiently than the conventional homogeneous $H_2SO_4$ catalyst.

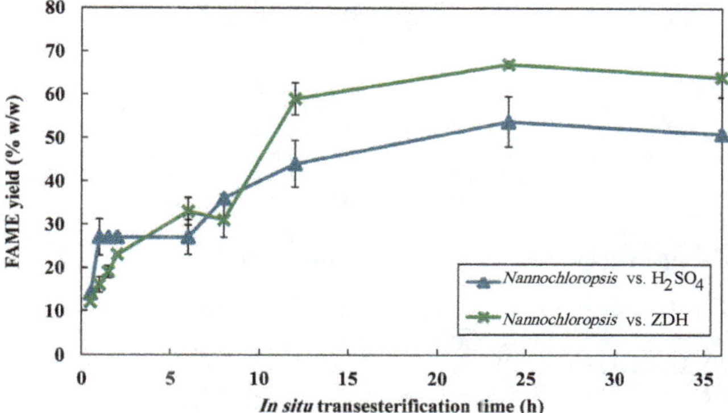


**Fig.3.** Reactively extracted FAME profile of *Nannochloropsis occulata* with $H_2SO_4$ vs ZDS. Process conditions: 600 mol methanol/ (mol lipids) = 0.47 mL methanol/(mg algae), 8.5 mol $H_2SO_4$/ (mol lipids) = 0.087 μL/ (mg algae), 100 % ZDS (w/w lipids), mass of microalgae = 100 mg, agitation rate = 450 rpm, temperature = 60°C.

### 3.4. *Mechanisms of enhancement of FAME yield by the ZDS catalyst*

The differences in the FAME production by the different catalysts used could be explained in terms of the chlorophyll extracts after the *in situ* transesterification, as shown in **Table 1**.

**Table 1.**
Chlorophyll content as a measure of cell disruption in *Nannochloropsis occulata.*

| Catalyst | Total chlorophyll (μg/mL) | Statistical analysis (P, *t* tests) |
|---|---|---|
| Control experiment | 0.59±0.02 | - |
| Acid | 2.68±0.12 | 0.01 |
| Acid+SDS | 2.74±0.19 | 0.03 |
| ZDS | 2.90±0.29 | 0.03 |

Total chlorophyll, i.e., chlorophyll A+C. Process conditions: 600 mol methanol/(mol lipids) = 0.47 mL methanol, 8.5 mol $H_2SO_4$ / (mol lipids) = 0.087 μL $H_2SO_4$/(mg algae), agitation = 450 rpm, temperature = 32°C mass of microalgae = 100 mg, mass of SDS = 9 mg, 100 % ZDS/(w/w lipids), reaction time = 24 h. The control experiment contained no catalyst but other conditions were the same.

Chlorophyll concentration has been positively correlated with cell wall disruption (Gerde et al., 2012). Based on this measurement, $H_2SO_4$, $H_2SO_4$+SDS, and ZDS significantly disrupted the cells (i.e. $p<0.05$) than the control experiment but there was no significant differences in cell wall disruption between $H_2SO_4$ and $H_2SO_4$/SDS even though there was a significant difference between the FAME yields as presented in **Table 2**. However, the highest chlorophyll extract was produced when using ZDS. Clearly, ZDS disrupted *N. occulata's* cell wall more effectively than $H_2SO_4$ which explains why it produced greater FAME yield than $H_2SO_4$ alone.

$H_2SO_4$ concentrations of 8.5 and 15 mol/(mol lipids) were equivalent to 0.326 and 0.578 mmol $H^+$, respectively. Increase in $H_2SO_4$ concentration from 8.5 to 15 mol/(mol lipids) resulted in increases in FAME production rate. The maximum FAME yield produced at 15 mol/(mol lipids) was greater than that produced by ZDS. However, 100 % ZDS (w/w lipids) used was equivalent to 0.0624 mmol $H^+$ indicating that ZDS was more efficient on a mass for mass basis than $H_2SO_4$ catalyst. The highest FAME yield (98%) was obtained using SDS+$H_2SO_4$ at 20% moisture content in the microalgae indicating that moisture did not adversely affect this process at this level when catalyst/surfactant was used.

**Table 2.**
Maximum FAME yields from *Nannochloropsis occulata.*

| Catalyst | FAME yield % (w/w) | Reaction time (h) |
|---|---|---|
| ${}^{a}H_2SO_4$ | 54±8 | 24 |
| ${}^{b}H_2SO_4$ | 87±2 | 24 |
| SDS + ${}^{a}H_2SO_4$ | 73±7.7 | 24 |
| SDS + ${}^{b}$H2SO4 | 98 ± 6.7 | 24 |
| ZDS | 67±1 | 24 |

${}^{a}$H2SO4 = 8.5 mol/(mol lipids); ${}^{b}H_2SO_4$ = 15 mol/(mol lipids); SDS + ${}^{a}H_2SO_4$ for dry algae; SDS + ${}^{b}H_2SO_4$ for wet algae at 20 % moisture (w/w dry algae). Process conditions: 600 mol methanol/(mol lipids), agitation rate = 450 rpm, temperature = 60°C, mass of microalgae = 100 mg, mass of SDS = 9 mg, 100 % ZDS (w/w lipids).

### 3.5. *Effect of inclusion of SDS in $H_2SO_4$ on water tolerance*

It has been shown that acid-catalysed direct transesterification exhibits higher water tolerance to microalgae-bound water (Velasquez-Orta et al., 2013) and free water (Wahlen et al., 2011). In order to investigate the level of water tolerance of $H_2SO_4$, with and without SDS, samples with 10, 20, and 30 % distilled water (w/w dry algae) were prepared and allowed to equilibrate for 1 h. Surprisingly, there was an increase in the FAME rate for $H_2SO_4$, with or without SDS, with increase in moisture content in the algae as shown in **Figure 4**.

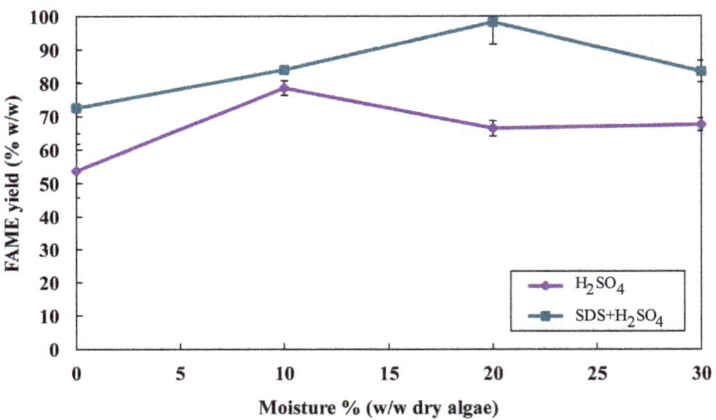


**Fig.4.** Reactively Extracted FAME produced from re-hydrated *Nannochloropsis occulata* with $H_2SO_4$ or $H_2SO_4$ + SDS. Process conditions: 600 mol methanol/(mol lipids) = 0.47 mL methanol, 8.5 mol $H_2SO_4$/(mol lipids) = 0.087 μL/ (mg biomass), agitation = 450 rpm, temperature = 60°C, mass of SDS = 9 mg, mass of microalgae = 100 mg.

The FAME production rates begin to decrease at 30 % moisture content. Cell wall lipids, such as phospholipids and glycolipids may be disrupted by polar organic solvents such as methanol, ethanol, other alcohols, and water (Cohen et al., 2012). However, the poor permeability of these solvents into the cells of completely dry oil-bearing biomass can significantly reduce their lipid extraction efficiency (Cohen et al., 2012). This can be counteracted to some extent by addition of a small quantity of water, as it swells the cell wall. The inclusion of water in extracting solvents including methanol or ethanol has been reported to increase extraction of phospholipids (Zhukov and Vereshchagrin, 1981). Removal of the cell wall lipids (phospholipids and glycolipids) from the algal cell walls compromises their integrity, i.e., it disrupts the cell wall to some degree thereby increasing accessibility of the solvent (methanol) to the internal body lipids (triglycerides). In addition, the interaction of water and methanol with cell wall proteins could compromise their integrity. The enhancement observed in the present study could be some combination of these two effects and the swelling effect. Therefore, the observed water tolerance in the re-hydrated microalgae was probably due to increased lipid extraction by moist methanol. This could be a key method of increasing the FAME yield in *in situ* transesterification of wet microalgae.

However, beyond 20 % moisture content, a drop in the FAME yield was observed, which showed that the water tolerance was exceeded for both catalysts. The amount of water tolerance achieved herein was greater than 10 % (w/w dry mass) obtained by Velasquez-Orta et al. (2013), perhaps because their moisture content was based on bound, rather than the free water used in this current investigation. However, the water tolerance achieved herein was lower than the 50 % (w/w dry mass) of free water during acid-catalysed *in situ* esterification of *C. gracilius* reported by Wahlen et al. (2011). It was also lower than the 80 % (w/w dry mass) of free water during acid-catalysed *in situ* transesterification of *N. gaditana* reported by Kim et al. (2015). It should be noted that Wahlen et al. (2011) used 0.04 mL methanol/(mg algae) while Kim et al. (2015) used 0.01 mL methanol/(mg algae). These methanol volumes/(mg algae) were significantly higher than the 0.0047 mL/(mg algae) used in this study. Therefore, their corresponding higher water tolerance than what observed herein is expected. *In situ* esterification of microalgae using $H_2SO_4$ as catalyst exhibited the same water tolerance, with or without SDS. However, the inclusion of SDS in $H_2SO_4$ produced greater FAME yields than $H_2SO_4$ alone at each moisture content as shown in **Figure 4**.

Park et al. (2014) has shown that the inclusion of sodium dodecyl benzene sulfonate (SDBS) in $H_2SO_4$ enhanced the extraction of FFAs and lipids from *Chlorella*. They also reported that SDS did not produce the same corresponding enhancement as SDBS (Park et al., 2014). It should be noted that their experiments were fundamentally different from what is reported herein. They investigated the effect of the inclusion of SDBS or SDS in $H_2SO_4$-catalysed hot water on the extraction of FFAs and lipids from *C. vulgaris*. They conducted additional experiments on the effect of including SDBS in $H_2SO_4$ for FAME production from the pre-extracted algal oil through a two-step transesterification. In better words, the approach used by Park et al. (2014) involved making biodiesel from pre-extracted algal oil which is fundamentally different from the single step transesterification ("*in situ* transesterification") reported in this study.

## 4. Conclusions

*In situ* transesterification has been shown to be technically feasible for FAME production from *N. occulata* using $H_2SO_4$, $H_2SO_4$/SDS (a surfactant), or ZDS (surfactant catalyst). ZDS produced a maximum 67±1 % FAME yield. SDS addition to $H_2SO_4$ enhanced the FAME yield and caused some levels of water tolerance. Addition of SDS in $H_2SO_4$ at 20 % moisture content produced a maximum FAME yield of 98.3±6.7 %. Finally, not only the process was more tolerant to water than transesterification-based routes, but the presence of a small quantity of external water increased the FAME yields in *in situ* transesterification, rather than inhibiting the reaction. This effect was apparent for all conditions up to 20-30 % water (w/w dry algae) which was significantly greater than the maximum of 0.5 % water (w/w oil) required in a two-step transesterification.

## References


[1] Brown, R.B, Audet, J., 2008. Current techniques for single-cell lysis. J. R. Soc. Interface. 5, S131-S138.

[2] Canakci, M., Gerpen, V.J., 1999. Biodiesel production via acid catalysis. Trans. ASAE. 42 (5),1203-1210.

[3] Chen L., Tianzhong L., Zhang, W., Chen X., Wang, J., 2012. Biodiesel production from algae oil high in free fatty acids by two-step catalytic conversion. Bioresour. Technol. 111, 208-214.

[4] Chisti Y., 2007. Biodiesel from microalgae. Biotechnol. Adv. 25(3), 294-306.

[5] Chisti, Y., 2013. Constraints to commercialization of algal fuels. J. Biotechnol. 167, 201-214.

[6] Cohen, Z., 1999. Chemicals from microalgae, CRC Press.

[7] Ehimen, E.A., Sun, Z.F., Carrington, C.G., 2010. Variables affecting the in situ transesterification of microalgae lipids. Fuel. 89 (3), 677-684.

[8] El-Shimi, H.I., Attia, N.K., El-Sheltawy, S.T., El-Diwani, G.I., 2013. Biodiesel production from Spirulina-platensis microalgae by in-situ transesterification process. J. Sust. Bioenerg. Syst. 3, 224-233.

[9] Standard UNE-EN 14103, 2003. Determination of ester and linolenic acid methyl ester contents. Issued by Asociación Española de Normalización y Certificación, Madrid.

[10] Eze, V.C., Phan, A.N., Harvey, A.P., 2014. A more robust model of biodiesel reaction, allowing identification of process conditions for enhanced rate and water tolerance. Bioresour. Technol. 156, 222-231.

[11] Folch, J., Lees, M., Stanley, G.H.S., 1956. A simple method for the isolation and purification of total lipids from animal tissue. J. Biol. Chem. 1, 497-509.

[12] Garces, R., Mancha, M., 1993. One-step lipid extraction and fatty acid methyl esters preparation from fresh plant tissues. Anal Biochem. 211, 139-143.

[13] Gerde, J.A., Montalbo-Lomboy, M., Yao, L., Grewell, D., Wanga, T., 2012. Evaluation of microalgae cell disruption by ultrasonic treatment. Bioresour. Technol. 125, 175-181.

[14] Gerken, H.G., Donohoe, B., Knoshaug, E.P., 2012. Enzymatic cell wall degradation of Chlorella vulgaris and other microalgae for biofuels production. Planta. 1, 239-253.

[15] Ghesti, G.F., Macedo, J.L., Parente, V.C.I., Dias, J.A., Dias, S.C.L., 2009. Synthesis, characterization and reactivity of Lewis acid/surfactant cerium trisdodecylsulfate catalyst for transesterification and esterification reactions. Appl. Catal, A. 355(1), 139 -147.

[16] Guldhe, A., Singh, B., Rawat, I., Ramluckan, K., Bux, F., 2014. Efficacy of drying and cell disruption techniques on lipid recovery from microalgae for biodiesel production. Fuel. 128, 46-52.

[17] Haas, M.J., Wagner, K., 2011. Simplifying biodiesel production:The direct or in situ transesterification of algal biomass. Eur. J. Lipid Sci. Technol. 113, 1219-1229.

[18] Hailegiorgis, S.M., Mahadzir, S., Subbarao, D., 2011. Enhanced *in situ* ethanolysis of *Jatropha curcas* L. in the presence of cetyltrimethylammonium bromide as a phase transfer catalyst. Renew. Energ. 36, 2502-2507.

[19] Jeffery, S.W., Humphrey, G.F., 1975. New spectrophotometric equations for determining chlorophylls *a*, *b*, $c_1$, and $c_2$ in higher plants, algae and natural phytoplankton. Biochem. Physiol. Pflanz. 167,191-194.

[20] Kaluzny, M.A., Duncan, L.A., Merritt, M.V., Epps, D.E., 1985. Rapid separation of lipid classes in high yield and purity using bonded phase columns. J. Lipid Res. 26, 135-140.

[21] Kasim, F.H., Harvey, A.P., 2011. Influence of various parameters on reactive extraction of *Jatropha curcas* L. for biodiesel production. Chem. Eng. J. 171, 1373-1378.

[22] Kim, B., Im, H., Lee, J.W., 2015. *In situ* transesterification of highly wet microalgae using hydrochloric acid. Bioresour. Technol. 185, 421-425.

[23] Li, Y., Lian, S., Tong, D., Song, R., Yang, W., Fan, Y., Qing, R., Hu, C., 2011. One-step production of biodiesel from *Nannochloropsis sp.* on solid base Mg-Zr catalyst. Appl. Energy. 88 (10), 3313-3317.

[24] Lotero, E., Liu, Y., Lopez, D.E., Suwannakarn, K., Bruce, D.A., Goodwin, J.G., 2005. Synthesis of biodiesel via acid catalysis. Ind. Eng. Chem. Res. 44, 5353-5363.

[25] Ma, F., Hanna, M.A., 1999. Biodiesel production: a review. Bioresour. Technol.70, 1-15.

[26] Meng, X., Yang, J., Xu, X., Zhang, L., Nie, Q., Xian, M., 2009. Biodiesel production from oleaginous microorganisms. Renew. Energy. 34, 1-5.

[27] Park, J.Y., Nam, B., Choi, S.A., Oh, Y.K., Lee, J.S., 2014. Effects of anionic surfactant on extraction of free fatty acid from *Chlorella vulgaris*. Bioresour. Technol. 166, 620-624.

[28] Tan, A., Ziegler, A., Steinbauer, B., Seelig, J., 2002. Thermodynamics of sodium decyl sulfate partitioning into lipid membranes. Biophys. J. 83, 1547-1556.

[29] Tuntiwiwattanapun, N., Tongcumpou, C., Haagenson, D., Wiesenborn, D., 2013. Development and Scale-up of Aqueous Surfactant-Assisted Extraction of Canola Oil for Use as Biodiesel Feedstock. J. Am. Oil Chem. Soc. 90(7), 1089-1099.

[30] Velasquez-Orta, S.B., Lee, J.G.M., Harvey, A., 2012. Alkaline *in situ* transesterification of Chlorella vulgaris. Fuel. 94, 544-550.

[31] Velasquez-Orta, S.B., Lee, J.G.M., Harvey, A.P., 2013. Evaluation of FAME production from wet marine and freshwater microalgae by *in situ* transesterification. Biochem. Eng. J. 76, 83-89.

[32] Wahlen, B.D, Willis, R.M, Seefeldt, L.C., 2011. Biodiesel production by simultaneous extraction and conversion of total lipids from microalgae, cyanobacteria, and wild mixed-cultures. Bioresour. Technol. 102(3), 2724-2730.

[33] Widjaja, A., Chien, C., Ju, Y., 2009. Study of increasing lipid production from fresh water microalgae *Chlorella vulgaris*. J. Taiwan Inst. Chem. Eng. 40, 13-20.

[34] Zeng, J., Wang, X., Zhao, B., Sun, J., Wang, Y., 2008. Rapid in situ transesterification of sunflower oil. Ind. Eng. Chem. Res. 48 (2), 850-856.

[35] Zhukov, A.V., Vereshchagrin, A.G., 1981. Current Techniques of extraction, purification and preliminary fractionation of polar lipids of natural origin. Adv. Lipids Res. 18, 247-282.

[36] Zolfigol, M.A., Salehi, P., Shiri, M., Tanbakouchian, Z., 2007. A new catalytic method for the preparation of bis-indolyl and tris-indolyl methanes in aqueous media. Catal. Commun. 8, 173-178.

# 21

# Improved lipid and biomass productivities in *Chlorella vulgaris* by differing the inoculation medium from the production medium

Shahrbanoo Hamedi, Mahmood A. Mahdavi*, Reza Gheshlaghi

*Department of Chemical Engineering, Ferdowsi University of Mashhad, Azadi Square, Pardis Campus, 91779-48944, Mashhad, Iran.*

## HIGHLIGHTS

➢Using different media for inoculums preparation and microalgae production improved lipid and biomass productivities.

➢When SH4 was selected for inoculation medium and N8 was selected for production medium, 130% increase in biomass productivity and 40% increase in lipid productivity was observed.

➢ Specific growth rate improved by differing inoculums preparation medium from production medium and changed from 0.0040/h to 0.0122/h.

## GRAPHICAL ABSTRACT

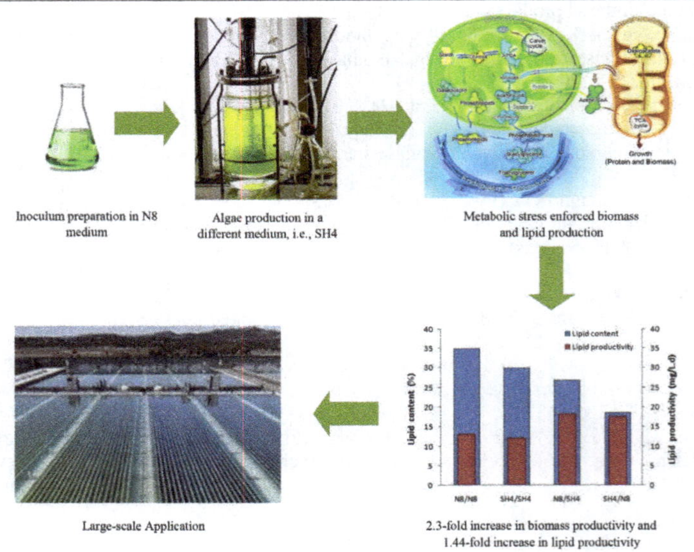


## ABSTRACT

Improvement of biomass and lipid productivities is now one of the main concerns in commercialization of microalgae cultivation as a feedstock for algal biofuel production. Conventional photoautotrophic processes using well-studied and rich in oil strain of *Chlorella vulgaris* are not able to meet such demands. A new strategy of inoculating algae production medium with cells grown in a different medium from the production medium was proposed herein. More specifically, when SH4 was used as production medium and N8 was used as inoculation medium, biomass and lipid productivities increased by 2.33 folds and 1.44 folds, respectively, compared with when the production and inoculation media were the same, such as SH4. The findings of the present investigation showed that this cultivation scheme resulted in 52% increase in cell number and 54% increase in dry weight leading to improved productivities. Although by even considering this improvement, photoautotrophic cultivation of algae can hardly compete with the heterotrophic cultivation, the high cost of hydrocarbon supply required in large-scale heterotrophic processes marks the technique proposed in the present study as a promising approach for commercialization of algal biofuel production.

**Keywords:**
Microalgae
Biomass productivity
Lipid productivity
Production medium
Inoculation medium
*Chlorella vulgaris*

* Corresponding author
E-mail address: mahdavi@um.ac.ir

## 1. Introduction

Microalgae has been at the center of attention during the past decades for its potential as a source of feedstock for biofuel production (Razon and Raymond, 2011) as well as other industrial applications (Carlsson et al., 2007). With the growing concern over carbon dioxide emission, sensible effects of global warming, and shortage of fossil fuel resources, this potential has been taken into serious consideration by governments and policy making agencies investing millions of dollars for research on commercialization of this technology (Sheehan et al., 1998). However, the feasibility of this technology as a future source of energy at a global scale requires significant increases to be achieved in algal productivity in order to generate sufficient feedstock to meet the increasing demands (US DOE, 2010).

Both biomass and lipid productivities are species-dependent at the first glance. However, environmental factors such as substrate type, nitrogen content of the media, metal elements, illumination scheme, cultivation strategy, and photobioreactor configuration could also influence the final productivities to a high extent. Numerous efforts have been reported on enhancement of biomass and lipid productivities through adjustment of these environmental factors (Mata et al., 2010; Pittman et al., 2011).

Changing environmental factors imply a "two-stage" cultivation in which conventional culture is perturbed by means of different strategies including nitrogen depletion, intense illumination, elemental supplements, etc; and consequently lipid accumulation in algal cells is stimulated under stress conditions. This approach has been adopted in many studies to improve lipid and biomass productivities. Total fatty acid was reported to increase from 86.2 mg/g cell to 137 mg/g cell in a 6-day batch culture when nitrogen content was reduced to 3% of dry weight (Richardson et al., 1969). Low-nitrogen media contributed to an overall increase in lipid content from 6% to 62% in different studies (Illman et al., 2000; Hu and Gao, 2006; Converti et al., 2009; Nigam et al., 2011; Uslu et al., 2011) while lipid productivity improved from 6 mg/L.d to 8 mg/L.d (Widjaja et al., 2009). In a two-stage marine microalgae cultivation, transferring from N-replete medium to N-free medium caused lipid content to increase to 20-26% due to metabolic stress (Jiang et al., 2012). Although continuous illumination increased algal cell concentration by 25% on average (Lee and Lee, 2001; Yusof et al., 2011) with no significant differences caused by the type of illumination source, changing light intensity from 100 $\mu E/m^2.s$ to 200 $\mu E/m^2.s$ under 15:9 light-dark cycles contributed to 50% increase in cell density from $0.8\times10^7$ cell/mL to $1.2\times10^7$ cell/mL (Tang et al., 2011). Controlled conditions at 60 μmol photons/$m^2$.s along with other parameters led to a biomass productivity of 40 mg/L.d (Lv et al., 2010). Optimization of media in terms of major components such as metal elements (Mandalam and Palsson, 1998), and specifically iron (Liu et al., 2008), potassium, and magnesium (Tran et al., 2010) increased lipid content to 56.6% and lipid and biomass productivities to 190 mg/L.d and 304 mg/L.d; respectively. In a recent study, medium optimization resulted in high lipid content of 59.6% and lipid productivity of 74 mg/L.d obtained under 750 mg/L nitrogen content and zero phosphorous content while iron was supplemented as well (Singh et al., 2015). Sufficient phosphorous concentration under nitrogen limitation condition could improve biomass production as high as nitrogen sufficient conditions (Chu et al., 2013).

Another aspect of the two-stage cultivation is switching from photoautotrophic algal growth to heterotrophic growth. The first studies on heterotrophic algal growth, with 10 g/L glucose supply, reported an increase in lipid content of *Chlorella protothecoides* from 15% to 55% (Miao and Wu, 2006). Under three organic carbon sources including glucose (1% w/v), acetate (1%w/v), and glycerol (1% w/v), *C. vulgaris* achieved biomass productivity of 254 mg/L.d and lipid productivity of 54 mg/L.d on glucose with light (mixotrophic) which was 19 folds and 14 folds higher than biomass and lipid productivities obtained in the same culture with merely air bubbling into the medium, respectively (Liang et al., 2009). However, lipid content decreased from 38% to 21% at the same condition. In another attempt on heterotrophic algal production, *C. vulgaris* was grown heterotrophically to reach higher cell densities, and then was diluted and subjected to a light environment for photoinduction. Using this strategy, after 12 h of illumination, biomass productivity reached 3333 mg/L.d and lipid productivity was 85 mg/L.d while lipid content of 25% was obtained (Fan et al., 2012). In a similar work, after seeding *C. vulgaris* in a heterotrophic medium, the production culture was inoculated resulting in biomass and lipid productivities 1.48 and 1.42 times higher than those with photoautotrophic seed, respectively (Han et al., 2012). Biomass productivity in seed culture was 25.2 times higher than those obtained by photoautotrophy.

The two-stage strategy was applied to 5-liter bioreactors and cell productivity increased from $24\times10^6$ cell/mL.d under phototrophic condition to $178\times10^6$ cell/mL.d under heterotrophic condition (Zheng et al., 2012). Although heterotrophic conditions improve biomass and lipid productivities significantly, this cultivation method does not appeal to algal industry too much as it requires supply of suitable feedstock such as lignocellulosic sugars which makes this technology costly (US DOE, 2010). Thus, autotrophic processes are still the main cultivation method particularly at large-scale operation.

In this study, as an alternative two-stage approach, a new strategy for improving biomass and lipid productivities of *C. vulgaris* based on a photoautotrophic cultivation scheme is proposed. This strategy relies on differing the inoculation (seed) culture from the production culture. Unlike the conventional cultivation scheme in which algal cells are grown in a medium and then inoculated in the same medium in higher volumes (usually 5-10% vol), in this proposal algal cells are grown in a medium and then are harvested and inoculated in a different medium in higher volumes with different composition. It is worth mentioning that the choice of the two media and their compositions is a key factor.

## 2. Materials and Methods

### *2.1 Algal Strain*

*Chlorella sp.* has been reported as the most suitable photoautotrophic microalgae for biofuel production due to its high productivity of fatty acids relevant to transesterification reaction (Hempel et al., 2012). These species are photosynthetic single cell green algae which are also used for body detox and human nutrition. Thereby, *C. vulgaris* was used as the working organism in this study. It was purchased from the Algae Culture Collection of Research Center for Basic and Applied Science at the University of Shahid Beheshti, Tehran, Iran.

### *2.2 Materials*

Chemicals used in this study were all obtained from Merck Chemicals (Germany). Two media were used in this study. The Shuisheng-4 medium (SH4) included chemical components (mg/L) as follows (Qian et al., 2008): $(NH_4)_2SO_4$ 200; $Ca(H_2PO_4)_2.H_2O$ 30; $MgSO_4.7H_2O$ 80; $NaHCO_3$ 100; KCl 25; $FeCl_3$ 1.5; $K_2HPO_4$ 10; and 1 ml elemental solution (g/L) containing: $H_3BO_3$ 2.86; $MnCl_2.4H_2O$ 1.81; $ZnSO_4.7H_2O$ 0.222; $Na_2MoO_4.2H_2O$ 0.391; $CuSO_4.5H_2O$ 0.079. The N8 medium contained the following components (mg/L) (Mandalam and Palsson, 1998): $KNO_3$ 1000; $KH_2PO_4$ 740; $Na_2HPO_4.2H_2O$ 260; $CaCl_2.2H_2O$ 13; FeEDTA 10; $MgSO_4.7H_2O$ 50; dissolved in distilled water and 1 mL of elemental solution (g/L) including: $MnCl_2.4H_2O$ 12.98; $ZnSO_4.7H_2O$ 3.2; $CuSO_4.5H_2O$ 1.83; $Al_2(SO_4)_3.18H_2O$ 3.58.

### *2.3 Culture conditions*

Cells were grown in 500 ml Erlenmeyer flasks containing 400 mL culture medium. The medium and flasks were sterilized in an autoclave for 15 min at 121 °C in order to prevent any contamination during stages of growth. Culture at 25 °C was illuminated with cool white fluorescent lamps at an intensity of 3500 lux (equivalent to 40 μmol photon/$m^2$.s) in 16:8 h light-dark cycles. Air and pure $CO_2$ (purity >99%) were mixed with approximately 1% $CO_2$ and passed through a filter for sterilization and then bubbled into the solution. The aeration rate was measured using a rotameter (LZB-4WB), setting on 0.5 L/min. The pH of both media was adjusted to 7 using NaOH and HCl at the beginning and monitored throughout the experiment. To prepare inoculants, algal growth continued until the midst of exponential phase, i.e., approximately 44 h (including lag phase) for SH4 medium and approximately 80 h (including lag phase) for N8 medium. Next, a volume equivalent to 10% of the production medium was centrifuged at 4300 rpm for 25 min to harvest cells. Then, the cell mass was washed twice with distilled water and re-suspended with production medium. In cases where inoculation and production media were the same, cell harvesting was ignored and inoculation was performed directly. In production medium, growth continued until late-exponential phase where optical density of the broth stayed nearly constant.

*2.4 Growth evaluation and biomass productivity*

After inoculation with 10% of the working volume of production medium, algal growth was determined by measuring optical density at 680 nm using a UNICO UV-VIS 2100 spectrophotometer (Unico Inc, Shanghai, China). One mL of sample was removed from culture broth in a sterile fashion and poured into a glass cuvette. Fresh medium was used as control. In order to confirm the relationship between optical density and growth, cell count was also performed. For each sample, 1 mL of algal suspension was removed through sampling tube and direct count was performed using Neubauer homocytometer under light microscope.

To determine the dry cell weight of the culture, 200 mL of culture broth was centrifuged at 4300 rpm for 25 min. Cells were washed twice with distilled water. Then, biomass was poured into an aluminum pre-weighted dish and dried at 80 °C for 24 h. Biomass productivity was calculated as the ratio of dried biomass per volume per incubation time.

*2.5 Lipid content and productivity*

Algal broth at the late-exponential growth phase was divided into two aliquot parts. One part was utilized for cell dry weight measurement and the other part for lipid extraction. It was performed gravimetrically based on a method adapted from Bligh and Dyer (1959). Then lipid content was calculated as per the following equation (**Eq. 1**):

$$C_{Lipid} = \frac{weight\ of\ extracted\ lipid}{weight\ of\ dried\ biomass} \times 100 \quad \text{Eq. 1}$$

Lipid productivity is a factor related to lipid content and biomass productivity. It was calculated as follows (**Eq. 2**):

$$P_{Lipid}(g/l.d) = \frac{C_{Lipid}(g/g) \times DCW(g/L)}{t(d)} \quad \text{Eq. 2}$$

Where $P_{Lipid}$ is lipid productivity, $C_{Lipid}$ is lipid content of cells, DCW is dry cell weight, and t is the cultivation period (d).

*2.6 Scheme of experiments*

Experiments were conducted in four different schemes as stated in **Table 1**. SH4 as production medium inoculated with seed cells grown in SH4 as inoculation medium (SH4/SH4), N8 as production medium inoculated with seed cells grown in N8 as inoculation medium (N8/N8), SH4 as production medium inoculated with seed cells grown in N8 as inoculation medium (SH4/N8), and N8 as production medium inoculated with seed cells grown in SH4 as inoculation medium (N8/SH4). The two schemes in which production and inoculation media were the same were utilized as control for the evaluation of the other two schemes.

**Table 1.**
Scheme of experiments.

| Cultivation scheme | SH4 inoculation medium | N8 inoculation medium | SH4 production medium | N8 production medium |
|---|---|---|---|---|
| Scheme 1 (SH4/SH4) | × | | × | |
| Scheme 2 (N8/N8) | | × | | × |
| Scheme 3 (SH4/N8) | | × | × | |
| Scheme 4 (N8/SH4) | × | | | × |

**3. Results and Discussion**

To investigate the effect of differing the inoculation medium from the production medium, two well-known culture media were adopted for growth of *C. vulgaris* in this study, i.e., N8 and SH4. The two media contained all the essential components required for the growth of *C. vulgaris* (Šoštarič et al., 2009). They were selected based on their quantitative elemental differences and their impacts on algal growth. Comparison of the media indicated that in N8, nitrogen source is in the form of nitrate while in SH4, it is in the form of ammonium. Carbon dioxide is the only carbon source in N8, whereas carbon dioxide and sodium bicarbonate are two carbon sources contained in SH4. Nitrogen content of N8 is 9.93 mM while it is 1.5 mM in SH4 that is 6.6 folds less than that of N8 due to the higher concentration of $KNO_3$ in N8 medium. Phosphorous content of the two media are significantly different as N8 has the phosphorous concentration of 6.8 mM while SH4 contains 0.18 mM of elemental phosphorous. In terms of essential ions, the two selected media are approximately the same. Differences in carbon and nitrogen sources directly affect growth and lipid production which in turn change biomass and lipid productivities.

Under the four different cultivation schemes, the growth curves were initially investigated as presented in **Figure 1**. As seen in this figure, the time length of growth phases for two media were different. Using SH4 as production medium, the log phase took 88 h to complete while it took 160 h when N8 was used as production medium. In cases where production and inoculation media were different, optical density at the end of the growth phase reached 1.5 in N8 medium.

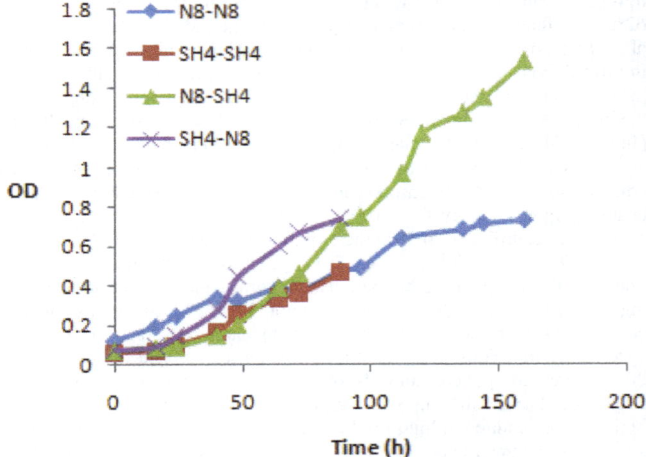


**Fig.1.** Growth curves of *Chlorella vulgaris* under different cultivation schemes.

The optical density reached 0.7 over the course of 160 h in N8 medium where inoculation and production media were the same. This showed a two-fold increase in optical density using a different inoculation culture. This was the case for SH4 production medium in which changing the inoculation medium from SH4 to N8 caused an increase in optical density from 0.45 to 0.7 in 88 h. According to **Table 2**, biomass and lipid

**Table 2.**
Measured dry cell weights and lipid concentrations under different cultivation schemes.

| Cultivation scheme | Dry Cell Weight (mg/L) | Lipid Concentration (mg/L) |
|---|---|---|
| SH4/SH4 | 150 | 45 |
| SH4/N8 | 350 | 65 |
| N8/N8 | 250 | 88 |
| N8/SH4 | 455 | 123 |

productivities increased by 130% (from 150 to 350 mg/L) and 44% (from 45 to 65 mg/L), respectively, using SH4/N8 in comparison with SH4/SH4. In cases where N8 was the production medium (comparison between N8/N8 and N8/SH4) the biomass and lipid productivities increased by 82% (from 250 to 455 mg/L) and 40% (from 88 to 123 mg/L), respectively. These figures indicated that when the production medium was different from the inoculation medium, biomass and lipid productions boosted significantly.

Close examination of the observations revealed that for the enhancement of biomass production, differing the production medium from the inoculation medium was an effective strategy. Specifically SH4/N8 scheme was the best choice where biomass productivity increased by 2.3 folds. For lipid production enhancement, the same strategy was efficient in both SH4/N8 and N8/SH4 schemes and the lipid productivity was improved by 40%-44%. Examination of the productivities shown in **Table 3** indicated that in N8/N8 and SH4/SH4 schemes of cultivation (identical production and inoculation media), lipid and biomass productivities were not significantly different; 13.2 *vs.* 12.3 mg/L.d for lipid productivity, and 37.5 *vs.* 40.9 mg/L.d for biomass productivity, respectively.

**Table 3.**
Biomass and lipid productivities and lipid contents at different cultivation schemes for *Chlorella vulgaris*.

| Cultivation scheme | Lipid Content (% dry weight biomass) | Lipid Productivity (mg/L.d) | Biomass Productivity (mg/L.d) |
|---|---|---|---|
| SH4/SH4 | 30.0 | 12.3 | 40.9 |
| SH4/N8 | 18.6 | 17.7 | 95.4 |
| N8/N8 | 35.2 | 13.2 | 37.5 |
| N8/SH4 | 27.0 | 18.5 | 68.3 |

In the two other schemes of cultivation, however, i.e., SH4/N8 and N8/SH4, the lipid and biomass productivities increased remarkably as in SH4/N8, biomass productivity reached 95.4 mg/L.d from base value of 40.9 mg/L.d and lipid productivity reached 17.7 mg/L.d from the base value of 12.3 mg/L.d. In N8/SH4 scheme, biomass productivity reached 68.3 mg/L.d from the base value of 37.5 mg/L.d and lipid productivity reached 18.5 mg/L.d from base value of 13.2 mg/L.d. These figures confirmed that when the production and inoculation media were different in algal cultivation, biomass and lipid productivities were improved (**Fig. 2**).

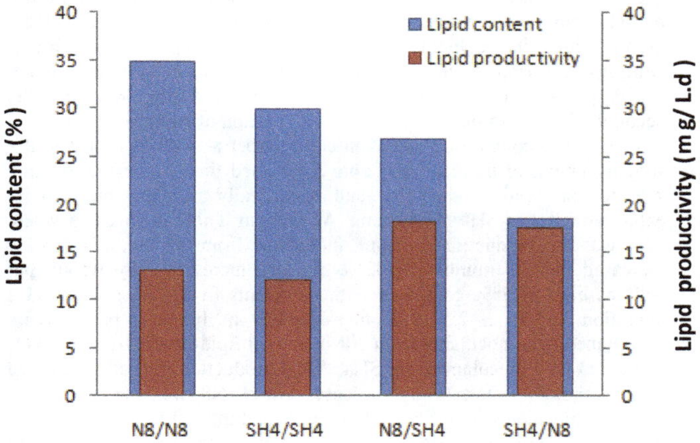


**Fig.2.** Comparison of lipid contents (% dry weight biomass) and lipid productivities in four cultivation schemes.

The above-mentioned strategy, however, caused adverse effect on lipid content as seen in **Figure 2**. It was observed that lipid production increased while lipid content decreased. Theoretically, it was because increases in biomass production outweighed increases in lipid production and consequently lipid content decreased. In SH4/N8 culture, for example, "cumulative lipid production" increased by 44% from 45 mg/L to 65 mg/L (see **Table 2**), whereas "lipid production per cell" decreased approximately by 5% compared with SH4/SH4. The relationship between optical density and cell number was measured according to **Equation 3**:

$$y = 5.917 \times OD + 0.203 \quad \text{Eq. 3}$$

Where y is the cell number when it is multiplied by $10^6$, and OD is optical density. Similarly, in N8/SH4 culture, "cumulative lipid production" increased by 40% from 88 mg/L to 123 mg/L (see **Table 2**) while "lipid production per cell" decreased approximately by 33% compared with N8/N8. Thus, increase in lipid production was due to the substantial increase in cell number and biomass rather than lipid production in a single cell. As a result, lipid content decreased while lipid productivity increased. The same observation was made in a study where growth under nutrient-rich conditions followed by cultivation under nitrogen starvation and controlled conditions of phosphate, light intensity, aeration, and carbon source was investigated (Mujtaba et al., 2012). In their study, lipid content decreased from 53% under nutrient-rich to 43% under nitrogen starvation while lipid productivity increased from 77.1 mg/L.d to 77.8 mg/L.d. It should be noted that in both cultivation schemes i.e., SH4/N8 and N8/SH4 lipid content decreased and lipid productivity increased. This observation indicates that nitrogen concentration of the media could not be the sole reason for lipid production alteration. Other factors such as phosphorous content, chemical form of nitrogen, i.e., ammonium *vs.* nitrate, and metal elements may also contribute to the improvement of lipid productivity and reduction of lipid content.

**Table 4** tabulates specific growth rates in the four cultivation schemes. As presented, in SH4/SH4, specific growth rate was 45% higher than in N8/N8 scheme (0.0058/h in SH4/SH4 compared with 0.0040/h in N8/N8). This was while growth phase in N8 medium was twice longer than that in SH4 medium. Growth rate in each medium depends on both carbon and nitrogen sources. Carbon dioxide was the only carbon source in N8 medium which was bubbled into the medium with air. The N8 is a nutritious medium that contains more nitrogen than SH4 but in the form of nitrate (1000 mg/L potassium nitrate in N8 *vs.* 200 mg/L ammonium sulphate in SH4). Basically, ammonium is the inorganic nitrogenous form that is easier to assimilate, since nitrate has to be reduced to ammonium prior to uptake (Richardson et al., 1969).

**Table 4.**
Specific growth rates of *Chlorella vulgaris* under four cultivation schemes.

| Cultivation scheme | Specific growth rates (/h ) |
|---|---|
| SH4/SH4 | 0.0058 |
| SH4/N8 | 0.0083 |
| N8/N8 | 0.0040 |
| N8/SH4 | 0.0122 |

This fact along with the continuous addition of carbon dioxide into the medium in excess with mass transfer limitation contributed to a longer growth phase in N8 (see **Fig. 1**). Suitable buffering capacity of N8 medium also might have contributed to the longer growth phase. The pH of the solution did not drop during growth phase (~7) indicating that all $CO_2$ supplementation was converted to carbohydrate rather than proton.
As shown in **Figure 3**, removing carbon dioxide from the process as the sole carbon source significantly reduced biomass production while pH did not change. This was the evidence that $CO_2$ was predominantly fixed for biomass production. In SH4 medium two carbon sources were available, i.e., carbon dioxide and sodium bicarbonate. Concentration of bicarbonate

**Table 5.**
Dry weight and lipid values per cell under four cultivation schemes.

| Production medium | Inoculation medium | OD | Cell number (y) (cell/mL)* | Lipid production per cell (mg Lipid/cell)** | Dry weight per cell (mg DCW/cell)*** |
|---|---|---|---|---|---|
| SH4 | SH4 | 0.45 | $2.8657 \times 10^6$ | $1.57 \times 10^{-8}$ | $5.23 \times 10^{-8}$ |
| | N8 | 0.7 | $4.3449 \times 10^6$ | $1.50 \times 10^{-8}$ | $8.06 \times 10^{-8}$ |
| N8 | N8 | 0.7 | $4.3449 \times 10^6$ | $2.03 \times 10^{-8}$ | $5.75 \times 10^{-8}$ |
| | SH4 | 1.5 | $9.0785 \times 10^6$ | $1.36 \times 10^{-8}$ | $5.01 \times 10^{-8}$ |

* Cell number was calculated based on **Equation 3**.
** Calculated by dividing total lipid (**Table 2**) by cell number.
*** Calculated by dividing total biomass (**Table 2**) by cell number.

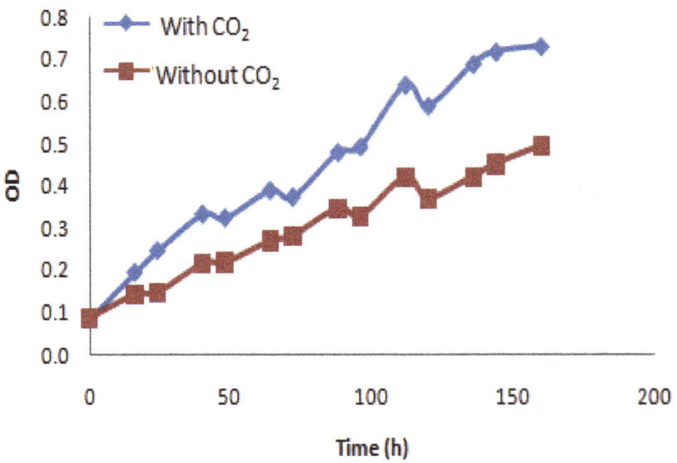


**Fig.3.** Growth curves of *Chlorella vulgaris* in N8 medium with/without $CO_2$.

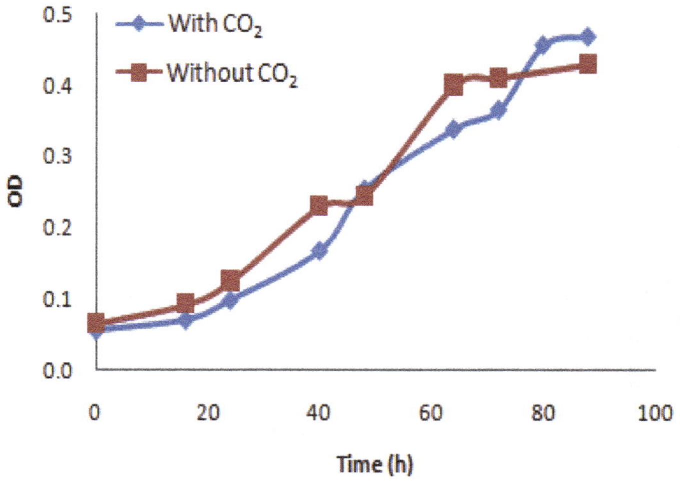


**Fig.4.** Growth curves of *Chlorella vulgaris* in SH4 medium with/without $CO_2$.

in this solution was 100 times more than that of $CO_2$ while pH was the final factor that determined which carbon source was consumed in the solution.

As seen in **Figure 4**, removing $CO_2$ from the medium did not noticeably change biomass concentration at the end of the growth phase due to the presence of bicarbonate but the growth levelled off at 65 h. However, the presence of $CO_2$ in the medium contributed to longer growth phase to 88 h with approximately the same biomass concentration.

Inoculation of N8 production medium with seed cells grown in SH4 significantly increased growth rate from 0.0040/h to 0.0122/h as well as biomass and lipid productivities. Cells already grown in SH4 were introduced in the high nitrogen content and nutritious N8 medium in which underwent a metabolic stress. This metabolic stress enforced protein and biomass production pathway, i.e., route 1 in **Figure 5**, resulting in higher biomass productivity while lipid production was suppressed due to higher nitrogen concentration. More biomass production compensated lower lipid content and consequently lipid productivity increased. It is indicated that under nutrient limitation stress condition there is a negative correlation between growth rate and lipid content (Roleda et al., 2013). Inoculation of SH4 medium with seed cells grown in N8 put cells in an environment with low nitrogen but in form of ammonium that was easier to uptake (Von Ruckert and Giani, 2004). This situation increased biomass production because of the change in metabolism and improved specific growth rate from 0.0058/h to 0.0083/h. Lipid production was inhibited in the presence of ammonia and metabolism was conducted toward biomass production through route 1 in **Figure 5**. Thus, biomass concentration reached higher values in a shorter period of time (approximately 88 h) resulting in higher lipid and biomass productivities due to higher cell concentration as reported in **Table 3**. During this period, biomass productivity underwent a 2.3-fold increase. The drop of pH from 7 to approximately 3.3 after this time, was due to the accumulation of protons as a result of assimilation of ammonium.

Close examination of changes in cell number as well as lipid and dry weight content of the cells in **Table 5** revealed that the best cultivation scheme for lipid productivity and specifically biomass productivity enhancement was SH4/N8 scheme. As seen in **Table 5**, in cases where SH4 was the production medium, inoculants from N8 culture not only increased the cell number by 52%, but also increased "dry weight per cell" as high as 54% compared with inoculants from SH4 culture. This alteration led to a 2.33-fold improvement in biomass productivity. Simultaneously, inoculants from N8 increased lipid productivity by 44% compared with inoculants from SH4, "lipid production per cell' decreased by 5% though. Interestingly, in cases where N8 was the production medium, inoculants from SH4 culture increased the cell number by 109%, while, both "lipid production per cell" and "dry weight per cell" decreased by 33% and 13%, respectively, compared with inoculants from N8 culture. Combination of these changes resulted in 82% improvement in biomass productivity as well as 40% improvement in lipid productivity.

Overall, this analysis indicated that inoculation of a medium such as SH4 or N8 with seed cells from another medium improved biomass and lipid productivities. The choice of inoculation medium, however, was

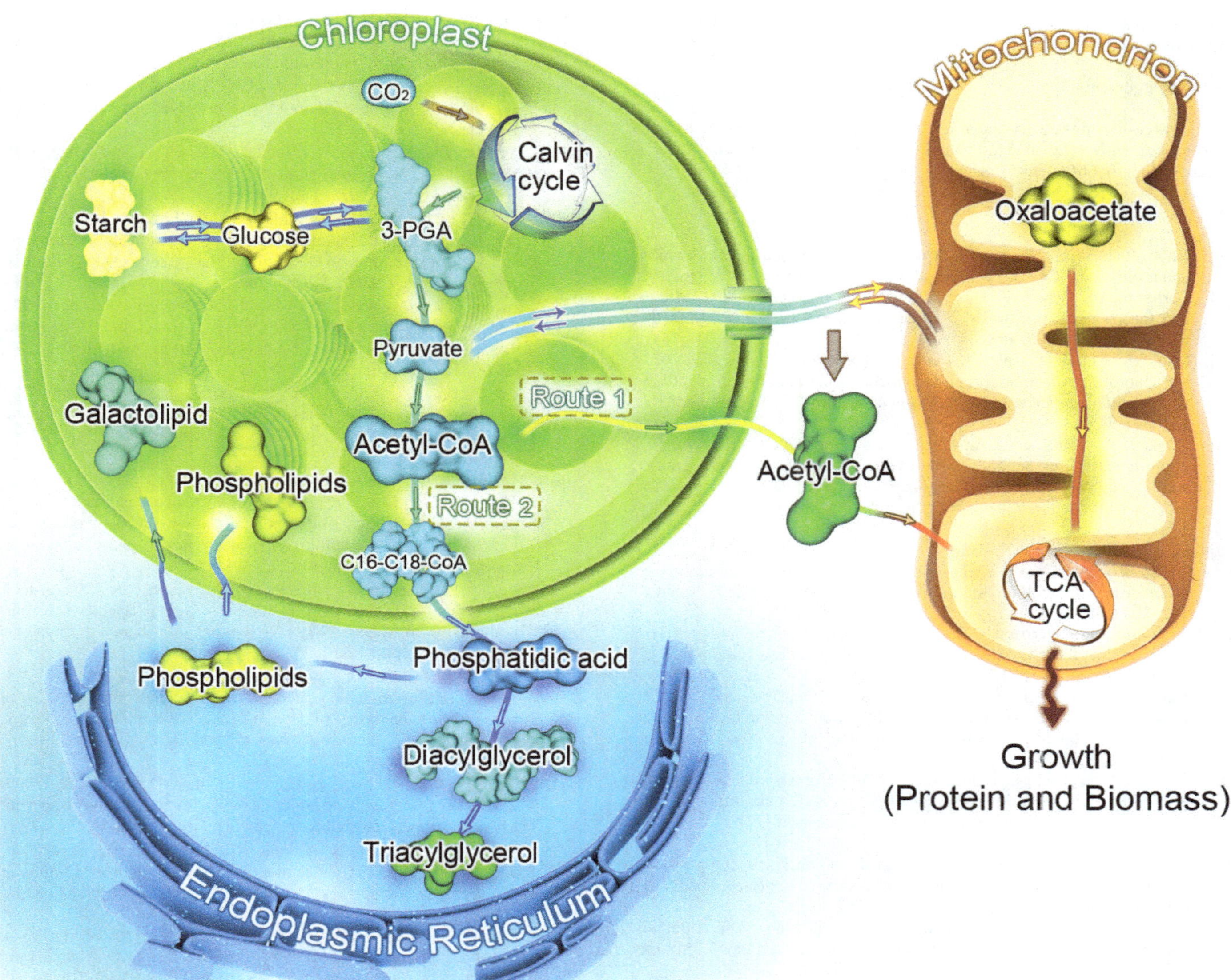


**Fig.5.** Different metabolic routes available to algae under various environmental conditions. Route 1 is the pathway for protein and biomass production under high nitrogen concentration while Route 2 is the pathway for lipid synthesis under low nitrogen concentration.

shown to be important as the SH4/N8 cultivation scheme improved biomass and lipid productivities to a higher extent compared with the N8/SH4 cultivation scheme. This method of algal cultivation has potential to be applied at commercial scale. On one hand, this technique is all about improvement of biomass and lipid productivities and it should be noted that lipid productivity is a more indicative index of total lipid production than lipid content from the commercial standpoint (Griffiths and Harrison, 2009). On the other hand, it imposes no additional cost on the large-scale operation expenses while the extent of improvement is significant in terms of lipid productivity (40-44%) and specifically biomass productivity (2.3 folds). Compared with the other two-stage approaches that provide stress conditions to increase lipid and biomass production, this technique may be too appealing to the industry to be ignored.

## 4. Conclusions

Production of microalgae *C. vulgaris* in a culture medium that was inoculated with cells grown in a different medium was proved to be an effective strategy to improve biomass and lipid productivities. This strategy was employed using two well-known algal media, i.e., SH4 and N8 and the results indicated that in SH4 production medium inoculated with cells grown in N8, biomass productivity increased by 2.3-folds and lipid productivity increased by 44% compared with the case where SH4 production medium inoculated with cells grown in the same medium. The extent of improvement in biomass and lipid productivities was remarkable with regard to the fact that no major changes were needed to be applied to a conventional photoautotrophic algal cultivation. Other algal media and their combinations for cultivation need to be examined for such a potential productivity improvement.

**Acknowledgements**

This work was financially supported by Graduate Students Research Fund at Ferdowsi University of Mashhad, Iran. Grant No. 11828. The authors declare no conflict of interest financially and personally.

**References**

[1] Bligh, E.G., Dyer, W.J., 1959. A rapid method of total lipid extraction and purification. Can. J. Biochem. Phys. 37, 911-917.

[2] Carlsson, A.S., van Beilen, J.B., Moller, R. Clayton, D., 2007. Micro- and macro-algae: utility for industrial applications: Outputs from the EPOBIO Project, CPL Press, UK.

[3] Chu, F.F., Chu, P.N., Cai, P.J., Li, W.W., Lam, P.K.S., Zeng, R.J., 2013. Phosphorous plays an important role in enhancing biodiesel productivity of *Chlorella vulgaris* under nitrogen deficiency. Bioresour. Technol. 134, 341-346.

[4] Converti, A., Cassaza, A.A., Ortiz, E.Y., Perego, P., Del Borghi, M., 2009. Effect of temperature and nitrogen concentration on the growth and lipid content of *Nannochloropsis oculata* and *Chlorella vulgaris* for biodiesel production. Chem. Process. Eng. Process Intensif. 48, 1146-1151.

[5] Fan, J., Huang, J., Li, Y., Han, F., Wang, J., Li, X., Wang, W., Li, S., 2012. Sequential heterotrophy-dilution-photoinduction cultivation for efficient microalgal biomass and lipid production. Bioresour. Technol. 112, 206-211.

[6] Griffiths, M.J., Harrison, S.T.L., 2009. Lipid productivity as a key characteristic for choosing algal species for biodiesel production. J. Appl. Phycol. 21, 493-507.

[7] Han, F., Huang, J., Li, Y., Wang, W., Wang, J., Fan, J., Shen, G., 2012. Enhancement of microalgal biomass and lipid productivities by a model of photoautotrophic culture with heterotrophic cells as seed. Bioresour. Technol. 118, 431-437.

[8] Hempel, N., Petrick, I., Behrendt, F., 2012. Biomass productivity and productivity of fatty acids and amino acids of microalgae strains as key characteristics of suitability for biodiesel production. J. Appl. Phycol. 24, 1407-1418.

[9] Hu, H., Gao, K., 2006. Response of growth and fatty acid compositions of *Nannochloropsis* sp. to environmental factors under elevated $CO_2$ concentration. J. Biotechnol. Lett. 28, 987-992.

[10] Illman, A.M., Scragg, A.H., Shales, S.W., 2000. Increase in Chlorella strains calorific values when grown in low nitrogen medium. J. Enzyme Microb. Technol. 27, 631-635.

[11] Jiang, Y., Yoshida, T., Quigg, A., 2012. Photosynthetic performance, lipid production and biomass composition in response to nitrogen limitation in marine microalgae. Plant Physiol. Biochem. 54, 70-77.

[12] Lee, K., Lee, C.G., 2001. Effect of light/dark cycles on wastewater treatments by microalgae. Biotechnol. Bioprocess Eng. 6, 194-199.

[13] Liang, Y., Sarkany, N., Cui, Y., 2009. Biomass and lipid productivities of *Chlorella vulgaris* under autotrophic, heterotrophic and mixotrophic growth conditions. Biotechnol. Lett. 31, 1043-1049.

[14] Liu, Z.Y., Wang, G.C., Zhao, B.C., 2008. Effect of iron on growth and lipid accumulation in *Chlorella vulgaris*. Bioresource. Technol. 99, 4717-4722.

[15] Lv, J.M., Cheng, L.H., Xu, X.H., Zhang, L., Chen, H.L., 2010. Enhanced lipid production of *Chlorella vulgaris* by adjustment of cultivation condition. Bioresour. Technol. 101, 6797-6804.

[16] Mandalam, R.K., Palsson, B.O., 1998. Elemental balancing of biomass and medium composition enhances growth capacity in high-density *Chlorella vulgaris* cultures. Biotechnol. Bioeng. 59, 605-611.

[17] Mata, T.M., Martins, A.A., Caetano, N.S., 2010. Microalgae for biodiesel production and other applications: a review. Renew. Sust. Energy Rev. 14, 217-232.

[18] Miao, X., Wu, Q., 2006. Biodiesel production from heterotrophic microalgal oil. Bioresource. Technol. 97, 841-846.

[19] Mujtaba, G., Choi, W., Lee, C.G., Lee, K., 2012. Lipid production by *Chlorella vulgaris* after a shift from nutrient-rich to nitrogen starvation conditions. Bioresour. Technol. 123, 279-283.

[20] Nigam, S., Rai, M.P., Sharma, R., 2011. Effect of nitrogen on growth and lipid content of *Chlorella pyrenoidosa*. Am. J. Biochem. Biotechnol. 7(3), 124-129.

[21] Pittman, J.K., Dean, A.P., Osundeko, O., 2011. The potential of sustainable algal biofuel production using wastewater resources. Bioresour. Technol. 102, 17-25.

[22] Qian, H., Chen, W., Sheng, G.D., Xu, X., Liu, W., Fu, Z., 2008. Effects of glufosinate on antioxidant enzymes, subcellular structure, and gene expression in the unicellular green algae *Chlorella vulgaris*. Aquat. Toxicol. 88, 301-307.

[23] Razon, L.F., Raymond, R.T., 2011. Net energy analysis of the production of biodiesel and biogas from the microalgae *Haematococcus pluvialis* and *Nannochloropsis*. Appl. Energy. 88, 3507-3514.

[24] Richardson, B., Orcutt, D.M., Schwertner, H.A., Martinez, C.L., Wickline, H.E., 1969. Effects of nitrogen limitation on the growth and composition of unicellular algae in continuous culture. Appl. Environ. Microbiol. 18, 245-250.

[25] Roleda, M.Y., Slocombe, S.P., Leakey, R.J.G., Day, J.G., Bell, E.M., Stanley, M.S., 2013. Effects of temperature and nutrient regimes on biomass and lipid production by six oleaginous microalgae in batch culture employing a two-phase cultivation strategy. Bioresour. Technol. 129, 439-449.

[26] Sheehan, J., Dunahay, T., Benemann, J., Roessler, P., 1998. A look back at the U.S. Department of Energy's Aquatic Species Program: biodiesel from algae. National Renewable Energy Laboratory, Golden, CO, Report NREL/TP-580.

[27] Singh, P., Guldhe, A., Kumari, S., Rawat, I., Bux, F., 2015. Investigation of combined effect of nitrogen, phosphorous and iron on lipid productivity of microalgae *Ankistrodesmus falcatus* KJ671624 using response surface methodology. Biochem. Eng. J. 94, 22-29.

[28] Šoštarič, M., Golob, J., Bricelj, M., Klinar, D., Pivec, A., 2009. Studies on the growth of *Chlorella vulgaris* in culture media with different carbon sources. Chem. Biochem. Eng. Q. 23(4), 471-477.

[29] Tang, H., Abunasser, N., Garcia, M.E.D., Chen, M., Simon Ng, K.Y., Salley, S.O., 2011. Potential of microalgae oil from *Dunaliella tertiolecta* as a feedstock for biodiesel. Appl. Energy. 88, 3324-3330.

[30] Tran, H.L., Kwon, J.S., Kim, Z.H., Oh, Y., Lee, C.G., 2010. Statistical optimization of culture media for growth and lipid production of *Botryococcus braunii* LB572. Biotechnol. Bioprocess Eng. 15, 277-284.

[31] US DOE, 2010. National algal biofuels technology roadmap. US Department of Energy, Office of Energy Efficiency and Renewable Energy, Biomass Program.

[32] Uslu, L., Isik, O., Koc, K., Goksan, T., 2011. The effects of nitrogen deficiencies on the lipid and protein contents of *Spirulina platensis*. Afr. J. Biotechnol. 10, 386-389.

[33] Von Ruckert, G., Giani, A., 2004. Effect of nitrate and ammonium on the growth and protein concentration of *Microcystis viridis* Lemmermann (Cyanobacteria). Braz. J. Bot. 27, 325-331.

[34] Widjaja, A., Chien, C.C., Ju, Y.H., 2009. Study of increasing lipid production from fresh water microalgae *Chlorella vulgaris*. J. Taiwan. Inst. Chem. Eng. 40, 13-20.

[35] Yusof, Y.A.M., Basari, J.M.H., Mukti, N.A., Sabuddin, R., Muda, A.R., Sulaiman, S., Makpol, S., Wan Ngah, W.Z., 2011. Fatty acids composition of microalgae *Chlorella vulgaris* can be modulated by varying carbon dioxide concentration in outdoor culture. Afr. J. Biotechnol. 10, 13536-13542.

[36] Zheng, Y., Chi, Z., Lucker, B., Chen, S., 2012. Two-stage heterotrophic and phototrophic culture strategy for algal biomass and lipid production. Bioresour. Technol. 103, 484-488.

# Recent updates on lignocellulosic biomass derived ethanol

Rajeev Kumar[1,*], Meisam Tabatabaei [2,3], Keikhosro Karimi[4,5], Ilona Sárvári Horváth[6]

[1] *Center for Environmental Research and Technology (CE-CERT), Bourns College of Engineering, University of California, Riverside, California, USA.*

[2] *Microbial Biotechnology Department, Agricultural Biotechnology Research Institute of Iran (ABRII), AREEO, Karaj, Iran.*

[3] *Biofuel Research Team (BRTeam), Karaj, Iran.*

[4] *Department of Chemical Engineering, Isfahan University of Technology, Isfahan 84156-83111, Iran.*

[5] *Microbial Industrial Biotechnology Group, Institute of Biotechnology and Bioengineering, Isfahan University of Technology, Isfahan 84156-83111, Iran.*

[6] *Swedish Centre for Resource Recovery, University of Borås, 501 90 Borås, Sweden.*

## HIGHLIGHTS

- Cellulosic biomass is the only source for sustainable fuels.
- Ethanol is a promising fuel candidate for near/long term applications.
- Ethanol can also serve as a precursor for other fuels and chemicals.
- However, processing cost for 2G ethanol is still high.
- Thus, urgent research efforts are needed to bring the cost down.

## GRAPHICAL ABSTRACT

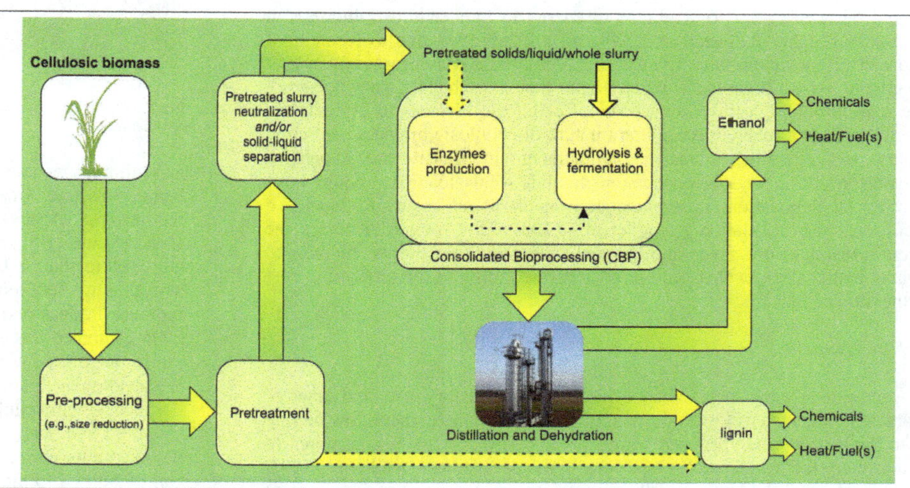


## ABSTRACT

Lignocellulosic (or cellulosic) biomass derived ethanol is the most promising near/long term fuel candidate. In addition, cellulosic biomass derived ethanol may serve a precursor to other fuels and chemicals that are currently derived from unsustainable sources and/or are proposed to be derived from cellulosic biomass. However, the processing cost for second generation ethanol is still high to make the process commercially profitable and replicable. In this review, recent trends in cellulosic biomass ethanol derived via biochemical route are reviewed with main focus on current research efforts that are being undertaken to realize high product yields/titers and bring the overall cost down.

**Keywords:**
Pretreatment
Cellulase
Fermentation
Consolidated bioprocessing

* Corresponding author
E-mail address: rkumar@cert.ucr.edu ; rajeev.dartmouth@gmail.com

**Contents**



## 1. Background

Lignocellulosic biomass, otherwise termed as cellulosic biomass, is the only sustainable feedstock for biorefineries to meet the ever increasing energy demand (Wyman, 2007; Lynd et al., 2008). Cellulosic biomass conversion into biofuels and chemicals has several advantages including greenhouse gas mitigation, near carbon neutrality, lesser dependence on fossil fuels, and improvement in nations' energy security (Wyman, 2007). Lignocellulosic biomass derived ethanol is often termed as "second generation" or "2G" as the "first generation" or "1G" ethanol is derived from sugar cane, corn, wheat, and other starchy feedstocks (Jordan et al., 2012). Studies suggest that the net energy return on 2G ethanol is much higher than ethanol derived from corn (Lynd et al., 2006; Schmer et al., 2008). In addition, 2G ethanol has much higher potential for greenhouse gas (GHG) emissions reduction than 1G ethanol (Hsu et al., 2010). The cost of energy in lignocellulosic biomass at $60/ton is roughly the same as $20/barrel oil; however, due to recalcitrant nature of cellulosic biomass (Lynd et al., 1999), the current processing cost of 2G ethanol is still high and is much higher than 1G ethanol (www.doe.gov). The reasons for high processing costs of cellulosic biomass to biofuels are several including inherent recalcitrant nature of cellulosic biomass than corn, energy and chemical intensive pretreatment, inefficient and expensive enzymes resulting in low conversion at high solids loadings required for commercial application, incomplete conversion of all sugars to fuels and chemicals, and distillation (Lynd et al., 2008). This review discusses the recent research efforts made in biological conversion of cellulosic biomass to ethanol and challenges that need to be addressed to bring the processing cost further down.

## 2. Why ethanol?

Among renewable fuels, ethanol due to its long history, use, and inherent characteristics, such as low toxicity to microbes and environment, low boiling point, high octane number, and comparable energy content, is considered to be a primary fuel candidate for near/long term applications (Lynd et al., 1991; Lynd et al., 2008). Although ethanol's energy content is roughly 2/3rd of gasoline and butanol, it has higher research octane number (RON; 107) than butanol (96) and gasoline (91-99) (Lynd, 1996). Research shows that ethanol can be used up to 85% (v/v) in vehicles without major modifications (Balat et al., 2008). Although, novel biochemical, thermo-catalytic, and hybrid routes are being developed to produce drop-in fuels and fuel additives to meet the infrastructure and other requirements (Huber et al., 2006; Anbarasan et al., 2012; Buijs et al., 2013; Caratzoulas et al., 2014; Harvey and Meylemans, 2014; Sreekumar et al., 2014). **Figure 1** shows that ethanol derived from cellulosic biomass can also be used to produce other fuel candidates such as butanol, gasoline, hydrogen, diesel, and others (Whitcraft et al., 1983; Costa et al., 1985; Deluga et al., 2004; Narula et al., 2015; Riittonen et al., 2015). Moreover, ethanol can also serve as a precursor for several other chemicals and intermediates that are currently derived from non-renewable resources (Angelici et al., 2013; Sun and Wang, 2014).

## 3. Cellulosic biomass

Lignocellulosic biomass includes forestry residues (e.g., hard & softwood), agricultural residues (e.g., corn stover, wheat straw, rice straw), herbaceous (e.g., switchgrass, miscanthus), and plants that grow in arid regions (e.g., Agave) (Somerville et al., 2010). The 2011 report from the United States (US) Department of Energy (DOE) suggests that in the US alone more than a billion ton of lignocellulosic biomass is potentially available at ~$60/ ton for conversion into >20 billion gallons of cellulosic biofuels (Perlack and Stokes, 2011). Whereas, a study published by Lal in 2005 estimated that total crop residue available is more than one billion ton in the US alone and more than 9 billion ton world-wide (Lal, 2005). Lignocellulosic biomass is primarily composed of cellulose (35-50 wt. %, dry basis), hemicelluloses (15-30%), pectin (2-5%), and lignin (12-35%). Cellulose and hemicelluloses that make more than 50% of total mass can be potentially converted to sugars for their conversion to ethanol. Lignin can be burned to meet the plants energy requirement and/or valorized to make fuels and chemicals (Ragauskas et al., 2014; Wyman and Ragauskas, 2015).

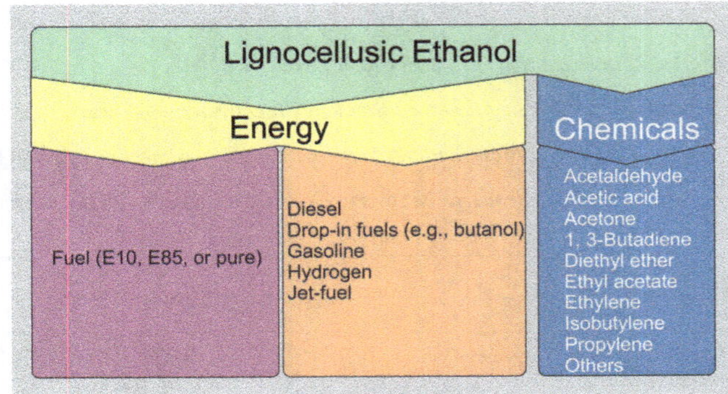


**Fig.1.** Promising applications of lignocellulosic ethanol for sustainable energy and chemicals.

## 4. Cellulosic biomass to ethanol

**Figure 2** shows the simple process flow diagram of converting cellulosic biomass to ethanol that is comprised of several steps: 1) biomass size reduction, 2) pretreatment, 3) enzymes production, 4) enzymatic hydrolysis of pretreated solids to fermentable sugars, 5) fermentation of sugars to ethanol, and 6) ethanol recovery. Steps 4 and 5 have several process configurations including separate hydrolysis and fermentation (SHF), simultaneous saccharification and fermentation (SSF), simultaneous saccharification and co-fermentation (SSCF) of hexose and pentose sugars, and consolidated bioprocessing (CBP), that combines enzymes production, enzymatic saccharification, and fermentation in a single step. Most pretreatments require some sort of size reduction to achieve better efficiency in terms of sugar release in pretreatment and/or biological conversion (Zhu and Pan, 2010). Nonetheless, pretreatment and enzymes are the most expensive

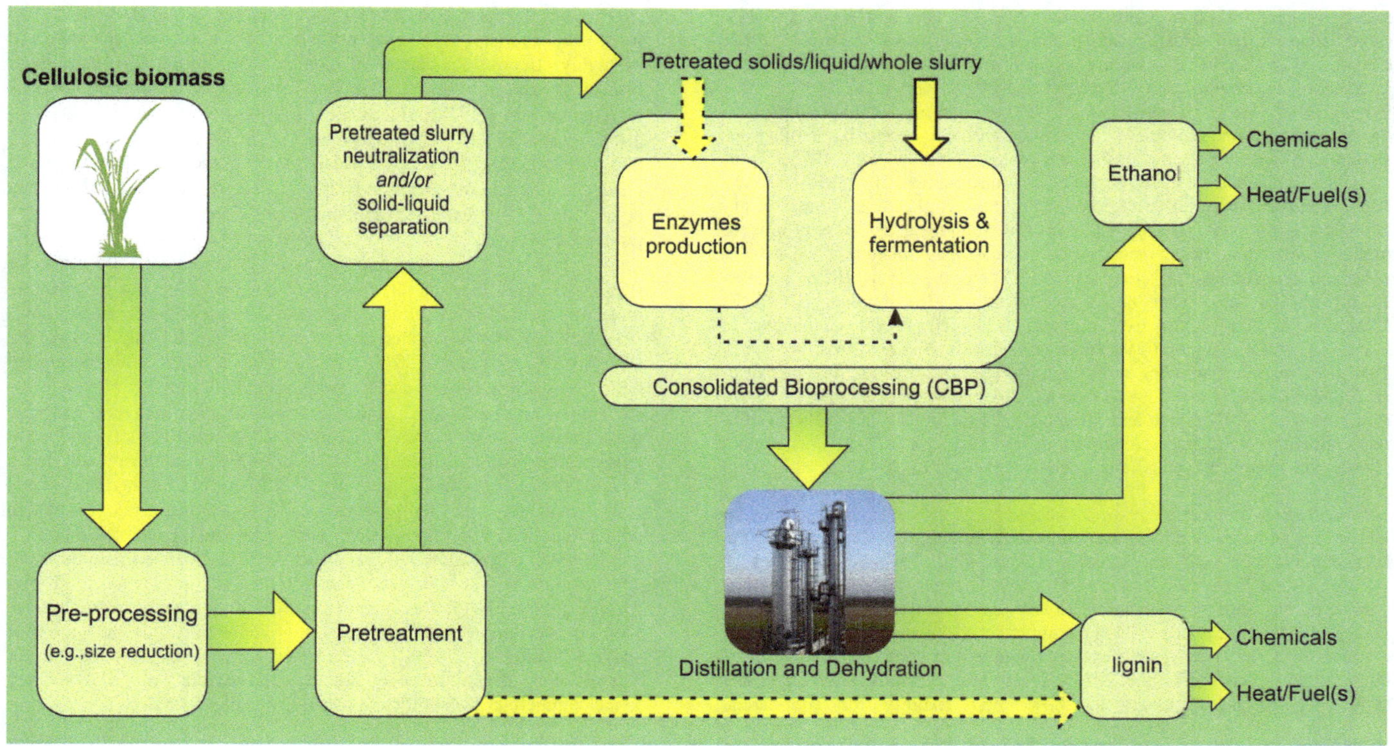


**Fig.2.** Schematic of second generation ethanol production process.

contributing factors to the 2G ethanol processing cost (Lynd et al., 2008), and, thus, have drawn a lot of attention in recent years.

## 5. Biomass recalcitrance

The inherent resistance of cellulosic biomass to pathogens, enzymes/microbes, and/or chemicals is called recalcitrance (Lynd et al., 1999), and is perceived to be majorly contributed by lignin including its amount, location, and type (syringyl vs. guacyl) (Studer et al., 2011; Ding et al., 2012). Other components such as pectin, hemicellulose, cellulose characteristics, and other biomass features are also believed to contribute to the plant's recalcitrance (Mohnen, 2008; Kumar et al., 2009b; Foston et al., 2011; Urbanowicz et al., 2012, Kumar and Wyman, 2013). However, it appears that all these features directly and/or indirectly contribute to the enzymes accessibility to plant's carbohydrates and/or enzymes effectiveness (Kumar and Wyman, 2010). Accessibility can further be divided into macro and micro-accessibility in that lignin, hemicellulose, and other components removal/relocation enhances macro-accessibility; whereas, changes in cellulose characteristics such as crystallinity and/or change in allomorph type (e.g., cellulose I to cellulose II and III) enhance micro-accessibility (Kumar and Wyman, 2013). However, for high sugar yields at low enzyme loadings, it is vital to enhance both macro and micro-accessibility and increase enzymes effectiveness (Kumar and Wyman, 2013), as enzymes are prone to deactivation and inhibition by their own end products and other components (Mandels and Reese, 1965; Reese and Mandels, 1980; Holtzapple et al., 1990; Kumar and Wyman, 2009b; Andrić et al., 2010; Ximenes et al., 2010; Kumar and Wyman, 2014).

Plants are being engineered to make them less resistant to break down, consequently requiring less harsher pretreatments and low enzyme loadings for high product yields (Funaoka et al., 1995; Chen and Dixon, 2007; Grabber et al., 2008; Sticklen, 2008; Fu et al., 2011). Since lignin is believed to be one of major impediments in low cost conversion of lignocellulosic biomass, the focus of most plant engineering studies is to alter the content, location, and type of lignin (syringyl: guaiacyl or S:G ratio) (Ding et al., 2012; Ragauskas et al., 2014; Wilkerson et al., 2014; Wagner et al., 2015). In this direction, an investigation showed that down regulating the hydroxycinnamoyl transferase (HCT) and caffeic acid 3-*O*-methyltransferase (COMT) genes in alfalfa (one of the energy grasses) resulted in decreased lignin content and enhanced sugar release (Chen and Dixon, 2007). However, in contrast to other studies (Davison et al., 2006; Studer et al., 2011), the changes in S: G ratios for these transgenic lines had no correlation with sugar release. Fu and coworkers recently showed that down regulating COMT gene in switchgrass resulted in a decrease in lignin content, reduction in the S:G lignin monomer ratio, improved forage quality, and an increase in the ethanol yield by up to 38% (Fu et al., 2011). On the other hand, changes in hemicelluloses, pectins, and other components in terms of backbone composition, chain length, branching, and content have also shown promise for reduction in plants recalcitrance (Daniel et al., 2006; Bindschedler et al., 2007; Dhugga, 2007; Mohnen, 2008; Foston et al., 2011; Cook et al., 2012; Urbanowicz et al., 2012; Doblin et al., 2014). However, more research efforts need to be directed to investigate the effect of hemicelluloses genetic engineering on plants recalcitrance (Pauly et al., 2013). Nonetheless, with genetically engineered plants, the question of their performance in field trials in terms of their growth, resistance to pathogens, and sugars yields often arises as most plant engineering studies are performed on model plants, such as *Arabidopsis thaliana*, grown in greenhouses. However, a recent study by researchers at the BioEnergy Science Center (BESC), one of the bioenergy research centers funded by the United States Department of Energy, showed that the field trials of switchgrass transgenic lines resulted in similar sugar and ethanol yields to those grown in greenhouses. In addition, the switchgrass grown in the fields was not susceptible to rust (Baxter et al., 2014).

## 6. Pretreatment

Pretreatment is a processing step to make lignocellulosic biomass more amenable to biological conversion at high yields that otherwise suffers from low yields and high processing costs (Wyman et al., 2013). The details on the type of earlier pretreatment technologies including liquid hot water or hydrothermal (Bobleter et al., 1976), dilute acid (Grethlein and Converse, 1991; Yang and Wyman, 2009; Trajano and Wyman, 2013), (non) aqueous and (near) critical ammonia (Dale and Moreira, 1982; Chundawat et al., 2013), ammonia recycled percolation (ARP), and soaking in aqueous ammonia (SAA) (Yoon et al., 1995; Kim et al., 2003), lime (Chang et al., 1997; Vincent et al., 1998), and others and their impact on biomass features and biological digestibility, and their economic viability are available in several previous and recent reviews (Millett et al., 1975; Lin et al., 1981; Ladisch et al., 1983; Knauf and Moniruzzaman, 2004; Mosier et al., 2005; Yang and Wyman, 2008; da Costa Sousa et al., 2009; Kumar et al., 2009a ; Karimi et al., 2013). It is worth mentioning a few new promising pretreatments that have recently been developed including co-solvent enhanced lignocellulosic fractionation (CELF) (Nguyen et al., 2015a; Nguyen et al., 2015b), co-solvent based lignocellulosic fractionation (COSLIF) (Zhang et al., 2007), extractive ammonia (EA) pretreatment (Chundawat et al., 2013), γ-valerolactone (GVL) pretreatment (Shuai et al., 2016; Wu et al., 2016), pretreatment applying ionic liquid(s) (Swatloski et al., 2002; Dadi et al., 2006; Seema et al., 2009; Li et al., 2010; Cheng et al., 2011; Perez-Pimienta et al., 2013; Singh and Simmons, 2013; Konda et al., 2014), sulfite pretreatment to overcome recalcitrance of lignocellulose (SPORL) (Zhu et al., 2009), and switchable butadiene sulfone pretreatment (de Frias and Feng, 2013).

Nonetheless, an ideal pretreatment should be feedstock agnostic, should be less energy and chemical intensive, and should generate highly reactive solids by enhancing their both macro and micro-accessibility (Kumar and Wyman, 2013) for their high yield conversion at low enzyme (biocatalyst) loadings with minimal sugars degradation (Yang and Wyman, 2008) and water demand (Kumar and Murthy, 2011). Although, as shown in **Figure 3**, some of the previously (and newly) developed pretreatment technologies meet some of these criteria (Dale and Ong, 2012), a rigorous techno-economic and life cycle analyses are necessary to show their viability for commercial applications (Mosier et al., 2005). For example, COSLIF pretreatment fractionates biomass at low temperatures (~50°C) and has been shown to be highly effective for variety of biomass types in terms of high sugar (especially, glucan to glucose) yields at low enzyme loadings (as low as 1 filter paper unit (FPU)/g glucan, i.e., ~ 2 mg of protein) (Rollin et al., 2011; Zhang et al., 2007 ). However, COSLIF requires concentrated phosphoric acid (>80 wt.%), which poses a recovery and recycling challenge, and doesn't appear to be highly effective for softwoods (Zhang et al., 2007). Pretreatments applying ionic liquids also appear highly promising and feedstock agnostic; however, the current cost of ionic liquids (>$3 per kg) makes this approach less commercially attractive (Klein-Marcuschamer et al., 2011). However, research efforts are underway at the Joint BioEnergy Institute, USA to develop ionic liquids from cellulosic biomass components (termed as bionic liquid) to drive the cost down (Socha et al., 2014). On the other hand, recently developed CELF uses a low boiling and renewable tetrahydrofuran (THF) as a co-solvent (boiling point-66°C), and fractionates all biomass types into three pure streams: highly reactive glucan enriched solids, xylose and other hemicellulose components in the liquid stream at near theoretical yields, and an ultra-pure stream of lignin, with >80% original lignin removed and recovered (Cai et al., 2013; Cai et al., 2014; Nguyen et al., 2015a). In addition, unlike most other pretreatment/fractionation technologies, CELF can be tuned to produce fuel precursors furfural, hydroxymethylfurfural, and levulinic acid at high yields for their catalytic conversion to drop-in fuels (Cai et al., 2013; Cai et al., 2014). CELF as a pretreatment defeats biomass recalcitrance and achieves high ethanol yields and titers at enzyme loadings as low as 2 mg protein/g glucan (Nguyen et al., 2015a; Nguyen et al., 2015b); however, recovery and recycling of THF is the key to the commercial scalability and feasibility of the technology.

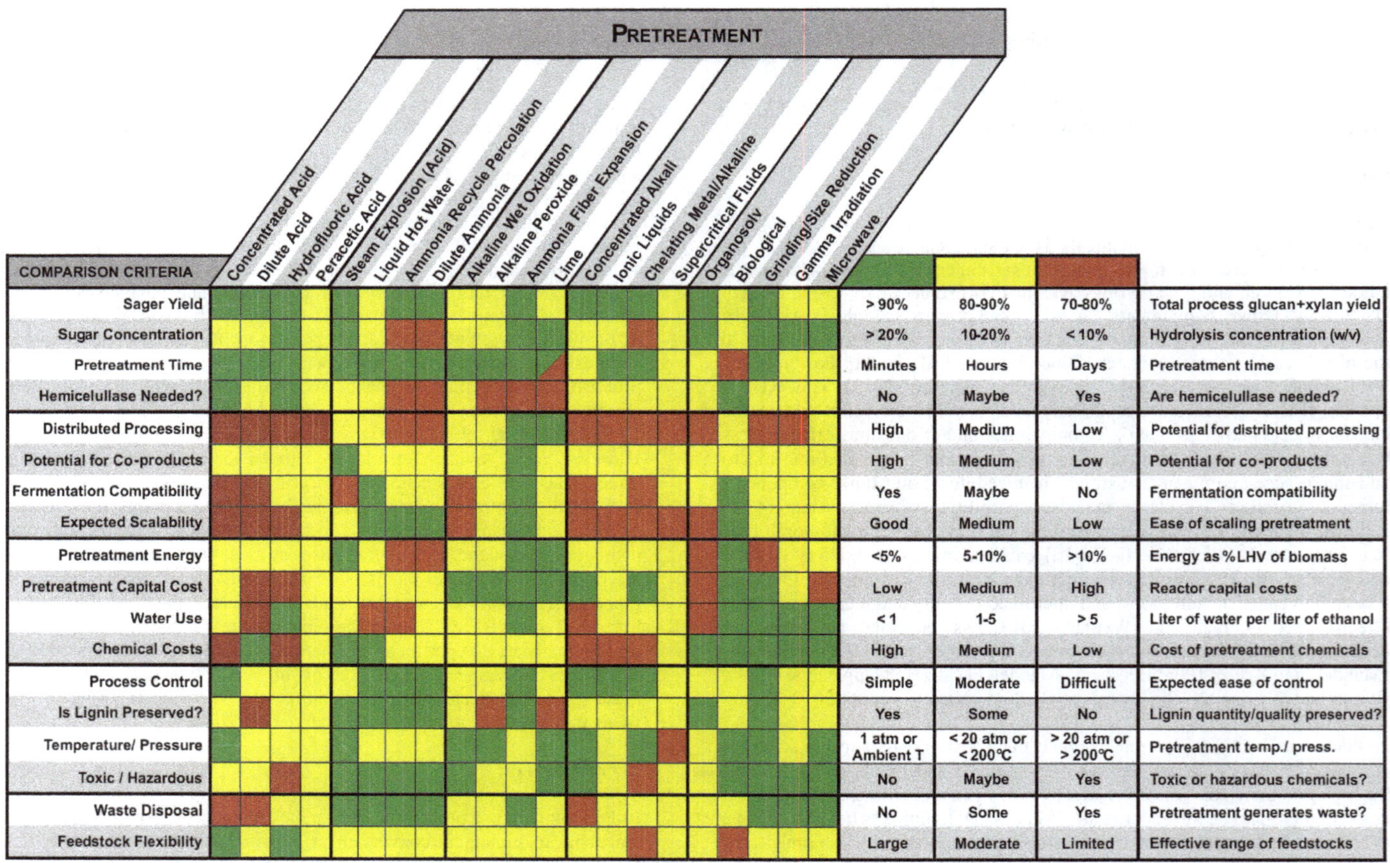


**Fig.3.** Comparison of various pretreatments for their characteristics and applicability in cellulosic biomass conversion to ethanol and other fuels and bio-based chemicals (taken with permission from Dale and Ong (2012)).

## 7. Enzymes

High cost of cellulase and other accessory enzymes required for biological conversion of pretreated lignocellulosic biomass into sugars is another major impediment in the commercialization of lignocellulosic biomass to fuels and chemicals (Culbertson et al., 2013; Hong et al., 2013). Although enhancement in enzymes stability, activity, and several fold decrease in cost have been reported in recent years, enzymes available at about $1.5-$2.0 per kg are still expensive (Stephen et al., 2012). Nonetheless, the enzymes cost per gallon would vary with the pretreatment applied, the extent of anhydrous polymers (cellulose and hemicellulose) conversion to sugars, and sugars conversion to ethanol (Klein-Marcuschamer et al., 2012).

In addition to enzymes low accessibility to (hemi) cellulose, their strong inhibition by components generated during pretreatment (e.g., phenols) (Ximenes et al., 2010; Kim et al., 2011) and enzymatic saccharification (Mandels and Reese, 1965; Halliwell and Griffin, 1973; Kumar and Wyman, 2008; Kumar and Wyman, 2009b; Qing et al., 2010; Kumar and Wyman, 2014) is one of the main reasons for high loading of enzymes required for commercially viable sugar yields. In addition, enzymes unproductive binding to lignin (Yang and Wyman, 2006; Selig et al., 2007; Kumar and Wyman, 2009a; Kumar and Wyman, 2009c; Kumar et al., 2012; Li et al., 2013) and pseudo-lignin (Hu et al., 2012; Kumar et al., 2013) also lowers the amount of enzymes available and affects their effectiveness. The rates and yields are also substantially lower at industrially relevant high solids loading than with low solids loadings (Kristensen et al., 2009; Di Risio et al., 2011) often applied and studied in laboratory settings. Although cellulase end-product inhibition by glucose can be alleviated in a process configuration called simultaneous saccharification and fermentation (SSF), and inhibition by cellobiose and hemicellulose oligomers can be alleviated by supplementing cellulase with accessory enzymes, low reaction rates at fermentation temperatures (32-37°C) (Alfani et al., 2000; Elia et al., 2008) and inhibition by ethanol still pose a challenge to high yields and titers at low enzyme loadings (Podkaminer et al., 2011; Podkaminer et al., 2012).

The discovery of novel non-hydrolytic enzymes like polysaccharide monooxygenases (LPMOs), appears to be highly promising in reducing cellulase and ultimately overall processing costs (Vaaje-Kolstad et al., 2010; Horn et al., 2012; Agger et al., 2014). Although the mechanism is not clear yet, these LPMOs are believed to oxidize the highly recalcitrant crystalline regions of cellulose and create more reducing/non-reducing ends for cellulase components to attack (Horn et al., 2012). In fact, a recent study with current generation of cellulase enzymes containing LPMOs (e.g., Cellic® Ctec2 from Novozymes) showed that it is possible to achieve higher rates and yields in SHF than SSF (Cannella and Jørgensen, 2014), which with older generation of enzymes was the other way around (Alfani et al., 2000; Lynd et al., 2002). This may be due to the fact that LPMOs require an electron donor, e.g., oxygen, for their effective action (Hu et al., 2014; Müller et al., 2015). Nonetheless, in addition to loss of some of the carbohydrates and requirement of different process configurations, the aldonic acids resulting from polysaccharides oxidation by LPMOs can be inhibitory to enzymes as well microbes (Cannella et al., 2012). In addition, it was recently shown that LPMOs can make cellulase cocktails less stable (Scott et al., 2015). Thus, it is still to be seen whether these new non-hydrolytic enzymes would be advantageous in the long run.

## 8. Fermentation

Incomplete utilization of all the sugars including hexoses (C6; glucose, galactose, and mannose) and pentoses (C5 sugars; xylose and arabinose) is another factor for high cost of 2G ethanol. In recent years, however, a lot more progress has been made in modifying various microbes including yeast (e.g., *Saccharomyces cerevisiae*, *Scheffersomyces (Pichia) stipites, Kluyveromyces marxianus)* and bacteria (e.g., *Zymomonas mobilis, Escherichia coli, Klebsiella oxytoca*) to make them capable of fermenting both hexoses and pentoses at comparatively high yields (metabolic (g ethanol/g sugar consumed) as well productive yield (g ethanol/g of total potential) (Hahn-Hagerdal et al., 1986; Jeffries and Jin, 2004; Jeffries, 2005; Kuhad et al., 2011; Fox et al., 2012; Laluce et al., 2012; Kim et al., 2013; Wang et al., 2013). The exhaustive details on the research efforts in making microbes capable of fermenting pentoses can be found in several recent reviews (Kuhad et al., 2011; Kim et al., 2012; Laluce et al., 2012; Balan, 2014; He et al., 2014). It is worth noting that in addition to making (mesophilic/thermophilic) microbes capable of fermenting pentoses together with hexoses, research efforts are also underway to make microbes metabolize cellobiose and higher cellodextrins directly to ethanol and other valuable metabolites. Although the concept is not new, as it was shown by (Spindler et al., 1989) that by directly fermenting cellobiose, it is possible to achieve higher conversion and ethanol yields, Galazka et al. (2010) recently reported a much higher conversion and yields by reconstituting the *Neurospora crassa* cellodextrins transporters system into *S.cerevisiae*. In another study, Ha et al. (2010) engineered a yeast strain to co-ferment cellobiose, glucose, and xylose together; however, high glucose concentrations expected after enzymatic saccharification of pretreated solids at high solids loading suppressed the metabolism of xylose. Although some of the engineered strains show great promises in metabolizing both hexose as well as pentose sugars, the incomplete pentose sugars utilization, low metabolic and productive yields and rates, low ethanol titers (<5wt% ethanol) than yeasts >10wt%, and inhibition by process-generated inhibitors (e.g., acetic acid , furfural) are still some of the challenges that must be overcome.

## 9. Consolidated bioprocessing

As shown in **Figure 4**, three main steps in lignocellulosic biomass conversion- enzymes production, biological hydrolysis of biomass to sugars and oligomers, and fermentative metabolites (e.g., ethanol) production- can be combined into a single bioprocessing system "Direct Microbial Conversion (DMC)" (Viljoen et al., 1926; Cooney et al., 1979; Demain et al., 2005) or lately known as "Consolidated Bioprocessing (CBP)" (Lynd, 1996). Studies have shown that CBP system combining three processing steps into one can save capital as well as operating costs (Lynd et al., 2008).

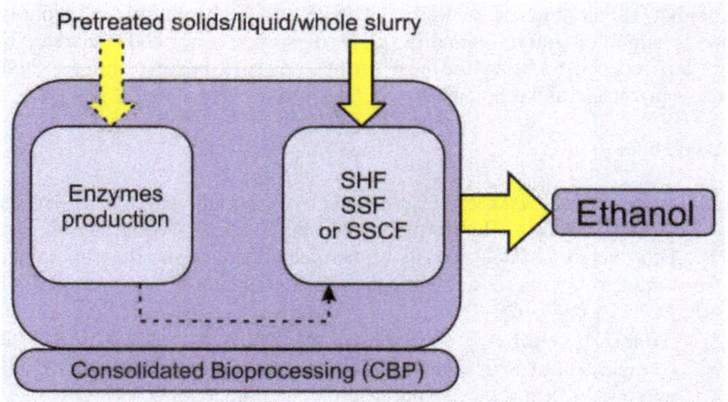


**Fig.4.** A simplified schematic of consolidated bioprocessing (CBP) system integrating three steps in lignocellulosic biomass conversion: enzymes production, hydrolysis (saccharification) of glucan and hemicelluloses in pretreated solids, and fermentation of sugars to ethanol (adapted from Lynd et al. (2002)). SHF, separate hydrolysis and fermentation; SSF, simultaneous saccharification and fermentation; and SSCF, simultaneous saccharification and co-fermentation.

There are several cellulolytic/non- cellulolytic and thermophilic/mesophilic candidate microorganisms for CBP including bacteria, e.g., *Clostridium thermocellum* (Lynd et al., 1989; Argyros et al., 2011; Shao et al., 2011), *Thermoanaerobacterium saccharolyticum* (Shaw et al., 2008), *Clostridium phytofermentans* (Jin et al., 2012), *Caldicellulosiruptor bescii* (Yang et al., 2009; Chung et al., 2014), and yeasts, e.g., *S. cerevisiae* and thermotolerant *K. marxianus* (Yamada et al., 2013). Thermophiles have an added advantage of higher hydrolysis rates and less probability of contaminations at fermentation temperatures of >60°C than mesophiles that usually operate at temperatures <50°C (Olson et al., 2012). However, most CBP organisms identified and developed, wild or genetically engineered, to date suffer from either low ethanol titer (<3wt %), low growth, or low metabolic yield and/or productive yield.

Among thermophiles, *C. thermocellum*- an anaerobe- is the most promising candidate due to its much faster degradation rates of crystalline cellulose than possible with free fungal enzymes (Shao et al., 2011), but it lacks the ability to metabolize pentoses. Another problem with *C. thermocellum* is its low metabolic yield (<0.30 g ethanol/ g sugar)- due to waste of carbon to undetectable and undesired products (Argyros et al., 2011; Deng et al., 2013; Yee et al., 2014)- and low ethanol tolerance (< 30 g/L)(Deng et al., 2013). A recent report, however, has shown that a titter of 38 g/L ethanol can be produced with *C. thermocellum* in a co-culture with *T. saccharolyticum* (Argyros et al., 2011). *T. saccharolyticum*- a thermophilic anaerobe- has been engineered to produce a high titer of ethanol (33-37 g/L) (Shaw et al., 2008), but it lacks the ability to hydrolyze cellulose and needs exogenous supplementation of cellulase.

*C. bescii* has recently been engineered to produce ethanol at high metabolic yield; however, the productive yields are too low for commercial application yet (Chung et al., 2014). It is important to note here that although all these microbes perform greatly with pure (hemi) cellulose compounds, their performance is comparatively not that great with real (unpretreated/pretreated) lignocellulosic biomass solids, most possibly due to inhibition of their free/cell-bound enzymes by lignin, hemicellulose, and/or other compounds in the unpretreated/pretreated solids (Shao et al., 2011; Brunecky et al., 2013; Resch et al., 2013). In addition to thermophilic and other bacteria, research is also underway in modifying yeasts to convert them into CBP organisms (Hasunuma and Kondo, 2012; Yamada et al., 2013). However, most of these genetically engineered strains still need some supplementation of exogenous enzymes for high ethanol yields.

## 10. Concluding remarks

In summary, a lot of progress has been made in recent years in terms of engineering plants, to make them less recalcitrant to breakdown, engineering microorganisms, to enhance their metabolic/productive yields and products/inhibitors tolerance, and developing novel pretreatments and improved enzyme cocktails to make lignocellulosic biomass derived ethanol commercially viable and profitable. Although several cellulosic ethanol plants are up and in operation around the globe; however, continued research efforts are still needed to bring the cost further down to make cellulosic ethanol plants profitable and replicable.

## References

[1] Agger, J.W., Isaksen, T., Várnai, A., Vidal-Melgosa, S., Willats, W.G.T., Ludwig, R., Horn, S.J., Eijsink, V.G.H., Westereng, B., 2014. Discovery of LPMO activity on hemicelluloses shows the importance of oxidative processes in plant cell wall degradation. Proc. Natl. Acad. Sci. 111(17), 6287-6292.

[2] Alfani, F., Gallifuoco, A., Saporosi, A., Spera, A., Cantarella, M., 2000. Comparison of SHF and SSF processes for the bioconversion of steam-exploded wheat straw. J. Ind. Microbiol. Biotechnol. 25(4), 184-192.

[3] Anbarasan, P., Baer, Z.C., Sreekumar, S., Gross, E., Binder, J.B., Blanch, H.W., Clark, D.S., Toste, F.D., 2012. Integration of chemical catalysis with extractive fermentation to produce fuels. Nature. 491(7423), 235-239.

[4] Andrić, P., Meyer, A.S., Jensen, P.A., Dam-Johansen, K., 2010. Reactor design for minimizing product inhibition during enzymatic lignocellulose hydrolysis: I. Significance and mechanism of cellobiose and glucose inhibition on cellulolytic enzymes. Biotechnol. Adv. 28(3), 308-324.

[5] Angelici, C., Weckhuysen, B.M., Bruijnincx, P.C.A., 2013. Chemocatalytic Conversion of Ethanol into Butadiene and Other Bulk Chemicals. Chem. Sus. Chem. 6(9), 1595-1614.

[6] Argyros, D.A., Tripathi, S.A., Barrett, T.F., Rogers, S.R., Feinberg, L.F., Olson, D.G., Foden, J.M., Miller, B.B., Lynd, L.R., Hogsett, D.A., Caiazza, N.C., 2011. High Ethanol Titers from Cellulose using Metabolically Engineered Thermophilic, Anaerobic Microbes. Appl. Environ. Microbiol. 77(23), 8288-8294.

[7] Balan, V., 2014. Current Challenges in Commercially Producing Biofuels from Lignocellulosic Biomass. ISRN Biotechnol 2014.

[8] Balat, M., Balat, H., Öz, C., 2008. Progress in bioethanol processing. Prog. Energy Combust. Sci. 34(5), 551-573.

[9] Baxter, H.L., Mazarei, M., Labbe, N., Kline, L.M., Cheng, Q., Windham, M.T., Mann, D.G.J., Fu, C., Ziebell, A., Sykes, R.W., Rodriguez, M., Davis, M.F., Mielenz, J.R., Dixon, R.A., Wang, Z.Y., Stewart, C.N., 2014. Two-year field analysis of reduced recalcitrance transgenic switchgrass. Plant Biotechnol. J. 12(7), 914-924.

[10] Bindschedler, L.V., Tuerck, J., Maunders, M., Ruel, K., Petit-Conil, M., Danoun, S., Boudet, A.M., Joseleau, J.P., Paul Bolwell, G., 2007. Modification of hemicellulose content by antisense down-regulation of UDP-glucuronate decarboxylase in tobacco and its consequences for cellulose extractability. Phytochemistry. 68(21), 2635-2648.

[11] Bobleter, O., Niesner, R., Röhr, M., 1976. The hydrothermal degradation of cellulosic matter to sugars and their fermentative conversion to protein. J. Appl. Polym. Sci. 20(8), 2083-2093.

[12] Brunecky, R., Alahuhta, M., Xu, Q., Donohoe, B.S., Crowley, M.F., Kataeva, I.A., Yang, S.J., Resch, M.G., Adams, M.W.W., Lunin, V.V., Himmel, M.E., Bomble, Y.J., 2013. Revealing Nature's Cellulase Diversity: The Digestion Mechanism of Caldicellulosiruptor bescii CelA. Science. 342(6165), 1513-1516.

[13] Buijs, N.A., Siewers, V., Nielsen, J., 2013. Advanced biofuel production by the yeast Saccharomyces cerevisiae. Curr. Opin. Chem. Biol. 17(3), 480-488.

[14] Cai, C.M., Nagane, N., Kumar, R., Wyman, C.E., 2014. Coupling metal halides with a co-solvent to produce furfural and 5-HMF at high yields directly from lignocellulosic biomass as an integrated biofuels strategy. Green Chem. 16(8), 3819-3829.

[15] Cai, C.M., Zhang, T., Kumar, R., Wyman, C.E., 2013. THF co-solvent enhances hydrocarbon fuel precursor yields from lignocellulosic biomass. Green Chem. 15(11), 3140-3145.

[16] Cannella, D., Hsieh, C.W., Felby, C., Jorgensen, H., 2012. Production and effect of aldonic acids during enzymatic hydrolysis of lignocellulose at high dry matter content. Biotechnol. Biofuels. 5(1), 26.

[17] Cannella, D., Jørgensen, H., 2014. Do new cellulolytic enzyme preparations affect the industrial strategies for high solids lignocellulosic ethanol production? Biotechnol. Bioeng. 111(1), 59-68.

[18] Caratzoulas, S., Davis, M.E., Gorte, R.J., Gounder, R., Lobo, R.F., Nikolakis, V., Sandler, S.I., Snyder, M.A., Tsapatsis, M., Vlachos, D.G., 2014. Challenges of and Insights into Acid-Catalyzed Transformations of Sugars. J. Phys. Chem. C. 118(40), 22815-22833.

[19] Chang, V.S., Burr, B., Holtzapple, M.T., 1997. Lime pretreatment of switchgrass. In Biotechnology for Fuels and Chemicals. Humana Press. pp, 3-19.

[20] Chen, F., Dixon, R.A., 2007. Lignin modification improves fermentable sugar yields for biofuel production. Nat. Biotechnol. 25(7), 759-761.

[21] Cheng, G., Varanasi, P., Li, C., Liu, H., Melnichenko, Y.B., Simmons, B.A., Kent, M.S., Singh, S., 2011. Transition of Cellulose Crystalline Structure and Surface Morphology of Biomass as a Function of Ionic Liquid Pretreatment and Its Relation to Enzymatic Hydrolysis. Biomacromolecules. 12(4), 933-941.

[22] Chundawat, S.P.S., Bals, B., Campbell, T., Sousa, L., Gao, D., Jin, M., Eranki, P., Garlock, R., Teymouri, F., Balan, V., Dale, B.E., 2013. Primer on Ammonia Fiber Expansion Pretreatment, in: Aqueous Pretreatment of Plant Biomass for Biological and Chemical Conversion to Fuels and Chemicals. John Wiley and Sons, Ltd, pp. 169-200.

[23] Chung, D., Cha, M., Guss, A.M., Westpheling, J., 2014. Direct conversion of plant biomass to ethanol by engineered *Caldicellulosiruptor bescii*. Proc. Natl. Acad. Sci. 111(24), 8931-8936.

[24] Cook, C.M., Daudi, A., Millar, D.J., Bindschedler, L.V., Khan, S., Bolwell, G.P., Devoto, A., 2012. Transcriptional changes related to secondary wall formation in xylem of transgenic lines of tobacco altered for lignin or xylan content which show improved saccharification. Phytochemistry. 74, 79-89.

[25] Cooney, C.L., Wang, D.I.C., Wang, S.D., 1979. Simultaneous cellulose hydrolysis and ethanol production by a cellulolytic anaerobic bacterium. Biotechnol. Bioeng. Symp. 8, 103-114.
[26] Costa, E., Uguina, A., Aguado, J., Hernandez, P.J., 1985. Ethanol to gasoline process: effect of variables, mechanism, and kinetics. Ind. Eng. Chem. Process Des. Dev. 24(2), 239-244.
[27] Culbertson, A., Jin, M., da Costa Sousa, L., Dale, B.E., Balan, V., 2013. In-house cellulase production from AFEXTM pretreated corn stover using Trichoderma reesei RUT C-30. RSC Adv. 3(48), 25960-25969.
[28] da Costa Sousa, L., Chundawat, S.P.S., Balan, V., Dale, B.E., 2009. `Cradle-to-grave' assessment of existing lignocellulose pretreatment technologies. Curr. Opin. Biotechnol. 20(3), 339-347.
[29] Dadi, A.P., Varanasi, S., Schall, C.A., 2006. Enhancement of cellulose saccharification kinetics using an ionic liquid pretreatment step. Biotechnol. Bioeng. 95(5), 904-910.
[30] Dale, B.E., Moreira, M.J., 1982. A freeze-explosion technique for increasing cellulose hydrolysis. Biotechnol. Bioeng. Symp. 12, 31-44.
[31] Dale, B.E., Ong, R.G., 2012. Energy, wealth, and human development: Why and how biomass pretreatment research must improve. Biotechnol. Prog. 28(4), 893-898.
[32] Daniel, G., Filonova, L., Kallas, A.M., Teeri, T.T., 2006. Morphological and chemical characterisation of the G-layer in tension wood fibres of Populus tremula and Betula verrucosa: labeling with cellulose-binding module CBM1HjCel7A and fluorescence and FE-SEM microscopy. Holzforschung. 60(6), 618-624.
[33] Davison, B.H., Drescher, S.R., Tuskan, G.A., Davis, M.F., Nghiem, N.P., 2006. Variation of S/G ratio and lignin content in a Populus family influences the release of xylose by dilute acid hydrolysis. Appl. Biochem. Biotechnol. 130, 427-435.
[34] de Frias, J.A., Feng, H., 2013. Switchable butadiene sulfone pretreatment of Miscanthus in the presence of water. Green Chem. 15(4), 1067-1078.
[35] Deluga, G.A., Salge, J.R., Schmidt, L.D., Verykios, X.E., 2004. Renewable Hydrogen from Ethanol by Autothermal Reforming. Science. 303(5660), 993-997.
[36] Demain, A.L., Newcomb, M., Wu, J.H.D., 2005. Cellulase, Clostridia, and Ethanol. Microbiol. Mol. Biol. Rev. 69(1), 124-154.
[37] Deng, Y., Olson, D.G., Zhou, J., Herring, C.D., Joe Shaw, A., Lynd, L.R., 2013. Redirecting carbon flux through exogenous pyruvate kinase to achieve high ethanol yields in Clostridium thermocellum. Metab. Eng. 15, 151-158.
[38] Dhugga, K.S., 2007. Maize Biomass Yield and Composition for Biofuels. Crop Sci. 47(6), 2211-2227.
[39] Di Risio, S., Hu, C.S., Saville, B.A., Liao, D., Lortie, J., 2011. Large-scale, high-solids enzymatic hydrolysis of steam-exploded poplar. Biofuels Bioprod. Biorefin. 5(6), 609-620.
[40] Ding, S.Y., Liu, Y.S., Zeng, Y., Himmel, M.E., Baker, J.O., Bayer, E.A., 2012. How Does Plant Cell Wall Nanoscale Architecture Correlate with Enzymatic Digestibility? Science. 338(6110), 1055-1060.
[41] Doblin, M.S., Johnson, K.L., Humphries, J., Newbigin, E.J., Bacic, A., 2014. Are designer plant cell walls a realistic aspiration or will the plasticity of the plant's metabolism win out? Curr. Opin. Biotechnol. 26, 108 -114.
[42] Elia, T.P., Jose, M.O., Mercedes, B., Lisbeth, O., 2008. Comparison of SHF and SSF processes from steam-exploded wheat straw for ethanol production by xylose-fermenting and robust glucose-fermenting Saccharomyces cerevisiae strains. Biotechnol. Bioeng. 100(6), 1122-1131.
[43] Foston, M., Hubbell, C.A., Samuel, R., Jung, S., Fan, H., Ding, S.Y., Zeng, Y., Jawdy, S., Davis, M., Sykes, R., Gjersing, E., Tuskan, G.A., Kalluri, U., Ragauskas, A.J., 2011. Chemical, ultrastructural and supramolecular analysis of tension wood in Populus tremula x alba as a model substrate for reduced recalcitrance. Energy Environ. Sci., 4(12), 4962-4971.
[44] Fox, J.M., Levine, S.E., Blanch, H.W., Clark, D.S., 2012. An evaluation of cellulose saccharification and fermentation with an engineered Saccharomyces cerevisiae capable of cellobiose and xylose utilization. Biotechnol. J. 7(3), 361-373.
[45] Fu, C., Mielenz, J.R., Xiao, X., Ge, Y., Hamilton, C.Y., Rodriguez, M., Chen, F., Foston, M., Ragauskas, A., Bouton, J., Dixon, R.A., Wang, Z.Y., 2011. Genetic manipulation of lignin reduces recalcitrance and improves ethanol production from switchgrass. Proc. Natl. Acad. Sci. 108(9), 3803-3808.
[46] Funaoka, M., Matsubara, M., Seki, N., Fukatsu, S., 1995. Conversion of native lignin to a highly phenolic functional polymer and its separation from lignocellulosics. Biotechnol. Bioeng. 46(6), 545-552.
[47] Galazka, J.M., Tian, C., Beeson, W.T., Martinez, B., Glass, N.L., Cate, J.H.D., 2010. Cellodextrin Transport in Yeast for Improved Biofuel Production. Science. 330(6000), 84-86.
[48] Grabber, J.H., Hatfield, R.D., Lu, F., Ralph, J., 2008. Coniferyl Ferulate Incorporation into Lignin Enhances the Alkaline Delignification and Enzymatic Degradation of Cell Walls. Biomacromolecules. 9(9), 2510-2516.
[49] Grethlein, H.E., Converse, A.O., 1991. Common aspects of acid prehydrolysis and steam explosion for pretreating wood. Bioresour. Technol. 36(1), 77-82.
[50] Ha, S.J., Galazka, J.M., Rin Kim, S., Choi, J.H., Yang, X., Seo, J.H., Louise Glass, N., Cate, J.H.D., Jin, Y.S., 2010. Engineered Saccharomyces cerevisiae capable of simultaneous cellobiose and xylose fermentation. Proc. Natl. Acad. Sci. 108(2), 504-509.
[51] Hahn-Hagerdal, B., Berner, S., Skoog, K., 1986. Improved ethanol production from xylose with glucose isomerase and *Saccharomyces cerevisiae* using the respiratory inhibitor azide. Appl. Microbiol. Biotechnol. 21, 173-175.
[52] Halliwell, G., Griffin, M., 1973. The nature and mode of action of the cellulolytic component C1 of Trichoderma koningii on native cellulose. Biochem. J. 135(4), 587-594.
[53] Harvey, B.G., Meylemans, H.A., 2014. 1-Hexene: a renewable C6 platform for full-performance jet and diesel fuels. Green Chem. 16(2), 770-776.
[54] Hasunuma, T., Kondo, A., 2012. Development of yeast cell factories for consolidated bioprocessing of lignocellulose to bioethanol through cell surface engineering. Biotechnol. Adv. 30(6), 1207-1218.
[55] He, M., Wu, B., Qin, H., Ruan, Z., Tan, F., Wang, J., Shui, Z., Dai, L., Zhu, Q., Pan, K., Tang, X., Wang, W., Hu, Q., 2014. Zymomonas mobilis: a novel platform for future biorefineries. Biotechnol. Biofuels. 7(1), 101.
[56] Holtzapple, M., Cognata, M., Shu, Y., Hendrickson, C., 1990. Inhibition of Trichoderma reesei cellulase by sugars and solvents. Biotechnol. Bioeng. 36(3), 275-287.
[57] Hong, Y., Nizami, A.S., Pour Bafrani, M., Saville, B.A., MacLean, H.L., 2013. Impact of cellulase production on environmental and financial metrics for lignocellulosic ethanol. Biofuels Bioprod. Biorefin. 7(3), 303-313.
[58] Horn, S., Vaaje-Kolstad, G., Westereng, B., Eijsink, V., 2012. Novel enzymes for the degradation of cellulose. Biotechnol. Biofuels, 5(1), 45.
[59] Hsu, D.D., Inman, D., Heath, G.A., Wolfrum, E.J., Mann, M.K., Aden, A., 2010. Life Cycle Environmental Impacts of Selected U.S. Ethanol Production and Use Pathways in 2022. Environ. Sci. Technol. 44(13), 5289-5297.
[60] Hu, F., Jung, S., Ragauskas, A., 2012. Pseudo-lignin formation and its impact on enzymatic hydrolysis. Bioresour. Technol. 117, 7-12.
[61] Hu, J., Arantes, V., Pribowo, A., Gourlay, K., Saddler, J.N., 2014. Substrate factors that influence the synergistic interaction of AA9 and cellulases during the enzymatic hydrolysis of biomass. Energy Environ. Sci. 7(7), 2308-2315.
[62] Huber, G.W., Iborra, S., Corma, A., 2006. Synthesis of Transportation Fuels from Biomass: Chemistry, Catalysts, and Engineering. Chem. Rev. 106(9), 4044-4098.
[63] Jeffries, T.W., 2005. Ethanol fermentation on the move. Nat. Biotechnol. 23(1), 40-41.
[64] Jeffries, T.W., Jin, Y.S., 2004. Metabolic engineering for improved fermentation of pentoses by yeasts. Appl. Microbiol. Biotechnol. 63(5), 495-509.
[65] Jin, M., Gunawan, C., Balan, V., Dale, B.E., 2012. Consolidated bioprocessing (CBP) of AFEX™-pretreated corn stover for ethanol

production using Clostridium phytofermentans at a high solids loading. Biotechnol. Bioeng. 109(8), 1929-1936.

[66] Jordan, D.B., Bowman, M.J., Braker, J.D., Dien, B.S., Hector, R.E., Lee, C.C., Mertens, J.A., Wagschal, K., 2012. Plant cell walls to ethanol. Biochem. J. 442(2), 241-252.

[67] Karimi, K., Shafiei, M., Kumar, R., 2013. Progress in Physical and Chemical Pretreatment of Lignocellulosic Biomass, in: Gupta, V.K., Tuohy, M.G. (Eds.), Biofuel Technologies. Springer Berlin Heidelberg, pp. 53-96.

[68] Kim, S.R., Ha, S.J., Wei, N., Oh, E.J., Jin, Y.S., 2012. Simultaneous co-fermentation of mixed sugars: a promising strategy for producing cellulosic ethanol. Trends Biotechnol. 30(5), 274-282.

[69] Kim, S.R., Park, Y.C., Jin, Y.S., Seo, J.H., 2013. Strain engineering of *Saccharomyces cerevisiae* for enhanced xylose metabolism. Biotechnol. Adv. 31(6), 851-861.

[70] Kim, T.H., Kim, J.S., Sunwoo, C., Lee, Y.Y., 2003. Pretreatment of corn stover by aqueous ammonia. Bioresour. Technol. 90(1), 39-47.

[71] Kim, Y., Ximenes, E., Mosier, N.S., Ladisch, M.R., 2011. Soluble inhibitors/deactivators of cellulase enzymes from lignocellulosic biomass. Enzyme Microb. Technol. 48(4-5), 408-415.

[72] Klein-Marcuschamer, D., Oleskowicz-Popiel, P., Simmons, B.A., Blanch, H.W., 2012. The challenge of enzyme cost in the production of lignocellulosic biofuels. Biotechnol. Bioeng. 109(4), 1083-1087.

[73] Klein-Marcuschamer, D., Simmons, B.A., Blanch, H.W., 2011. Techno-economic analysis of a lignocellulosic ethanol biorefinery with ionic liquid pre-treatment. Biofuels Bioprod. Biorefin. 5(5), 562-569.

[74] Knauf, M., Moniruzzaman, M., 2004. Lignocellulosic biomass processing: a perspective. Int. Sugar J. 106(1263), 147-150.

[75] Konda, N., Shi, J., Singh, S., Blanch, H., Simmons, B., Klein-Marcuschamer, D., 2014. Understanding cost drivers and economic potential of two variants of ionic liquid pretreatment for cellulosic biofuel production. Biotechnol. Biofuels. 7(1), 86.

[76] Kristensen, J., Felby, C., Jorgensen, H., 2009. Yield-determining factors in high-solids enzymatic hydrolysis of lignocellulose. Biotechnol. Biofuels. 2(1), 11.

[77] Kuhad, R.C., Gupta, R., Khasa, Y.P., Singh, A., Zhang, Y.H.P., 2011. Bioethanol production from pentose sugars: Current status and future prospects. Renew. Sust. Energy Rev. 15(9), 4950-4962.

[78] Kumar, D., Murthy, G.S., 2011. Impact of pretreatment and downstream processing technologies on economics and energy in cellulosic ethanol production. Biotechnol. Biofuels. 4, 27.

[79] Kumar, L., Arantes, V., Chandra, R., Saddler, J., 2012. The lignin present in steam pretreated softwood binds enzymes and limits cellulose accessibility. Bioresour. Technol. 103(1), 201-208.

[80] Kumar, P., Barrett, D.M., Delwiche, M.J., Stroeve, P., 2009a. Methods for Pretreatment of Lignocellulosic Biomass for Efficient Hydrolysis and Biofuel Production. Ind. Eng. Chem. Res. 48(8), 3713-3729.

[81] Kumar, R., Hu, F., Sannigrahi, P., Jung, S., Ragauskas, A.J., Wyman, C.E., 2013. Carbohydrate derived-pseudo-lignin can retard cellulose biological conversion. Biotechnol. Bioeng. 110(3), 737-753.

[82] Kumar, R., Mago, G., Balan, V., Wyman, C.E., 2009b. Physical and chemical characterizations of corn stover and poplar solids resulting from leading pretreatment technologies. Bioresour. Technol. 100(17), 3948-3962.

[83] Kumar, R., Wyman, C.E., 2009a. Effect of additives on the digestibility of corn stover solids following pretreatment by leading technologies. Biotechnol. Bioeng. 102(6), 1544-1557.

[84] Kumar, R., Wyman, C.E., 2009b. Effect of enzyme supplementation at moderate cellulase loadings on initial glucose and xylose release from corn stover solids pretreated by leading technologies. Biotechnol. Bioeng. 102(2), 457-467.

[85] Kumar, R., Wyman, C.E., 2009c. Effects of cellulase and xylanase enzymes on the deconstruction of solids from pretreatment of poplar by leading technologies. Biotechnol. Prog. 25(2), 302-314.

[86] Kumar, R., Wyman, C.E., 2008. An improved method to directly estimate cellulase adsorption on biomass solids. Enzyme Microb. Technol. 42(5), 426-433.

[87] Kumar, R., Wyman, C.E., 2010. Key features of pretreated lignocelluloses biomass solids and their impact on hydrolysis, in: Waldon, K. (Ed.), Bioalcohol production : Biochemical conversion of lignocellulosic biomass. Woodhead publishing limited, Oxford, pp. 73-121.

[88] Kumar, R., Wyman, C.E., 2013. Physical and Chemical Features of Pretreated Biomass that Influence Macro-/Micro-Accessibility and Biological Processing. in: Aqueous Pretreatment of Plant Biomass for Biological and Chemical Conversion to Fuels and Chemicals. John Wiley and Sons, Ltd, pp. 281-310.

[89] Kumar, R., Wyman, C.E., 2014. Strong cellulase inhibition by Mannan polysaccharides in cellulose conversion to sugars. Biotechnol. Bioeng. 111(7), 1341-1353.

[90] Ladisch, M.R., Lin, K.W., Voloch, M., Tsao, G.T., 1983. Process considerations in the enzymatic hydrolysis of biomass. Enzyme Microb. Technol. 5(2), 82-102.

[91] Lal, R., 2005. World crop residues production and implications of its use as a biofuel. Environ. Int. 31(4), 575-584.

[92] Laluce, C., Schenberg, A., Gallardo, J., Coradello, L., Pombeiro-Sponchiado, S., 2012. Advances and Developments in Strategies to Improve Strains of Saccharomyces cerevisiae and Processes to Obtain the Lignocellulosic Ethanol - a review. Appl. Biochem. Biotechnol. 166(8), 1908-1926.

[93] Li, C., Knierim, B., Manisseri, C., Arora, R., Scheller, H.V., Auer, M., Vogel, K.P., Simmons, B.A., Singh, S., 2010. Comparison of dilute acid and ionic liquid pretreatment of switchgrass: Biomass recalcitrance, delignification and enzymatic saccharification. Bioresour. Technol. 101(13), 4900-4906.

[94] Li, H., Pu, Y., Kumar, R., Ragauskas, A.J., Wyman, C.E., 2013. Investigation of lignin deposition on cellulose during hydrothermal pretreatment, its effect on cellulose hydrolysis, and underlying mechanisms. Biotechnol. Bioeng. 111(3), 485-492.

[95] Lin, K.W., Ladisch, M.R., Schaefer, D.M., Noller, C.H., Lechtenberg, V., Tsao, G.T., 1981. Review of effect of pretreatment on digestibility of cellulosic materials. AICHE Symp. Ser. 207(77), 102-106.

[96] Lynd, L., Greene, N., Dale, B., Laser, M., Lashof, D., Wang, M., Wyman, C., 2006. Energy returns on ethanol production. Science (New York, N.Y.), 312(5781), 1746-1748.

[97] Lynd, L.R., 1996. Overview and evaluation of fuel ethanol from cellulosic biomass :Technology, economics, the environment, and policy. Annu. Rev. Energy Env. 21, 403-465.

[98] Lynd, L.R., Cushman, J.H., Nichols, R.J., Wyman, C.E. 1991. Fuel ethanol from cellulosic biomass. Science (Washington, DC, United States), 251(4999), 1318-1323.

[99] Lynd, L.R., Grethlein, H.E., Wolkin, R.H., 1989. Fermentation of cellulosic substrates in batch and continuous culture by Clostridium thermocellum. Appl. Environ. Microbiol. 55(12), 3131-9.

[100] Lynd, L.R., Laser, M.S., Bransby, D., Dale, B.E., Davison, B., Hamilton, R., Himmel, M., Keller, M., McMillan, J.D., Sheehan, J., Wyman, C.E., 2008. How biotech can transform biofuels. Nat. Biotechnol. 26(2), 169-72.

[101] Lynd, L.R., Weimer, P.J., van Zyl, W.H., Pretorius, I.S., 2002. Microbial cellulose utilization: fundamentals and biotechnology. Microbiol. Mol. Biol. Rev. 66(3), 506-77.

[102] Lynd, L.R., Wyman, C.E., Gerngross, T.U., 1999. Biocommodity Engineering. Biotechnol. Prog. 15(5), 777-793.

[103] Mandels, M., Reese, E.T., 1965. Inhibition of cellulases. Annu. Rev. Phytopathol. 3, 85-102.

[104] Millett, M.A., Baker, A.J., Satter, L.D., 1975. Pretreatments to enhance chemical, enzymatic, and microbiological attack of cellulosic materials. Biotechnol. Bioeng. Symp. (5), 193-219.

[105] Mohnen, D., 2008. Pectin structure and biosynthesis. Curr. Opin. Plant Biol. 11(3), 266-277.

[106] Mosier, N., Wyman, C., Dale, B., Elander, R., Lee, Y.Y., Holtzapple, M., Ladisch, M., 2005. Features of promising technologies for pretreatment of lignocellulosic biomass. Bioresour. Technol. 96(6), 673-686.

[107] Müller, G., Várnai, A., Johansen, K.S., Eijsink, V.G.H., Horn, S.J., 2015. Harnessing the potential of LPMO-containing cellulase cocktails poses new demands on processing conditions. Biotechnol. Biofuels. 8(1), 1-9.

[108]Narula, C.K., Li, Z., Casbeer, E.M., Geiger, R.A., Moses-Debusk, M., Keller, M., Buchanan, M.V., Davison, B.H., 2015. Heterobimetallic Zeolite, InV-ZSM-5, Enables Efficient Conversion of Biomass Derived Ethanol to Renewable Hydrocarbons. Sci. Rep. 5, 16039.

[109]Nguyen, T.Y., Cai, C.M.Z., Osman, O., Kumar, R., Wyman, C.E., 2015a. CELF Pretreatment of Corn Stover Boosts Ethanol Titers and Yields from High Solids SSF with Low Enzyme Loadings. Green Chem. DOI: 10.1039/C5GC01977J.

[110]Nguyen, T.Y., Cai, C.M., Kumar, R., Wyman, C.E., 2015b. Co-solvent Pretreatment Reduces Costly Enzyme Requirements for High Sugar and Ethanol Yields from Lignocellulosic Biomass. Chem. Sus. Chem. 8(10), 1716-1725.

[111]Olson, D.G., McBride, J.E., Joe Shaw, A., Lynd, L.R., 2012. Recent progress in consolidated bioprocessing. Curr. Opin. Biotechnol. 23(3), 396-405.

[112]Pauly, M., Gille, S., Liu, L., Mansoori, N., de Souza, A., Schultink, A., Xiong, G., 2013. Hemicellulose biosynthesis. Planta. 238(4), 627-642.

[113]Perez-Pimienta, J.A., Lopez-Ortega, M.G., Varanasi, P., Stavila, V., Cheng, G., Singh, S., Simmons, B.A., 2013. Comparison of the impact of ionic liquid pretreatment on recalcitrance of agave bagasse and switchgrass. Bioresour. Technol. 127, 18-24.

[114]Perlack, R.D., Stokes, B.J., 2011. U.S. Billion-Ton Update: Biomass Supply for a Bioenergy and Bioproducts Industry. U.S. Department of Energy, Oak Ridge National Laboratory, Oak Ridge, TN.

[115]Podkaminer, K., Kenealy, W., Herring, C., Hogsett, D., Lynd, L., 2012. Ethanol and anaerobic conditions reversibly inhibit commercial cellulase activity in thermophilic simultaneous saccharification and fermentation (tSSF). Biotechnol. Biofuels. 5(1), 43.

[116]Podkaminer, K.K., Shao, X., Hogsett, D.A., Lynd, L.R., 2011. Enzyme inactivation by ethanol and development of a kinetic model for thermophilic simultaneous saccharification and fermentation at 50 °C with *Thermoanaerobacterium saccharolyticum* ALK2. Biotechnol. Bioeng. 108(6), 1268-1278.

[117]Qing, Q., Yang, B., Wyman, C.E., 2010. Xylooligomers are strong inhibitors of cellulose hydrolysis by enzymes. Bioresour. Technol. 101(24), 9624-9630.

[118]Ragauskas, A.J., Beckham, G.T., Biddy, M.J., Chandra, R., Chen, F., Davis, M.F., Davison, B.H., Dixon, R.A., Gilna, P., Keller, M., Langan, P., Naskar, A.K., Saddler, J.N., Tschaplinski, T.J., Tuskan, G.A., Wyman, C.E., 2014. Lignin Valorization: Improving Lignin Processing in the Biorefinery. Science. 344(6185), 1246843.

[119]Reese, E.T., Mandels, M., 1980. Stability of the cellulase of Trichoderma reesei under use conditions. Biotechnol. Bioeng. 22(2), 323-335.

[120]Resch, M.G., Donohoe, B.S., Baker, J.O., Decker, S.R., Bayer, E.A., Beckham, G.T., Himmel, M.E., 2013. Fungal cellulases and complexed cellulosomal enzymes exhibit synergistic mechanisms in cellulose deconstruction. Energy Environ. Sci. 6(6), 1858-1867.

[121]Riittonen, T., Eränen, K., Mäki-Arvela, P., Shchukarev, A., Rautio, A.R., Kordas, K., Kumar, N., Salmi, T., Mikkola, J.P., 2015. Continuous liquid-phase valorization of bio-ethanol towards bio-butanol over metal modified alumina. Renew. Energ. 74, 369-378.

[122]Rollin, J.A., Zhu, Z., Sathitsuksanoh, N., Zhang, Y.H.P., 2011. Increasing cellulose accessibility is more important than removing lignin: A comparison of cellulose solvent-based lignocellulose fractionation and soaking in aqueous ammonia. Biotechnol. Bioeng. 108(1), 22-30.

[123]Schmer, M.R., Vogel, K.P., Mitchell, R.B., Perrin, R.K., 2008. Net energy of cellulosic ethanol from switchgrass. Proc. Natl. Acad. Sci., 105(2), 464-469.

[124]Scott, B.R., Huang, H.Z., Frickman, J., Halvorsen, R., Johansen, K.S., 2015. Catalase improves saccharification of lignocellulose by reducing lytic polysaccharide monooxygenase-associated enzyme inactivation. Biotechnol. Lett. 1-10.

[125]Seema, S., Blake, A.S., Kenneth, P.V. 2009. Visualization of biomass solubilization and cellulose regeneration during ionic liquid pretreatment of switchgrass. Biotechnol. Bioeng. 104(1), 68-75.

[126]Selig, M.J., Viamajala, S., Decker, S.R., Tucker, M.P., Himmel, M.E., Vinzant, T.B., 2007. Deposition of Lignin Droplets Produced During Dilute Acid Pretreatment of Maize Stems Retards Enzymatic Hydrolysis of Cellulose. Biotechnol. Prog. 23(6), 1333-1339.

[127]Shao, X., Jin, M., Guseva, A., Liu, C., Balan, V., Hogsett, D., Dale, B.E., Lynd, L., 2011. Conversion for Avicel and AFEX pretreated corn stover by Clostridium thermocellum and simultaneous saccharification and fermentation: Insights into microbial conversion of pretreated cellulosic biomass. Bioresour. Technol. 102(17), 8040-8045.

[128]Shaw, A.J., Podkaminer, K.K., Desai, S.G., Bardsley, J.S., Rogers, S.R., Thorne, P.G., Hogsett, D.A., Lynd, L.R., 2008. Metabolic engineering of a thermophilic bacterium to produce ethanol at high yield. Proc. Natl. Acad. Sci. 105(37), 13769-13774.

[129]Shuai, L., Questell-Santiago, Y.M., Luterbacher, J.S., 2016. A mild biomass pretreatment using [gamma]-valerolactone for concentrated sugar production. Green Chem. 18, 937-943.

[130]Singh, S., Simmons, B.A., 2013. Ionic Liquid Pretreatment: Mechanism, Performance, and Challenges, in: Aqueous Pretreatment of Plant Biomass for Biological and Chemical Conversion to Fuels and Chemicals, John Wiley and Sons, Ltd, pp. 223-238.

[131]Socha, A.M., Parthasarathi, R., Shi, J., Pattathil, S., Whyte, D., Bergeron, M., George, A., Tran, K., Stavila, V., Venkatachalam, S., Hahn, M.G., Simmons, B.A., Singh, S., 2014. Efficient biomass pretreatment using ionic liquids derived from lignin and hemicellulose. Proc. Natl. Acad. Sci. 111(35), E3587-E3595.

[132]Somerville, C., Youngs, H., Taylor, C., Davis, S.C., Long, S.P., 2010. Feedstocks for Lignocellulosic Biofuels. Science. 329(5993), 790-792.

[133]Spindler, D.D., Wyman, C.E., Grohmann, K., 1989. Evaluation of thermotolerant yeasts in controlled simultaneous saccharifications and fermentations of cellulose to ethanol. Biotechnol. Bioeng. 34(2), 189-195.

[134]Sreekumar, S., Baer, Z.C., Gross, E., Padmanaban, S., Goulas, K., Gunbas, G., Alayoglu, S., Blanch, H.W., Clark, D.S., Toste, F.D., 2014. Chemocatalytic Upgrading of Tailored Fermentation Products Toward Biodiesel. Chem. Sus. Chem. 7(9), 2445-2448.

[135]Stephen, J.D., Mabee, W.E., Saddler, J.N., 2012. Will second-generation ethanol be able to compete with first-generation ethanol? Opportunities for cost reduction. Biofuels Bioprod. Biorefin. 6(2), 159-176.

[136]Sticklen, M.B., 2008. Plant genetic engineering for biofuel production: towards affordable cellulosic ethanol. Nat. Rev. Genet. 9(6), 433-443.

[137]Studer, M.H., DeMartini, J.D., Davis, M.F., Sykes, R.W., Davison, B., Keller, M., Tuskan, G.A., Wyman, C.E., 2011. Lignin content in natural Populus variants affects sugar release. Proc. Natl. Acad. Sci. 108(15), 6300-6305.

[138]Sun, J., Wang, Y., 2014. Recent Advances in Catalytic Conversion of Ethanol to Chemicals. ACS Catal. 4(4), 1078-1090.

[139]Swatloski, R.P., Spear, S.K., Holbrey, J.D., Rogers, R.D., 2002. Dissolution of Cellose with Ionic Liquids. J. Am. Chem. Soc. 124(18), 4974-4975.

[140]Trajano, H.L., Wyman, C.E., 2013. Fundamentals of Biomass Pretreatment at Low pH. in: Aqueous Pretreatment of Plant Biomass for Biological and Chemical Conversion to Fuels and Chemicals, John Wiley and Sons, Ltd, pp. 103-128.

[141]Urbanowicz, B.R., Pena, M.J., Ratnaparkhe, S., Avci, U., Backe, J., Steet, H.F., Foston, M., Li, H., O'Neill, M.A., Ragauskas, A.J., Darvill, A.G., Wyman, C., Gilbert, H.J., York, W.S., 2012. 4-O-methylation of glucuronic acid in Arabidopsis glucuronoxylan is catalyzed by a domain of unknown function family 579 protein. Proc. Natl. Acad. Sci. 109(35), 14253-14258.

[142]Vaaje-Kolstad, G., Westereng, B., Horn, S.J., Liu, Z., Zhai, H., Sorlie, M., Eijsink, V.G.H., 2010. An Oxidative Enzyme Boosting the Enzymatic Conversion of Recalcitrant Polysaccharides. Science. 330(6001), 219-222.

[143]Viljoen, J.A., Fred, E.B., Peterson, W.H., 1926. The fermentation of cellulose by thermophilic bacteria. J. Agric. Sci. 16(01), 1-17.

[144] Vincent, C., Nagwani, S., Holtzapple, M., Mark T., 1998. Lime pretreatment of crop residues bagasse and wheat straw. Appl. Biochem. Biotechnol. 74(3), 135-159.

[145] Wagner, A., Tobimatsu, Y., Phillips, L., Flint, H., Geddes, B., Lu, F., Ralph, J., 2015. Syringyl lignin production in conifers: Proof of concept in a Pine tracheary element system. Proc. Natl. Acad. Sci. 112(19), 6218-6223.

[146] Wang, X., Yomano, L.P., Lee, J.Y., York, S.W., Zheng, H., Mullinnix, M.T., Shanmugam, K.T., Ingram, L.O., 2013. Engineering furfural tolerance in Escherichia coli improves the fermentation of lignocellulosic sugars into renewable chemicals. Proc. Natl. Acad. Sci. 110(10), 4021-4026.

[147] Whitcraft, D.R., Verykios, X.E., Mutharasan, R., 1983. Recovery of ethanol from fermentation broths by catalytic conversion to gasoline. Ind. Eng. Chem. Process Des. Dev. 22(3), 452-457.

[148] Wilkerson, C.G., Mansfield, S.D., Lu, F., Withers, S., Park, J.Y., Karlen, S.D., Gonzales-Vigil, E., Padmakshan, D., Unda, F., Rencoret, J., Ralph, J., 2014. Monolignol Ferulate Transferase Introduces Chemically Labile Linkages into the Lignin Backbone. Science. 344(6179), 90-93.

[149] Wu, M., Yan, Z.Y., Zhang, X.M., Xu, F., Sun, R.C., 2016. Integration of mild acid hydrolysis in γ-valerolactone/water system for enhancement of enzymatic saccharification from cotton stalk. Bioresour. Technol. 200, 23-28.

[150] Wyman, C.E., 2007. What is (and is not) vital to advancing cellulosic ethanol. Trends Biotechnol. 25(4), 153-157.

[151] Wyman, C.E., Dale, B.E., Balan, V., Elander, R.T., Holtzapple, M.T., Ramirez, R.S., Ladisch, M.R., Mosier, N.S., Lee, Y.Y., Gupta, R., Thomas, S.R., Hames, B.R., Warner, R., Kumar, R., 2013. Comparative Performance of Leading Pretreatment Technologies for Biological Conversion of Corn Stover, Poplar Wood, and Switchgrass to Sugars, in: Aqueous Pretreatment of Plant Biomass for Biological and Chemical Conversion to Fuels and Chemicals, John Wiley and Sons, Ltd, pp. 239-259.

[152] Wyman, C.E., Ragauskas, A.J., 2015. Lignin Bioproducts to Enable Biofuels. Biofuels Bioprod. Biorefin. 9(5), 447-449.

[153] Ximenes, E., Kim, Y., Mosier, N., Dien, B., Ladisch, M., 2010. Inhibition of cellulases by phenols. Enzyme Microb. Technol. 46 170-176.

[154] Yamada, R., Hasunuma, T., Kondo, A., 2013. Endowing non-cellulolytic microorganisms with cellulolytic activity aiming for consolidated bioprocessing. Biotechnol. Adv. 31(6), 754-763.

[155] Yang, B., Wyman, C.E., 2006. BSA treatment to enhance enzymatic hydrolysis of cellulose in lignin containing substrates. Biotechnol. Bioeng. 94(4), 611-617.

[156] Yang, B., Wyman, C.E., 2009. Dilute acid and autohydrolysis pretreatment. Methods Mol. Biol. 581, 103-114.

[157] Yang, B., Wyman, C.E., 2008. Pretreatment: the key to unlocking low-cost cellulosic ethanol. Biofuels Bioprod. Biorefin. 2(1), 26-40.

[158] Yang, S.J., Kataeva, I., Hamilton-Brehm, S.D., Engle, N.L., Tschaplinski, T.J., Doeppke, C., Davis, M., Westpheling, J., Adams, M.W.W., 2009. Efficient Degradation of Lignocellulosic Plant Biomass, without Pretreatment, by the Thermophilic Anaerobe "*Anaerocellum thermophilum*" DSM 6725. Appl. Environ. Microbiol. 75(14), 4762-4769.

[159] Yee, K., Rodriguez Jr, M., Thompson, O., Fu, C., Wang, Z.Y., Davison, B., Mielenz, J., 2014. Consolidated bioprocessing of transgenic switchgrass by an engineered and evolved *Clostridium thermocellum* strain. Biotechnol. Biofuels. 7(1), 75.

[160] Yoon, H., Wu, Z., Lee, Y., 1995. Ammonia-recycled percolation process for pretreatment of biomass feedstock. Appl. Biochem. Biotechnol. 51(1), 5-19.

[161] Zhang, Y.H.P., Ding, S.Y., Mielenz, J.R., Cui, J.B., Elander, R.T., Laser, M., Himmel, M.E., McMillan, J.R., Lynd, L.R., 2007. Fractionating recalcitrant lignocellulose at modest reaction conditions. Biotechnol. Bioeng. 97(2), 214-223.

[162] Zhu, J.Y., Pan, X.J., 2010. Woody biomass pretreatment for cellulosic ethanol production: Technology and energy consumption evaluation. Bioresour. Technol. 101(13), 4992-5002.

[163] Zhu, J.Y., Pan, X.J., Wang, G.S., Gleisner, R., 2009. Sulfite pretreatment (SPORL) for robust enzymatic saccharification of spruce and red pine. Bioresour. Technol. 100(8), 2411-2418.

# Growth and characterization of deposits in the combustion chamber of a diesel engine fueled with B50 and Indonesian biodiesel fuel (IBF)

M Taufiq Suryantoro*, Bambang Sugiarto, Fariz Mulyadi

*Departement of Mechanical Engineering, University of Indonesia, 16424 Depok West Java, Indonesia.*

## HIGHLIGHTS

➢Deposits formation as one of the main challenges for biodiesel utilization was investigated.

➢Deposits structure and compositions were compared and found to be different for Indonesian biodiesel fuel (IBF) and B50.

➢ Higher inclusion rate of biodiesel increased deposits formation on engine components, especially on the valves and injector tip.

## GRAPHICAL ABSTRACT

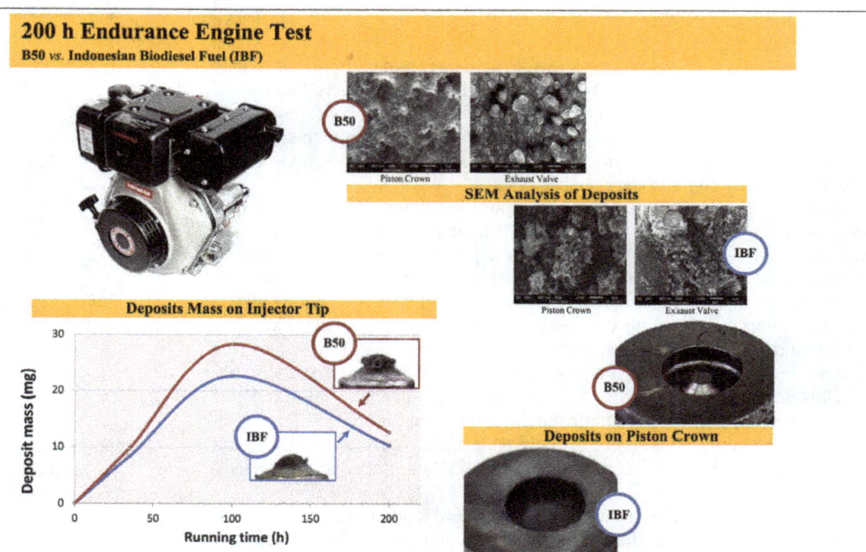


## ABSTRACT

Although used since 1893, biodiesel still faces problems that must be overcome before it can fully replace petroleum diesel. Existing literature shows that continuous use of biodiesel could lead to higher growth of deposits on critical engine components, contributing to lots of problems that could ultimately decrease engine performance. In this context, endurance tests were performed to compare the impacts of B50 and Indonesian biodiesel fuel (IBF: diesel fuel containing 10% palm oil biodiesel) on engine durability. More specifically, deposits growth as well as deposits structure and composition in response to the application of the above-mentioned fuel blends were investigated over 200 h. The results revealed that B50 produced relatively larger amounts of deposits especially on the valves and injector tip while also increased the risk of ring sticking. In addition, the structure and the elemental composition of the deposits formed on engine important components, i.e., injector tips, piston crown, intake/exhaust valves, cylinder head, and piston grooves when B50 was used were quite different compared with the IBF. Overall, more deposits formation was observed by increasing biodiesel inclusion rate especially on the valves and injector tip, while deposits tended to be wet and brittle as well.

**Keywords:**
Biodiesel
Indonesian biodiesel fuel (IBF)
Endurance test
Deposit growth
Deposit composition
Deposits structure

* Corresponding author
E-mail address: taufiq_suryo@yahoo.co.id

## 1. Introduction

Owing to the growing industrialization and the expanding transportation sector around the globe, demands have largely increased for various fossil-oriented energy carriers. However, the limited amount and the uneven distribution of these nonrenewable energy resources have led to a worldwide search for renewable or alternative energy resources. To achieve economic viability, it is important to try to develop renewable energies whose feedstocks are available and locally-abundant. Indonesia and the other tropical countries have focused on the production of renewable energies such as biodiesel or bioethanol using both edible and non-edible plants such as palm, coconut, *Jatropa curcas*, *Calophyllum inophyllum*, rubber seed, *Schleicheera oleosa*, etc.

In the year 2016, Indonesia committed itself to increase the use of biodiesel from 10 to 20% (http://www.migas.esdm.go.id/) through the implementation of a program called "Peraturan Menteri EDM No. 12 Tahun 2015". Accordingly, Indonesian should use Fatty Acid Methyl Esters (FAMEs) produced from palm oil to substitute petroleum diesel fuel while the replacement rate is supposed to be increased slightly every year in order to reduce diesel fuel imports. Nevertheless, such implementation is not without any problems either. One of the most crucial problems associated with the growing application of biodiesel is the excessive formation of carbon deposits in the combustion chamber. This is ascribed to the fact that biodiesel blends have less thermal stability as well as higher density and viscosity values compared with typical diesel fuel, thus increasing the possibility of deposit formation in engines (Reksowardojo et al., 2010; Liaquat et al., 2014). These deposits could negatively impact heat transfer in the combustion chamber (Yamada et al., 2002; Kalam and Masjuki, 2004), hydrocarbon emissions (Caceres et al., 2003; Gopal and Karupparaj, 2015), and the overall quality of the combustion process and the components' lifetime which could ultimately result in increased maintenance cost. It is worth highlighting that various biodiesel blends vary in terms of their physicochemical properties and consequently their combustion characteristics.

Therefore, researchers in various countries have been conducting experiments in order to scrutinize the use of biodiesel both at laboratory and industrial scales. Agarwal (2007) in a review paper concluded that based on the long-term endurance tests, biodiesel can be used as a substitute for mineral diesel. In spite of such efforts, the usability of biodiesel and its blends especially in accordance with engine durability is still in question. There are various editions of the European (EN) and American (ASTM) standards for different biodiesel inclusion rates in petroleum diesel. For instance, EN 590:2004 and ASTM D975 allow only 5% biodiesel to be blended into diesel fuel. In general, most car manufacturers recommend maximum biodiesel inclusion rate of only 5% in order to avoid undesirable impacts on unmodified engines such as deposits formation (Joint FIE Manufacturers Statement, 2009).

To address this challenge, further studies and comprehensive research activities with a focus on fuels' characteristics and their deposit formation capability in different engine parts are strongly required. It is worth quoting that research on deposit formation in injector and combustion chamber basically includes measuring the thickness of the deposits in the engine componentd in response to the application of test and reference fuels. Deposits can contain various materials and residues which gradually grow or accumulate on critical parts of internal combustion engines (Güralp et al., 2006; Arifin and Arai, 2009). Therefore, it is essential to analyze the chemical and physical attributes of deposits in order to figure out the causes. Deposit characterization generally includes structural and compositional analyses through scanning electron microscopy (SEM) and Energy Dispersive Spectroscopy (EDS). Having considered both diesel fuel and biodiesel as references, the deposits formed can be compared for different fuel compositions.

The objective of the present study was to investigate the formation of deposits in the combustion chamber of a diesel engine and to characterize them when Indonesian biodiesel fuel (IBF) containing 10% palm oil biodiesel and B50 fuel were used. The investigated parameters included the rate of deposit growth on several diesel engine components, the composition of the deposits, the morphological features of the deposits formed on different engine componentd, and how the rate of deposit growth affected engine's performance.

## 2. Materials and Methods

### 2.1. Experimental set-up and deposit analysis procedure

Descriptive results on deposit growth in different parts of a diesel engine were obtained by conducting 200 h endurance tests for each of the investigated fuel blends. Deposits are more likely be formed when engines reach maximum power or maximaum torque at low combustion chamber temperature, combined with engine cooling. Therefore, the present study was conducted mostly at maximum torque.

As mentioned earlier, the duration of the endurance test was 200 h. After 35, 100, and 200 h of the test, the engine was dismantled in order to analyze deposit growth by photography and mass weighing in different engine components, i.e., piston crown, piston grooves, cylinder head, injector tip, intake valve, and exhaust valve. A digital single lens camera was used to capture the images. Since the deposits were formed in micro scale thickness, and as a result of the complexity of the engine components, it was difficult to get deposit thickness for every engine component and, therefore, non-destructive method (photographic method) was used. The images obtained were further processed by using a photo editing software and MATLAB program (MATLAB, 2013) to obtain coordinate points, visualizing the injector in the form of graphs, and to estimate the deposits volume through the integration process (**Table 1**).

**Table 1.**
The deposit mass calculation approach.

| Taking images at 4 different angles | Processing images | Production of output 3-D images |
|---|---|---|

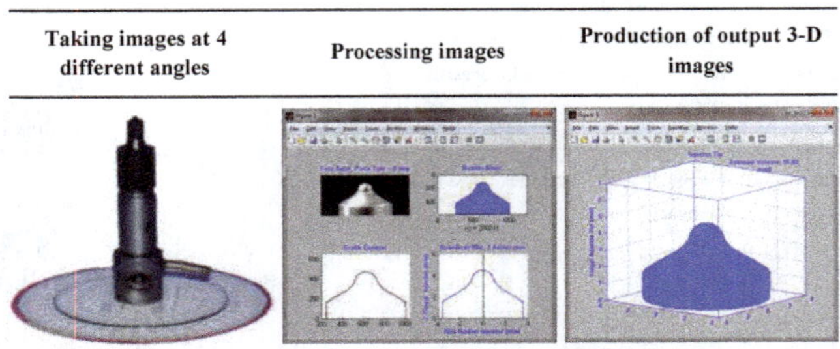

Engine off mode was used to stimulate cooling condition as used in real field. Moreover, the structure and chemical composition of the deposits were also analyzed by using a JEOL JSM-6510 LA SEM-EDS (3000× magnification). Smoke samples from the smoke filter were also taken and analyzed in order to investigate their relation with the deposits formed in the valves.

A single cylinder Yanmar L48 diesel engine (**Table 2**) was used in the experiments. A single-phase Dongdong generator with a 3.5 kW capacity set with bulbs installed in series was used to gain a constant power of 1.7 kW (75 % of the engine maximum load capacity) during the 200 h endurance test. Oil and exhaust temperatures were recorded every 2 h, while smoke and fuel flow rates were logged in every 4 h.

**Table 2.**
The specifications of the engine set-up used.

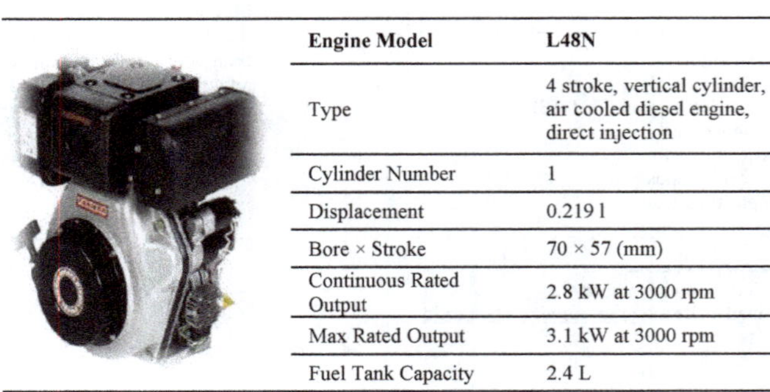

| Engine Model | L48N |
|---|---|
| Type | 4 stroke, vertical cylinder, air cooled diesel engine, direct injection |
| Cylinder Number | 1 |
| Displacement | 0.219 l |
| Bore × Stroke | 70 × 57 (mm) |
| Continuous Rated Output | 2.8 kW at 3000 rpm |
| Max Rated Output | 3.1 kW at 3000 rpm |
| Fuel Tank Capacity | 2.4 L |

## 3. Results and Discussion

Deposits or as they are often called carbon deposits in the combustion chamber of diesel engines are heterogenic mixtures consisting of carbon ash, soot, and the residues of oxygenated materials. In a comprehensive study, Diabya et al. (2009) argued that down-sizing of the engine as well as extreme conditions of the engine operation, e.g., operational temperature and pressure of the combustion chamber, oil sump, or cylinder liner temperature could accelerate the deposits formation process in engines. These deposits remain attached to the critical parts of the combustion chamber such as piston (crown, rings, and crevices), cylinder head, intake and exhaust valves and injector. Deposits on the injector often occur on the injector tip and in the interior of the fuel oulet. It is worth quoting that deposits gradually grow and accumulate in these critical parts and will negatively affect engine durability and consequently the maintenance cost.

Besides, deposits accumulation on the wall of the combustion chamber could interfere with heat transfer in the combustion chamber leading to increased $NO_x$ emission (Ra et al., 2006). It has been reported that the thickness of deposits could be estimated by measuring the temperature of the combustion chamber wall and the area of fuel impingement (Goldsworthy, 2006; Güralp et al., 2006). The combination of low wall temperature and unburned fuel could cause more deposit formation in the engine chamber. Overall, deposit formation in the combustion chamber is a complex process that depends on several interrelated parameters, i.e., fuel type, the material used in the combustion chamber wall, temperature, compression, etc. Among those, the wall is the most important parameter in the formation of deposits (Arifin and Arai, 2009).

Spilners and Hedenburg (1985), who studied carbon deposit formation in a single cylinder test engine, observed that the olefin content of the fuel had the greatest effect on the deposit formation tendency. Fuel injectors in diesel engines are precision instruments and their function is to control engine combustion, fuel economy, and engine noise by injecting exactly the right amount of fuel into the combustion chamber on each compression stroke. Carbon deposits formed on the jnjector may cause operational problems, such as excessive smoke emissions, loss of power, poor fuel economy, degraded emissions, excessive engine noise, rough engine operation, and poor drivability. It is worth quoting that the jnjector carbon deposits, usually appearing as lacquers on the piston, are believed to result from the reactions of the oxidation products of unstable fuel blends, e.g., when high biodiesel inclusion rates are practiced (Goldsworthy, 2006).

### *3.1. Deposit growth*

Deposits in the combustion chamber of a diesel engine are classified based on their location into 5 categories: injector tips, piston crown, intake/exhaust valves, cylinder head and piston grooves. Deposit growth was measured by gravimetric and photographic methods. Extent of deposit growth on the injector tips can be seen in **Figure 1**. More specifically, **Figure 1A** shows deposit growth on the injector tip by time in the engine running on the IBF fuel blend, while **Figure 1B** presents the trend of deposit growth when the B50 fuel was utilized. Visual inspection of the injector nozzles at the end of each running test for various fuel blends i.e., diesel containing coconut oil, was also conducted previously by Kalam et al. (2003) who indicated that there was no significant carbon deposit on the nozzle tips. In a different study on analyzing the wear of various vital engine parts, Agarwal et al. (2003) reported up to 30% wear reduction in response to additional lubricity properties of linseed biodiesel added to the fuel (B20). Similar observations made in the present study, i.e., wear reduction using both B50 and IBF (**Fig. 1**). From the data presented in **Figure 1A**, it could be concluded that the IBF fuel led to a more efficient combustion during the test. Moreover, an increase in deposit mass after 100 h of the endurance test was observed but it was slightly decreased by the end of experiment, i.e., after 200 h. Physically, the deposits were found harder and drier, but more brittle compared with the deposits found on the cylinder head or piston crown. As a result, they could be easily shed off from the injector tip surface.

Deposit growth in the engine fueled by B50 showed a different trend (**Fig. 1B**) as after the first 35 h, the thickness of the deposits increased sharply. Moreover, the deposits were relatively soft and wet/oily. The wet nature of the deposits is ascribed to the presence of lots of fuel residues and therefore, these deposits are flammable.

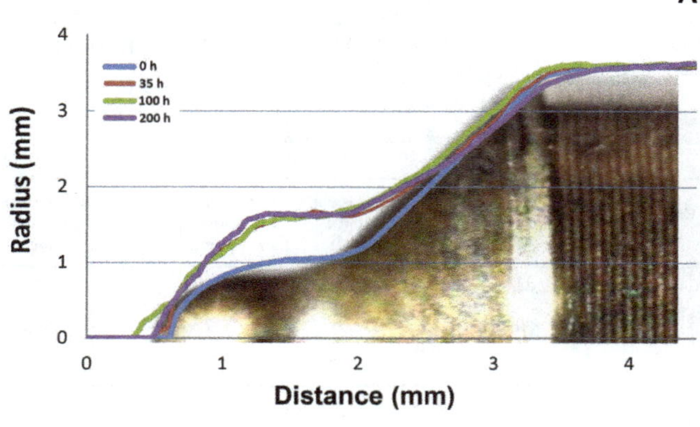


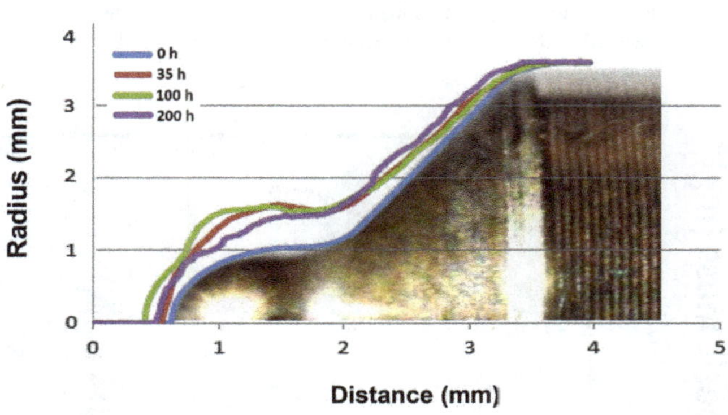


**Fig.1.** Deposit growth on the injector tip in the engines running on, A) IBF and B) B50.

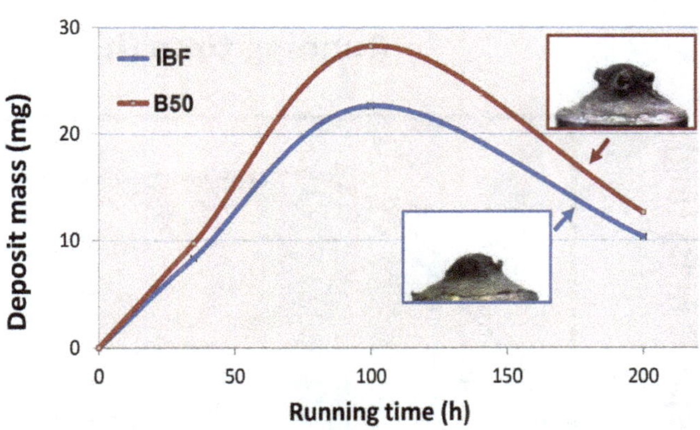


**Fig.2.** Deposit growth on the injector tip using IBF and B50 after 100 h of engine operation.

**Figure 2** also shows the deposit mass generated on the injector tip. As presented B50 led to higher deposit mass than IBF. Furthermore, the deposits were found wetter, thicker and more brittle in the hole of the nozzle. In general, deposit growth in the engine running on B50 was considerably higher than in the engine running on IBF. These results were in agreement with those of Reksowardojo et al. (2010) and Liaquat et al. (2014). **Table 3** tabulates the data obtained on deposit volume and mass on the injector tip using IBF and B50. As shown, after 100 h, a decrease in deposit volume occurred using both fuel types. Measurements were conducted by using both photography and mass weighing in order to minimize errors. For instance, weighing has its own shortcoming of taking the true mass stuck in an engine part into account owing to the presence of non-deposit mass, i.e., oil and fuel residues.

**Table 3.**
Deposit volume and mass in injector tip uses IBF and B50.

| Endurance test (h) | Deposit vol. using IBF ($mm^3$) | Deposit vol. B50 ($mm^3$) | Deposit mass using IBF (mg) | Deposit mass using B50 (mg) |
|---|---|---|---|---|
| 35 | 44.586 | 45.201 | 8,30 | 9.70 |
| 100 | 50.884 | 53.371 | 22,59 | 28,23 |
| 200 | 45.463 | 46.497 | 10.30 | 12,60 |

The trend observed in deposits formation by time in the intake and exhaust valves are shown in **Figure 3A and B**, respectively. As presented, the deposit mass measured in the intake valve using B50 was higher than that of IBF.

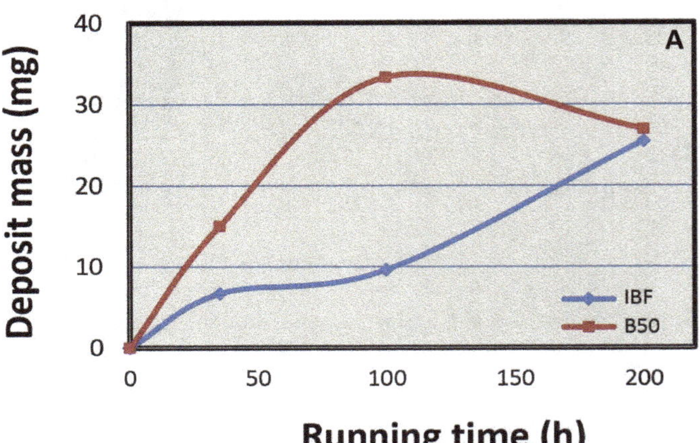


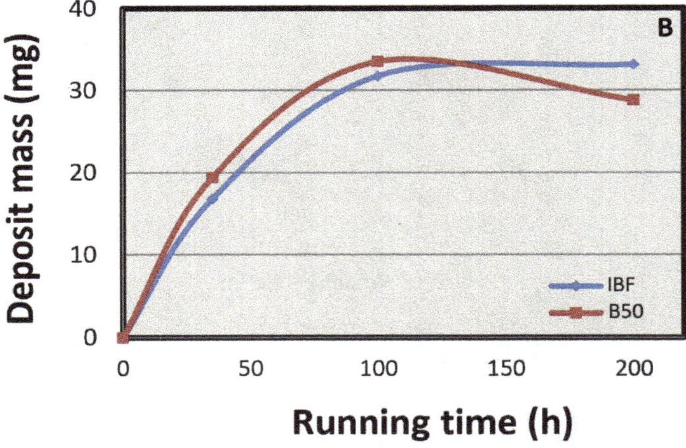


**Fig.3.** Deposit mass growth trends on, A) the intake valve and B) the exhaust valve using IBF and B50.

This could be attributed to the swelling of rubbers as a result of high biodiesel inclusion as in B50 and the consequent oil leakage to the combustion chamber. Rubber swelling did not happen when IBF was used during 200 h of endurance test while the application of B50 resulted in the swelling of the intake valve rubber after only 100 h of operation (**Fig. 4**). As for the exhaust valve, the mass of deposits formed were relatively similar using both fuels after 200 h of engine operation. Moreover, the deposits formed in the exhaust valve using both fuels were brittle and could be easily detached from the valve surface. In a study, Ziejewski et al. (1984) formulated, characterized, and evaluated a nonionic sunflower oil-aqueous ethanol microemulsion as fuel in a diesel engine during a 200 h endurance test. After each durability test, the carbon, sludge and varnish deposits were rated to directly measure the wear of the engine components. Contrary with the findings of the present study, no differences in deposits formation in the intake and exhaust valves were detected for the tested fuels. More specifically, in their study, all valve seats as well as all valve faces showed light peening caused by hard particles released from the combustion chamber deposits while light carbon residue and amber lacquer were noted on the intake valve faces and on the valve stems, respectively.

It is worth mentioning that deposit growth in the other engine components such as cylinder head, piston crown, and piston ring could not be quantified herein by using the photography method while the deposits growth on the cylinder head and piston ring could be measured by the weighing method through shedding off all the deposits stuck on the component's wall. None of the methods were capable of measuring the mass of deposits stuck in the piston ring accurately. Simillar studies were conducted by Ziejewski et al. (1983 and 1984) in which all pistons

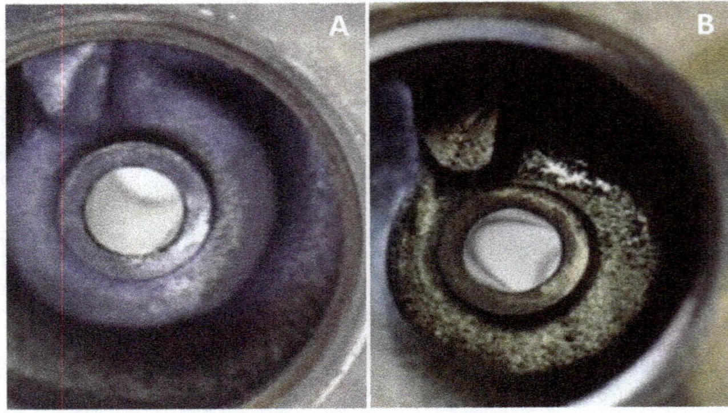


**Fig.4.** Intake valve seat condition using, A) IBF and B) B50 after 100 h of engine operation.

showed significantly greater carbon and lacquer buildups in the ring grooves and on the piston lands, as well as on the piston undercrowns and skirts. Moreover, in a comparative study, Agarwal et al. (2011) in a qualitative analysis of soot formation on engine components under EGR operation condition found that higher soot deposits were formed on the cylinder head, injector tip, and piston crown when EGR was used compared with the condition where the engine was operated without EGR. More specifically, the wear of top compression ring in the engine operated with EGR was found to be slightly lower than the engine operated without EGR that may be due to the lower combustion temperature. But the wear of the second and third compression rings as well as the oil ring in the engine operated with EGR was far more than those of the engine operated without EGR.

### *3.2. Deposit structure*

The deposits formed in different components, i.e., piston, valve, cylinder head, and injector were quite different in structure (**Fig. 5**). As presented, the deposits particles formed as a result of B50 were bigger in comparison with IBF (**Fig. 5**). Moreover, the shapes of the deposits

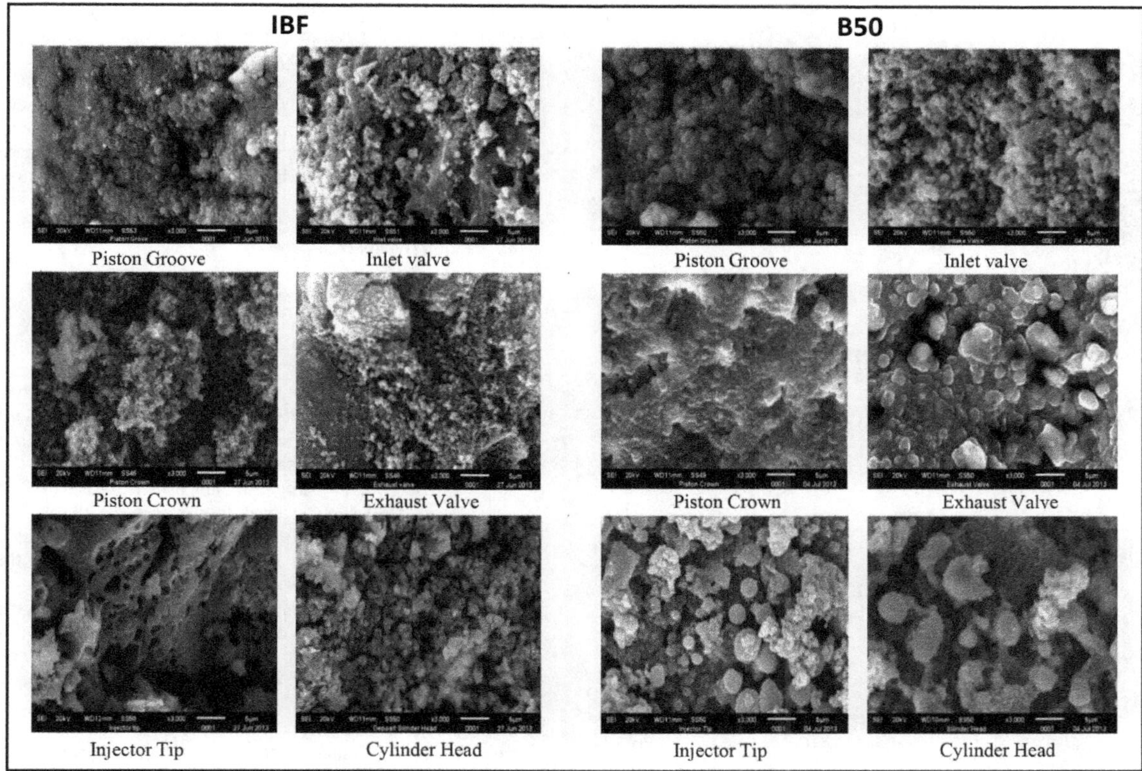


**Fig.5.** SEM monographs showing deposit structure on different engine components (3000×).

obtained were also different. More specifically, the deposits particles formed in the B50-fueled engine had a relatively soft shape, while in the IBF-fueled engine deposits were sharp. This could be attributed to the wetness of the deposits and the fact that were not evaporated completely. The spherical structure of the deposits in the B50-fueled engine was likely ascribed to the presence of long chain hydrocarbons and unstable compounds in the fuel. Long chain hydrocarbons cause the fuel harder to evaporate, whereas unstable compounds in fuel lead to fuel oxidization and the formation of polymers. The formation of deposit in the piston groove was not impacted by fuel type and the changes observed were not significant. Similarly, deposit structures in intake valve using IBF and B50 were also not significantly different.

**Figure 6** revealed that after 200 h of the endurance test, the piston surface was significantly different using the investigated fuels. More specifically, IBF-derived deposits were relatively dry while B50-derived deposits were wet/oily. This may be caused by the unburned fuel articles which failed to diffusely combust after being sprayed by the injector, possibly because of insufficient air fuel ratio (AFR). It has been mentioned that the deposits caused by unburned fuel, and thus wetter than the other deposits on the wall, were thicker (Arifin and Arai, 2009). This was also observed in the present study on the piston crown and especially on the injector tip.

### *3.3. Deposit composition*

Besides morphological variations in the deposits formed, elemental composition of the deposits in different engine components were also analyzed herein. As mentioned earlier, deposits are a layer of various materials and residues gradually grown or accumulated on critical parts of internal combustion engines. More specifically, deposits are a pile of soot and solid particles that stick on a thin liquid layer in the combustion chamber and is compressed by the high compression present in the combustion chamber. The theory of deposits formation was formulated by Lepperhoff and Houben (1993) and was further developed in the present study by identifying the

**Fig.6.** Piston crown condition after dismantling the engines running on IBF (dry) and B50 (wet).

elements found in the deposits. The results obtained by EDS revealed that the deposits in the piston crown of the B50-fueled engine contained a high level of calcium (**Fig. 7**). This element can also be found in oil lubricants as detergent or corrosion inhibitors. While in the IBF-fueled engine, calcium content was considerably lower. In addition to calcium, zinc was also found in the deposits. Zinc is known as additive for anti-wear and oxidant in most oil lubricants (Nicholls et al., 2005). Elemental comparison also showed that O/C content of the deposits formed on the injector tip was higher when IBF was used compared with B50. These findings were similar to those of Liaquat et al. (2014) who also argued that although B50 fuel contained more oxygen but the resultant deposits contained less oxygen. This could probably indicate that the deposits formed on the B50-fueled engine were easier to be oxidized or burned.

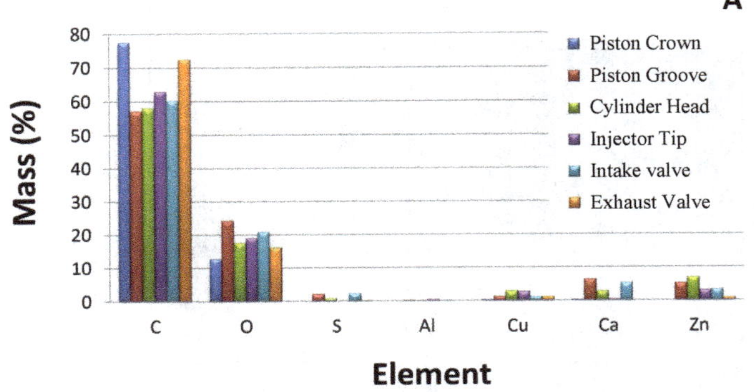


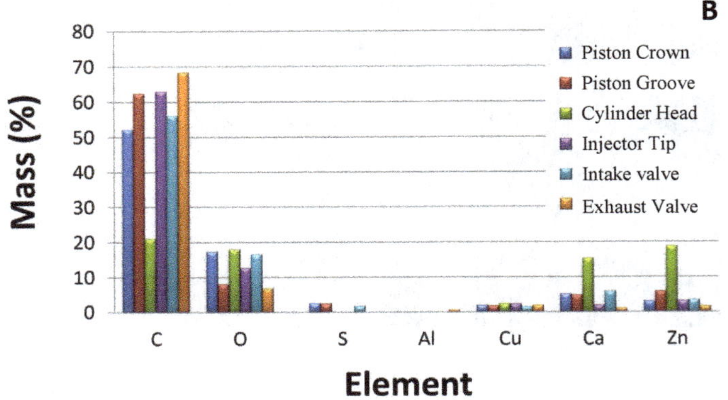


**Fig.7.** Elemental composition of the deposits formed on different engine components using, A) B50 and B) IBF.

As for the deposits stuck in the piston groove of both IBF and B50 fueled engines, calcium and zinc were found indicating that the piston groove was loaded with lubricant, i.e., biodiesel. In general, the zinc found in wear debris could be because of additive depletion, extra additives being added due to addition of make-up oil, wear of bearings, brass components, neoprene seals, etc. (Agarwal et al., 2003). An important observation was made by Agarwal et al. (2003) who claimed that the application of B20 led to approximately 65% less increase in zinc content indicating less wear of zinc-containing components and lower consumption of the lubricating oil in comparison with neat diesel fuel.

Overall, the deposits morphology and elemental composition in the piston groove were similar using both fuel blends, i.e., B50 and IBF (**Figs. 5 and 7**). Traces of potassium and calcium were also identified in the deposits proving that lubricant oil was present in the deposits. In the intake and exhaust valves, the percentage of carbon mass and oxygen deposited in IBF-fueled engine was higher than in B50-fueled engine. Moreover, in the intake valve, the deposits found contained Fe, Mg, and Mo, which are the elements of the wear debris (Liaquat et al., 2014). On the whole, the elemental compositions obtained in the present study revealed lubricant leakage caused by B50 as a result of rubber swelling. Moreover, the deposits in the exhaust value when B50 was used contained a trace of aluminum indicating the occurrence of metal wear process in the piston ring as a sign of ring sticking. The copper measured in the wear debris basically originates from wear of bearings, and bushings (**Fig. 7**). Agarwal et al. (2003) also argued that the lubricating impacts of B20 resulted in approximately 25% lower copper content revealing less wear when compared with neat diesel fuel.

The formation processes of carbonaceous deposits in the first piston ring grooves of direct injection diesel engines were studied by Diaby et al. (2009). Their analyses showed that the deposits, with a structure of cracked varnishes, mainly resulted from the degradation of lubricants. These deposits contained a noticeable proportion of metallic elements resulting from the wear of metallic parts of the engine. According to their results, the degradative oxidation of the lubricant apparently induced polymerization reactions which in turn led to the formation of a varnish acting as a binder. The formed binder bridged between carbon particles and metallic particles of wear and worsened the cycle of lubricant degradation (Diaby et al., 2009).

## 4. Conclusions

Based on the findings of the present study, the following conclusions can be drawn:

- The higher inclusion rate of biodiesel increased deposits formation on engine components, especially on the valves and injector tip.
- By increasing biodiesel inclusion rate, a tendency toward increased ring sticking was observed.
- B50 fuel led to a large amount of deposits. More specifically, the use of B50 showed excessive wetness in the piston crown. This could be ascribed to the fact that the injected fuel hitting on the piston crown wall would fail to dry, thus making the B50-caused deposits tend to be wet and brittle.
- Based on the deposits morphology, in every engine component, different shapes could be observed even by using the same fuel. By using different fuels, different deposit structures were detected in similar parts of the engine, especially in the injector tip, piston crown, and cylinder head. This was not observed in the exhaust valve and ring groove.
- In the B50-caused deposits, spherical patterns dominated the structures observed. This could be mainly attributed to the unburned fuel that failed to properly combust, and would later dominate the deposits structure. This deposit structure is similar to deposit structure observed in the piston groove.

## Acknowledgements

The authors would like to especially thank BTMP-BBPT for their utmost support by providing facilities and funds. We also would like to extend our appreciation to BATAN who helped with Tomography and SEM analyses.

## References

[1] Adams, C., Peters, J.F., Rand, M.C., Schroer, B.J., Ziemke, M.C., 1983. Investigation of soybean oil as a diesel fuel extender: endurance tests. JAOCS. 60(8), 1574-1579.

[2] Agarwal, A.K., Bijwe, J., Das, L. M., 2003. Effect of biodiesel utilization of wear of vital parts in compression ignition engine. J. Eng. Gas Turb. Power. 125, 604-611.

[3] Agarwal, A.K., 2007. Biofuels (alcohols and biodiesel) applications as fuels for internal combustion engines. Prog. Energy Combust. Sci. 33, 233–271.

[4] Agarwal, D., Singh, S.K., Agarwal, A.K., 2011. Effect of Exhaust Gas Recirculation (EGR) on performance, emissions, deposits and durability of a constant speed compression ignition engine. Appl. Energ. 88, 2900–2907.

[5] Arifin, Y.M., Arai, M., 2009. Deposition characteristics of diesel and bio-diesel fuels. Fuel 88(11), 2163-2170.

[6] Caceres, D., Reisel, J.R., Sklyarov, A., Poehlman, A., 2003. Exhaust emission deterioration and combustion chamber deposits composition over the life cycle of small utility engine. J. Eng. Gas Turbines Power. 125(1), 358-364.

[7] Diabya, M., Sablier, M., Negrateb, A. L., El Fassi, M., Bocquet, J., 2009. Understanding carbonaceous deposit formation resulting from engine oil degradation. Carbon 47, 355-366.

[8] Goldsworthy, L., 2006. Computational fluid dynamics modeling of residual fuel oil combustion in the context of marine diesel engines. Int. J. Engine Res. 7(2), 181-199.

[9] Gopal, K.N., Karupparaj, R.T., 2015. Effect of pongamia biodiesel on emission and combustion characteristics of DI compression ignition engine, Ain Shams Eng. J. 6(1), 297-305.

[10] Güralp, O., Hoffman, M., Assanis, D.N., Filipi, Z., Kuo, T.W., Najt, P. Rask, R., 2006. Characterizing the effect of combustion chamber deposits on a gasoline HCCI engine. SAE technical paper, No. 2006-01-3277.

[11] Joint FIE Manufacturers Statement, 2009, Fuel requirements for diesel fuel injection systems.

[12] Kalam, M.A., Husnawan, M., Masjuki, H.H., 2003. Exhaust emission and combustion evaluation of coconut oil-powered indirect injection diesel engine. Renew. Energ. 28, 2405–2415.

[13] Kalam, M.A., Masjuki, H.H., 2004. Emissions and deposits characteristics of a small diesel engine when operated on preheated crude palm oil. Biomass Bioenergy. 27(3), 289-297.

[14] Lepperhoff, G., Houben, M., 1993. Mechanisms of deposit formation in internal combustion engines and heat exchangers. SAE technical paper, No. 931032.

[15] Liaquat, A.M., Masjuki, H.H., Kalama, M.A., Fazal, M.A., Faheem Khan, A., Fayaz, H., Varman, M., 2013. Impact of palm biodiesel blend on injector deposit formation. Appl. Energ. 111, 882–893.

[16] Liaquat, A.M., Masjuki, H.H., Kalam, M.A., Fattah, I.R., 2014. Impact of biodiesel blend on injector deposit formation. Energy. 72, 813-823.

[17] Nicholls, M.A., Do, T., Norton, P.R., Kasrai, M. Bancroft, G.M., 2005. Review of the lubrication of metallic surfaces by zinc dialkyl-dithiophosphates. Tribol. Int. 38(1), 15-39.

[18] Ra, Y., Reitz, R.D., Jarrett, M.W., Shyu, T.P., 2006. Effects of piston crevice flows and lubricant oil vaporization on diesel engine deposits. SAE Technical Paper. No. 2006-01-1149.

[19] Reksowardojo, I.K., Hung, B.N., Sok, R., Kilgour, A.J., Brodjonegoro, T.P., Soerawidjaja, T.H., Mai, P.X., Shudo, T., Arismunandar, W., 2010. The Effect of biodiesel fuel from Rubber seed oil (*Hevea Brasiliensis*) on a direct injection (DI) diesel engine, The 3rd Regional Conference on New and Renewable Energy, 13-14th, October, Penang, Malaysia.

[20] Spilners, I.J., Hedenburg, J.F., 1985. Effect of Fuel and Lubricant Composition on Engine Deposit Formation, in: Ebert, L.B. (Ed.), Chemistry of Engine Combustion Deposits, Plenum Press, pp. 289-302.

[21] Yamada, Y., Emi, M., Ishii, H., Suzuki, Y., Kimura, S., Enomoto, Y., 2002. Heat loss to the combustion chamber wall with deposit in D.I. diesel engine: variation of instantaneous heat flux on piston surface with deposit. JSAE Rev. 23(4), 415-421.

[22] Ziejewski, M., Kaufman, K.R., 1983. Laboratory endurance test of a sunflower oil blend in a diesel engine. JAOCS. 60(8), 1567-1573.

[23] Ziejewski, M., Kaufman, K. R., Schwab A.W., Pryde, E.H., 1984. Diesel engine evaluation of a nonionic sunflower oil-aqueous ethanol microemulsion. JAOCS. 61(10), 1620-1626.

# 24

# Lipase immobilized on polydopamine-coated magnetite nanoparticles for biodiesel production from soybean oil

Marcos F.C. Andrade, Andre L.A. Parussulo, Caterina G.C.M. Netto, Leandro H. Andrade, Henrique E. Toma *

*Instituto de Química, Universidade de São Paulo, 05508-000, São Paulo, Brazil.*

## HIGHLIGHTS

➢Lipase immobilized onto polydopamine magnetite nanoparticle converted soybean oil into biodiesel with high efficiency.
➢Polydopamine film allowed direct binding of the enzyme.
➢Polydopamine film led to immobilization of a large amount of enzymes onto the magnetic nanoparticles.
➢The enzyme could be magnetically recycled.

## GRAPHICAL ABSTRACT

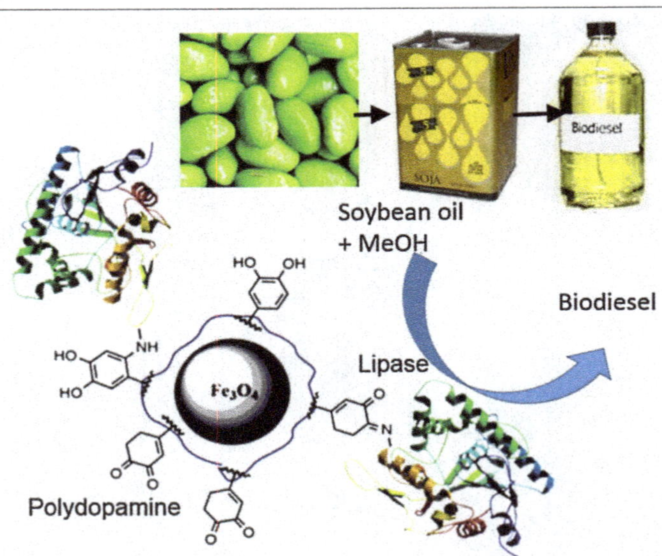


## ABSTRACT

Lipase from *Pseudomonas cepacia* was covalently attached to magnetite nanoparticles coated with a thin polydopamine film, and employed in the enzymatic conversion of soybean oil into biodiesel, in the presence of methanol. The proposed strategy explored the direct immobilization of the enzyme *via* Michael addition and aldolic condensation reactions at the catechol rings, with no need for using specific coupling agents. In addition, a larger amount of enzymes could be bound to the magnetic nanoparticles, allowing their efficient recycling with the use of an external magnet. For biodiesel production, the transesterification reaction was carried out directly in soybean oil by stepwise addition of methanol, in order to circumvent its inactivation effect on the enzyme. A better yield of 90% was achieved at 37 °C compared with the free enzyme. However, the immobilized biocatalyst became gradually less effective after the third cycle, due to its prolonged exposition to the denaturating methanol medium.

**Keywords:**
Biodiesel
Soybean oil
Lipase
Magnetic nanoparticles
Polydopamine

* Corresponding author
E-mail address: henetoma@iq.usp.br

## 1. Introduction

Biodiesel is an attractive substitute to conventional diesel, conveying significant environmental advantages originated from its renewable feedstocks, i.e., vegetable oils and animal fats. Other benefits include the lower emissions of sulfur-based pollutants, aerosols, and carbon monoxide compared with the petrochemical products (Sharp et al., 2000; Shrivastava et al., 2000). Currently, production of biodiesel is carried out by means of the transesterification reaction, employing corrosive chemical catalysts (strong acids or bases) at relatively high temperatures (Moser, 2011). Hence, the downstream processing costs and environmental issues have led to a search for alternative eco-friendly biodiesel production methods (Kim et al., 2006; Hama et al., 2007; Bisen et al., 2010; Verma et al., 2013).

In this context, lipases have been proposed as a feasible alternative to sodium hydroxide or sulfuric acid (Moser, 2011). Enzyme-mediated processes are considered more advantageous over the chemical routes, since they proceed at room temperature and do not practically away triger environmental concerns (Tan et al., 2010; Li et al., 2011; Atabani et al., 2012; Verma et al., 2013). In addition, enzymes can achieve high catalytic efficiencies under brand reaction conditions, and moreover, due to substrate specific, they are regarded as very attractive alternatives to the chemical processes. However, enzymes are also easily denatured entities and their short catalytic lifespan and high prices inevitably limit their applications in large-scale reactors. For this reason, there is a great interest in immobilizing enzymes on solid supports, allowing recyclability while substantially reducing the catalyst cost (Katchalski-Katzir, 1993; Straathof et al., 2002; Mateo et al., 2007; Hanefeld et al., 2009; Bose et al., 2010; Rebelo et al., 2010; Garcia-Galan et al., 2011; Wang et al., 2011; Yucel et al., 2011; Rodrigues et al., 2013). Re-usage is not the only advantage of enzyme immobilization and it has been previously reported that immobilized enzymes can exhibit higher reaction rates and better thermal stability compared with their free counterparts (Verma et al., 2013).

The advantages of nanomaterials as support for enzyme immobilization have opened new perspectives in biocatalysis arena. Such nanomaterials should be engineered by controlling the size, shape, and functionalization, in order to improve their suitability for applications in nanobiocatalysis. In fact, the nanoscale dimension of the nanoparticles ensures a very good partnership with enzymes, allowing their association and performance as new modified species, keeping their mobility and mass transfer characteristics in solution, but incorporating better qualities in terms of stability and activity.

In particular, the use magnetic nanoparticles is especially rewarding, for allowing easy enzyme recovery by applying an external magnet. Supermagnets of $Nd_2Fe_{14}B$ displaying strong magnetic fields are quite available in the market, and are suitable for this purpose. However, one has to carefully consider the biomolecule linking procedure to the nanoparticle beforehand, since the coupling method can impact the enzyme catalytic activity and its immobilization efficiency (Rebelo et al., 2010; Netto et al., 2009, 2011, 2012, 2013, 2015). A large variety of methods has already been employed (Netto et al., 2013); however, they have led to very contrasting results.

In this regard, polydopamine was recently introduced as a very promising material for the immobilization of biomolecules in inorganic substrates (Xu et al., 2004; Lee et al., 2007; 2009; Ren et al., 2011; Black et al., 2013). This bio-inspired polymer is easily formed by the partial oxidation of dopamine in a mild alkaline medium, and exhibits high affinity for most inorganic surfaces. Its great attractiveness offers not only a fast and easy way to coat magnetite nanoparticles, but also the possibility of direct attachment of the biomolecules, by means of Michael addition or aldolic condensation reactions at its exposed catechol rings.

Among the inorganic supports used for the attachment of proteins, many interesting advantages could be achieved by using magnetite nanoparticles (Netto et al., 2013). Superparamagnetic materials allow a convenient separation of the catalyst from the reaction medium by simply using an external magnetic field. In addition, their large surface area allows the adsorption of high amounts of biomolecules, while the nanoparticulate nature provides a good mobility for catalysis. Other important aspects such as higher stability and environmental compatibility should also be mentioned. However, as Verma et al. (2013) pointed out, in the case of nanomaterials bound enzymes used in catalyzed biofuel production, there are yet many aspects to be explored and improved, as their large-scale use is still in their infancy.

In this sense, a successful association of lipase with a versatile coating polymer such as polydopamine and a superparamagnetic particle (magnetite), has already been reported in the literature (Xu et al., 2004; Lee et al., 2007; 2009; Ren et al., 2011; Black et al., 2013), but to the best of our knowledge, its application in biofuel production has never been attempted. This is rather surprising, since lipases are well-known candidates in bio-catalyzed biodiesel synthesis (Kim et al., 2006; Hama et al., 2007; Bisen et al., 2010; Li et al., 2011; Wang et al., 2011; Atabani et al., 2012; Verma et al., 2013). As a matter of fact, the immobilization of lipase on silver nanoparticles *via* adhesive polydopamine for biodiesel production has already been successfully reported (Dumri et al., 2014). However, in this case, the enzyme recycling was found not feasible, and moreover, the generation of silver contaminants could raise serious environmental concerns. Therefore, the use of *Pseudomonas cepacia* lipase immobilized on polydopamine coated $Fe_3O_4$ nanoparticles to perform the bio-catalytical synthesis of biodiesel was explored in the present study. Biodiesel was produced using the immobilized nano-biocatalyst, soybean oil, and methanol under environmentally-compatible conditions. Such initiative is in line with the modern trends in sustainable biotechnology, and could be particularly of interest for soybean biodiesel producing countries.

## 2. Materials and Methods

Lipase from *P. cepacia* (powder), $FeCl_3.6H_2O$, $FeCl_2.4H_2O$, NaOH, $NaH_2PO_4.H_2O$, dopamine hydrochloride, and Bradford reagent solution were obtained from Sigma-Aldrich (Germany). Methanol was purchased from Synth (Brazil). A commercially-available refined soybean oil produced by Lisa Company (Brazil) was used in all transesterification reactions.

Bradford essays were performed using a HP8453 diode array (190-1100 nm) spectrophotometer. Transmission electron microscopy (TEM) images were obtained using a JEOL JEM 2100 equipment employing a $LaB_6$ emission filament, with 200 kV maximum acceleration tension. A Bruker Eco-ATR with a Ge crystal was used to obtain the FT-IR spectra. For atomic force microscopy, a Nanoscope E from Digital Instruments was used.

### *2.1. Magnetite nanoparticles synthesis*

The magnetite nanoparticles were synthesized using a previously described method (Yamaura et al., 2004) involving the coprecipitation of $Fe^{3+}$ and $Fe^{2+}$ hydroxides, in an alkaline medium. More specifically, a system containing 200 mL of a 1 mol $L^{-1}$ NaOH aqueous solution was initially deoxygenated, and was mechanically stirred at 1100 rpm. Then, 50 mL of 0.2 mol $L^{-1}$ $FeCl_3$ and 0.1 mol $L^{-1}$ $FeCl_2$ solution were added drop-wise. The reaction was allowed to proceed for 30 min at room temperature. Subsequently, the black precipitate was magnetically separated and washed 5 times with nano-pure water.

### *2.2. Nanoparticles functionalization with polydopamine*

The freshly-prepared magnetite nanoparticles were dispersed in 100 mL of distilled water, and 250 mg of dopamine hydrochloride was added. The pH was adjusted to 8.4 using a 1 mol $L^{-1}$ NaOH aqueous solution. The dispersion was air-bubbled for 3 h, at 30 min intervals. The particles were again magnetically separated and rinsed 3 times with distilled water.

### *2.3. Immobilization of lipase from P. cepacia on $Fe_3O_4$@polydopamine*

For lipase immobilization, the general procedure described by Ren et al. (2011) was used. The polydopamine coated magnetite nanoparticles ($Fe_3O_4$@PD) were dispersed in 100 mL of 10 mM PBS at pH 7. Then, 4 mL of this dispersion was cooled at 4 °C, 200 mg of the lipase powder was added, and the system was stirred for 1 h at 4 °C. The black powder was magnetically separated, washed with distilled water, and dried overnight under reduced pressure.

*2.4. Biodiesel synthesis catalyzed by lipase from P. cepacia immobilized on magnetic nanoparticles*

The transesterification reaction was carried out directly in soybean oil and methanol, with no need for additional solvents, as expected for a green strategy. Initially, 200 mg of the nanocatalyst was added to 1 g of a 1:1 (mol/mol) mixture of methanol and soybean oil, and the system was agitated at 37 °C for 12 h. Then, 3 equivalents of methanol (relative to the initial amount of soybean oil) were added in two steps, after 150 and 300 min of the reaction. Finally, the particles were magnetically separated, washed with tert-butanol, and dried under reduced pressure for 10 min.

*2.5. ATR-FTIR biodiesel quantification in soybean oil / biodiesel mixtures*

The amount of biodiesel present in mixture with soybean oil was determined by adapting a fast and reliable method proposed in the literature (Mahamuni et al., 2009; Zhang et al., 2013). More specifically, the method was based on monitoring the intensities $CH_3$ asymmetric stretching signals (characteristic of fatty acids methyl ester infra-red spectra) as a function of biodiesel to soybean oil mass ratio. For this purpose, a highly pure biodiesel reference material was employed and standard biodiesel/soybean oil mixtures were prepared in order to construct a calibration curve correlating the intensity (area) of 1427–1450 $cm^{-1}$ band and the biodiesel percentage as shown in **Figure 1**.

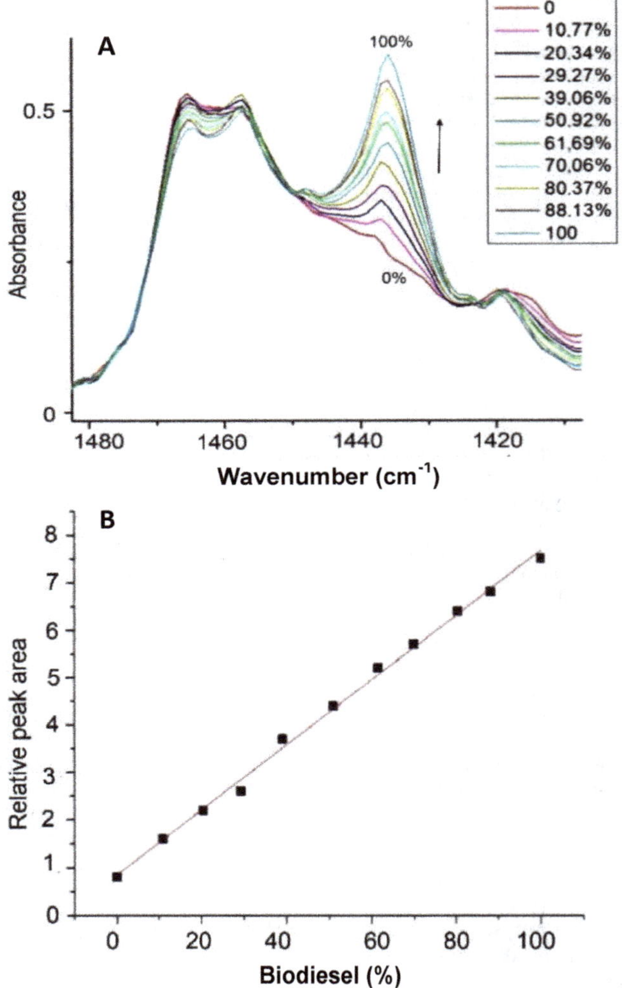


**Fig.1.** (A) FTIR spectra of soybean oil / biodiesel mixtures. The box on the top right indicates the mass percentage of biodielsel in the sample. (B) Calibration curve made by correlating the area under 1427-1450 $cm^{-1}$ band and the relative mass of biodiesel in the soybean oil / biodiesel solutions.

### 3. Results and discussion

Dynamic light scattering measurements for the $Fe_3O_4$ nanoparticles in aqueous solution revealed a typical hydrodynamic radius distribution of around 86 nm (**Fig. 2A**). Such average distribution actually involves aggregates of smaller nanoparticles which are of interest in enzymatic catalysis. This is ascribed to the fact that they exhibit a rather strong magnetization behavior, thus responding more rapidly to the applied magnetic fields. After the treatment with dopamine, the average hydrodynamic radius was increased to 112 nm (**Fig. 2B**), reflecting the polymeric coating around the $Fe_3O_4$ nanoparticles.

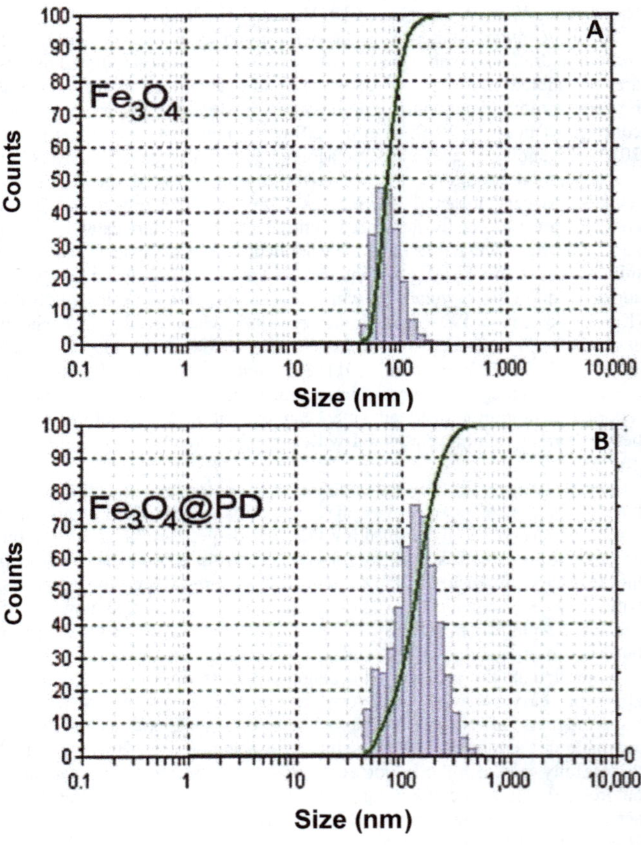


**Fig.2.** DLS profiles for the $Fe_3O_4$ (A) and $Fe_3O_4$@PD. (B) nanoparticles in aqueous solution.

The AFM images of $Fe_3O_4$@PD nanoparticles, in the contact mode and in the phase contrast mode, can be seen in **Figures 3A and 3B**, respectively, corroborating the results obtained from the DLS measurements. In the phase contrast image, it is possible to detect the aggregated clusters core surrounded by a soft material attributed to the polydopamine film coating. In the corresponding TEM images, a large amount of $Fe_3O_4$ aggregates comprising typical 11 nm magnetic cores could be observed (**Fig. 4 A and B**). As shown in **Figure 4B,** it can be concluded that the sample preparation caused the magnetite nanoparticles be surrounded by an amorphous and almost transparent film, which is completely absent in **Figure 4A**.

The FTIR spectra of the $Fe_3O_4$ and $Fe_3O_4$@PD nanoparticles can be seen in **Figure 5**. By comparing these images, magnetite functionalization with polydopamine represented by the presence of three sharp peaks, i.e., aromatic C-C stretching (1486 $cm^{-1}$), $NH_3$ in-plane bending (1422 $cm^{-1}$), and C-O-H symmetric bending (1266 $cm^{-1}$) could be detected. In fact, the presence of these bands represents the presence of polydopamine, however, it is worth noticing that at the present time there is no general agreement about the real structure of this particular coating material (Liebscher et al., 2013).

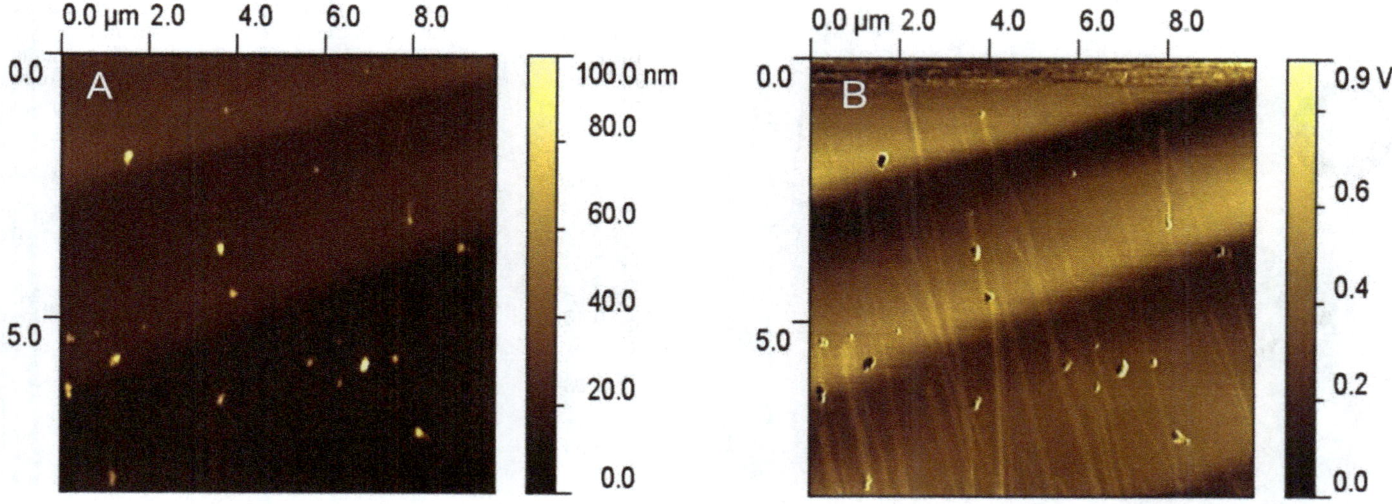


**Fig.3.** (A) Contact mode AFM and (B) Phase contrast AFM of magnetite nanoparticles coated with polydopamine.

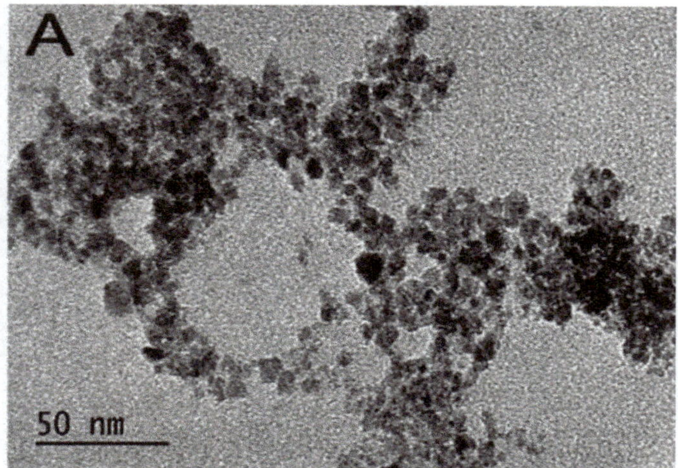


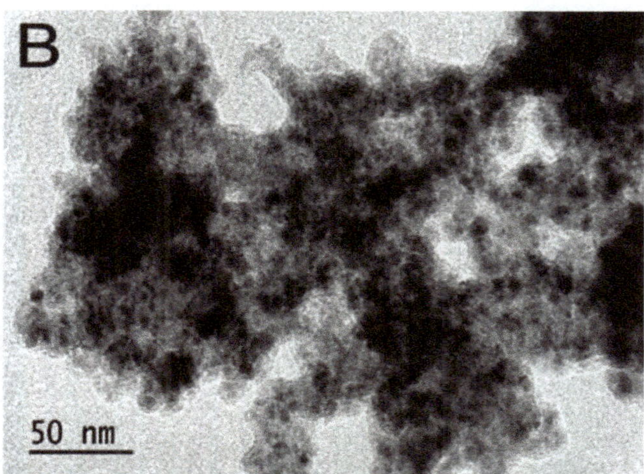


**Fig.4.** TEM images of (A) $Fe_3O_4$ and (B) $Fe_3O_4$@PD nanoparticles.

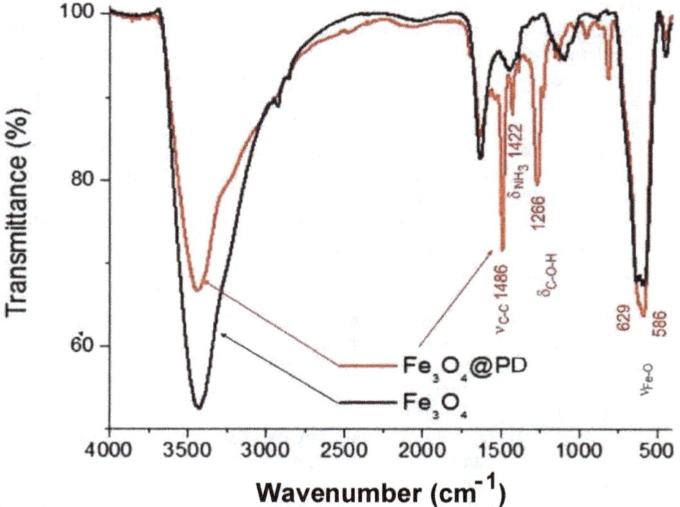


**Fig.5.** FTIR spectra spectra of bare and polydopamine - coated magnetite nanoparticles.

Polydopamine allows an easy immobilization of biomolecules by means of Michael addition or aldolic condensation reactions at the catecholic ring. This possibility entirely dispenses the use of linking agents such as EDC and glutaraldehyde (**Fig. 6**). Moreover, it also readily polymerizes over a wide range of materials in the presence of molecular oxygen.

Hence, lipase from *P. cepacia* was easily immobilized on $Fe_3O_4$@PD without using any additional linking agents and previous treatments of the lipase powder. The reaction was accomplished within 10 min and such fast chemical binding has a great advantage, i.e., decreased the risk of enzyme denaturation as it is the case when more drastic procedures are employed.

Distinct differences in the FTIR spectra from the vibrational bands in the 1100 – 1700 $cm^{-1}$ region can be noticed before and after the protein immobilization procedure (**Figs. 5 and 7**). These differences were ascribed to the covalent attachment of the enzyme to the nanomaterial. Nevertheless, the assignment of the vibrational peaks was not feasible, because of the strong superimposition of the polydopamine and lipases spectra in this region. However, all non-adsorbed enzymes in the final material was completely discarded by rinsing three times with PBS, pH 7.

The main interest to covalently attach the lipases on magnetite nanoparticles surfaces is attributed to the easy separation of the catalyst from the solution. Another advantage of this attractive material is its recyclability. Therefore, both the catalytic efficiency of the immobilized

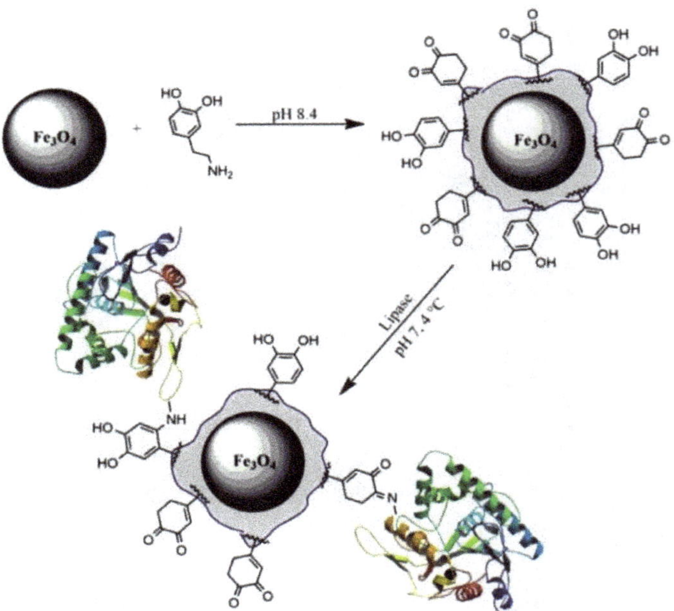


Fig.6. Schematic illustration of the magnetite coating by *in-situ* polymerization of dopamine at the nanomaterial surface, and the binding of lipase by means of the Michael addition or aldolic condensation involving the amino groups of the exposed catechol groups. The irregular form (in gray) around the $Fe_3O_4$ spheres represents the polydopamine film and the enzyme is represented by colored structures.

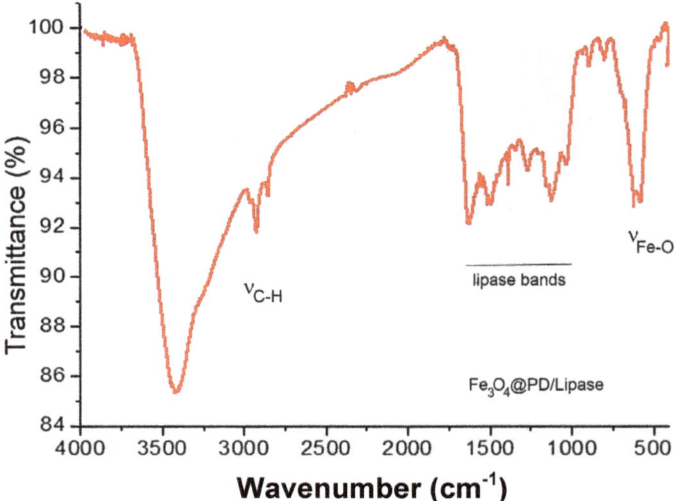


Fig.7. FTIR spectrum of $Fe_3O_4$@PD-Lipase showing the characteristic enzyme vibrational peaks in the 1100-1700 $cm^{-1}$ range.

enzyme and its activity during 5 successive reaction cycles were investigated herein.

Prior to biodiesel synthesis, the conditions for lipase immobilization at the polydopamine shell around the magnetite nanoparticles were optimized. More specifically, the effect of the dopamine/magnetite mass ratio on the amount of lipase adsorbed on the functionalized nanomaterial was first investigated. Accordingly, as shown in **Figure 8**, the amount of immobilized lipase did not increase linearly with increasing the dopamine/magnetite mass ratio, but exhibited a saturation behavior when the ratio was above 2. For this reason, this condition was selected to coat magnetite nanoparticles with polydopamine. At the second step, the concentration of $Fe_3O_4$@PD in the medium containing lipase was changed to maximize both enzyme adsorption efficiency and the protein relative mass in the nanocatalyst. These experiments were carried at a constant initial mass of enzymes of 0.077 mg, 10 mmol $L^{-1}$ (pH 7, PBS solution) at 4 °C and by using different $Fe_3O_4$@PD concentrations.

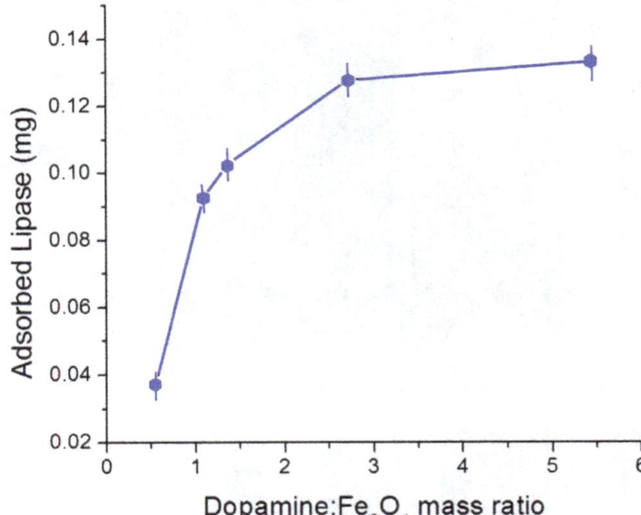


Fig.8. Maximum amount of lipase adsorbed as a function of dopamine to magnetite mass ratio.

The analysis of the data presented in **Figure 9** readily indicates that although higher nanoparticle to lipase ratios resulted in higher adsorption of the protein, for a constant mass of enzyme, the relative amount of lipase in the final nanocatalyst decreased by increasing the $Fe_3O_4$@PD concentration, probably because of the redistribution effect. So, in the next sequence of the experiments, 2% enzyme load was used as a good compromise between adsorption efficiency and the relative amount of lipase in the material.

After the optimization of lipase chemical adsorption onto the functionalized nanomaterials, the enzyme activity was evaluated in the transesterification of soybean oil with methanol in a solvent-free reaction. A major concern with this reaction is the irreversible denaturation of lipases by insoluble methanol in the reaction media, as equimolar proportions of soybean oil and methanol are used. Unfortunately, this problem is inherent in the lipase-based biodiesel production systems and represents a critical challenge to be overcome, regardless of the immobilization process employed. As an attempt to circumvent the problem, several experiments under variable flow conditions were performed in the present study, in order to minimize the exposition of lipase to the denaturating condition. Since such experiments could be intrinsically more complicated by reproducibility problems and by introducing more variables in the process. Hence, an alternative procedure was adopted herein by stepwise addition of methanol to the soybean oil. The immobilized lipase was initially added in an equimolar mixture of methanol and soybean oil (near the solubility limit of methanol in SBO). The remaining equivalents of methanol were equally added in two steps, after 150 and 300 min. A higher solubility of methanol in biodiesel compared with its solubility in soybean oil validates this approach. The proposed protocol is illustrated in **Figure 10**, where the dashed lines and arrows indicate the points when methanol was added to the reaction media.

During the first reaction cycle at 37 °C, the immobilized enzyme converted 93% of soybean oil into biodiesel after 12 h reaction, which was better than the result obtained using the free enzyme (86%) under the same reaction conditions. This result corroborates the general observation in immobilized enzyme catalysis, that the enzyme activity is improved when compared with its free counterpart (Netto et al., 2009; 2010; 2011).

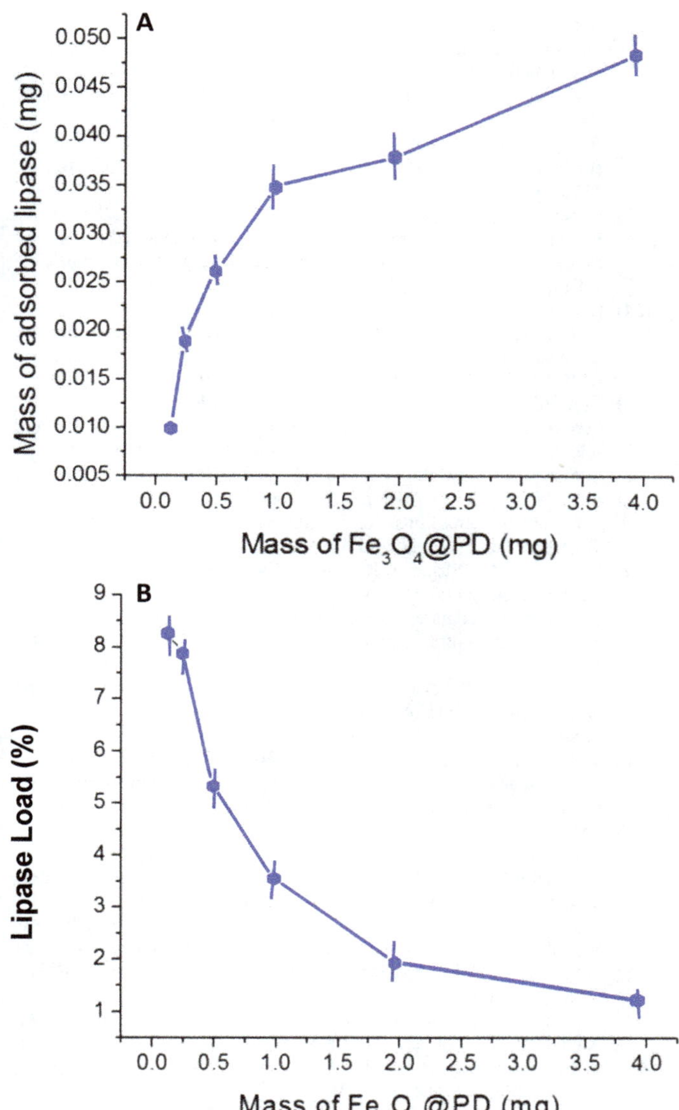


**Fig.9.** (A) Mass of adsorbed lipase and (B) relative mass of lipase on the nanocatalyst as a function of the initial mass of the functionalized nanomaterial. Experiments were carried in 10 mM (pH 7, PBS) at 4°C and lipase initial mass of 0.077 mg.

Although the catalyst achieved a high soybean oil into biodiesel conversion during the first reaction, the efficiency of the transesterification reaction gradually decreased in the subsequent cycles, particularly after the third cycle (**Fig. 11**). This might be due to the prolonged exposition of the enzyme to the medium containing high methanol concentration. In fact, both temperature and methanol concentration can be responsible for slow denaturation of the immobilized protein, and this aspect should be more deeply investigated. In spite of this, the use of the superparamagnetic nanoparticles functionalized with polydopamine during the first three cycles was very successful in terms of the yield achieved.

## 4. Conclusions

Lipase from *P. cepacia* was readily adsorbed onto magnetite nanoparticles coated with a thin polydopamine film, with no need for using additional linking agents, thus ensuring a green procedure. The immobilized lipase led to a higher soybean oil conversion of 93% into biodiesel within 12 h than the free enzyme (86%). The reaction could be carried out directly in soybean oil and methanol as reactants, without using any additional solvents, encompassing another relevant green aspect to be mentioned. The magnetic nanocatalyst could be recycled at least three times by coupling the process with a stepwise addition of methanol during the 12 h solvent-free reaction at 37 °C. The findings of the present study revealed that enzymatic catalysis could be effectively applied for biodiesel synthesis fulfilling the expectations considered for a green chemistry strategy by employing a recyclable natural catalyst under environmentally-friendly conditions, in contrast with the industrial processes currently in-use.

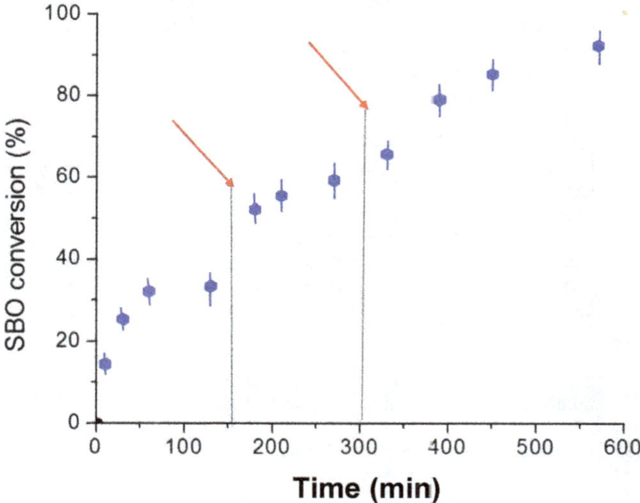


**Fig.10.** Percentage of soybean oil conversion into biodiesel using 20% of nanocatalyst mass relative to the initial mass of the reactants. Dashed lines and arrows indicate the stepwise addition of 3 mole equivalents of methanol relative to soybean oil.

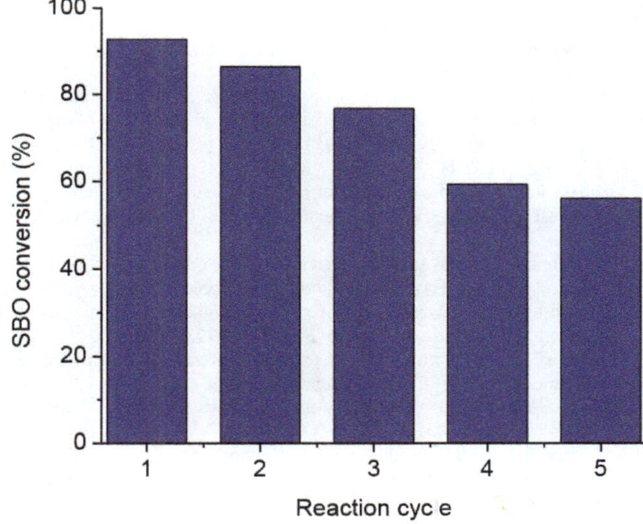


**Fig.11.** Total amount of SBO converted into biodiesel after 12 h reaction at 37 °C in a solvent free reaction. The immobilized enzyme was recycled up to 5 times and the amount of SBO converted at each cycle is presented.

**Acknowledgements**

The support from Fundação de Amparo à Pesquisa Científica e Tecnologica, FAPESP – Grant 2013/24725-4, and Conselho Nacional de Desenvolvimento Científico e Tecnológico, CNPq, Grant 482383/2013-5 is gratefully acknowledged.

## References


[1] Atabani, A.E., Silitong, A.S., Badruddin, I.A., Mahlia, T.M.I., Masjuki, H.H., Mekhilef, S.A., 2012. Comprehensive review on biodiesel as an alternative energy source and its characteristics. Renew. Sust. Energy Rev. 16, 2070-2093.

[2] Bisen, P.S., Sanodiya, B.S., Thakur, G.S., Baghed, R.K., Prasad, G.B., 2010. Biodiesel production with special emphasis on lipase-catalised transesterification. Biotechnol. Lett. 32, 1019-1030.

[3] Black, K.C., Yi, J., Rivera, J.G., Zelasko-Leon, D.C., Messersmith, P.B., 2013. Polydopamine-enabled surface functionalization of gold nanorods for cancer cell-targeted imaging and phothermal therapy. Nanomedicine. 8, 17-28.

[4] Bose, S., Armstrong, D.W., Petrich, J.W., 2010. Enzyme-catalyzed hydrolysis of cellulose in ionic liquids: a green approach toward the production of biofuels. J. Phys. Chem. B. 114, 8221-8227.

[5] Dumri, K., Hung-Anh, D., 2014. Immobilization of lipase on silver nanoparticles via adhesive polydopamine for biodiesel production. Enzyme Res. Article ID 389739.

[6] Garcia-Galan, C., Berenguer-Murcia, A., Fernandez-Lafuente, R., Rodrigues, R.C., 2011. Potential of different enzyme immobilization strategies to improve enzyme performance. Adv. Synth. Catal. 353, 2885-2904.

[7] Hama, S., Yamaji, H., Fukumizu, T., Numata, T., Tamalampudi, S., Kondo, A., Noda, H., Fukuda, H., 2007. Biodiesel-fuel production in a packed-bed reactor using lipase-producing *Rhizopus oryzae* cells immobilized within biomass support particles. Biochem Eng. J. 34, 273-278.

[8] Hanefeld, U., Gardossi, L., Magner, E., 2009. Understanding enzyme immobilization. Chem. Soc. Rev. 38, 453-468.

[9] Katchalski-Katzir, E., 1993. Immobilized enzymes learning from past successes and failures. Trends Biotechnol. 11, 471-478.

[10] Kim, J., Jia, J., Wang, P., 2006. Challenges in biocatalysis for enzyme based biofuel cells. Biotechnol. Adv. 24, 296-308.

[11] Lee, H., Dellatore, S.M., Miller, W.M., Messersmith, P.B., 2007. Mussel inspired surface chemistry for multifunctional coatings. Science. 318, 426-430.

[12] Lee, H., Rho, J., Messersmith, P.B., 2009. Facile conjugation of biomolecules onto surfaces via mussel adhesive protein inspired coatings. Adv. Mat. 21, 431-434.

[13] Liebscher, J., Mrowczynski, R., Scheidt, H.A., Filip, C., Hadade, N.D., Turcu, R., Bende, A., 2013. Structure of polydopamine: a never-ending story?. Langmuir. 29, 10539-10548.

[14] Li, S.F., Fan, Y.H., Hu, R.F., Wu, W.T., 2011. *Pseudomonas cepacia* lipase immobilized onto the electrospun pan nanofibrous membranes for biodiesel production from soybean oil. J. Mol. Catal. B: Enzym. 72, 40-45.

[15] Mahamuni, N.N., Adewuyi, Y.G., 2009. Fourier transform infrared spectroscopy (FTIR) method to monitor soy biodiesel and soybean oil in transesterification reactions, petrodiesel-biodiesel blends and blend adulteration with soy oil. Energy Fuels. 23, 3773-3782.

[16] Mateo, C., Palomo, J.M., Fernandes-Lorente, G., Guisan, J.M., Fernanes-Lafuente, R., 2007. Improvement of enzyme activity, stability and selectivity via immobilization techniques. Enzyme Microb. Tech. 40, 1451-1463.

[17] Moser, B.R., 2011. Biodiesel production, properties, and feedstocks, in: Tomes, D., Lakshmanan, P., Sonstad, D. (Eds.), Biofuels. New York, Springer, pp. 285-347.

[18] Netto, C.G.C.M., Andrade, L.H., Toma, H.E., 2009. Enantioselective transesterification catalysis by *Candida antarctica* lipase immobilized on superparamagnetic nanoparticles. Tetrahedron: Asymmetry. 20, 2299-2304.

[19] Netto, C.G.C.M., Nakamatsu, E.H., Netto, L.E.S., Novak, M.A., Zuin, A., Nakamura, M., Araki, K., Toma, H.E., 2011. Catalytic properties of thioredoxin immobilized on superparamagnetic nanoparticles. J. Inorg. Biochem. 105, 738-744.

[20] Netto, C.G.C.M., Nakamura, M., Andrade, L.H., Toma, H. E., 2012. Improving the catalytic activity of formate dehydrogenase from *Candida boidinii* by using magnetic nanoparticles. J. Mol. Catal. B: Enzym. 84, 139-143.

[21] Netto, C.G.C.M., Toma, H.E., Andrade, L.H., 2013. Superparamagnetic nanoparticles as versatile and supporting materials for enzymes. J. Mol. Catal. B: Enzym. 85-86, 71-92.

[22] Netto, C.G.C.M., Andrade, L.H., Toma, H.E., 2015. Association of *Pseudomonas putida* formaldehyde dehydrogenase with superparamagnetic nanoparticles: an effective way of improving the enzyme stability, performance and recycling. New J. Chem. 39, 2162-2167.

[23] Rebelo, L.P., Netto, C.G.C.M., Toma, H.E., Andrade, L.H., 2010. Enzymatic kinetic resolution of (rs)-1-(phenyl)ethanols by *Burkholderia cepacia* lipase immobilized on magnetic nanoparticles. J. Braz. Chem. Soc. 21, 1537-1542.

[24] Ren, Y., Rivera, J.G., He, L., Kulkarni, H., Lee, D.K., Messersmith, P.D., 2011. Facile, high efficiency immobilization of lipase enzyme on magnetic iron oxide nanoparticles via a biomimetic coating. BMC Biotech. 11, 63. DOI: 10.1186/1472-6750-11-63.

[25] Rodrigues, R.C., Ortiz, C., Berenguer-Murcia, A., Torres, R., Fernandez-Lafuente, R., 2013. Modifying enzyme activity and selectivity by immobilization. Chem. Soc. Rev. 42, 6290-6307.

[26] Sharp, C.A., Howell, S.A., Jobe, J., 2000. The effect of biodiesel fuels on transient emissions from modern diesel engines, Part II, unregulated emissions and chemical characterization. SAE Technical Paper. No. 2000-01-1968.

[27] Srivastava, A., Prasad, A.R., 2000. Triglycerides-based diesel fuels. Renew. Sust. Energy Rev. 4, 111-133.

[28] Straathof, A.J., Panke, S., Schmid, A., 2002. The production of fine chemicals by biotransformations. Curr. Opin. Biotechnol. 13, 548-556.

[29] Tan, T., Lu, J.K., Nie, K.L., Deng, L., Wang, F., 2010. Biodiesel production with immobilised lipase: a review. Biotechnol. Adv. 26, 628-634.

[30] Verma, M.L., Barrow, C.J., Puri, M., 2013. Nanobiotechnology as a novel paradigm for enzyme immobilisation and stabilisation with potential applications in biodiesel production. Appl. Microbiol. Biotechnol. 97, 23-39.

[31] Wang, X., Liu, X., Zhao, C., Ding, Y., Xu, P., 2011. Biodiesel production in packed-bed reactors using lipase-nanoparticle biocomposite. Bioresour. Technol. 102, 6352-6355.

[32] Xu, C., Xu, K., Gu, H., Zheng, R., Liu, H., Zhang, X., Guo, Z., Xu, B., 2004. Dopamine as a robust anchor to immobilize functional molecules on the iron oxide shell of magnetic nanoparticles. J. Am. Chem. Soc. 126, 9938-9939.

[33] Yamaura, M., Camilo, R.L., Sampaio, L.C., Macedo, M.A., Nakamura, M., Toma, H.E., 2004. Preparation and characterization of (3-aminopropyl) triethoxysilane coated magnetite nanoparticles. J. Magn. Magn. Mater. 279, 210-217.

[34] Yücel, Y., Demir, C., Dizge, N., Keskinler, B., 2011. Lipase immobilization and production of fatty acid methyl esters from canola oil using immobilized lipase. Biomass Bioenergy. 35, 1496-1501.

[35] Zhang, F., Adachi, D., Tamalampudi, S., Kondo, A., Tominaga, K., 2013. Real-time monitoring of the transesterification of soybean oil and methanol by fourier transform infrared spectroscopy. Energy Fuels. 27, 5957-5961.

# Permissions

All chapters in this book were first published in Biofuel Research Journal, by Green Wave Publishing of Canada; hereby published with permission under the Creative Commons Attribution License or equivalent. Every chapter published in this book has been scrutinized by our experts. Their significance has been extensively debated. The topics covered herein carry significant findings which will fuel the growth of the discipline. They may even be implemented as practical applications or may be referred to as a beginning point for another development.

The contributors of this book come from diverse backgrounds, making this book a truly international effort. This book will bring forth new frontiers with its revolutionizing research information and detailed analysis of the nascent developments around the world.

We would like to thank all the contributing authors for lending their expertise to make the book truly unique. They have played a crucial role in the development of this book. Without their invaluable contributions this book wouldn't have been possible. They have made vital efforts to compile up to date information on the varied aspects of this subject to make this book a valuable addition to the collection of many professionals and students.

This book was conceptualized with the vision of imparting up-to-date information and advanced data in this field. To ensure the same, a matchless editorial board was set up. Every individual on the board went through rigorous rounds of assessment to prove their worth. After which they invested a large part of their time researching and compiling the most relevant data for our readers.

The editorial board has been involved in producing this book since its inception. They have spent rigorous. hours researching and exploring the diverse topics which have resulted in the successful publishing of this book. They have passed on their knowledge of decades through this book. To expedite this challenging task, the publisher supported the team at every step. A small team of assistant editors was also appointed to further simplify the editing procedure and attain best results for the readers.

Apart from the editorial board, the designing team has also invested a significant amount of their time in understanding the subject and creating the most relevant covers. They scrutinized every image to scout for the most suitable representation of the subject and create an appropriate cover for the book.

The publishing team has been an ardent support to the editorial, designing and production team. Their endless efforts to recruit the best for this project, has resulted in the accomplishment of this book. They are a veteran in the field of academics and their pool of knowledge is as vast as their experience in printing. Their expertise and guidance has proved useful at every step. Their uncompromising quality standards have made this book an exceptional effort. Their encouragement from time to time has been an inspiration for everyone.

The publisher and the editorial board hope that this book will prove to be a valuable piece of knowledge for researchers, students, practitioners and scholars across the globe.

# List of Contributors

**Nurul Aini Edama, Alawi Sulaiman and Siti Noraida Abd. Rahim**
Tropical Agro-Biomass Research Group, Faculty of Plantation and Agrotechnology, Universiti Teknologi MARA, 40450 Shah Alam, Selangor, Malaysia

**Awais Bokhari, Lai Fatt Chuah, Suzana Yusup, Junaid Ahmad, Muhammad Rashid Shamsuddin and Meng Kiat Teng**
Biomass Processing Laboratory, Centre of Biofuel and Biochemical Research (CBBR), Chemical Engineering Department, Universiti Teknologi PETRONAS, Bandar Seri Iskandar, 32610 Seri Iskandar, Perak, Malaysia

**Dave Barchyn and Stefan Cenkowski**
Department of Biosystems Engineering, University of Manitoba, Winnipeg, Manitoba, Canada

**Gopalakrishnan Kumar**
Center for Materials Cycles and Waste Management Research, National Institute for Environmental Studies, Tsukuba, Japan

**Péter Bakonyi, Nándor Nemestóthy and Katalin Bélafi-Bakó**
Research Institute on Bioengineering, Membrane Technology and Energetics, University of Pannonia, Egyetem u. 10, 8200 Veszprém, Hungary

**Periyasamy Sivagurunathan**
Department of Environmental Engineering, Daegu University, Republic of Korea

**Chiu-Yue Lin**
Department of Environmental Engineering and Science, Feng Chia University, 40724 Taichung, Taiwan

**G.R. Moradi, M. Ghanbari, M.J. Moradi, Sh. Hosseini and Y. Davoodbeygi**
Catalyst Research Center, Chemical Engineering Department, Faculty of Engineering, Razi University, Kermanshah, Iran

**M. Mohadesi**
Chemical Engineering Department, Kermanshah University of Technology, Kermanshah, Iran

**Mebrahtu Haile**
Land resource management and environmental protection department, college of dry land agriculture and natural resource, Mekelle University, Ethiopia

**Mateus S. Amaral, Patrícia C.M. Da Rós, Heizir F. de Castro and Sara A. Machado**
Engineering School of Lorena-University of São Paulo, Lorena, SP, Brazil

**Carla C. Loures**
Engineering Faculty of Guaratingueta - State University Julio de Mesquita Filho-UNESP, Guaratingueta, SP, Brazil

**Cristiano E. R. Reis**
Department of Bioproducts and Biosystems Engineering, College of Food, Agricultural and Natural Resource Sciences, University of Minnesota, USA

**Messias B. Silva**
Engineering School of Lorena-University of São Paulo, Lorena, SP, Brazil
Engineering Faculty of Guaratingueta - State University Julio de Mesquita Filho-UNESP, Guaratingueta, SP, Brazil

**Aarón Millán-Oropeza, Luis G. Torres -Bustillos and Luis Fernández-Linares**
Departamento de Bioprocesos, Unidad Profesional Interdisciplinaria de Biotecnología, Instituto Politécnico Nacional (UPIBI - IPN), Av. Acueducto s/n Col. Barrio la Laguna Ticomán, 07340, Mexico City, Mexico

**Maryam M. Kabir, Mohammad J. Taherzadeh and Ilona Sárvári Horváth**
Swedish Centre for Resource Recovery, University of Borås, 501 90, Borås, Sweden

**Abiodun Aladetuyi , Gabriel A. Olatunji and Stephen O. Oguntoye**
Department of Chemistry, University of Ilorin, P.M.B 1515, Ilorin, Nigeria

**David S. Ogunniyi and Temitope E. Odetoye**
Department of Chemical Engineering, University of Ilorin, P.M.B 1515, Ilorin, Nigeria

**Mandana Akia, Farshad Yazdani and Elahe Motaee**
Chemistry & Chemical Engineering Research Center of Iran (CCERCI), P.O. Box 4335-186, Tehran, Iran
Nano/catalysts Research Group, Biofuel Research Team (BRTeam), Karaj, Iran

**Dezhi Han**
Key Laboratory of Biofuels, Qingdao Institute of Bioenergy and Bioprocess Technology, Chines Academy of Sciences, Qingdao 266101, PR China

**Hamidreza Arandiyan**
State Key Joint Laboratory of Environment Simulation and Pollution Control (SKLESPC), School of Environment, Tsinghua University, Beijing 100084, China

**A.I. Gavrisheva, E.S. Shastik, T.V. Laurinavichene and A.A. Tsygankov**
Institute of Basic Biological Problems, Russian Academy of Sciences, Pushchino, Moscow Region, 142290, Russia

**B.F. Belokopytov**
Institute of Physiology and Biochemistry of Microorganisms, Russian Academy of Sciences, Pushchino, Moscow Region, 142290, Russia

**V.I. Semina**
Institute of Basic Biological Problems, Russian Academy of Sciences, Pushchino, Moscow Region, 142290, Russia
LLC "Ecoproject", Mytishchi, 141014, Moscow Region, Russia

**A.E. Atabani**
Department of Mechanical Engineering, Faculty of Engineering, University of Malaya, 50603 Kuala Lumpur, Malaysia
Department of Mechanical Engineering, Erciyes University, 38039 Kayseri, Turkey
Erciyes Teknopark A.Ş, Yeni Mahalle Aşıkveysel Bulvarı Erciyes Teknopark Tekno3 Binası 2. KatNo: 28, 38039 Kayseri, Turkey

**M. Mofijur, H.H. Masjuki, Irfan A. Badruddin and W.T. Chong**
Department of Mechanical Engineering, Faculty of Engineering, University of Malaya, 50603 Kuala Lumpur, Malaysia

**S.F. Cheng and S.W. Gouk**
Unit of Research on Lipids (URL), Department of Chemistry, Faculty of Science, University of Malaya, 50603 Kuala Lumpur, Malaysia

**Ahmad Farhad Talebi**
Semnan university, Semnan, Iran
Energy Crops Genetic Engineering Group, Biofuel Research Team (BRTeam), Karaj, Iran

**Masoud Tohidfar and Meisam Tabatabaei**
Energy Crops Genetic Engineering Group, Biofuel Research Team (BRTeam), Karaj, Iran
Microbial Biotechnology and Biosafety Dept, Agricultural Biotechnology Research Institute of Iran (ABRII), Karaj, Iran

**Abdolreza Bagheri**
Biotechnology and Plant Breeding Dept., College of Agriculture, Ferdowsi University of Mashhad, Mashhad, Iran

**Stephen R. Lyon**
AlgaXperts, LLC, Milwaukee, Wisconsin, USA

**Kourosh Salehi-Ashtiani**
Division of Science and Math, and Center for Genomics and Systems Biology (CGSB), New York University Abu Dhabi, P.O. Box 129188, Abu Dhabi, UAE

**B. Ndaba, I. Chiyanzu and S. Marx**
School of Chemical and Minerals Engineering, North-West University (Potchefstroom Campus), Potchefstroom, South Africa

**Marc Y. Menetrez**
United State Environmental Protection Agency, Office of Research and Development, National Risk Management Research Laboratory, Air Pollution Prevention and Control Division, Research Triangle Park, NC 27711, USA

**Jundika C. Kurnia**
Mechanical Engineering Department, Universiti Teknologi PETRONAS, 32610 Bandar Seri Iskandar, Perak Darul Ridzuan, Malaysia

**Sachin V. Jangam**
Department of Chemical and Biomolecular Engineering, National University of Singapore, 4 Engineering Drive 4, 117575 Singapore

**Saad Akhtar and Agus P. Sasmito**
Department of Mining and Materials Engineering, McGill University, 3450 University Street, Frank Dawson Adams Bldg, Montreal Quebec H3A 2A7, Canada

**Arun S. Mujumdar**
Department of Chemical and Biomolecular Engineering, National University of Singapore, 4 Engineering Drive 4, 117575 Singapore
Department of Mining and Materials Engineering, McGill University, 3450 University Street, Frank Dawson Adams Bldg, Montreal Quebec H3A 2A7, Canada

**Stuart Wilkinson and David J. Cook**
International centre for Brewing Science (ICBS), Division of Food Sciences, The University of Nottingham, Sutton Bonington Campus, Loughborough, Leicestershire LE12 5RD, U.K

**Katherine A. Smart and Sue James**
SABMiller Plc, SABMiller House, Church Street West, Woking, Surrey, GU21 6HS, U.K

**Stuart Wilkinson , Darren Greetham and Gregory A. Tucker**
Brewing Science Section, Division of Food Sciences, The University of Nottingham, Sutton Bonington Campus, Loughborough, Leicestershire LE12 5RD, U.K

**Kamoru A. Salam, Sharon B. Velasquez-Orta and Adam P. Harvey**
School of Chemical Engineering and Advanced Materials (CEAM), Newcastle University, NE1 7RU, United Kingdom

**Shahrbanoo Hamedi, Mahmood A. Mahdavi and Reza Gheshlaghi**
Department of Chemical Engineering, Ferdowsi University of Mashhad, Azadi Square, Pardis Campus, 91779-48944, Mashhad, Iran

**Rajeev Kumar**
Center for Environmental Research and Technology (CE-CERT), Bourns College of Engineering, University of California, Riverside, California, USA

**Meisam Tabatabaei**
Microbial Biotechnology Department, Agricultural Biotechnology Research Institute of Iran (ABRII), AREEO, Karaj, Iran
Biofuel Research Team (BRTeam), Karaj, Iran

**Keikhosro Karimi**
Department of Chemical Engineering, Isfahan University of Technology, Isfahan 84156-83111, Iran
Microbial Industrial Biotechnology Group, Institute of Biotechnology and Bioengineering, Isfahan University of Technology, Isfahan 84156-83111, Iran

**Ilona Sárvári Horváth**
Swedish Centre for Resource Recovery, University of Borås, 501 90 Borås, Sweden

**M Taufiq Suryantoro and Bambang Sugiarto, Fariz Mulyadi**
Departement of Mechanical Engineering, University of Indonesia, 16424 Depok West Java, Indonesia

**Marcos F.C. Andrade, Andre L.A. Parussulo, Caterina G.C.M. Netto, Leandro H. Andrade and Henrique E. Toma**
Instituto de Química, Universidade de São Paulo, 05508-000, São Paulo, Brazil

# Index

www.ingramcontent.com/pod-product-compliance
Lightning Source LLC
Chambersburg PA
CBHW050443200326
41458CB00014B/5050
* 9 7 8 1 6 3 2 3 9 9 9 1 5 *